Natural Water Treatment Systems for Safe and Sustainable Water Supply in the Indian Context

Natural Water Treatment Systems for Safe and Sustainable Water Supply in the Indian Context

Saph Pani

Edited by
Thomas Wintgens, Anders Nättorp, Lakshmanan Elango
and Shyam R. Asolekar

PUBLISHING

Published by **IWA Publishing**
Alliance House
12 Caxton Street
London SW1H 0QS, UK
Telephone: +44 (0)20 7654 5500
Fax: +44 (0)20 7654 5555
Email: publications@iwap.co.uk
Web: www.iwapublishing.com

First published 2016
Revised edition with addition of Chapter 19, 2016

British Library Cataloguing in Publication Data
A CIP catalogue record for this book is available from the British Library

Original Edition

ISBN: 9781780407104 (Paperback)
ISBN: 9781780407111 (eBook)

This Revised Edition

ISBN: 9781780408385 (Paperback)
ISBN: 9781780408392 (eBook)

Contents

About the Editors

Thomas Wintgens holds a PhD in Chemical Engineering from RWTH Aachen University and is Professor in Environmental Technologies at the University of Applied Sciences and Arts Northwestern Switzerland since September 2008. He is leading a research team working on water and wastewater treatment technologies in municipal and industrial applications as well as on management of scarce water resources. In the last decade he has been involved in many international research projects on water treatment and resources management. He chaired the water reuse task force of the European Water Supply and Sanitation Technology Platform. Thomas Wintgens was the coordinator of Saph Pani.

Anders Nättorp holds a PhD in Chemical Reaction Engineering from Ecole Polytechnique Fédérale de Lausanne. He worked 10 years in development and production departments in industry and joined the University of Applied Sciences and Arts Northwestern Switzerland in 2006. As a senior researcher he leads projects and work packages, complementing technical results with evaluation of cost and system analysis in the fields of water treatment technologies and resource recovery, in particular phosphorus. Anders Nättorp was the project manager of Saph Pani and responsible for the scientific quality assurance of the outputs of the project.

Lakshmanan Elango is a Professor in Geology at Anna University, Chennai, India with a PhD in Hydrogeology. He has over 28 years of experience, and gained research training from the University of Birmingham, Swiss Federal Institute of Technology, University of New Castle and Ruhr University in Europe. He has edited seven books, authored several book chapters, technical reports and research articles. He is the President of Association of Global Groundwater Scientists, Vice President of the International Association of Hydrological Sciences, Vice President of Indian National Committee of International Association of Hydrogeologists. He has organised advanced training workshops for the World Bank funded Hydrology Project and UNESCO's International Hydrology Program. Lakshmanan Elango was leader of a work package and Co-Chair of Saph Pani.

Dr. Shyam R. Asolekar is currently a Professor at the Centre for Environmental Science and Engineering at the Indian Institute of Technology Bombay (also served as the Head during 2006–2009). He is author of three books, six patents as well as several policy documents, training manuals, chapters of books and research papers. He has been a member (since, 1997) of the "Dahanu Taluka Environmental Protection Authority" constituted by the Honorable Supreme Court of India. His current research and teaching areas include: 1. reuse of treated wastewater for achieving near-zero emissions especially by combining advanced tertiary treatment technologies with low cost eco-centric natural treatment systems, 2. rejuvenation of ponds, lakes, rivers and wetlands and 3. development and application of 'Decision Support Tools' based on "life cycle" approach, minimization of carbon footprint and sustainability criteria. Dr. Shyam R. Asolekar was a work package leader of Saph Pani.

Foreword

Rossella Riggio and Dr. Panagiotis Balabanis (European Commission)

Water is a precondition for human, animal and plant life on Earth. Deforestation, pollution, over-exploitation of water resources, damage to aquatic ecosystems, climate change and security issues are challenging the sustainability of water systems. In parallel, population increase, economic development, urbanization, and land use or natural geomorphic changes also challenge the sustainability of resources by decreasing water supply or increasing demand.

Still too many people around the world do not have access to safe drinking water or basic sanitation. At the same time, three billion people will join the global consumer class over the next two decades, accelerating the degradation of natural resources and escalating competition for them.

If we continue business as usual, global demand for water will exceed viable resources by 40 percent by 2030.

In this context, access to safe drinking water and sanitation, integrated water management, including water efficiency, can clearly contribute to manage the challenges of climate change water scarcity and global equality. Water is also an indispensable resource for the economy and has a high strategic and economic importance. Water crisis have been recognized as the 1st highest risk that could undermine economic growth according to the 2015 Global Risk Report of the World Economic Forum.

Research and innovation plays an important role in providing solutions to major water challenges. Over the past decades, EU research funding has Framework Programme dedicated over EUR 1 billion to water research and Horizon 2020 will continue to support fundamental and applied research to address this complex and cross- cutting societal challenge. Water is also a very important area for international research cooperation with non-EU countries for promoting sustainable development in the context of the on-going Sustainable Development Goals discussion.

Within the context of the Environment (including climate change) Theme of the FP7 Cooperation Programme, a dedicated research topic on water systems and treatment technologies to cope with water shortages in urbanised areas in India was launched in 2011. SAPH PANI "Enhancement of natural water systems and treatment methods for safe and sustainable water supply in India" was selected for funding following the evaluation of that call. Since then a more strategic cooperation on water purification and wastewater reclamation, and reuse issues was built between the European Commission and the Indian Department of Science and Technology that gave rise to a joint coordinated call for proposals in 2012 and the emergence of a strong network of European and Indian researchers working together. In this context, SAPH PANI could be considered as a precursor of such cooperation.

This book summarises the key achievement of the SAPH PANI EU funded project.

On behalf of the European Commission and its Directorate General for Research, we would like to express our appreciation to the SAPH PANI partners for their efforts. We are confident that the book shall constitute a state of art knowledge experience which can find its way to contribute solving in practice the water challenge in India.

Rossella Riggio and Dr. Panagiotis Balabanis

European Commission
DG Research & Innovation
I2 Eco-Innovation

Foreword

P. Rajendra Prasad (Saph Pani Advisory Board)

Water a natural resource and essential component of human survival, contributes significantly to sustained economy and hence it is naturally in demand of multiple stakeholders. Though the principle of mass conservation indicates the quantum of water to remain the same over time, decrease in usable water resources day-by-day is witnessed globally. In recent times, the vast spread of uneven distribution of the resource in time and space supplemented by many anthropogenic interventions has induced a high degree of complexity making it more susceptible to even minor marginal changes. As a result the increasing gap between demand and supply constrained by diminished hope and scope to augment new resources has brought in a paradigm shift from development to management of the resource.

The average annual rainfall in India is quite reasonable and is around 1200 mm. However, its uneven distribution in time and space supplemented by frequent monsoon failures and ineffective management of the resource make it even scarce. So further research needs to be carried out for augmenting new sources of water and maintain its quality. Adoption of efficient renovation and recycling approaches can bring in balance in addressing the quantity and quality issues associated with the requirements.

India, being an agrarian nation, 85% of its usable water resources is spent in the farming sector leaving the rest for industrial, domestic and recreational purposes. The recent spurt in agricultural activity, industrial development and urbanization supplemented by natural disasters and liberalized policies of Government and financial institutions have been mounting stress on ground water resources in India, especially in hard rock terrain. Added to this people started realizing the importance and need of maintaining 'water quality' only in recent times.

The multi fold increase in irrigational needs associated with excess application of fertilizers and pesticides and lack of suitable technologies at affordable costs for solid and liquid waste disposal have led to unprecedented pollution of the water bodies. In addition, the impinging threat of sea water intrusion into coastal aquifer regime renders the water resources in the coastal regions more fragile and vulnerable to anthropogenic as-well-as natural hazards. Further the changing climatic conditions are yet another factor in drastically effecting the hydrological cycle. At present, in many emerging nations, water policies are driven mostly by short term economic and political concerns rather than scientific perceptions in which India is not an exception.

As per many studies done by national and international agencies, more than 50% of the deceases are found to be waterborne in India. The need to supply adequate safe drinking water with easy access has become the top national priority. The country in a bid to address the issue of meeting the growing demands for water initiated a number of programmes like rainwater harvesting, replenishing ground water from surface water resources and creating large dams and reservoirs mainly to help irrigation. Some of them do address quantity and quality aspects but incidentally they also brought in undesirable social and environmental impacts. In addition improper planning in managing sewage and irrigation waste waters has become a major issue in recent times.

The Govt. of India, in its endeavour to meet the national water needs has created huge dams and associated distribution systems. However, the ever increasing demands could not be met from available resources, forcing the system to look for alternate methods of conservation and management. This has been the guiding factor to prompt the ministries of water resources and the Department of Science and Technology (DST) to initiate major national programs like WAR for water Rajiv Gandhi drinking water mission etc., to initiate innovative research and management strategies. Many programmes supported by European Union like India- EU water projects, Indigo, FP7 etc. have been in place in support of this cause.

The Saph Pani initiative supported by the FP7 programme of European Union is not only timely but also evolved as a model in Natural Treatment Systems in water resources treatment and management in India. Among many EU supported programmes Saph Pani initiative has the unique distinction on more than one count. It is a unique programme that dealt with treatment of fresh water, waste water and treatment of natural systems. It dealt with treatment of natural systems in different geographic, geomorphic and climatic environments. It addressed issues related to normal and extreme events. It also dealt with three major components of water cycle viz., atmospheric, surface and groundwater with a balanced approach. In addition it could also create awareness among stakeholders.

The Saph Pani Programme is focussed around natural treatment systems for safe and sustainable water supply in India. The Programme is built around three major components dealing with bank filtration (BF), management of aquifer recharge and soil aquifer treatment (SAT) including constructed wetlands. The project aims at adopting a comprehensive approach synergizing the European experts and the Indian resources with application to Indian field conditions. The project demonstrates its potential in the Indian context in developing and implementing cost effective innovative scientific technologies and also contributes to capacity building to replicate in other parts of the country. The different technologies were implemented in different geographic hydro climatic and hydro geological settings with different field conditions and varying uses.

Bank Filtration was adopted as a natural field treatment technology for treating the surface water from lakes and rivers to make them potable. As a part of the programme, a number of bank filtration wells and systems have been developed and monitored. Apart from keeping their turbidity, the systems have demonstrated its efficacy in the removal of pathogens, colour, dissolving organic carbon and reduced coliform count. The studies also briefed to be effective even under high abstraction rates both during monsoon and non-monsoon periods. The studies have led to the design and construction of flood-proved wells and prevention of bank erosion. A standard protocol has also been developed for adoption in different hydro climatic and hydro geological conditions.

Managed Aquifer Recharge (MAR) has been implemented not only to inject treated storm water and surface water into the aquifer regime but also to improve the water quality including sea water intrusion. Techniques were developed to adopt and manage highly varying flows to recharge the aquifer regime. The project mainly dealt with the design and development of various wetland systems for treatment of waste water. Different types of wetlands constructed worldwide have been modified and engineered to suit the local waste water characteristics and the environmental conditions.

The project could achieve low operational maintenance costs. The results have demonstrated many reuse options acceptable at community level. It also demonstrated that systems can be operated with the skills of rural folks with suitable training and it can contribute to supplement the shortfall between the water supply and sewage treatment.

The project was able to produce state of the art technology that can easily be adopted to suit the local conditions in improving the quality of water and contribute to augment additional resources. The quality research publications resulted out of the project is a testimony to the high degree of professionalism, industry and academia interaction and its societal application.

P. Rajendra Prasad
Saph Pani Advisory Board

Acknowledgements

This book is based on collaborative Indo-European research activities in the project Saph Pani on Enhancement of natural water systems and treatment methods for safe and sustainable water supply in India. This project has received funding from the European Union's Seventh Programme for research, technological development and demonstration under grant agreement No 282911 (www.saphpani.eu). The Saph Pani project started in October 2011 and ran till September 2014. All project partners and advisory board members are acknowledged for their dedication to the project and the inputs received for this publication. Lena Breitenmoser and Julia Plattner are thanked for their dedicated editorial support. The project officers at the European Commission DG Research and Innovation, Unit I.2 Eco-Innovation, are acknowledged for project guidance.

Partners of the Saph Pani project:

FACHHOCHSCHULE NORDWESTSCHWEIZ FHNW Switzerland, Thomas Wintgens, Anders Nättorp, Julia Plattner, Liang Yu, Linda Stamm, Jeremias Brand.

UTTARAKHAND JAL SANSTHAN UJS India, P. C. Kimothi, S. K. Sharma, Ramesh Chandra, R. K. Rauhela, Manish Semwal, P. K. Saini, Pooran Singh Patwal.

NATIONAL INSTITUTE OF HYDROLOGY NIH India, R. D. Singh; N. C. Gosh, V. C. Goyal, C. K. Jain, Sudheer Kumar, A. K. Lohani, Surjeet Singh, Anupama Sharma, Sumant Kumar, Shashi Indwar, Biswajit Chakravorty, Y. R. S. Rao, B. Venkatesh, T. Thomas, B. K. Puarendra, Sanjay Mittal, Rakesh Goyal, Biswajiit Das, Saroj Khatania.

INDIAN INSTITUTE OF TECHNOLOGY ROORKEE IITR India, Pradeep Kumar, Indu Mehrotra, Ankush Gupta, Medalson Ronghang, Soma Kumari, Himashu Singh, Fuzail Ahmed, Laxmi Das, Anand Bharti.

VEOLIA WATER (INDIA) PVT LTD VEOLIA India, S. V. K. Babu, Priyanka Bhat, Vikas Gupta, Anuj Goel, Brune Poirson, Mélanie Grignon, Bodhisattwa Dasgupta, Naresh Kumar, Ashok Parashar, Bharat Bhushan Chadha.

ANNA UNIVERSITY CHENNAI ANNA India, Lakshmanan Elango, Parimala Renganayaki, K. Brinda, Rajesh Rajendran, Rajaveni Sundarapandian, M. C. Raicy, Jagadesan Gunalan, Indu S. Nair, G. Gowrisankar.

SPT CONSULTANCY SERVICES PARTNERSHIP SPT India, Thirunavukkarasu Munuswamy, Shenbaganandam Ganapathy, Saravanan Janakiraman, Arulprakasm Subramanian, Balaji Karuppaiah, Parimala Renganayaki Sundaram, Balasubramanian Krishnan.

MUNICIPAL CORPORATION OF RAIPUR RMC India, A. K. Malwe.

ARUN GULATI AJD India, Arun Gulati, Manish Gupta, Siddarth Kimothi.

COUNCIL OF SCIENTIFIC AND INDUSTRIAL RESEARCH NGRI India, Shakeel Ahmed, Ramaswamy Rangarajan, Vilasrao Somvanshi, Subash Chandra, Nepal Mondal, Tanvi Arora, Sahebrao Sonkamble, Sarah Sarah, Farooq A. Dar, Naziya Jamal, Deepa Negi Kapardar, P. Raghavendra, Tarun K. Gaur, Rekha Kumari, Rakesh K. Tiwari, Adeyuppu Pratyusha, Satyajit Raut, Akoju Ramadevi, Vikram Kumar, Deepak Kumar, Napasani Veerababu, Taufique Warsi, Md. Wajihuddin, Satya Chari.

INDIAN INSTITUTE OF TECHNOLOGY BOMBAY IITB India, Shyam R. Asolekar, Dinesh Kumar, Anana Hiremath, Pradip Kalbar, Richa Singh, Rahul Sutar, Shruti Ranjan, Sachin Pandey, Ankit Srivastava, Ashish Kumar, Deepak Vishawakarma, Lohit Reddy, Sanjeev Yendamui.

DHI – (INDIA) WATER & ENVIRONMENT PVT LTD DHI India, Bertram Moninkoff, Mohamed Fahimuddin.

KOMPENTENTZZENTRUM WASSER BERLIN GEMEINNUTZIGE GMBH KWB Germany, Bodo Weigert, Christoph Sprenger, David Stevens, Gesche Grützmacher, Hella Schwarzmüller, Maike Gröschke, Michael Rustler.

BUREAU DE RECHERCHES GEOLOGIQUES ET MINIERES BRGM France, Marina Alazard, Stéphanie Aulong, Alexandre Boisson, Abdel Majiit Bouzit, Alain Chevalier, Céline Cosson, Benoît Dewandel, Christine Flehoc, Wolfram Kloppmann, Thierry Laurioux, Jean-Christophe Maréchal, Jérôme Perrin, Marie Pettenati, Géraldine Picot-Colbeaux, Géraldine Quarton, Benjamin Tellier, Dominique Thierry, Matthieu Basset.

ZENTRUM FÜR UMWELTMANAGEMENT UND ENTSCHEIDUNGSTHEORIE CEMDS Austria, Markus Starkl.

HOCHSCHULE FÜR TECHNIK UND WIRTSCHAFT DRESDEN HTWD Germany, Thomas Grischek, Cornelius Sandhu, Ulrike Feistel, Rico Bartak, Thomas Voltz, Stephanie Fischer.

UNITED NATIONS EDUCATIONAL, SCIENTIFIC AND CULTURAL ORGANIZATION – UNESCO IHE The Netherlands, Saroj Sharma, Maria Kennedy, Haziz Mutabuzi, Richard Missa, Charles Nyongo.

INTERNATIONAL WATER MANAGEMENT INSTITUTE IWMI Sri Lanka, Priyanie Amerasinghe, Mahesh Jampani.

COMMONWEALTH SCIENTIFIC AND INDUSTRIAL RESEARCH ORGANISATION CSIRO Australia, Declan Page, Peter Dillon, Joanne Vanderzalm, Jatinder Sindhu.

FREIE UNIVERSITAET BERLIN FUB Germany, Christoph Sprenger, Thomas Taute, Lutz Thomas, Maike Gröschke, Theresa Frommen, Mario Eybing, Kolja Bosch, Florian Brückner.

Glossary

Anicut A dam made in a stream for maintaining and regulating irrigation.

Artificial Storage and Recovery Injection of water into a well for storage and recovery from the same well.

Artificial Storage, Transfer and Recovery Injection of water into a well for storage, and recovery from a different well.

Bank filtration A process whereby the subsurface serves as a natural filter and also biochemically attenuates potential contaminants present in the surface water. Bank filtration can occur naturally due to higher surface water levels compared to groundwater levels, or it can be induced by lowering groundwater levels by pumping from wells.

Caisson well A well of a comparatively larger diameter (1–10 m) and shallow depth (5–10 m) that has a circular concrete, reinforced concrete or brick-lined caisson as a casing. The well is constructed by building the casing on the ground surface and subsequently sinking it (using weights or jacks) as an open caisson by excavating the interior. On account of their large diameter, caisson wells usually have a high water storage capacity.

Check dam Structure constructed across the river to harvest run-off water for groundwater recharging and regulating irrigation.

Constructed wetlands Treatment systems that use natural processes involving wetland vegetation, soils, and their associated microbial assemblages to improve water quality.

Contour trench Structures used to break the slope at intervals and reduce the velocity of surface run-off. The water retained in the trench helps to increase the soil moisture content and ground water recharge.

Disinfection Removal or inactivation of pathogenic microorganisms.

Disinfection by-products A chemical compound formed by the reaction of a water disinfectant with a precursor (e.g. natural organic matter) in a water supply system.

Duckweed pond Pond used for wastewater treatment in which thin mat of duckweed grows at the surface of water which maintains anaerobic conditions in the pond.

Escherichia coli (*E. coli*) Coliform bacteria of faecal origin used as an indicator organism in the determination of (waste) water pollution.

Gravity injection well Ordinary bore wells and dug wells used for pumping may also be alternatively used as recharge wells.

Infiltration Downward movement of water in unsaturated zone.

Injection well Structure similar to a tube well but with the purpose of augmenting the groundwater storage of a confined aquifer by pumping in treated surface-water under pressure.

Karnal Technology A method of land-treatment/disposal of wastewater which involves growing tree on ridges 1m wide and 50 cm high and disposing of the treated effluent in furrows.

Managed Aquifer Recharge Intentional storage of water into the aquifers for subsequent recovery or environmental benefits.

Nalahs bund Structures constructed across streams (Nalah or Nala) to check the flow of surface water in the stream channel and to retain water for longer durations in the previous soil or rock surface.

Natural Treatment Systems Multi-objective treatment systems employing natural processes and components (soil/aquifer, vegetation and sunlight) to improve water quality.

Open well Dugwell, commonly used at household level.

Organic micropollutants organic contaminants which are present in water, soil and environment in the range of mg/L to ng/L (also known as trace organics).

Oxidation pond Also known as lagoons, are stabilization ponds generally used to treat primary effluents (from septic tanks) by heterotrophic bacteria.

Pathogen indicators Bacteria like faecal coliform and E. coli. Their presence indicates that water may be contaminated by human or animal faecal matter.

Percolation tank Artificially created surface water body submerging a highly permeable land area so that the surface run-off is made to percolate and recharge the ground water storage.

Pre-treatment Treatment steps to improve quality of source water before employing natural treatment systems to enhance their performances.

Polishing ponds Pond systems are used to improve the quality of effluents from efficient anaerobic sewage treatment plants like UASB reactors, so that the final effluent quality becomes compatible with legal or desired standards.

Post-treatment Treatment steps to further improve quality of water after the natural treatment systems to meet the water quality standards/guidelines and regulations.

Recharge shaft Structure constructed to increase recharge into unconfined aquifers where water levels are much deeper or into confined aquifers, which are overlain by strata having low permeability.

Recharge pit Used in artificial recharge of phreatic aquifer from surface-water sources. They are excavated of variable dimensions that are sufficiently deep to penetrate less permeable strata.

Reclaimed water Wastewater that has been treated to a level that allows for its reuse for a beneficial purpose.

Risk Assessment Identification, evaluation, and estimation of the levels of risks involved in a situation (with or without certain intervention), their comparison against benchmarks or standards, and determination of an acceptable level of risk.

Soil Aquifer Treatment Artificial recharge of wastewater treatment plant effluents or storm water for further polishing its quality (in soil and aquifer) aiming at reuse.

Subsurface dam System for storing groundwater by a "cut-off wall" (dam body) set up across a groundwater channel.

Talabs Mainly natural ponds.

Tube well The typical name for vertical production wells in rural India with small diameters of 125 to 200 mm used mainly for irrigation.

Waste stabilization pond Large, shallow basins in which raw sewage is treated entirely by natural processes involving both algae and bacteria; generally consists of series of anaerobic, facultative or maturation ponds.

Wastewater Any water that has been adversely affected in quality by human influence. Wastewater can originate from e.g. domestic, industrial, commercial or agricultural activities and is often a combination thereof (often mixed with storm-water or surface run-off). Municipal wastewater is also known as sewage comprising faecal matter and urine.

Wastewater treatment plant Treatment systems consisting of different processes for improving quality of domestic and industrial wastewater (also known as sewage treatment plant).

Water table Level below which the formation is saturated with water.

Water reuse The general term for the beneficial use of treated or reclaimed (waste) water.

List of Abbreviations

AFTM	Audio frequency telluric method
AOP	Advanced oxidation process
ARR	Aquifer Recharge and Recovery
ASL	Above sea level
ASP	Activated sludge process
ASR	Aquifer Storage and Recovery
ASTR	Aquifer Storage Transfer and Recovery
ASTER	Advanced Spaceborne Thermal Emission and Reflection Radiometer
$BaCl_2$	Barium chloride
BC	Boundary condition
BCM	Billion cubic metres
BF	Bank filtration
BIS	Bureau of Indian Standards
BOD	Biological oxygen demand
BOM	Biodegradable matter
CEC	Cation exchange capacity
CGWB	Central Ground Water Board
COD	Chemical oxygen demand
CMW	Chennai Metro Water
CMWSSB	Chennai Metro Water Supply and Sewerage Board
CPCB	Central Pollution Control Board
CSP	City Sanitation Plan
CURE	Centre for Urban & Regional Excellence
CW	Constructed wetland
CWC	Central Water Commission
DC	Direct current
DEM	Digital Elevation Model
DGPS	Differential Global Positioning System
DJB	Delhi Jal Board
DO	Dissolved oxygen
DOC	Dissolved organic carbon
DP	Duckweed pond
EC	Electrical conductivity
ERT	Electrical Resistivity Tomography
EUR	Euro
FAP	Facultative anaerobic ponds
FC	Faecal coliforms
FF-CW	Free floating constructed wetland

FTIR	Fourier transform infrared spectroscopy
GAC	Granular activated carbon
GCW	Groundwater circulation wells
GHMC	Greater Hyderabad Municipal Corporation
GoI	Government of India
GMWL	Global meteoric water line
GPS	Global Positioning System
HMWSSB	Hyderabad Metropolitan Water Supply and Sewerage Board
HRT	Hydraulic retention time
HSSF-CWs	Horizontal sub-surface flow constructed wetlands
IFCGR	Indo French Center for Groundwater Research
IGP	Indo Gangetic Plain
INR	Indian Rupees
IWDP	Integrated Wastelands Development Programme
IWMP	Integrated Watershed Management Programme
JMRPL	Jalmahal Resorts Pvt. Ltd.
JNNURM	Jawaharlal Nehru National Urban Renewal Mission
KSMWS	Kachiwani Singaram micro-watershed
KT-CW	Karnal type-constructed wetlands
LBF	Lake bank filtration
LMWL	Local meteoric water line
m a.s.l.	Meters above sea level
m bgl	Meters below ground level
MAR	Managed Aquifer Recharge
MF	Microfiltration
MGNREGA	Mahatma Gandhi National Rural Employment Guarantee Act
MHT	Monitoring boreholes
MLD	Million litres per day
MoEF	Ministry of Environment and Forests, Government of India
MoWR	Ministry of Water Resources, Government of India
MP	Maturation pond
MPN	Most probable number
NARBAD	National Bank for Agriculture and Rural Development
NCT	National Capital Territory
NERC	Natural Environment Research Council
NF	Nanofiltration
NGRI	National Geophysical Research Institute
NH_3	Ammonia
NH_3-N	Ammonia nitrogen
NH_4^+	Ammonium
NIH	National Institute of Hydrology
NO_2–	Nitrite
NO_3–	Nitrate
NRMMC-EPHC-NHMRC	Natural Resource Management Ministerial Council, Environment Protection Heritage Council, Australian Health Ministers Conference
NTS	Natural Treatment Systems
NTU	Nephelometric Turbidity Unit
ODEX	Overburden drilling with excentric bit
O&M	Operation & Maintenance
OMP	Organic micropollutants
PP	Polishing ponds
PPP	Public-private partnership
PVC	Polyvinyl chloride
QMRA	Quantitative Microbial Risk Assessment
RBF	River bank filtration
RCW	Radial collector well
RO	Reverse osmosis
RRR	Repair, Renovation and Restoration scheme

RSF	Rapid sand filtration
RWH	Rainwater harvesting
SAT	Soil aquifer treatment
SGT	Subsurface groundwater treatment
SEM	Scanning electron microscopy
SME	Small and medium enterprise
SSF	Slow sand filtration
SSP	State Sanitation Plan
SUVA	Specific Ultraviolet Absorbance
SWOT	Strengths, Weaknesses, Opportunities, Threats
TDEM	Time Domain Electromagnetic Methods
TDS	Total dissolved solids
TKN	Total kjeldahl nitrogen
TLERT	Time lapse Electrical Resistivity Tomography
TN	Total nitrogen
TOC	Total organic carbon
TP	Total phosphorous
TSS	Total suspended solids
TTC	Thermotolerant coliforms
UASB	Up-flow anaerobic sludge blanket
UGC	Upper Ganga Canal
UJS	Uttarakhand Jal Sansthan
ULB	Urban local body
UNDP	United Nations Development Programme
UNESCO	United Nations Educational, Scientific and Cultural Organization
USAID FIRE D	USAID Financial Institutions Reform and Expansion (FIRE), D stands for debt
UV	Ultraviolet
VES	Vertical Electrical Soundings
VF-CW	Vertical flow constructed wetlands
VFW	Vertical filter well
VLF	Very low frequency
VSMOW	Vienna Standard Mean Ocean Water
VSSF-CW	Vertical sub-surface flow constructed wetlands
WHO	World Health Organization
WP	Work package
WSP	Waste stabilization ponds
WTP	Water treatment plant
WWTP	Wastewater treatment plant

FURTHER INFORMATION

Open access publication of the eBook version of this handbook was made possible through the OpenAIRE-FP7-Post-Grant-OA-Pilot. Underlying research material can be accessed at https://zenodo.org/. The names of all related files mention Saph Pani. Additional information can also be requested from the lead author of the chapter using one of the mail addresses in the table below.

Thomas Wintgens	thomas.wintgens@fhnw.ch (Editor)
Cornelius Sandhu	sandhu@htw-dresden.de
Ankush Gupta	ag343@cornell.edu
Maike Gröschke	maike.groeschke@fu-berlin
Anders Nättorp	anders.naettorp@fhnw.ch (Editor)
M. C. Raicy	raicygeo@gmail.com
Alexandre Boisson	a.boisson@brgm.fr
Dinesh Kumar	dinesh.poswal0197@gmail.com
Priyanie Amerasinghe	P.Amerasinghe@cgiar.org
Saroj Sharma	s.sharma@unesco-ihe.org
Christoph Sprenger	christoph.sprenger@kompetenz-wasser.de
Wolfram Kloppman	w.kloppmann@brgm.fr
Peter Dillon	pdillon500@gmail.com
Markus Starkl	markus.starkl@boku.ac.at
Shakeel Ahmed	shakeel.ngri@gmail.com
Elango Lakshmanan	elango@annauniv.edu (Editor)
Shyam R. Asolekar	asolekar@gmail.com (Editor)

Chapter 1

Introduction to natural water treatment systems in the Indian context

Thomas Wintgens, Julia Plattner, Lena Breitenmoser, Lakshmanan Elango, Shyam Aselokar, Cornelius Sandhu and Anders Nättorp

1.1 INTRODUCTION TO SAPH PANI

It is well known that groundwater is the largest source of fresh water in the world. Domestic, agricultural and industrial activities require water, and in most parts of the world the requirement is met by pumping groundwater. Even though it is a renewable resource, today, the availability of fresh water has become a vital issue in almost all parts of the world. Due to increased population, the demand for water is greater than the recharge of the groundwater sources. The demand for water is reported to increase even more by 2050 as the population is expected to rise to 9 billion by this time (FAO, 2007). In this context, water conservation needs priority attention to manage the existing water resources and to ensure sufficient availability of water for future generations. With these aspects in mind, a multi-disciplinary and multi-institutional pilot study based project was conceived. This project, given the acronym Saph Pani, on the 'Enhancement of natural water systems and treatment methods for safe and sustainable water supply in India' was co-funded by the European Commission within the 7th Framework Programme. The project was carried out at several study sites in India during the period 2011–2014. The project partners include academic institutions, research institutes, water supply agencies, and small and medium level companies. The institutions are located in Australia, Austria, France, Germany, India, Sri Lanka, The Netherlands, and Switzerland. The project consortium consisted of investigators with varied level of expertise and disciplines of study including economists, social scientists, geologists, hydrogeologists, chemical engineers, biologists, microbiologists, hydrologists, civil engineers, chemists, geophysicists, environmental engineers etc. The project highlights are field based pilot level studies in northern, central, western and southern parts of India. Aspects related to natural water treatment systems such as Managed Aquifer Recharge (MAR) by ponds and check dams, Soil Aquifer Treatment (SAT) and River Bank Filtration (RBF) where closely investigated at those sites.

1.1.1 Water resources in India

India possesses 18% of the world's population i.e., 1.25 billion, but accounts for only 4% of the world's water resources. There are different estimates reported about India's water budget in terms of available resources and current use. Table 1.1 gives a comparison of different estimates considering two evapotranspiration scenarios (UNICEF *et al.* 2013).

Depending on the scenario it becomes more or less obvious how strained the given resources are. However, not only is the total country perspective relevant but also the geographical and temporal distribution as well as variability of water availability. This can be illustrated by the fact that annual rainfall can range from around 3,700 mm in Karnataka to around 500 mm in Rajasthan (CWC, 2014). In many river basins 80% of the run-off occurs during the few months of monsoon and cannot be captured effectively but rather can lead to flooding and destruction (Mishra *et al.* 2009). The agricultural sector is the biggest water user in India with around 90% use of the total water resources. 80% of the irrigation water in agriculture is abstracted from groundwater, due to water pollution in rivers (World Bank Report, 2005). Although domestic (8%) and industrial supply (2%) are currently smaller water users (Aquastat, 2010), strong growth is expected based on the increasing population. This problem is particularly acute for India's urban population which makes up more than 28% of the total population. The potable water supply increased from 3×10^9 m^3/year in 1990 to 56×10^9 m^3/year in 2010 and is estimated to reach 110×10^9 m^3/year by 2050 according to FAO data. Similarly, the "India Vision 2020" (2002) report predicts that total water consumption in India is expected to rise by 20–40% until 2020. Wherever urbanization has been growing, the quantity of water infiltrating from the rains has decreased due to construction of impervious surfaces such as buildings, cemented pavements and roads.

Table 1.1 General water management data for India (UNICEF *et al.* 2013).

	Analysis Based on MoWR* (Values in BCM)**	**Analysis Based on MoWR* Estimates Based on Worldwide Comparison (Values in BCM**)**
Annual rainfall	3,840	3,840
Evapotranspiration	3,840 – (1,869 + 432) = 1,539 (40%)	2,500 (65%) World-wide comparison
Surface run-off	1,869 (48.7%)	Not used in estimate
Groundwater recharge	432 (11.3%)	Not used in estimate
Available water	2,301 (60%)	1,340 (35%)
Utilisable water	1,123 (48.8% of 2,301)	654 (48.8% of 1,340)
Current water use	634	634

*Ministry of Water Resources
**BCM = Billion Cubic Meters

In many cases these pressures result in a scarcity of safe drinking water supply, sanitation problems, urbanisation of wetlands, and encroachment of urban areas onto swamp and floodplain areas of rivers. The people living in these suburban areas generally have neither organized potable water supply systems nor adequate sewerage facilities. The National Water Policy of India (NWP, 2002), among other government mandates and regulations, emphasizes drinking water (section 8) and water quality (section 14) as important priorities to be addressed. The Central Pollution Control Board (2009) estimates only a third of the collected urban wastewater is actually treated in India's class I cities (the biggest 498 cities in India) and class II towns (CPCB, 2009).

Climate change, causing frequent failure of monsoons and resulting in limited surface water resources, has led to increased dependence on groundwater resources (Lal, 2001). Hence, the health of populations that depend entirely on groundwater as their sole source of drinking water (Ragunath, 1987) are particularly vulnerable due to groundwater depletion and water quality degradation. Thus, future supplies of water for urban areas will depend on the quantity and quality of the groundwater available (McIntosh, 2003).

Faced with poor water supply services, farmers and urban dwellers alike have resorted to pumping groundwater through tube wells. For example, as many as 10.2 million dug wells, 5.4 million private production (tube) wells, and 60,000 deep production wells became operational in the country in the four decades before 1999 (Nagaraj, 1999).

Groundwater levels are rapidly declining in most states in India. In the regions south of the Ganga and the Sutlej Rivers (the Indo-Gangetic alluvial plain) up to the southern margin of the Deccan Plateau (Nagaraj, 1999), which includes the nations bread basket Punjab (Brown, 2005), levels declined by more than 0.2 m/year between 1981 and 2000. In Tamil Nadu, a state with more than 62 million people in southern India, falling water tables are very common which has resulted in many wells drying up (The Economic Times, 2007). Associated crop yields are also decreasing. Declining water tables lead to land subsidence and make surface water flows less reliable since many rivers are maintained in the dry season by the groundwater discharge. Furthermore, groundwater helps to maintain other surface water bodies such as ponds, lakes and wetlands, which are also threatened by excess pumping. In coastal cities like Chennai ungoverned pumping has led to seawater intrusion, which forces the abandonment of well fields progressively inwards, away from the sea (Elango, 2009).

Groundwater management today involves many facets – it is not just the science of groundwater hydrology and water quality, but includes other issues such as sustainability and integration of social, economic and ecological aspects (Campana, 2007).

1.1.2 The role of natural treatment technologies in mitigating water scarcity in India

There is a growing need to address the water scarcity problems not only in India but all over the world. Natural treatment processes utilising the attenuation and buffering capacity of natural soil-aquifer and plant-root systems are particularly suited for this as they can be implemented by using locally available resources both in terms of the initial investment needed and operational effort (e.g., in terms of energy and chemicals). Natural treatment processes are considered most relevant for the Indian context and have huge, undeveloped potential to address water scarcity issues (UN Water, 2006).

MAR is a very suitable technique which includes rainwater harvesting, construction of infiltration wells, percolation tanks, recharge pits and shafts and managing run-off water to facilitate recharge of suitable aquifers. Some of the advantages of MAR include floodwater mitigation, control of saltwater intrusion, storage of water to reduce pumping and piping costs, temporary regulation of groundwater abstraction, and water quality improvements (Asano, 1985). A subset of the MAR concept is SAT, RBF or bank filtration (BF, common usage for rivers or lakes), which enables the utilisation of surface water sources after passing through the natural porous sub-surface (aquifer) to the production well. The porous media serves as a natural filter and also biochemically attenuates potential contaminants present in the surface water. BF is also recognised by the WHO (World Health Organisation) as a natural treatment method for drinking water (WHO, 2011) Similarly, constructed and natural wetland systems can be used for water resources protection and provision of non-potable water, e.g., for irrigation applications both in the agricultural and urban context.

Several studies have shown that MAR can be effective in raising groundwater levels (Athavale *et al.* 1992; CGWB, 2000; Karanth & Prasad, 1979; Muralidharan *et al.* 2007; Parthasarathi & Patel, 1997; Raju, 1998; Shivanna *et al.* 2004). Effective groundwater recharge through the construction of subsurface barriers was also studied by Elango and Senthilkumar (2006) in Tamil Nadu, India. These studies demonstrated that MAR is a potent solution to overcoming water scarcity. Gale (2005) states that in India, MAR in rural areas is typically performed as part of a package of measures aimed at developing or rehabilitating watersheds. Such watershed development programmes combine a range of land development/protection, soil moisture conservation, afforestation, pasture development and horticultural activities, as well as explicit water resource conservation/augmentation measures. Those measures additionally emphasize the potential role of RBF as an element of MAR and groundwater banking (Sandhu *et al.* 2011).

While numerous studies on BF exist for Europe and North America, there have been very few holistic and comprehensive studies on RBF in India. For example, the quality of bank filtrate abstracted for domestic use was studied in Nainital, Haridwar, Delhi and Mathura. Bank filtrate was found to have lower concentrations of turbidity, coliforms (especially during the monsoon) and dissolved organic carbon compared to directly abstracted surface water (Dash *et al.* 2008, 2010; Gupta *et al.* 2014b; Lorenzen *et al.* 2010; Pekdeger *et al.* 2008; Singh *et al.* 2010; Sprenger *et al.* 2008). A summary of preliminary studies of RBF systems within a description of the potential for RBF in India (Sandhu *et al.* 2011), demonstrates that such systems already play a significant role in providing drinking water to urban areas. RBF (along with more limited post-treatment chlorination) is generally accepted as the sole drinking water treatment at some sites such as in Haridwar, Nainital and Srinagar in the State of Uttarakhand. Bank filtrate monitored in recent years has shown a significantly higher quality compared to water abstracted directly from surface sources (JIWWA, 2012; Sandhu & Grischek, 2012; Bartak *et al.* 2015; see also Chapters 2 and 3). With these techniques there is great potential for improved treatment of surface water. Hence, due to the need to augment ground water quantity in order to sustainably meet India's future water needs it is sensible and essential to adopt and implement natural water systems and treatment technologies.

1.1.3 Saph Pani project objectives

The Saph Pani project aimed to improve natural water treatment systems such as BF, MAR and wetlands in India. The project considered particularly water resources and water supply in water- stressed urban and peri-urban areas in different parts of the Indian sub-continent.

The objective was to strengthen the scientific understanding of the performance-determining processes occurring in the root, soil and aquifer zones of the relevant regions and considered the removal and fate of important water quality parameters such as pathogenic microorganisms and respective indicators, organic substances, nutrients and metals. The hydrologic characteristics (infiltration and storage capacity) were also investigated to strengthen local or regional water resources management strategies (e.g., by providing buffering of seasonal variations in supply and demand). The socio-economic value of enhanced utilisation of the treatment and storage capacity of natural systems was evaluated taking into account sustainability and system risk management.

1.1.4 Saph Pani approach and methodology

The project focused on a set of case studies in India (see Section 1.1.5). These included a range of natural water systems and treatment technologies investigated by different work-packages (WP) such as BF (WP1), MAR (WP2) and constructed wetlands (CWs) (WP3) as illustrated in Figure 1.1.

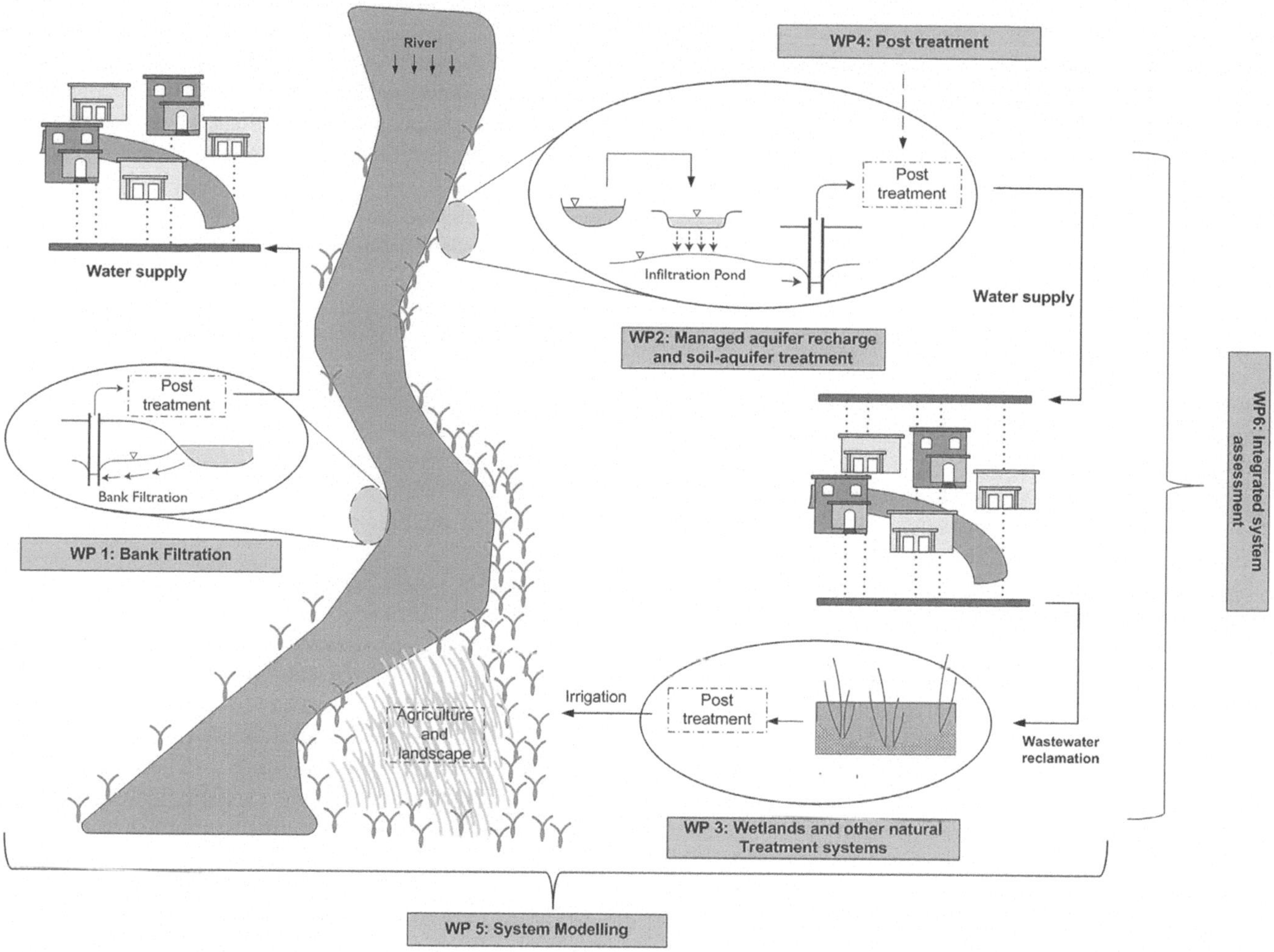

Figure 1.1 Overall Saph Pani approach showing main technology areas and relation to work package.

The field site investigations included hydrogeological, hydrological and geochemical characterisation and in some cases water quality monitoring or pre-feasibility studies for new treatment schemes. In addition, recommendations regarding appropriate pre- and post-treatment (WP4) to produce potable water quality and to avoid clogging of the sub-surface structures was given. The experimental and conceptual studies were supported by modelling (WP5) which improved the theoretical understanding of the sites and enhanced the transferability of results. Furthermore, a sustainability assessment was performed, covering health, environmental, economical, institutional and social aspects (WP6).

1.2 SAPH PANI CASE STUDY SITES

Saph Pani has performed case studies on eight different sites. In WP1 BF as a natural water system and (pre-)treatment technology was addressed in four case studies in urban areas in North India, namely Haridwar, Srinagar (Uttarakhand), Nainital and Delhi (see Figure 1.2, Table 1.2 and Chapter 2 and 4). In WP 2 MAR was investigated in the rural site Maheshwaram and the urban sites Chennai and Raipur (see Table 1.3). In work package 3 a natural wetland with SAT near Hyderabad (Musi River Watershed) and a specially designed and CW in Mumbai have delivered "hands-on" knowledge and data for further interpretation. The case study sites are situated in different geographic and climactic situations, addressing natural phenomena such as drought, flood and saltwater infiltration. The main characteristics of the study sites are briefly described in the paragraphs below.

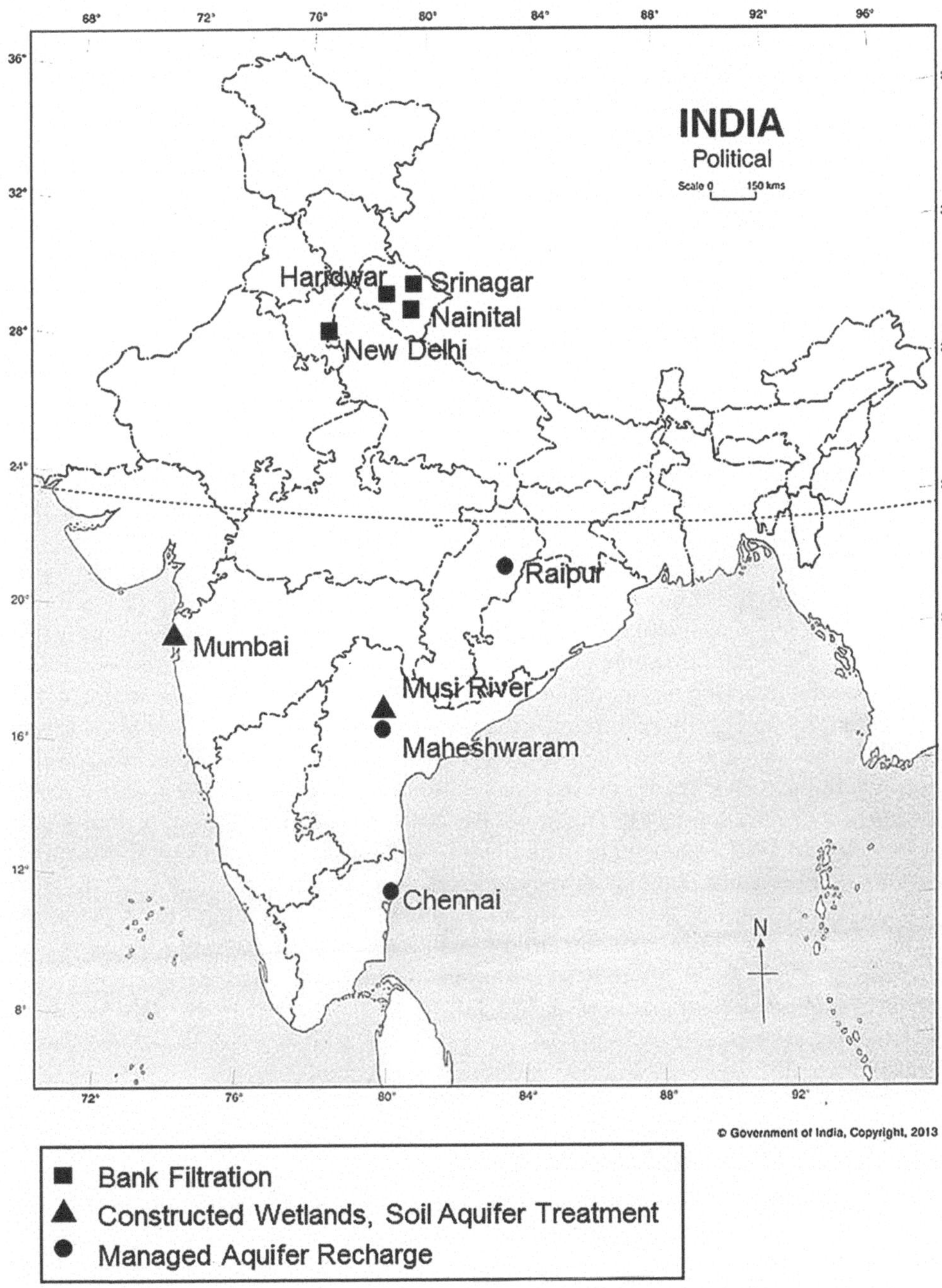

Figure 1.2 Location of the Saph Pani case study sites (Source Map: © Government of India, 2013).

Table 1.2 Features of case studies on bank filtration (BF) in Uttarakhand and NCT Delhi.

Case Study Site	Haridwar	Srinagar	Nainital	Delhi
Population	>225,000 (plus 5,000 pilgrims daily & occasionally 8.2 million)	~80,000 (Gupta *et al.* 2014)	>41,000 (plus 5,000 tourists in summer months)	~16,000,000 (Census of India, 2011)
Source-water quality	**Good**	**Good**	**Moderate**	**Poor**
Current raw water source	Bank filtrate & groundwater	Surface water, groundwater & bank filtrate	Bank filtrate & groundwater	Surface water, groundwater & bank filtrate
Issues relevant to natural water systems & treatment technologies addressed in the case studies on BF	• Removal of pathogens and turbidity • Reliable drinking water disinfection • Flood protection of wells • High nitrate concentration in land-side groundwater (in Srinagar) • Adaptation of management of RBF schemes (Srinagar) • Catchment area and well head protection of wells		• Removal of pathogens and organics • Sustainability of existing bank filtration system • Discharge of partially treated to untreated urban wastewater via sewer canals entering lake	• High Ammonium in groundwater • High pollution from wastewater in Yamuna • Identification of suitable remediation options for existing RBF wells

1.2.1 Field site in Haridwar by Ganga River

Haridwar is located in the state of Uttarakhand next to the Ganga River and is one of the most important Hindu pilgrimage sites in the world. In addition to the permanent residents of around 225,235 residing in the main part of the city (Census of India, 2011), about 50,000 pilgrims come to Haridwar daily with up to 8.2 million people during specific religious festivals such as the Kumbh Mela (Gangwar & Joshi, 2004) for the ritual of bathing in the Ganga. Hence the city's drinking water supply, managed by the Uttarakhand State Water Supply and Sewerage Organisation – Uttarakhand Jal Sansthan (UJS), both quantitatively and qualitatively, is difficult because of the highly varying demand. As of 2013, around 59,000 to 67,000 m^3/d of water for drinking purposes is abstracted from 22 RBF wells (Figure 1.3), which account for around two-thirds of the total drinking water supply to Haridwar city; the remainder is abstracted from groundwater production wells. The abstracted water only requires disinfection by chlorination, and can provide safe drinking water even when facing high variations in water demand and during monsoons. However, in September 2010 during one of the severest monsoons in North India, the Ganga River rose to an unprecedented level of 296 m above sea level (CWC, 2014) inundating the ground surface around some of the nearby RBF wells. Consequently, high turbidity and coliform counts were observed in these wells. In WP1, water quality and hydrogeological site data from Haridwar was collected and analysed. Subsequently, and supported by column experiments, the need for flood-proofing RBF wells was elucidated and guidelines for mitigation of the risk of floods on RBF wells were formulated (Saph Pani D1.2, 2013; see also chapter 2). In conjunction with WPs 4, 5 and 6 the current disinfection (post-treatment) of RBF water was assessed and suggestions for improvement were made, a numerical groundwater flow modelling study was conducted to obtain an improved understanding of the travel time, flow path and portion of bank filtrate abstracted by the RBF wells, and an integrated assessment of technical and socio-economic aspects (Essl *et al.* 2014) as well as a risk-based assessment of the RBF site were conducted (Bartak *et al.* 2015).

1.2.2 Field site in Srinagar by Alaknanda River

The town of Srinagar is located along the road to the important Hindu shrine of Badrinath in the Himalayas. It is the main commercial and administrative centre of the district of Pauri in Uttarakhand, and is one of the largest towns along the Alaknanda River. However, the production of drinking water has not been able to match the growing demand. Around 90% of the drinking water is conventionally treated surface water from Alaknanda River (produced from directly abstracted Alaknanda River water followed by conventional treatment), with the remainder being supplied by groundwater and RBF wells. During the severe Monsoons of 2010, 2011 and 2013 the surface water supply had to be discontinued for several days due to excessive turbidity and damage to water abstraction structures. To determine whether RBF could serve as an alternative to surface water treatment for the year-round production of drinking water and to address the deficit between demand and supply, UJS constructed an RBF well in 2010 (Kimothi *et al.* 2012; Figure 1.3). However, extreme events like the highest ever-recorded flood of June 2013 resulted in the RBF well being completely inundated and a fine sand layer of more than 1 m thickness was deposited over the well, even though the site was considered to have an elevation above the previous highest recorded flood level (see also Chapter 2). In WP1 the water quality of the RBF system was investigated at this site (Saph Pani D1.1, 2012). Nitrate in excess of the Indian drinking water limit of 45 mg/L was found in the water abstracted by the RBF well. The source might be the leaching of nitrate from geogenic phyllite present in the bedrock (Gupta *et al.* 2015a) and as such is an extremely rare case for an RBF site. Using the primary field data collected in WP1, a groundwater flow modelling study was conducted in WP5 to optimise the operation of the RBF site in order to lower the nitrate concentration in the abstracted water (Saph Pani D5.2, 2012). Furthermore, the breakthrough of bacteriological indicators was assessed to mitigate the risk of floods and a sanitary sealing was constructed for the RBF well (Saph Pani D1.2, 2013; see also Chapter 2).

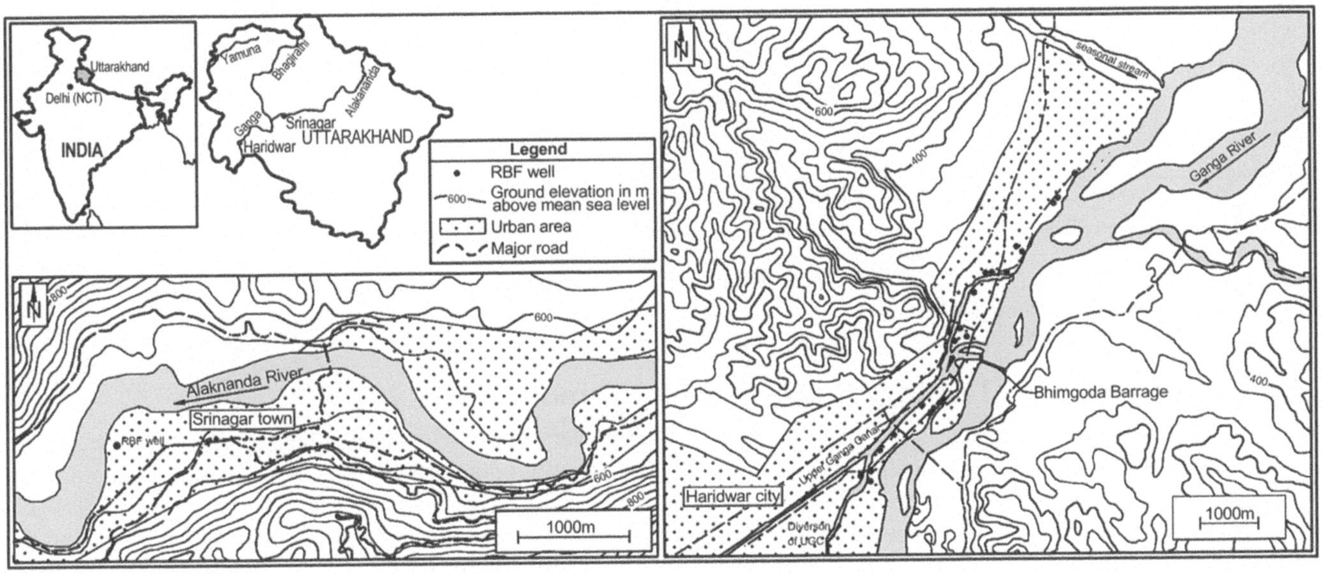

Figure 1.3 Location of riverbank filtration (RBF) wells at Srinagar (left) and Haridwar (right), Uttarakhand (HTW Dresden and IIT Roorkee, 2014).

1.2.3 Nainital by Nainital Lake

Nainital, located around the Naini Lake in the foothills of the Himalayas in the Kumaon region of Uttarakhand, is the administrative headquarters of the district of Nainital and is a popular tourist destination especially in summer and by virtue of the lake. From 1955 up to around 2007, lake water was directly abstracted and conventionally treated to supplement the existing water supply from springs (Dash *et al.* 2008). The BF system was subsequently developed in stages at the bank of Lake Naini (Figure 1.4) starting in 1990 and by around 2007 had completely replaced the direct abstraction of surface water on account of the superior treatment by BF (Dash *et al.* 2008). However, over the past decades anthropogenic activities in Nainital led to severe degradation of the Naini Lake, which became increasingly eutrophic (Pant *et al.* 1981). Since 2007, several measures have been taken to conserve the lake and check its eutrophication, including upgradation of the town's sewer system to prevent wastewater from entering the lake, cleaning and prohibiting solid waste disposal into the lake and introduction of hypolimnetic aeration in the lake (Gupta *et al.* 2015b). In WP1, the water quality of the BF wells, lake and groundwater was investigated by Gupta *et al.* (2015b; see also Chapter 3) to assess BF performance after recent changes in the Naini Lake and its catchment (see chapter 3). Stable isotope analysis was conducted to assess the portion of groundwater and bank filtrate in each well. The results were compared with previous data (Dash *et al.* 2008) to assess the long-term changes in water quality. This study also revealed important insights into the groundwater hydrology of the BF well field in

Nainital and showed that bank filtrate has better water quality than groundwater in terms of inorganic ion concentrations throughout the year. The water quality data was used to determine options for post-treatment in conjunction with WP4.

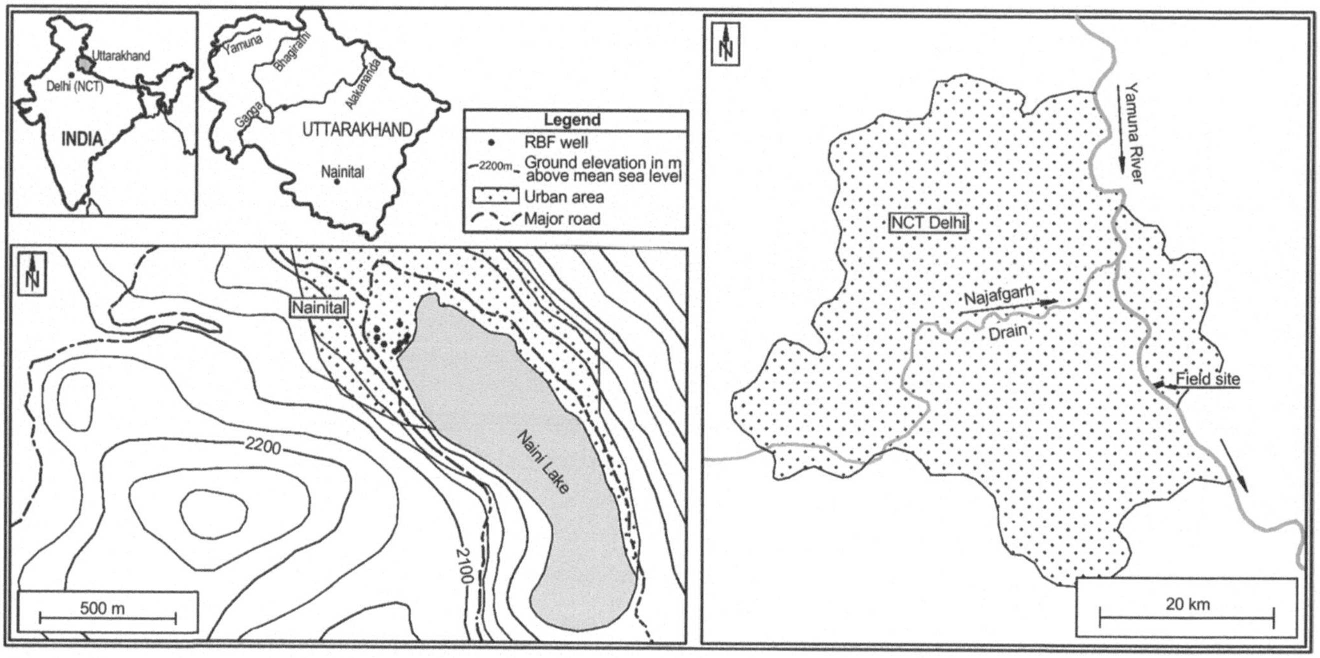

Figure 1.4 Location of bank filtration (BF) wells at Nainital (left) and study area in National Capital Territory (NCT) Delhi (right) (HTW Dresden and IIT Roorkee, 2014).

1.2.4 National Capital Territory (NCT) Delhi by Yamuna River

The study site is located on the eastern bank of the Yamuna River in Delhi opposite the Nizamuddin area on the western bank of the river (Figure 1.4). Here, the Yamuna River is heavily polluted by poorly treated and untreated urban wastewater and the existing drinking water production wells close to the river produce bank filtrate of low quality. The high concentration of nitrogen (mostly ammonium) makes the abstracted groundwater unfit for drinking water purposes and one of the production wells is not used for the water supply most of the time (Groeschke, 2013). Furthermore, the groundwater is used by urban dwellers and small scale farmers. Before distribution, the bank filtrate is mixed with surface water to reduce the ammonium concentration. If the ammonium concentration is too high the distribution is stopped. In order to understand the behaviour of ammonium in aquifers at BF sites with surface waters highly polluted by untreated sewage, field data have been collected and laboratory column studies have been conducted with aquifer material from Delhi. Results of the analyses and experiments were used as the basis for recommendations about the application of BF in nitrogen contaminated aquifers (see also Chapter 4). The removal of pollutants (nitrogen species and trace organic compounds) was studied and appropriate post-treatment was suggested in WP4.

1.2.5 Maheshwaram

One of the main experimental watersheds relevant for MAR studies in WP2 is located around the town of Maheshwaram (Figure 1.5) near Hyderabad. With a total area of 54 km^2, the watershed is located in a semi-arid hard-rock context typical of the entire region where the saprolite, a chemically weathered rock, layer (10–20 m thick) is usually unsaturated. It is a watershed with a high density of groundwater production wells (>700) mostly for paddy irrigation. Changes in land use have occurred since 2006 because of the new Hyderabad international airport located less than 10 km away. It is expected to become a peri-urban area within the coming years as significant housing projects are planned. Over the last decade MAR has been implemented throughout the watershed in the form of percolation tanks, check dams, etc. (see Chapter 7). Intensive groundwater exploitation for irrigation and domestic use has resulted in aquifer over-exploitation and deterioration of groundwater quality. The fluoride level is above the maximum permissible limit of 1.5 mg/L and salinization and agricultural inputs have been observed. MAR is an attractive concept for groundwater augmentation and enhanced groundwater quality.

The objective of the tasks in WP2 for the case study site was to investigate the potential of percolation tanks to enhance recharge and groundwater quality in the over-exploited hard-rock aquifer by implementing a sophisticated monitoring strategy for groundwater levels and quality, conducting hydro-geochemical analysis, and investigating the hydrodynamics with the support of a conceptual groundwater balance model developed in cooperation with WP5.

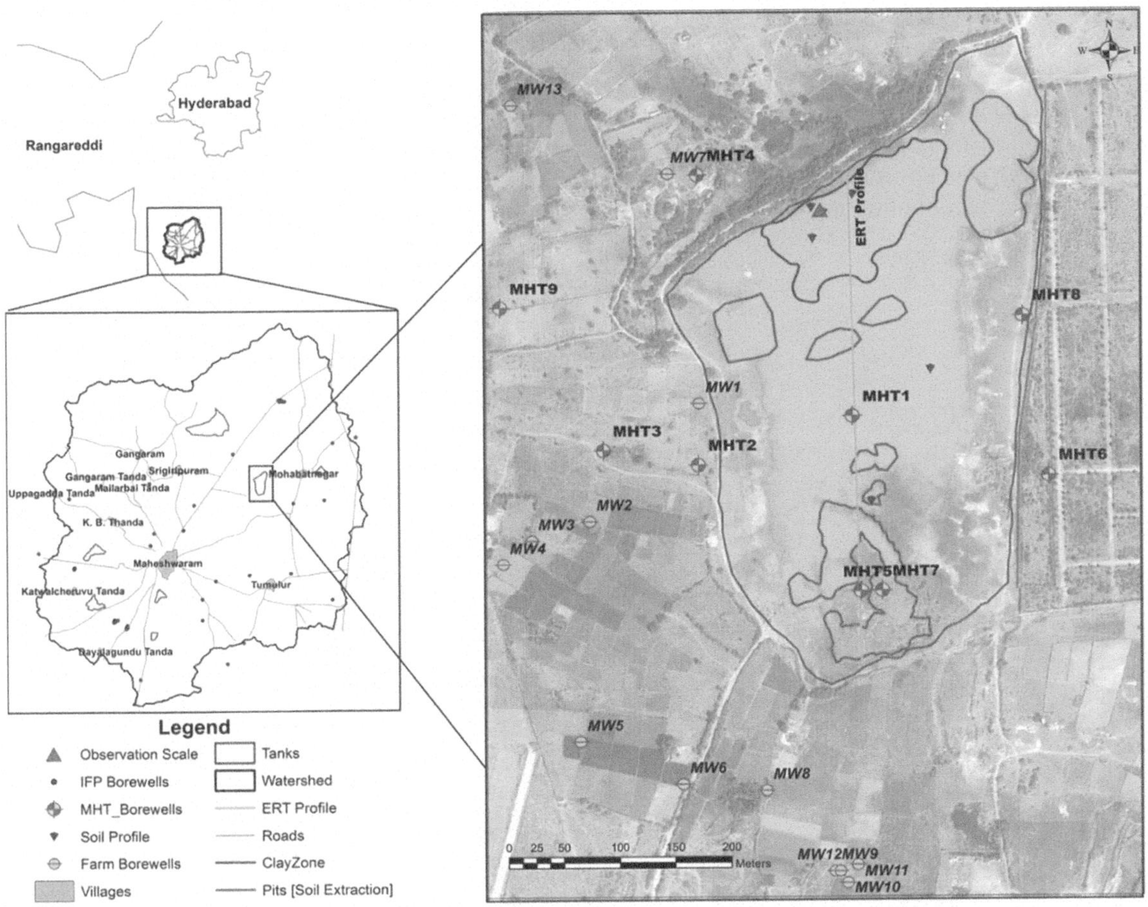

Figure 1.5 Geological map of Maheshwaram watershed (see Figure 1.2 for location of study site within India) (Boisson *et al.* 2014c).

1.2.6 Chennai

Chennai is the largest city in South India located in the eastern coastal plains. Water supply to the Chennai city is met from reservoirs and by pumping groundwater. The total surface water basin area of Chennai region is 7,282 km^2, divided between Andhra Pradesh and Tamil Nadu. The north east monsoon is significant as it replenishes the surface reservoirs and also recharges the groundwater. Water is stored in reservoirs and used cautiously. The reservoirs are virtually empty before the next annual rainfall and the shortfall in surface water storage systems is met from groundwater resources (Elango, 2009). More than one third of the water demand is met by groundwater from three wellfields known as Minjur, Panjetty and Tamaraipakkam situated about 40 km north of Chennai (Figure 1.6). The average rainfall on the basin is 7–9 billion m^3/year, which corresponds to 950–1,250 mm/year. Even though the annual rainfall is moderate, extreme cases of very high daily rainfall were recorded in past in the Chennai basin. Severe rainfall during short periods of time combined with high percentage of impervious areas in this region is the major source of flooding (D5.2, Saph Pani). The city has two major rivers, Adyar and Cooum. Both rivers usually carry sewage and only heavy rain events can flush them. The city itself has meager groundwater resources due to very little recharge. However, a considerable amount of groundwater is pumped to the city from well fields located in the Arani and Korttalai river basins north of Chennai. Severe pumping from these regions for Chennai's water supply and for local irrigational needs has resulted in seawater intrusion (D5.2, Saph Pani). The Minjur wellfield lies nearest to the coast (9 km) and is hydraulically connected with the sea. Since 1969 this wellfield has been intruded by seawater up to a distance of 15km inland due to extensive extraction of groundwater for agricultural, industrial and domestic uses for prolonged periods. Thus, on one hand the Chennai region is affected by floods and on the other hand by severe shortage of water. The objective of the tasks in WP2 for the Chennai case study site was to undertake a comprehensive assessment of MAR for coping with sea-water intrusion and groundwater over-exploitation and to develop recommendations and a management plan for implementing MAR systems that utilize excess monsoon water to counteract seawater intrusion by:

- Using existing data in close collaboration with WP5 (modelling).
- Conducting a water quality monitoring programme, hydro-geochemical analyses and geophysical investigations.
- Identifying a pilot area or an existing recharge structure to determine its efficiency in preventing seawater intrusion.
- Development of a decision support system for identifying suitable measures to counteract seawater intrusion by utilizing artificial recharge in cooperation with WP5.
- Assessing the impact of infiltration through temple tanks on groundwater quality development.

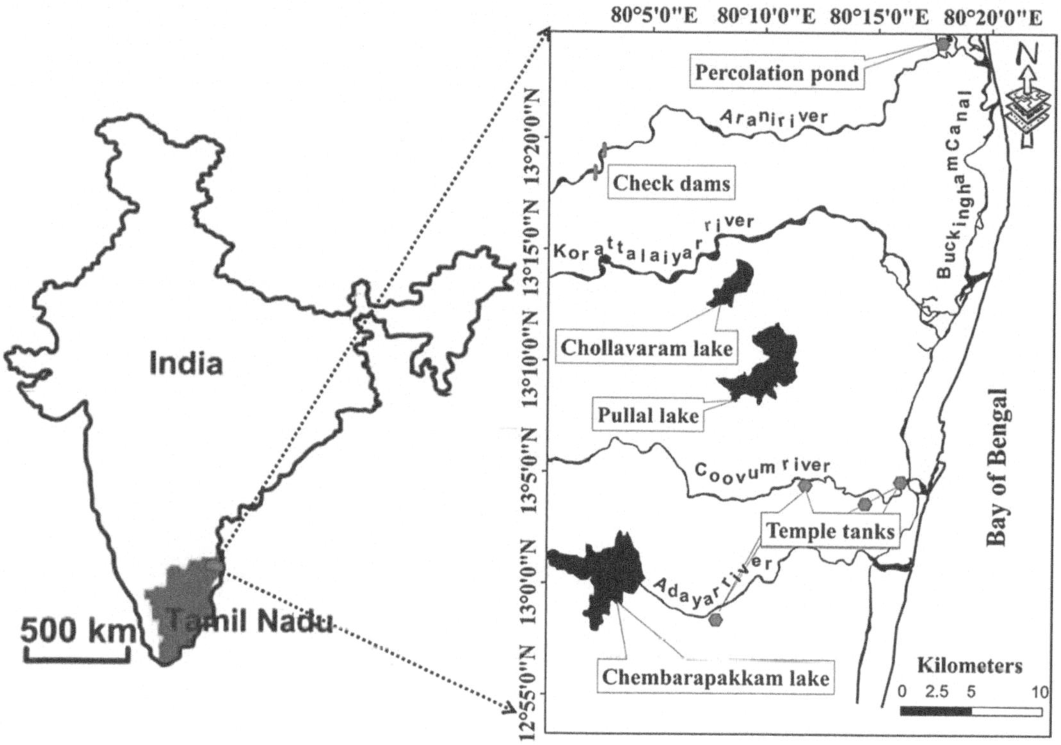

Figure 1.6 Location of the Check dams and the Percolation pond north of Chennai city (Raicy *et al.* 2015).

1.2.7 Raipur

Raipur city is situated in the central Indian state of Chhattisgarh in the fertile plains of the Mahanadi River basin. According to a projection, the present population of the city is nearly 1,200,000 (Agrawal, 2013). Currently about 65% of the city's water supply is covered by pipeline distribution of surface water from Kharun River, a tributary of Mahanadi that has been dammed to create a reservoir. About 35% is supplied by local hand pumps which tap the aquifers that are sparsely distributed throughout the city and consist mainly of deeply weathered baseline rocks. Growth of population led to excessive withdrawal and depletion of the old dug wells and hand pumps. At present any private individual interested in having a bore well has to seek a series of permits and 'No Objection Certificates' from the Government.

Initially Raipur had 154 natural and man-made water bodies, locally called talabs. These water bodies were usually connected by storm water channels. Therefore, despite a history of heavy rainfall of more than 1,200 mm during the monsoon period of 4–5 months per year, the city never had any problems with water logging and flooding. Because of the unplanned city development only 85 talabs have survived to the present. As a result the excess storm water leads to water logging even during short but intensive rainfall events. The excess of water during monsoon time is in strong contrast to the periods of water scarcity throughout the rest of the year. Furthermore, the large scale deterioration of lake water quality throughout the city due to the discharge and dumping of municipal wastes, both liquid and solid, is a big problem. Some highly contaminated lakes have become so malodorous and intense breeding sites for mosquitos that nearby residents are considering the option of converting these talabs into landfilling or drying them up. This would, however, further complicate storm water management and increase the problem of water logging.

In lieu of population growth and the resulting pressure to the already inadequate infrastructure, there is a need for reintegration of lakes in the overall water management in order to address the problems of ground water depletion, buffering of flood waters, and prevention of water logging. The objective of the tasks in WP2 for Raipur was to make an assessment of the potential of storm water infiltration from existing talabs by selecting test sites for infiltration studies, hydrogeological investigations and groundwater quality investigations on the basis of an existing storm water management model, and to derive recommendations for measures to enhance lake water quality and improve infiltration capacity if possible.

1.2.8 Mumbai

The Mumbai case study in WP3 focused on the setup and evaluation of a CW. The study site is located in Mumbai on the campus of IIT Bombay which is set in the northern suburb of the city. Currently, about 15,000 people reside on the campus including students, staff and family of faculties. The campus area is about 200 hectares and a wide range of biodiversity can be found. A pilot scale horizontal subsurface flow CW was commissioned for demonstration and research purposes. Local available wastewater from the campus was used as a feed for the wetland. Tracer tests were performed to investigate the flow pattern in the wetland and biochar was produced from the plant material. The objectives of the case study were to determine suitable plant species to be applied in the wetland, the assessment of performance of the system and to explore innovative ways to utilise the harvested plant material. The schematic setup of the pilot-scale CW-facility is displayed in Figure 1.7.

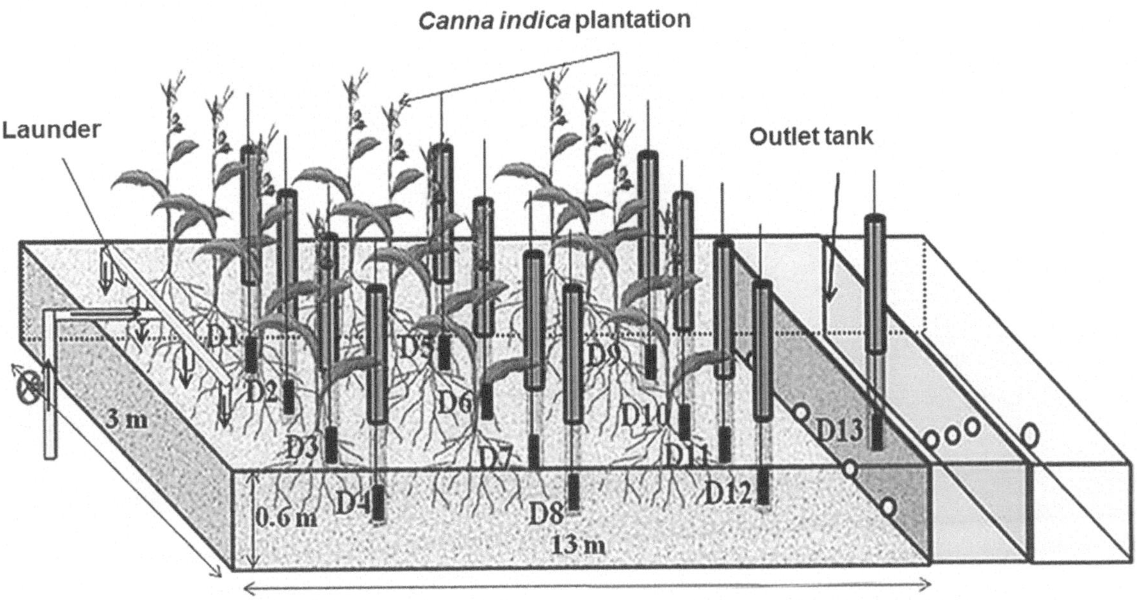

Figure 1.7 Scheme of the constructed wetland, (D1, D2, D3, . . . , D13 are the detectors).

1.2.9 Hyderabad, Musi River watershed

The Musi River flows through the city of Hyderabad which is located in the state of Telangana and is associated with an ancient irrigation system comprising a series of natural and engineered wetlands (Figure 1.8). Over 1.2 million m^3/day of partially treated domestic and industrial wastewater is discharged into the river from the city.

The river water is used for irrigation, either directly via a system of irrigation canals or after storage in sedimentation ponds/tanks (Ensink *et al.* 2010). The wastewater is a significant resource in this semi-arid periurban environment where the cultivation of fodder grass, paddy and vegetables provides economic benefits to many inhabitants of the area.

Year round cultivation, which generates large return flows from irrigated fields contributes to a large share of the hard rock aquifer recharge. Shallow groundwater is also pumped locally for irrigation on terrains where canal water is not accessible, or where it is too polluted for certain crops, especially paddy rice. A micro-watershed in the Musi River catchment, near Hyderabad was hydrodynamically characterised (WP3) and modelled (WP5) and the natural cycles of water, salt, nutrients and selected pollutants where described. This model was used to optimise the performance of a combination of passive and engineered forms of treatment.

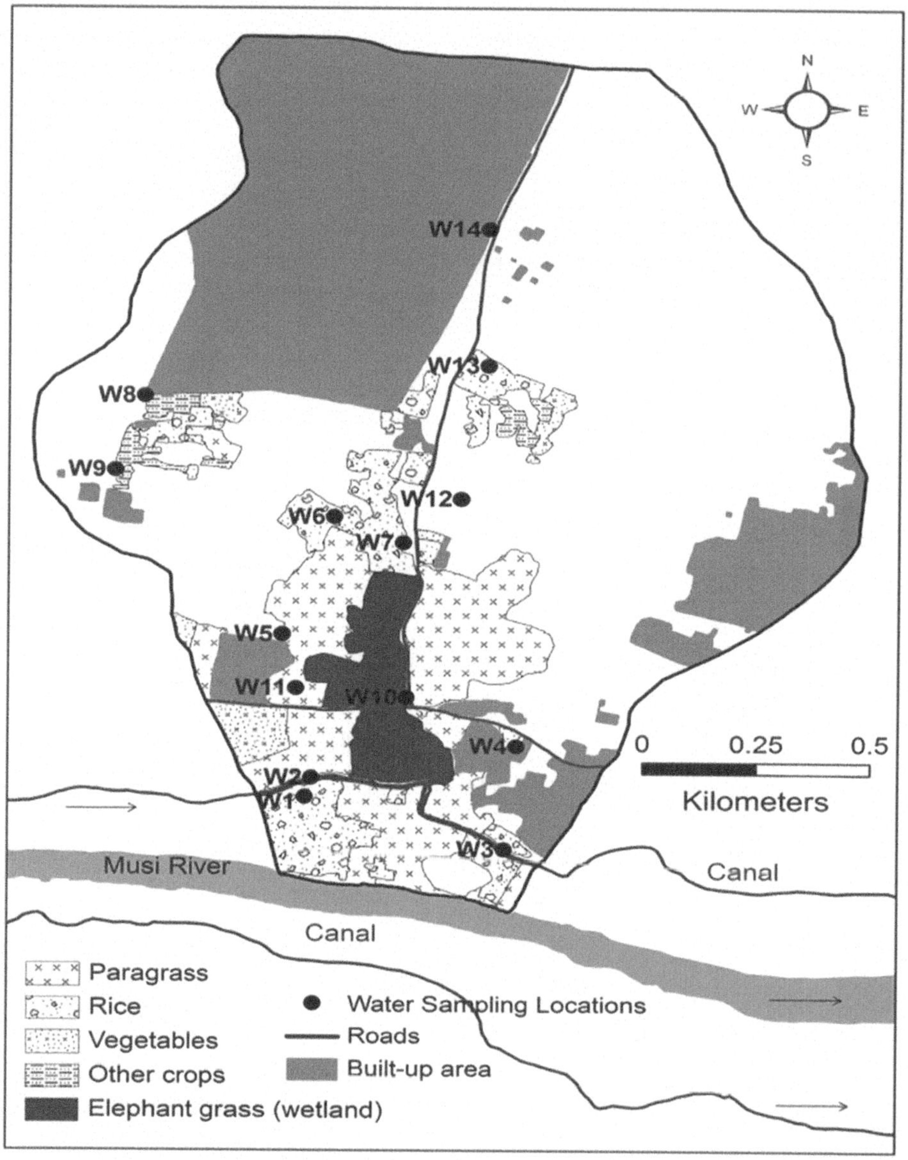

Figure 1.8 Musi River micro-watershed downstream of Hyderabad city. *Source*: IWMI (see chapter 11).

1.2.10 MAR and SAT Case study summary

Some basic features pertaining to study sites investigating MAR or wetland approaches are presented in Table 1.3. Three of the sites investigated MAR and are intended to augment the groundwater resources for drinking water (two cases) and irrigation water (one case) supply. In the Musi River Watershed wetlands could make a contribution to improve water resources quality for irrigation water.

Table 1.3 Features of case studies on MAR and wetlands in Maheshwaram Chennai, Hyderabad and Raipur.

Urban Case Study Site	Maheshwaram	Chennai	Raipur	Hyderabad (Musi River Watershed)
Natural water systems & treatment technologies considered	MAR	MAR	MAR	Wetland
Permanent population (Census of India, 2011)	4,000	6,560,250	1,200,000 (2010)	5,742,000
Rainfall [mm/a]	700–1,000	950–1,250	1,400	300–500
Current major raw water source	Groundwater	Groundwater (supplied from well fields of Arani and Korttalai river basins north of city) & surface water	Surface & Groundwater	Surface water & shallow groundwater
Uses of raw water	Irrigation	Drinking water	Drinking water	Irrigation, ground water recharge, nutrient recycling, pathogen removal
Water quality	Poor (high concentration of salts & nutrients)	Varying degrees (salt water intrusion due to over abstraction of groundwater)	Poor, due to eutrophication of local talabs of varying sizes (disposal of partially treated to untreated domestic sewage)	Poor due to domestic sewage, industrial pollutants and agriculture run-off
Issues relevant to natural water systems	• Overexploitation of groundwater • Deterioration of groundwater quality • Fluoride exceeds maximum permissible limit (1.5 mg/L) • Salinisation	• Low exploitable groundwater resources due to low recharge • Floods due to inadequate storm water run-off in monsoon • Salt water intrusion • Hydro-geochemical evolution – cation exchange	• Contamination of surface water • Eutrophication • Waterlogging • Flooding from storm water discharge • Risk for future contamination of groundwater with nitrate	• Contaminants from sewers and industries; treatment potential not quantified • Minimal institutional support
Treatment technologies to be addressed in the project	• Groundwater aug-mentation by MAR • Enhanced ground-water quality by infiltrating water into the aquifer through percolation tanks and defunct dugwells	• Check-dams • Pre-treatment i.e. settling ponds • treatment: sand filtration, infiltration ponds	Using city's numerous existing lakes and talabs as: • Infiltration ponds • For sand filtration • Integrating storm water drainage into an integrated MAR approach	Based on the status of existing surface water quality, enhancement of natural treatment technologies to reach water quality standards for drinking and/or agriculture purposes
Relevant references	Boisson *et al.* (2014a, 2014b); Pettenati *et al.* (2014); Picot-Colbeaux *et al.* (2014); Dewandel *et al.* (2006); Maréchal *et al.* (2004, 2006); Wyns *et al.* (2004); Perrin *et al.* (2011)	Parimalarenganayaki and Elango (2014), Elango (2009)	CGWB (2007), Mukherjee *et al.* (2011)	Amerasinghe *et al.* (2009); Schmitt *et al.* (2010)

1.3 STRUCTURE OF THE BOOK

The book follows the project structure and depicts the various case studies and the accomplished investigations in detail in the different chapters. The first sections cover the BF studies, first giving an overview on the relevance of that technology for India and then reporting on case study specific work. Similarly, the book section on MAR will start with a review chapter on the current MAR practice in India. Maheshwaram and Chennai will then be detailed as case studies and the methodologies and results will be depicted.

Natural and CW systems are in the focus of the third book section, which is opened by a status on natural water treatment including wetlands for wastewater treatment in India. The experiences gained with the experimental studies utilizing the CW system on the IIT Bombay campus including complementary small scale studies are described subsequently. The Musi river basin is the second case study on wetland systems which is illustrated.

The book goes further to analyse the role of pre- and post-treatment systems in combination with natural treatment components, particularly wetlands and MAR. A methodology to present adequate pre- and post-treatment systems to comply with water quality requirements is presented. Dedicated chapters detail the role of hydrological and hydro-geochemical modelling approaches to facilitate analysis, design and operation of natural water treatment systems. Examples are provided related to the Saph Pani case studies which show how the fundamental understanding of the conditions can help in system development. A section on integrated assessment of natural water treatment systems concludes the book. The basic methodology followed is outlined and application cases are described. Specific attention is paid to health risk assessment and the basic structure of MAR recharge risk assessment guidelines is outlined.

1.4 REFERENCES

Agrawal V. (2013). Raipur as a new capital: Impact on population. *International Journal of Engineering Research Science & Technology*, **2**(1), 41–46. ISSN: 2319–5991

Amerasinghe P., Weckenbrock P., Simmons R., Acharya S. and Blummel M. (2009). An atlas of water quality, health and agronomic risks and benefits associated with "wastewater" irrigated agriculture an atlas of water quality and agronomic risk and benefits associated with "wastewater" irrigated agriculture. A study from the banks of the Musi River, India. http://www.freidok.uni-freiburg.de/volltexte/6963 (accessed 15 January 2014).

Asano T. (1985). Artificial Recharge of Groundwater. Butterworth Publishers, Boston, United States.

Athavale R. N., Rangarajan R. and Muralidharan D. (1992). Measurement of natural recharge in India. *Journal of Geological Society of India*, **39**(3), 235–244.

Aquastat (2010). FAO's global water information system. http://www.fao.org/nr/aquastat/ (accessed 8 September 2014).

Bartak R., Page D., Sandhu C., Grischek T., Saini B., Mehrotra I., Jain C. K. and Ghosh N. C. (2015). Application of risk-based assessment and management to riverbank filtration sites in India. *Journal of Water and Health*, **13**(1), 174–189.

Boisson A., Baisset M., Alazard M., Perrin J., Villesseche D, Kloppmann W., Chandra S., Dewandel B., Picot-Colbeaux G., Ahmed S. and Maréchal J. C. (2014a). Comparison of surface and groundwater balance approaches in the evaluation of managed aquifer recharge structure: Case of a percolation tank in hard rocks aquifer in India. *Journal of Hydrology*, **519**, 1620–1633. doi:10.1016/j.jhydrol.2014.09.022.

Boisson A., Villesseche D., Baisset M., Perrin J., Viossanges M., Kloppmann W., Chandra S., Dewandel B., Picot-Colbeaux G., Rangarajan R., Maréchal J. C. and Ahmed S. (2014b). Questioning the impact and sustainability of percolation tanks as aquifer recharge structures in semi-arid crystalline context. *Environmental Earth Sciences*, doi 10.1007/s12665–014–3229–2.

Boisson A., Guihéneuf N., Perrin J., Maréchal J. C., Bour O., Dewandel B., Dausse A., Viossanges M. and Chandra S. (2014c). Estimation of the vertical evolution of hydrodynamic parameters in weathered and fractured crystalline rock aquifers: Insights from a detailed study on an instrumented site. *Submitted to Hydrogeology Journal*.

Brown L. R. (2005). Outgrowing the Earth: The Food Security Challenge in an Age of Falling Water Tables and Rising Temperatures. W. W. Norton & Co, New York, United States.

Census of India (2011). Cities having population 1 lakh and above. Provisional Population Totals. Office of The Registrar General and Census Commissioner, Ministry of Home Affairs, Government of India, India. In: Saph Pani D1.2 (2013). Guidelines for flood-risk management of bank filtration schemes during monsoon in India. Saph Pani Project Deliverable. http://www.saphpani.eu/downloads (accessed 29 September 2014).

CGWB (Central Ground Water Board) (2000). Manual on Artificial Recharge of groundwater, Ministry of Water Resources, Government of India, India. http://www.cgwb.gov.in/documents/Guide_on_ArtificialRecharge.pdf (accessed 30 September 2014).

CGWB (2007). Central Ground Water Board North Central Chhattisgarh Region. www.cgwb.gov.in/NCCR/index.html (accessed 8 September 2014).

CWC (2014). River stage gauge and flood forecasting station Haridwar. Central Water Commission (CWC), Ministry of Water Resources, Government of India, India. http://www.india-water.gov.in/ffs/data-flow-list-based/flood-forecasted-site (accessed 28 August 2014).

Campana M. E. (2007). Sustainable Management of Groundwater in Mexico. In: Proceedings of a workshop, Series: Strengthening Science-Based Making in Developing Countries, pp. 26–35.

CPCB (Central Pollution Control Board) (2009). Status of water supply, wastewater generation and treatment in class-I cities and class-II towns of India. Control of Urban Pollution series: CUPS/70/2009–10, CPCB, Ministry of Environment and Forests, Government of India, New Delhi, India.

Dash R. R., Mehrotra I., Kumar P. and Grischek T. (2008). Lake bank filtration at Nainital, India: Water-quality evaluation. *Hydrogeol. J.*, **16**(6), 1089–1099.

Dash R. R., Bhanu Prakash E. V. P., Kumar P., Mehrotra I., Sandhu C. and Grischek T. (2010). River bank filtration in Haridwar, India: Removal of turbidity, organics and bacteria. *Hydrogeology Journal*, **18**(4), 973–983.

Dewandel B., Lachassagne P., Wyns R., Maréchal J. C. and Krishnamurthy N. S. (2006). A generalized 3-D geological and hydrogeological conceptual model of granite aquifers controlled by single or multiphase weathering. *J. Hydrol.*, **330**, 260–284.

Elango L. (2009). Groundwater modeling to assess the feasibility of pumping seawater from a beach well for Chennai desalinization plant. *Journal of Applied hydrology*, **22**(1), 84–92.

Elango L. and Senthilkumar M. (2006). Modelling the effect of subsurface barrier on groundwater flow regime. In: MODFLOW and More, Managing groundwater systems, Colorado School of Mines, pp. 806–810.

Ensink J. H., Scott C. A., Brooker S. and Cairncross S. (2010). Sewage disposal in the Musi-River, India: Water quality remediation through irrigation infrastructure. *Irrigation and Drainage Systems*, **24**(1–2), 65–77.

Essl L., Starkl M., Kimothi P. C., Sandhu C. and Grischek T. (2014). Riverbank filtration and managed aquifer recharge as alternative water supply technologies for India: Strengths–weaknesses–opportunities–threats analysis. *Water Science and Technology: Water Supply*, **14**(4), 690–697.

FAO (2007). The state of the world's land and water resources and for food and agriculture, managing systems at risk, Food and Agriculture Organization of the United States and Earthscan, http://www.fao.org/docrep/017/i1688e/i1688e.pdf (accessed 21 January 2015).

Gale I. and Dillon P. (2005). Strategies for Managed Aquifer Recharge (MAR) in semi-arid areas. UNESCO-IHP, Paris, France.

Gangwar K. K. and Joshi B. D. (2004). A preliminary study on solid waste generation at Har Ki Pauri, Haridwar, around the Ardh-Kumbh period of sacred bathing in the river Ganga in 2004. *Environmentalist*, **28**(3), 297–300.

Groeschke M. (2013). Challenges to riverbank filtration in Delhi (India): Elevated ammonium concentrations in the groundwater of an alluvial aquifer. *Zentralblatt für Geologie und Paläontologie*, **1**(1), 1–9.

Gupta A., Ronghang M., Kumar P., Mehrotra I., Kumar S., Grischek T., Sandhu C. and Knoeller K. (2015a). Nitrate Contamination of Riverbank Filtrate at Srinagar, Uttarakhand, India: A Case of Geogenic Mineralization. *Submitted to Journal of Hydrology.*

Gupta A., Singh H., Ahmed F., Mehrotra I., Kumar P., Kumar S., Grischek T. and Sandhu C. (2015b). Lake bank filtration in landslide debris: Irregular hydrology with effective filtration. *Sustain. Water Resour. Manag.* doi:10.1007/s40899–015–0001-z

HTW Dresden and IIT Roorkee (2014). Case study of nitrate pollution in groundwater of Srinagar and investigations about the water quality in Haridwar. Report prepared by R. Mink and S. Paufler, Division of Water Sciences, Faculty of Civil Engineering & Architecture, Dresden University of Applied Sciences (HTW Dresden), Dresden, Germany in cooperation with Environmental Engineering Section, Department of Civil Engineering, Indian Institute of Technology Roorkee (IIT Roorkee), Roorkee, India.

India Vision 2020 (2002). Planning Commission Report, Government of India, New Delhi, India. http://www.google.ch/url?sa=t&rct=j&q=&esrc=s&source=web&cd=2&ved=0CCoQFjAB&url=http%3A%2F%2Fwww.teindia.nic.in%2FFiles%2FReports%2FCCR%2Fpl_vsn2020.pdf&ei=GrwNVMKaDsay7AafhYHoCg&usg=AFQjCNEJF3UivByv5IjPIrXbZPG-NQWj8Q&bvm=bv.74649129,d.ZGU (accessed 8 September 2014).

Journal of Indian Water Works Association (JIWWA) (2012). Special Issue on River Bank Filtration. Journal of Indian Water Works Association, Mumbai, India. ISSN 0970–275X.

Karanth K. R. and Prasad P. S. (1979). Some studies on hydrologic parameters of groundwater recharge in Andhra Pradesh, *Journal of Geological Society of India*, **20**, 404–414.

Kimothi P. C., Dimri D. D., Adlakha L. K., Kumar S., Rawat O. P., Patwal P. S., Grischek T., Sandhu C., Ebermann J., Ruppert M., Dobhal R., Ronghang M., Kumar P., Mehrotra I. and Uniyal H. P. (2012). Development of Riverbank Filtration in Uttarakhand. *Journal of Indian Water Works Association*, Special Issue on River Bank Filtration (December, 2012), 13–18.

Lal M. (2001). Climatic Change—Implications for India's Water Resources. *Journal of Social and Economic Development*, **3**, 58–87.

Lorenzen G., Sprenger C., Taute T., Pekdeger A., Mittal A. and Massmann G. (2010). Assessment of the potential for bank filtration in a water-stressed megacity (Delhi, India). *Environmental Earth Sciences*, **61**(7), 1419–1434.

Maréchal J. C., Dewandel B. and Subrahmanyam K. (2004). Use of hydraulic tests at different scales to characterize fracture network properties in the weathered-fractured layer of a hard rock aquifer. *Water Resour. Res.*, **40**(11), W11508. doi:10.1029/2004WR003137.

Maréchal J. C., Dewandel B., Ahmed S., Galeazzi L. and Zaidi F. K. (2006). Combined estimation of specific yield and natural recharge in a semi-arid groundwater basin with irrigated agriculture. *J. Hydrol.*, **329**, 281–293.

McIntosh A. C. (2003). Asian Water Supplies: Reaching the Urban Poor. IWA Publ., London, United Kingdom.

Mishra B. K., Takara K., Asce M. and Tachikawa Y. (2009). Integrating the NRCS Runoff Curve Number in Delineation of Hydrologic Homogeneous regions. *Journal of hydrologic engineering*, 10.1061/ASCEHE.1943–5584.0000101.

Mukherjee R., Sahoo M. and Naik K. C. (2011). Raipur city, Chhattisgarh. In: Groundwater scenario in major cities of India. Report of Central Ground Water Board, Government of India, pp. 188–195.

Muralidharan D., Andrade R. and Rangarajan R. (2007). Evaluation of check-dam recharge through water-table response in ponding area, *Current Science*, **92**(10), 1350–1352.

Nagaraj N. (1999). Institutional management regimes for pricing of irrigation water: The French model—lessons for India, *Agricultural Systems*, **61**(3), 191–205. doi:10.1016/S0308-521X(99)00052-9.

NWP (National Water Policy) (2002). Ministry of Water Resources, Government of India, New Delhi, India.

Pant M. C., Gupta P. K., Pande J., Sharma P. C. and Sharma A. P. (1981). Aspects of water pollution in Lake Nainital, U.P., India. *Environmental Conservation*, **8**(2), 113–117.

Parimalarenganayaki S. and Elango L. (2014). Impact of recharge from a check dam on groundwater quality and assessment of suitability for drinking and irrigation purposes. *Arabian Journal of Geosciences*, **7**(8), 3119–3129.

Parthasarathi G. S. and Patel A. S. (1997). Groundwater recharge through people's participation in Jamnagar region. *Journal of Indian Water Works Association*, 51–56.

Pekdeger A., Lorenzen G. and Sprenger C. (2008). Preliminary report on data of all inorganic substances and physicochemical parameters listed in the Indian and German Drinking Water Standards from surface water and groundwater at the 3 (+1) field sites. TECHNEAU integrated project deliverable D5.2.2. http://www.techneau.org (accessed 30 September 2014).

Pettenati M., Picot-Colbeaux G., Thiéry D., Boisson A., Alazard M., Perrin J., Dewandel B., Maréchal J. C., Ahmed S. and Kloppmann W. (2014). Water quality evolution during managed aquifer recharge (MAR) in India crystalline basement aquifers: Reactive transport modeling in critical zone. *GES - Procedia: Earth and Planetary Science*, **10**, 82–87. doi:10.1016/j.proeps.2014.08.016.

Perrin J., Mascre C., Pauwels H. and Ahmed S. (2011). Solute recycling: An emerging threat to groundwater quality in southern India? *J. Hydrol.*, **398**(1–2), 144–154.

Picot-Colbeaux G., Thiéry D., Pettenati M., Boisson A., Perrin J., Sarah S., Dewandel B., Maréchal J. C., Ahmed S. and Kloppmann, W. (2014). Modeling managed aquifer recharge capacity of crystalline aquifers in semi-arid context (South India): Implementing natural percolation tank dynamics into MARTHE code. *Submitted to Hydrogeology Journal.*

Ragunath H. M. (1987). Groundwater, 2nd edn, New Age, International (p) Limited Publisher, New Delhi, India.

Raicy M. C., Renganayaki S. P., Thirunavukkarasu M. and Elango L. (2015). Understanding the surface water-ground water inter-actions by geochemical measurements in three different groundwater re-charge structures. *Submitted to Journal of Environmental Science and Management.*

Raju K. C. B. (1998). Importance of recharging depleted aquifers: State of the art of artificial recharge in India. *Journal of Geological Society of India*, **51**, 429–454.

Sandhu C. and Grischek T. (2012). Riverbank filtration in India – using ecosystem services to safeguard human health. *Water Science and Technology: Water Supply*, **12**(6), 783–790.

Sandhu C., Grischek T., Kumar P. and Ray C. (2011). Potential for riverbank filtration in India. *Clean Technologies and Environmental Policy*, **13**(2), 295–316.

Schmitt R. (2010). Wastewater reuse in Indian peri-urban agriculture: Assessment of irrigation practices in a small, peri-urban catchment in Hyderabad, Andhra Pradesh, India. MSc report, Swiss Federal Institute of Technology, Zurich, Switzerland.

Saph Pani D1.1 (2012). Database of relevant pollutants in urban areas and their attenuation at RBF sites. Saph Pani Project Deliverable. http://www.saphpani.eu/downloads (accessed 19 August 2014).

Saph Pani D1.2 (2013). Guidelines for flood-risk management of bank filtration schemes during monsoon in India. Saph Pani Project Deliverable. http://www.saphpani.eu/downloads (accessed 19 August 2014).

Saph Pani D5.2 (2012). Preliminary models and system design. Saph Pani Project Deliverable. http://www.saphpani.eu/downloads (accessed 19 August 2014).

Shivanna K., Kulkarni U. P., Joseph T. B. and Navada S. V. (2004). Contribution of storms to groundwater recharge in the semi-arid region of Karnataka, India. *Hydrological Processes*, **18**, 473–485.

Singh P., Kumar P., Mehrotra I. and Grischek T. (2010). Impact of riverbank filtration on treatment of polluted river water. *J. Environ. Manage.*, **91**(5), 1055–1062.

Sprenger C., Lorenzen G. and Pekdeger A. (2008). Occurrence and fate of microbial pathogens and organic trace compounds at riverbank filtration sites in Delhi, India. TECHNEAU integrated project: Deliverable D5.2.6. http://www.techneau.org (accessed 30 November 2014).

The Economic Times (2007). http://economictimes.indiatimes.com/articleshow/msid-2234576,prtpage-1.cms (accessed 24 October 2010).

UNICEF, FAO and SaciWATERs (2013). Water in India: Situation and Prospects, Report UNICEF; FAO and SaciWATERS, New Delhi, India.

UN Water (2006). Coping with water scarcity: A strategic issue and priority for system-wide action. UN Water thematic initiatives Report, New York, United States.

World Bank (2005). India's Water Economy: Bracing for a Turbulent Future. World Bank Report, Washington DC, United States.

Wyns R., Baltassat J. M., Lachassagne P., Legtchenko A. and Vairon J. (2004). Application of proton magnetic resonnace soundings to groundwater reserves mapping in weathered basement rocks (Brittany, France). *Bull SocGéol France*, **175**, 21–34.

WHO (2011). Guidelines for drinking-water quality. 4th edn, World Health Organization, Geneva, Switzerland, ISBN 978–9241541851.

Chapter 2

Overview of bank filtration in India and the need for flood-proof RBF systems

Cornelius Sandhu, Thomas Grischek, Medalson Ronghang, Indu Mehrotra, Pradeep Kumar, Narayan C. Ghosh, Yellamelli Ramji Satyaji Rao, Biswajit Chakraborty, Pooran Patwal and Prakash C. Kimothi

2.1 INTRODUCTION

In some towns and cities in India, existing riverbank filtration (RBF), or simply bank filtration (BF), systems (mainly on rivers, but also at some lakes) currently serve as sustainable alternatives and supplements to existing surface water and groundwater sources for public water supply (Sandhu & Grischek, 2012). The application of BF has certain regionally specific advantages. For instance, most urban drinking water production systems in the hilly parts of the north Indian state of Uttarakhand, but also in other hilly regions of India, are typically supplied by directly abstracted surface water from springs, rivers and streams with a highly variable seasonal discharge. These surface water abstraction systems face two major recurring problems with respect to quantity and quality for drinking water production:

- In the pre-monsoon, especially the hot-dry summer season (March–May), the discharge of spring-fed streams and small rivers decreases considerably, thereby significantly reducing the quantity of drinking water produced and making such schemes drought-prone resulting in a drinking water production-deficit. During this period, the BF systems continue to operate sustainably by abstracting the sub-surface flow in the riverbeds, e.g. in Satpuli (Kimothi *et al.*, 2012; Ronghang *et al.*, 2012).
- During monsoon (June–September), the rapid sand filters used currently in the conventional drinking water treatment plants are unable to remove the turbidity from the raw water and the subsequent conventionally-applied disinfection by chlorination does not guarantee the elimination of pathogens. Furthermore, silting of water supply pipes and damage to water abstraction structures from surface water treatment plants is common, especially by floods due to washing away of abstraction pipes, structural damage, and inundation of pumps and electrical installations. As a result surface water abstraction and treatment plants are rendered inoperable with consequent interruptions in water supply from 2–3 days to a few weeks. This is a ubiquitous issue experienced in many regions in India. In comparison, investigations of various BF systems in India have demonstrated their efficiency in the pre-treatment of raw water for drinking and especially the removal of pathogens and turbidity during monsoon (Dash *et al.*, 2008, 2010; Ronghang *et al.*, 2012; Sandhu & Grischek, 2012; Saph Pani D1.1, 2012, D1.2, 2013; see also Chapter 3).

The first aim of this chapter is to elucidate the importance of BF systems for potable water supply in India as a sustainable alternative to the direct abstraction of surface water followed by conventional post-treatment. In this context, an overview of the hydrogeology, design and water quality of some BF systems in India is given. BF systems also have weak points, being vulnerability to breakthrough of pathogens and contamination to wells from floods. Thus, the second aim is to evaluate the risk from floods to BF systems using the example of Haridwar and Srinagar in Uttarakhand. An assessment of the removal of bacteriological indicators in the context of floods was conducted by column studies under laboratory and field conditions and from field data collected from some RBF sites.

2.2 OVERVIEW OF BANK FILTRATION SYSTEMS IN INDIA

2.2.1 Summary of design-parameters of bank filtration systems in India

Data-collection field surveys and scientific studies from 2005 onwards (as cited in Table 2.1) of sites of drinking water supply organisations in the states of Uttarakhand, Bihar, Jharkhand, Andhra Pradesh and Gujarat have revealed the presence of various BF systems (Figure 2.1). These systems are located in different hydro-climatic and -geological settings, whose features and suitability for BF are summarised in Table 2.2. Most BF sites are located in regions having a substantial vertical and horizontal extent of alluvium, most prominently in the Indo-Gangetic Plain (IGP) (Figure 2.1). However, some BF sites are also located in hilly, hard rock and coastal areas, such as in the states of Uttarakhand, Jharkhand and coastal Andhra Pradesh. In these areas the alluvium is mainly confined to the river course and nearby areas and is also of limited thickness (3–7 m in Jharkhand and up to 20 m in Uttarakhand).

The hydro-climatic variations range from arid and semi-arid in the northwest, west and central India, to humid in parts of the Himalayas (north), the northeast and coastal regions. Hence, due to the topography and tropical monsoon climate influencing the country, major temporal (seasonal) and spatial variations in rainfall occur which significantly influence the surface run-off regime of all rivers. Most rivers that have a significant year-round discharge and flow through the IGP from the Himalayas are glacier-fed. In addition, these rivers also receive water from solely spring-fed tributaries from the north and south from the central Indian highlands (Deccan Plateau and Peninsular India) with significant seasonal fluctuations in discharge. According to Kale (2003), the southwest summer monsoon accounts for >80% of the annual rainfall over a major part of the Indian region, whereas the southeastern part of the Indian Peninsula and the northwest Himalaya receive <60%. Consequently, the average monsoon flows tend to be at least 1–2 orders of magnitude more that in the non-monsoon season. This variability of river flows affects the availability (quantity and quality) of raw surface water for treatment plants producing drinking water, especially for larger towns and cities in the IGP where the terrain is levelled, gradients are low and significant volumes of partially-to-untreated industrial and domestic wastewater are discharged into the rivers (e.g. Delhi, Mathura, Agra, Kanpur). In comparison, the BF systems in Uttarakhand and the IGP (Figure 2.1 and Table 2.1) are able to abstract relatively constant volumes of water year-round.

On the other hand, south-north flowing spring-fed tributaries of the Ganga in south Bihar and Jharkhand and rivers originating in the central highlands flowing towards coastal Andhra Pradesh and Gujarat experience peak flows during monsoon and minimum flow (in some cases no flow) in the dry pre-monsoon period. As the rainfall over the northern part of the Indian Peninsula including the Deccan Plateau and south IGP is strongly concentrated in the monsoon season, the rivers in this part have a seasonal discharge related to the monsoon (Kale *et al.*, 1997) and the rivers are generally incised in rock or alluvium and have stable channels (Kale, 2003). One notable feature is that at the locations of the BF schemes by these rivers (Figure 2.1 and Table 2.1), the riverbeds consist of medium to coarse sand and gravel and thereby exhibit a suitable hydraulic conductivity for RBF. Due to this and similar to spring-fed rivers in Uttarakhand, most of these rivers have a substantial sub-surface flow in their beds even during the summer pre-monsoon, when no or only negligible flow is visible on the surface. This feature allows the RBF systems to operate during the relatively dry non-monsoon period also, albeit with lower discharges and reduced operating hours. Nevertheless, considering the predominant hard-rock geology of the region with limited alluvium confined to the riverbed, in nearby river areas and in the plains of northern Jharkhand, the existing BF schemes are suitable means for production of drinking water, and in some areas the only viable options to obtain water compared to direct surface water or even groundwater. In this context, BF buffers the quantity of water required through bank-/bed-storage and can thus be considered as an element of managed aquifer recharge and integrated water resources management.

It has been observed that most of the older BF systems constructed in India prior to 2000 (Table 2.1) have been designed primarily to fulfil the target of obtaining water of improved quality (compared to direct surface water abstraction) in sufficiently high volumes. Considering this and the local hydrogeological conditions, the wells of the BF systems have been accordingly designed. Typically these are classical vertical production wells, large-diameter caisson wells, radial collector wells (RCW) and one or more RCW with shorter radials at shallower depths connected to a single collector well (Figure 2.2a–d). Vertical wells have been constructed in relatively thick-layered homogeneous fine to medium alluvium (Table 2.1 and Figure 2.2a; sites built in 2010 in Uttarakhand, Nainital, Bhimtal, Patna, Gaya and Delhi).

Table 2.1 Summary of design parameters of bank filtration systems in India listed in ascending order of production capacity.

Location (State)	Source Water Body	Well Type (Number of Wells)	Production Capacity [m³/d]	Depth [m]	Distance from Source Water [m]	Travel-time of Bank Filtrate	References
Muzaffar Nagar* (UP)	Kali	VFW (1)	29–300	8–15	68	n.d.	*a*
Dandeli (KA)**	Kali	VFW	55–220	20–23.7	52	~9 days	*b*
Dehradun, Sahaspur (UK)	Swarna	RCW(s)	210–570	laterals 3–4 m beneath riverbed		>150 min	*c*
Ray Bazaar (JH)	Saphi Nadi (n.p.)	CW (1)	225	3–6	within riverbed	minutes–hours	*own data*
Agastmuni* (UK)**	Mandakini	VFW (1)	>280	30	33	n.d.	*c, d*
Bhimtal (UK)	Lake Bhimtal	VFW (1)	>320	48	16	n.d.	*c*
Karnaprayag* (UK)**	Alaknanda	VFW (1)	>700	20	53	n.d.	*c, d*
Satpuli* (UK)**	East Nayar	VFW (1)	756	26	43–45	2 days–2 weeks	*c, d, e*
Srinagar* (UK)**	Alaknanda	VFW (1)	852–937	18	170	>2 years	*own data, d*
Japla, Hussainabad (JH)	Son	RCW (1)	900	laterals 1 m beneath riverbed		minutes–hours	*own data*
Gumla (JH)	Nagpheri (n.p.)	CW (2)	>1,800 (max. 12,000)	3–6	within riverbed	minutes–hours	*own data*
Mathura (UP)	Yamuna	RCW (1)	2,400	laterals 15.5 & 18 m beneath riverbed		1.5–3 days	*f*
Patna (Bihar)	Ganga	VFW (6)	>3,500	150–300	9–236	n.d.	*g*
Anakapalli (AP)	Sarada	RCW (4)	>4,000	10	within riverbed	minutes–hours	*own data*
Daltonganj (JH)	North Koel (n.p.)	RCW (3)	4,000–5,000 (max. 7,000)	1–6	within riverbed	minutes–hours	*own data*
Rishikesh (UK)	Ganga	CW (2)	7,200	13–16	15–25	n.d.	*own data*
Gaya, Dandi Bagh (Bihar)	Falgu	VFW (12)	~10,000	25	5–10	n.d.	*own data*
Nainital (UK)	Lake Nainital	VFW (9)	12,000–16,000	22–37	4–94	8 ≥ 30 days	*h*
Medinipur (WB)	Kangsabati	RCW (1)	15,900	laterals 6 & 11 m beneath riverbed		n.d.	*i*
Kharagpur (WB)	Kangsabati	RCW (1)	22,700	laterals 6 8 m beneath riverbed		n.d.	*i*
Visakhapatnam, Boni (AP)	Gosthani (n.p.)	RCW (5)	27,300	10	within riverbed	hours–days	*own data*
Haridwar (UK)	Ganga and UGC	CW (22)	59,000–67,000	7–10	4–110	2 ≥ 100 days	*j, own data*
Delhi, Palla	Yamuna	VFW (~90)	~100,000 (in, 2007)	45–54	few m to 600 m	few weeks	*k, l*
Ahmedabad (Gujarat)	Sabarmati	RCW (7)	110,000	laterals 10 & 11 m beneath riverbed		n.d.	*i*

*Water used primarily for irrigation. **constructed in 2008. ***constructed in 2010. UP: Uttar Pradesh; KA: Karnataka; UK: Uttarakhand; JH: Jharkhand; AP: Andhra Pradesh; WB: West Bengal; UGC: Upper Ganga Canal; CW: large-diameter (5–10 m) caisson well; VFW: vertical filter well; RCW: radial collector well; RCW(s): Koop well or small-scale radial collector well; n.d.: not determined; n.p.: non-perennial – extreme low flow in pre-monsoon season; *a*: Thakur *et al.* (2010); *b*: Boving *et al.* (2012); *c*: Sandhu and Grischek (2012); *d*: Kimothi *et al.* (2012); *e*: Ronghang *et al.* (2012); *f*: Singh *et al.* (2010); *g*: Sandhu *et al.* (2011a); *h*: see Chapter 3; *i*: Sandhu *et al.* (2011b); *j*: Bartak *et al.* (2015); *k*: Rao *et al.* (2007); *l*: Lorenzen *et al.* (2010).

This is invariably related to the economical and fast construction of such wells compared to the other types. Nevertheless, the presence of large cobbles and boulders in foot-hill and mountainous regions (e.g. of Uttarakhand; Kimothi *et al.*, 2012) can hinder the construction of vertical wells using conventional rotary borehole drilling techniques that allow more accurate grain size distribution and classification of the sub-surface material. While in the absence of large cobbles and boulders rotary borehole drilling techniques permit borehole diameters of up to 500 mm, field-experience from Uttarakhand has revealed that it is a comparatively time- and cost intensive technique. Furthermore, the availability of rotary drilling rigs capable of drilling larger diameter (>300 mm) boreholes is limited. To circumvent these practical difficulties, the "overburden drilling with excentric bit (ODEX)" method (Murphy, 1991) is favoured in practise and frequently used. Although the ODEX technique is comparatively faster and more economical than the rotary drilling technique, it can reduce the actual size of the grains of the aquifer material through the elliptical grinding motion of the drilling bit thereby causing an underestimation of the grain size and an inaccurate interpretation of the sub-surface. The diameter of the boreholes that can be drilled using the ODEX technique is also limited to a maximum of 200 mm. This highlights the importance of conducting a sub-surface geophysical exploration during the investigation of new RBF sites for an accurate interpretation of the sub-surface lithology. Additionally, the construction of an exploratory well using a rotary drilling technique is advantageous for accurately determining particle grain sizes required for dimensioning the well filter-screens.

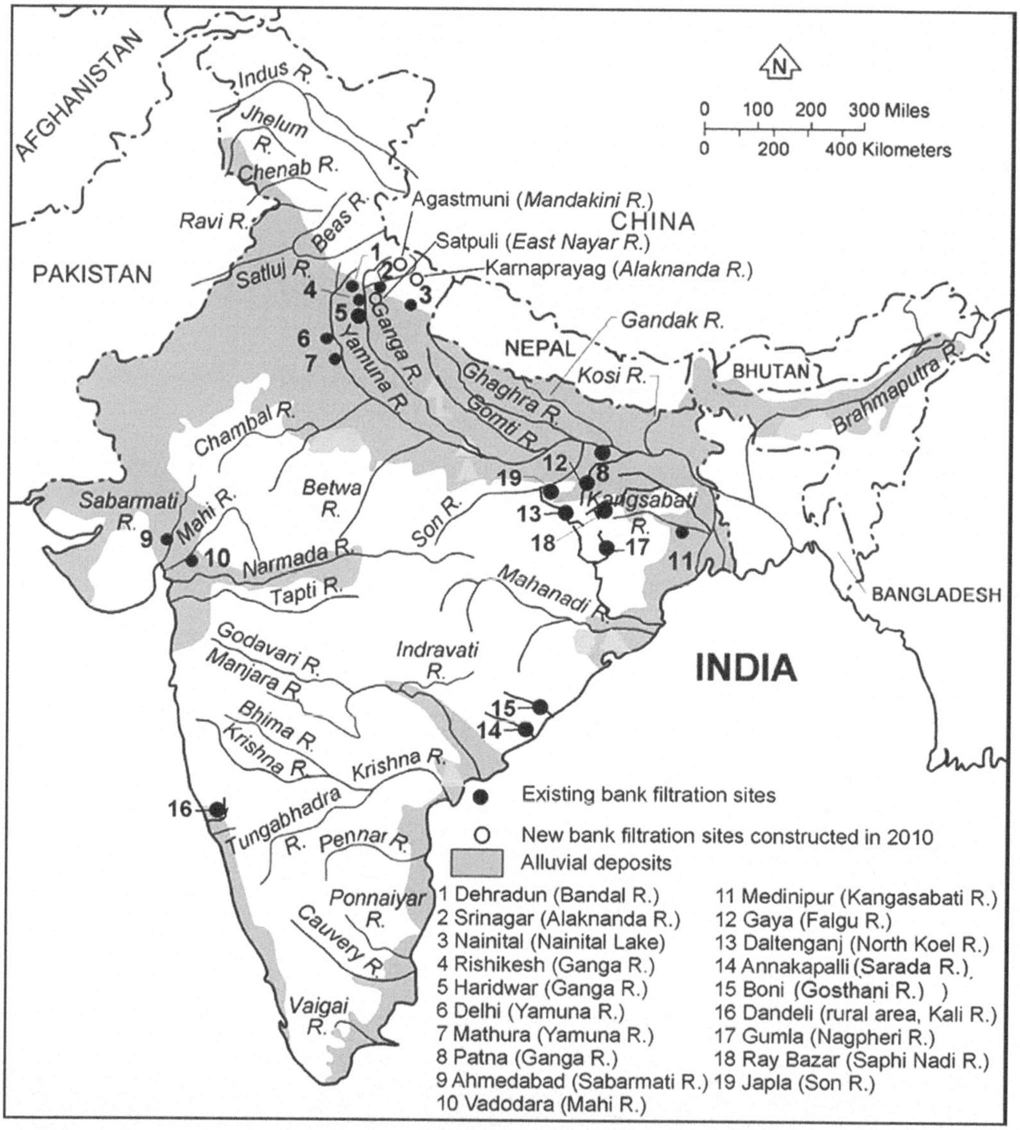

Figure 2.1 Existing bank filtration sites in India (adapted from Sandhu *et al.*, 2011b).

Table 2.2 Summary of hydrogeological features, advantages and issues of bank filtration sites in India.

BF site in Figure 2.1	General Features of BF Site location	Aquifers	Wells	Advantages of BF	Main Issues for BF
1–5	Hilly or foothill regions by perennial surface water bodies (snow-melt and spring-fed)	• Mostly shallow (up to ~20 m) • Medium to coarse sand & gravel • Presence of large fluvial boulders common (influences choice of drilling technique)	• Large-diameter caisson wells (7–10 m) • Vertical filter wells • Koop wells	• Removal of pathogens & turbidity • Year-round abstraction of water during monsoon & dry non-monsoon periods	• Construction of flood-proof wells • High sediment transport and turbidity in rivers during monsoon
6–8	Middle-lower courses of rivers in Indo-Gangetic Plain	• Mostly shallow to deep (>20 m) • Medium to fine alluvium	• Vertical filter wells typically 200–400 mm diameter • Radial collector wells located on riverbank and in riverbed	• Removal of pathogens, turbidity & organics • Year-round abstraction of water during monsoon & dry non-monsoon periods	• Partly extremely polluted surface water (sites 6 & 7) • Regulated river flow (sites 6 & 7) • Ambient landward groundwater contamination • Fine sediments in lower courses of rivers may impede surface water – groundwater interaction*
9–19	Peninsular India, east coast, semi-arid western India (Gujarat) and parts of South Ganga Plain	• Mainly hard rock aquifers in peninsular India. Limited alluvium deposits, partly confined to river courses • Medium to coarse sand and gravel riverbeds with thickness 3–20 m	• Radial collector wells in riverbed • Vertical wells (site 12,400 mm diameter) • Caisson well (site 11)	• Removal of pathogens & turbidity • Year-round abstraction of water during monsoon & dry non-monsoon periods • Abstraction during monsoon and post-monsoon is generally higher compared to dry pre-monsoon summer	• Construction of flood-proof wells • Relatively short travel time of bank filtrate to radial collector wells located in riverbeds, thereby possibility of breakthrough of turbidity and pathogens

*Possibility of presence of low-hydraulic conductivity layer which only partly cuts through the riverbed, local deposition of fines (e.g. site 8 in Patna).

Large-diameter (~10 m) caisson wells are used for BF systems designed to meet high water demands in areas with shallow groundwater tables (≤3 m below ground level) having medium to coarse alluvium containing cobbles and boulders (Table 2.1 and Figure 2.2b; Rishikesh and Haridwar). The caisson wells allow a significant storage capacity on account of their large-diameter. While simple in concept, caisson well construction requires specialised work and specific techniques involving significant manual labour and time (Herrick, 1996).

The BF sites in the cities of Ahmedabad, Baroda and Mathura are designed to abstract very large volumes of water using RCWs sited within the riverbed (Table 2.1 and Figure 2.2c). According to Kumar *et al.* (2012), the practise in India of siting the RCW within the riverbed is quite successful where groundwater is saline as there is little or no mixing of the filtered river water with the groundwater. However, the travel-time is relatively short and correspondingly moderate purification in terms of organics and microorganisms occur (Kumar *et al.*, 2012).

Examples of vertical filter wells and RCW systems at various BF sites in North America and Europe have been discussed in detail by Hunt *et al.* (2002) and Grischek *et al.* (2002). Accordingly, each well-type has its advantages and disadvantages.

Therefore the choice of the well-type for the BF system is site-specific and must consider the hydrogeological conditions of the aquifer and hydraulic conditions of the river, especially riverbed clogging.

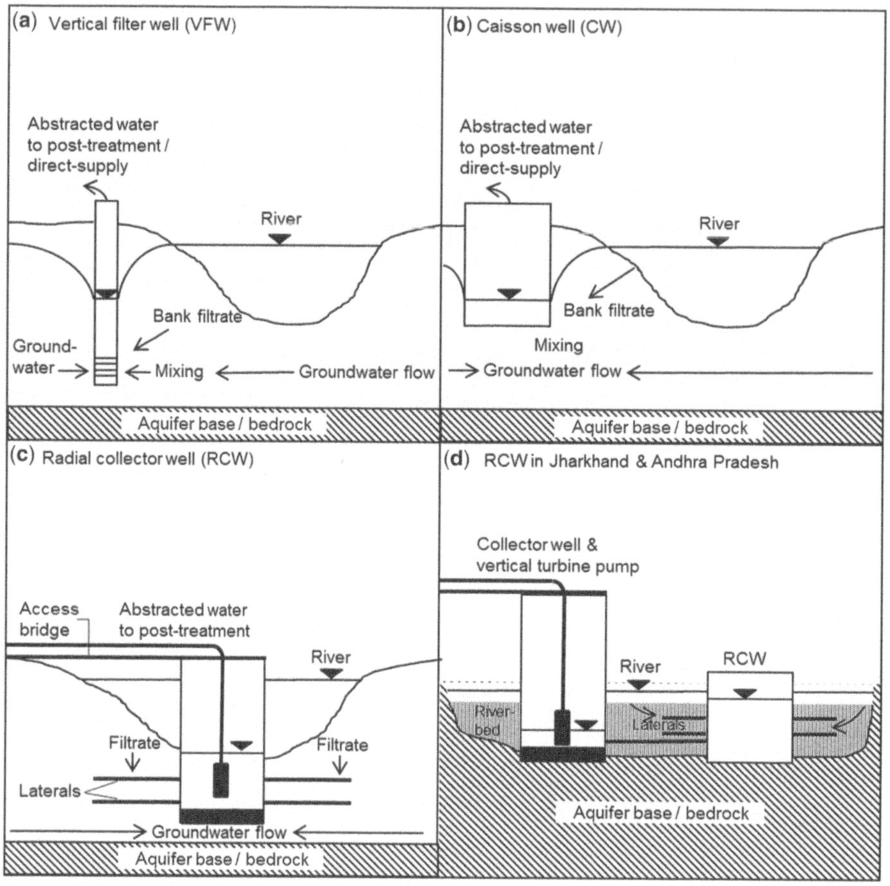

Figure 2.2 Schematic designs of typical wells at bank filtration sites in India.

2.2.2 Overview of water quality aspects at bank filtration sites

Water quality aspects for many BF sites in Europe and North America have been widely investigated. Consequently, while most BF sites in Europe are optimized for the removal of trace organic compounds, in North America BF is more often used to remove pathogens such as Cryptosporidium and, in limited settings, to minimize frequent clogging of costly membrane filters (Sandhu *et al.*, 2011b). In light of the discharge of untreated to partially treated sewage into surface water, bathing of livestock and defecation along riverbanks, as well as the very high turbidity during monsoon in India, BF is considered to serve as an extremely important pre-treatment for the removal of pathogenic microorganisms from drinking water derived from surface water.

At the BF sites of Srinagar, Haridwar and Nainital, the $\log_{10}$ removal of total thermotolerant coliforms (TTC) is observed to be 2.1 to 4.4 (Table 2.3). At these sites, and also at the BF systems in Karnaprayag, Agastmuni and Satpuli in Uttarakhand, a very good removal of TTC is attained over short flow-paths of 4–170 m of bank filtrate despite short travel times of 2–8 days. This is also due to the relatively superior surface water quality in the hilly regions, in contrast especially to the Yamuna River between central Delhi and Mathura. This stretch of the river is infamous for its pollution (CSE, 2007) and 85% of the total pollution of the Yamuna is attributed to the discharge of partially to entirely untreated domestic sewage (CPCB, 2006). Nevertheless, according to Singh *et al.* (2010) and Kumar *et al.* (2012) the BF system in Mathura effectively attenuates organic contaminants, colour, UV-absorbance and TTC by around 50% and is thus a vital pre-treatment step to the necessary post-treatment by aeration, filtration and disinfection. Compared to the direct abstraction of river water followed by conventional treatment but with pre-chlorination in Mathura, the BF system reduces or eliminates the need for pre-oxidation or pre-chlorination (Kumar *et al.*, 2012). Thus, according to Kumar *et al.* (2012) by using BF as a pre-treatment step, adsorbable organic halogenes, ammonia-chlorine complexes and disinfection by-products do not build up in the treated water.

Table 2.3 Removal of thermotolerant coliforms (TTC; *E. coli*) by BF at some sites in North India (Table 2.1).

BF Site (Reference)	TTC (*E. coli*) counts [MPN/100 mL]		Removal of TTC (*E. coli*)		Monitoring Period (n)
	Surface Water	BF Well	Log_{10}	[%]	
Srinagar (1)	104–6,570 (1,388)	<1	>3.4 (mean)	>99.96	Sep.–Nov. 2012 (5)
Haridwar (2,3)	4,298[a]–48,650[b]	1[a]–18[b]	3.5–4.4	99.97 ≥ 99.99	2005–2013 (129[a], 113[b])
Nainital (4)	141[c]	<2[c]	>2.1	>99.29	2011–2013 (18[SW], 116[well])
Mathura (5)	150–230,000	43–9,300	0.5–1.4	70–95	Nov. 2006–May 2007 (15[SW], 11[well])

1: Saph Pani D1.2 (2013); 2: Bartak *et al.* (2015); 3: Dash *et al.* (2010); 4: see Chapter 3 ; 5: Singh *et al.* (2010); MPN: most probable number; n: number of samples; [a]: Mean values for non-monsoon season; [b]: Mean values for monsoon season; [c]: median; [SW]: surface water, [well]: well water.

A study on the removal of human adenoviruses and noroviruses during the sub-surface passage of infiltrating Yamuna River water in central Delhi from December 2007 to March 2008 observed that although both viruses were present in the river water in the range of 10^5 genomes/100 mL, they were not detected in an observation well located 50 m from the river after a residence time of approximately 119 days (Sprenger *et al.*, 2014). In the same study, the $\log_{10}$ removal of somatic coliphages between a transect of three observation wells each at a distance (from the river) of 1 m, 2.4 m and 3.8 m was determined to be around 3.3, 0.7 and 0.7 respectively.

At the other BF sites of Karnaprayag, Agastmuni and Satpuli in the Uttarakhand hills, sampling on three occasions at each site from November 2011 to May 2012 showed total coliform counts to be relatively low in the river water in the range of 23–900 MPN/100 mL. Some river water samples showed a positive presence for *E. coli*. In the water from the wells, total coliforms were not detected (detection limit ≤2 MPN/100 mL) and *E. coli* was absent in all the samples. The relatively low TTC counts at the hilly BF sites in Uttarakhand are due to the low population living upstream accompanied by enhanced biodegradation due to the relatively high dissolved oxygen content in the rivers and high river gradient allowing for enhanced aeration of the water. During the non-monsoon period, the turbidity in the river water was found to be below the 5 Nephelometric Turbidity Unit (NTU) drinking water limit (BIS 10500, 2012), except in Srinagar where it was 12 NTU. Under such favourable surface water quality conditions, including extremely low dissolved organic carbon concentrations of mostly ≤1 mg/L, the surface water treatment plants are able to produce potable water conventionally by flocculation, rapid sand filtration and disinfection. However, BF is advantageous for the year-round supply of water even in monsoon when the turbidity rises to up to 200 NTU (Dash *et al.*, 2010), which conventional water treatment plants are unable to remove. The removal of TTC during monsoon has been observed to be greater than, if not equal to, the removal in non-monsoon due to greater TTC counts in surface water in monsoon at some BF sites such as Haridwar.

In hilly and sub-montane rural areas of Uttarakhand, these monsoon-associated problems necessitated the development of an economical and robust small-scale BF well called Koop (in Hindi: "well") as an alternative to direct surface water abstraction structures built on smaller seasonal streams. The Koop assembly (made of mild steel), consisting of a vertical collector cylinder (1–2 m in length) and four perforated 0.05 m diameter radial pipes each around 0.5–1 m long, is installed 3–4 m below or beside the bed or bank of a stream. Bacteriological indicator counts in the filtrate abstracted by the conventionally built Koops are observed to be significantly lower but not completely eliminated on account of the very short travel time and flowpath (Dash *et al.*, 2007; Sandhu & Grischek, 2012). While the travel time of the filtrate for a few conventionally built Koops was determined to be between 2 and 4 minutes only, it increased to more than 150 minutes for an experimental Koop constructed in 2008 using a sorted filter media and the additional installation of an extremely low permeability geotextile with the aim of increasing the length of the flow path and thereby the travel time (Sandhu & Grischek, 2012).

The BF wells in Jharkhand and coastal Andhra Pradesh, that are conceptually similar in design to Koop wells and are likewise located within the riverbed, also experience short travel time of the filtrate especially during monsoon (Table 2.1, Figure 2.2d, Section 2.2.1). Consequently a breakthrough of total coliforms was observed in these wells in monsoon 2014. However, at all these sites the abstracted filtrate is post-treated by aeration, flocculation, rapid sand filtration and finally disinfection.

Additional advantages of BF may also be seen principally in the removal of colour and organic compounds measured as UV absorbance and dissolved organic carbon (DOC). All sites produce bank filtrate that meets the acceptable limits of most of the general ionic water quality parameters of the Indian Standards (BIS 10500, 2012). An exception is found in Srinagar, where nitrate in the abstracted water from the BF well exceeds the 45 mg/L limit (BIS, 10500, 2012). There the abstracted water is diluted prior to distribution. At one location in Jharkhand, in Gaya and Mathura, the abstracted filtrate showed elevated manganese and iron concentrations as observed during a snap-shot screening of water quality at the peak of the dry pre-monsoon season in early June 2014 and in monsoon in late June 2013.

BF, however, does not present an absolute barrier to other substances of concern (e.g. ammonium) and some inorganic trace elements may even be mobilized. This has been observed in Delhi, which has poor surface water quality, and where extensive post-treatment is applied to remove high levels of ammonium (Groeschke, 2013). Elevated manganese concentrations were also observed in the water abstracted by the wells during the pre-monsoon in May 2014.

2.2.3 Mitigation of risks to bank filtration sites in India

Minimizing health-risks due to the use of BF systems and a commitment to protect the environment (aquifer or ambient groundwater resources) from any undesirable effects through induced BF are important considerations when using BF. In this context, water safety plan measures consistent with the WHO have been developed using the BF site in Haridwar as a case-study (Bartak *et al.*, 2015) to manage risks associated with BF sites in India. After the risk assessment of the BF site, it was concluded that risks from inorganic chemicals, salinity, nutrients and turbidity were acceptable in Haridwar. Furthermore, the quantitative microbial risk to human health from bacterial pathogens is below the reference risk used in this study. However, the quantitative microbial risk assessment was limited due to lack of data on virus and protozoa concentrations in source water (Bartak *et al.*, 2015). However, high removal capabilities even for viral and protozoan pathogens are reported in BF literature (e.g. in Sprenger *et al.*, 2014) albeit these risks need improved characterisation, in a longer-term assessment. General recommendations include the need for well head protection as a sanitary measure and to safeguard the well from floods, characterization of both source and groundwater quality, and management of monsoon effects (Bartak *et al.*, 2015; Saph Pani D1.2, 2013).

2.3 RISKS FROM MONSOON FLOODS TO BANK FILTRATION SYSTEMS IN INDIA

2.3.1 The effect of the monsoon on drinking water production

The monsoon and consequent dynamic river flows, including floods that cause widespread inundation of adjacent low-lying areas, are an annual event in the Indian subcontinent. As described in section 2.1, the disruption of drinking water production during monsoon is also common, especially for those towns and cities dependent on raw water that is directly abstracted from rivers. Unprecedented floods in June 2013 and other preceding extreme events as observed in August-September 2010 and 2011 in Uttarakhand and other parts of North India, accompanied with simultaneous inundation of the floodplain caused widespread deposition of sediments and structural damage to the drinking water production units and water-pipe distribution networks. Such an event potentially results in faecal contamination of drinking water. Faecal contamination is one of the most common causes of viral hepatitis (type E caused by the hepatitis E virus) and other waterborne disease outbreaks in Delhi and many parts of India (Hazam *et al.*, 2010). Various incidences of waterborne disease outbreaks in India from 1990–2011 that can be directly linked to the drinking water supply reveal that viral hepatitis, gastroenteritis, typhoid and diarrhoea were frequent, thereby underlining the fact that viruses even in small concentrations can be highly infectious. Viral hepatitis outbreaks in Kanpur in 1990–1991 (Naik *et al.*, 1992) and Baripada in 2004 (Swain *et al.*, 2010) can be linked to the faecal contamination in directly abstracted surface water and subsequent insufficient removal of viruses by the post-treatment, which is usually disinfection by chlorination. Gastroenteritis, typhoid, cholera and diarrhoea outbreaks across India are linked to the faecal contamination in the drinking water distribution system (leakages and low pressure in pipelines) resulting from wastewater, overland run-off due to extreme rainfall and flood water coming into contact with drinking water (Bhunia *et al.*, 2009; Sailaja *et al.*, 2009; Bhunia & Ghosh, 2011; Shah *et al.*, 2012).

2.3.2 Risks to riverbank filtration sites from floods

In India, wells used for the production of drinking water are at risk of microbial contamination during floods. There is also a risk of interruption of power supply, both of which lead to disruptions in drinking water supply. This occurs in Europe too, where floods are already the most common natural hazard and are expected to increase in frequency and severity resulting in rise in damages (Rambags *et al.*, 2011). Despite sparse data sets and temporal inconsistencies in the methodology used to investigate raw water quality at RBF schemes along some rivers in Germany (Elbe and Rhine), a correlation between flood events and a temporary influence on the bank filtrate quality in terms of an increase in coliform counts and turbidity of the bank filtrate has been observed. Increased numbers of bacteria and viruses have been reported in bank filtrate at the Rhine River during floods (Schubert, 2000). A high probability of a breakthrough of pathogenic microorganisms in bank filtrate was identified for a sand and gravel aquifer in the Netherlands (Medema *et al.*, 2000), whereby the residence time of the bank filtrate in the aquifer decreased from 45–65 days to 10–14 days with a rapid increase in the surface water level. Numerical simulation studies on virus transport during RBF for flood scenarios in large rivers connected to shallow unconfined sandy gravel and gravel fluvial aquifers indicate that for river level increases of 1 to 5 m, 2- to 4-log higher virus concentrations in groundwater can be found accompanied with up to 30% shorter travel times that are attributed to higher advection during rising river levels and more dispersion (Derx *et al.*, 2013). It is thus hypothesized that the risk to raw water contamination by pathogenic microorganisms during floods is likely to be a result of:

- Damage to the biologically active clogging layer (effective filter layer) on the riverbed as a result of increased shear stress during floods.
- Infiltration of river water along areas of the riverbank where no protective clogging layer exists; thereby causing transport of water in the upper part of the aquifer that was unsaturated before the flood resulting in poorer filtration.
- Changes in hydraulic pressure and shorter travel times of the bank filtrate towards the well until confining conditions are reached.
- Seepage of the flood water into the upper sub-surface and unsaturated zone, whereby microbial pathogen loading in the bank filtrate may be observed even many months after the flood has subsided.
- Direct contamination through unsealed/unprotected well heads and observation wells.

Observations at different RBF sites along the rivers Elbe and Rhine showed that the changes in the hydraulic pressure as a result of the higher flood water levels could lead to a release of already existing microorganisms in the sub-surface, which break through into the well before the younger bank filtrate from the flood reaches the well.

2.3.3 Flood-risk identification at the RBF case study sites of Haridwar and Srinagar

Description of RBF site and extreme flood event in Haridwar

The potable water to the main city of Haridwar is supplied by 22 RBF caisson wells that have a total daily capacity of 59,000–67,000 m^3 (Table 2.1). The RBF wells are located on the west-bank of the Ganga River in the north, on Pant Dweep Island and on a narrow stretch of land between the Upper Ganga Canal (UGC) and the Ganga River in the southern part of the city (Figure 2.3). The abstracted water is disinfected with sodium hypochlorite at the well prior to being distributed to the consumer directly or being pumped into storage reservoirs that are also disinfected.

Along the area where the RBF wells are located, the unconfined aquifer comprises fluvial deposits of fine to coarse sand mixed with pebbles and boulders (Saph Pani D1.2, 2013). Beneath these deposits, the aquifer-base consists of a clay layer. This lithology is generally consistent with the more detailed lithological information available for Pant Dweep Island, where according to Dash *et al.* (2010) a 19 m thick unconfined alluvial aquifer comprising mainly poorly graded sand (0.0075–4.75 mm) in the upper 14 m, is followed by a 5 m thick silty sand layer. The hydraulic conductivity of the aquifer varies from 16 to 59 m/d. The aquifer is in hydraulic contact with the adjoining Ganga and UGC, whose bed sediments are made of silt (mainly within Bhimgoda Barrage reservoir and New Supply Channel; Figure 2.3), fine sand, coarse gravel, cobbles and boulders.

The highest extreme monsoon flood event that was ever recorded in Haridwar occurred on 19 September 2010 (CWC, 2014). Water levels recorded at the gauging station of the Central Water Commission (CWC) located approximately 1–2 km upstream of Pant Dweep Island in Haridwar reached 296.3 m above sea level. During August–September 2010, most of the abstraction structures that pump water directly out of rivers in Uttarakhand were also submerged. From 18–21 September 2010, the area around some of the RBF wells in Haridwar (from North to South: IW 31, 27, 42, 43, 25, 24) was inundated by the flood waters of the Ganga (Figure 2.3). During inundation the wells ceased operation. This lead to an interruption of the water supply for at least 2–3 days as the well operators were forced to abandon the wells and shut down the pumps (due to the severe danger from the approaching flood water). After the flood water had receded, a visual inspection by Uttarakhand State Water Supply and Sewerage Organisation (Uttarakhand Jal Sansthan – UJS) revealed some damage to the base of the wells. It was also visually observed that the water in the wells had become turbid, presumably due to direct seepage of the flood water down the well shaft, or through cracks and fissures in the wall of the caisson. The turbid water was pumped out of the wells via a bypass, until no more turbidity was visible.

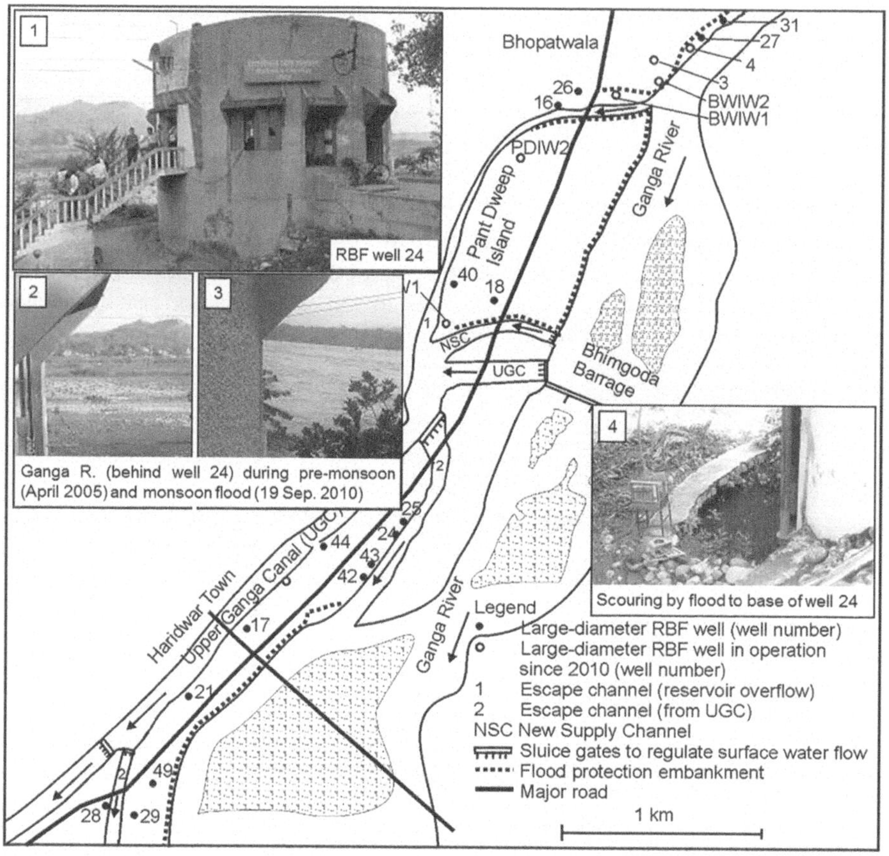

Figure 2.3 Layout of RBF system, flood protection embankment and flood-damage to RBF wells in Haridwar (Base map: Sandhu, 2015; Photo credits: (1) D. Schoenheinz, 2005; (2) T. Grischek, 2005; (3 & 4) S. Kumar, 2010).

Description of RBF site and extreme flood event in Srinagar

The direct abstraction of water from the Alaknanda River followed by conventional treatment accounts for around 80% of the potable water production for Srinagar and the town of Pauri (Saph Pani D1.2, 2013). The remainder is supplied by one RBF well (Table 2.1; Figure 2.4; designated as PW-DST) and some groundwater abstraction wells. The RBF well is located 170 m from the riverbank and is 18 m deep (Kimothi *et al.*, 2012). At the location of the RBF well, the aquifer is around 21 m deep and consists of medium-coarse sand. The mean seasonal discharge of the Alaknanda river increases nearly 10 times from a minimum of around 121 m^3/s in the winter months (January-March) to a maximum of nearly 1,200 m^3/s in monsoon, with peak monsoon discharges of up to 1,815 m^3/s (Chakrapani & Saini, 2009).

During the unprecedented flood from 15–17 June 2013 in the Alaknanda River catchment, the river level rose to around 15 m above its mean level in Srinagar thereby submerging the RBF site in nearly 8 m deep flood water (Figure 2.4). The adjacent area was also submerged. After the flood subsided, the receding water deposited a 1.5–3 m thick fine sand layer over the RBF site, the access road to the site, the adjacent Srinagar-Rishikesh main road and numerous buildings. The RBF site remained inaccessible for several weeks for vehicles and heavy equipment until the sand was cleared from the connecting roads.

In order to get an estimation of the bacteriological contamination to the wells that were inundated by the flood, two water samples (diluted to different concentrations with sterile water for improved accuracy) were collected on 2 July 2013 each from the production well Silk Farm located around 100 m to the landward (East) side of the RBF well PW-DST (Figure 2.4) and from a hand pump located around 100 m to the South of PW-DST. The samples were analysed in the laboratory of UJS in Srinagar for total coliforms and *E. coli* using the Colilert-18 method of IDEXX (Table 2.4).

From Table 2.4 it is evident that the water abstracted by the production well Silk Farm and the hand pump, both in the vicinity of the PW-DST, show a significant and high bacteriological contamination. The magnitude of bacteriological indicator counts is similar in the samples from the identical source indicating a high confidence in the results of the analyses. However, no turbidity was visible in any of the samples.

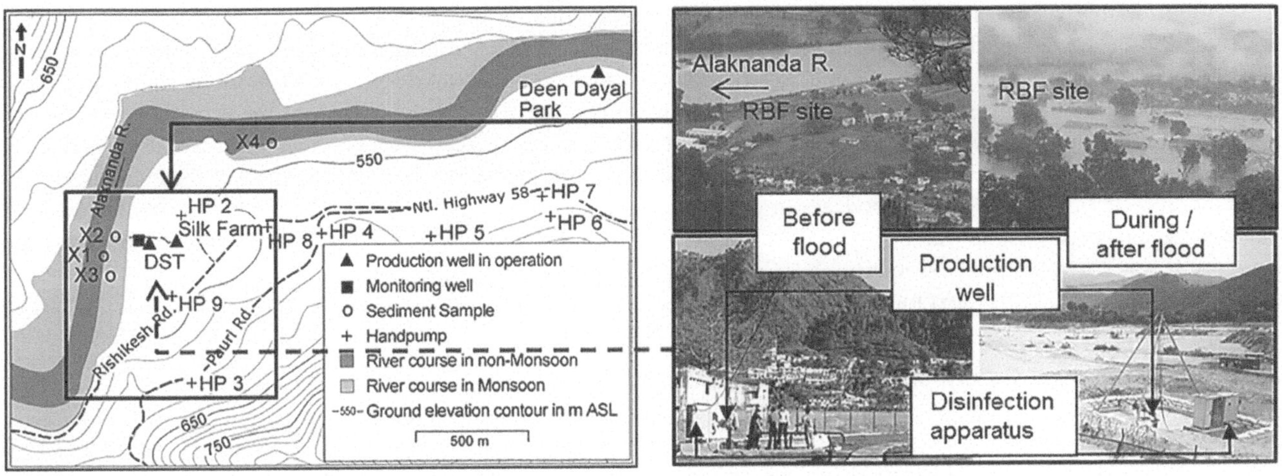

Figure 2.4 Location of RBF site in Srinagar before and after the unprecedented flood in June 2013 (Sandhu, 2015).

Table 2.4 Bacteriological indicator counts in well water in Srinagar on 2 July 2013 after the flood (Sandhu, 2015).

Location	Total Coliform Count [MPN/100 mL]		*E. coli* Count [MPN/100 mL]	
	Sample 1	Sample 2	Sample 1	Sample 2
Well Silk Farm	2,586	2,613	375	495
Hand pump	959	1,439	107	189

Summary of identifiable risks and existing flood protection measures

Taking the highest ever recorded flood event in September 2010 in Haridwar and the unprecedented flood of June 2013 in Srinagar as references, the risks to the RBF sites are summarised in Table 2.5. It is evident that most risks associated with the location of the RBF wells and their designs are applicable.

Table 2.5 Summary of risks from floods to RBF sites in Haridwar and Srinagar (Sandhu, 2015).

Risks	Haridwar	Srinagar
Risks associated to location of RBF site		
– Unconfined aquifer	X	X
– Inundation of land around RBF well and direct contamination	X	X
– Seepage of the flood water into the upper sub-surface and unsaturated zone	X	X
– Shorter travel times of the bank filtrate towards the well	X	X
Risks associated with RBF well design above ground level		
– Insufficient geodetic elevation of well head	X	X
– Inappropriate sealing of well head (Srinagar) or area around caisson well (Haridwar)	X	X
– Direct entry of flood water through improperly sealed well head and fissures in well caisson → direct contamination of well	X	X
– Inaccessibility to wells due to inundation of area around wells → difficulty to start backup power supply (e.g. generators)	X	X
Location of control system for pump operation	n/a	X
Design below ground level		
– Insufficient sealing immediately below well head chamber (uppermost part of borehole)	X	n/a[1]
– Insufficient sealing of annulus (area between casing and sub-surface material) where casing penetrates through confining layer of aquifer at ground level	n/a	n/a

X risk applicable; n/a risk not applicable; [1]Sanitary sealing measures were implemented after the August 2011 flood.

As a rule in many parts of India, the banks of rivers that experience or are at risk of serious flooding, are fortified by flood-protection measures. Such measures include stone and boulder filled galleries reinforced with wire-mesh, concrete blocks and permanently constructed stone and concrete embankments as well as dykes. As such, along the Ganga River's West bank in Haridwar, there is a flood protection embankment (Figure 2.3). The elevation of the top of the embankment ranges from south to north between 279 and 302 m above sea level and is thereby largely above the normal ground surface elevations where the 22 RBF wells are located. While 15 RBF wells are located to the west (behind) of the embankment and are thereby protected from an extreme flood, seven wells are unprotected. The extreme flood in 2010 inundated the ground at the base of the caissons of these wells (Figure 2.3). The RBF site in Srinagar is also protected by an embankment. However a portion of it was damaged by the flood in 2011. Consequently embankments have to be inspected after each monsoon and repaired before the onset of the next monsoon.

Failure of main power supply and contingency measures

On Pant Dweep Island in Haridwar, only the RBF well 18 has a generator that provides backup electricity to the pumps and disinfection system. Of the remaining wells, only 24, 25, 42 and 43 are provided with backup electricity by diesel generators. However, as these wells are not protected by a flood embankment, during the extreme flood in September 2010 these generators could not be accessed/operated because they were also inundated. This highlights the fact that extreme flood events and subsequent direct contamination and inaccessibility of the wells are a risk for some wells in Haridwar. Accessibility to the wells during a flood in Haridwar and Srinagar, as well as at all other RBF sites in India, is important because currently there are no known examples of on-line systems installed to monitor microbial contamination and turbidity peaks in time and to ensure uninterrupted disinfection.

2.4 ASSESSMENT OF RISKS TO BANK FILTRATION WELLS

2.4.1 Design of wells and direct contamination

There is a significant difference in the design of the RBF wells in Haridwar and Srinagar (Saph Pani D1.2, 2013). The caisson well design of the wells in Haridwar implies that the well head or the ceiling of the caisson on top of which the vertical turbine pumps and associated armatures, valves and electrical installations are installed is at a sufficient elevation above ground level so that the entry of flood water from directly above would not be expected. However, if the wellheads are insufficiently protected such that flood water inundates the area around the wells or enters the wells directly through cracks/fissures present in the caisson wall around or below ground level, then these provide a pathway for contamination of the well (Figure 2.5).

In case of some of the RBF wells in Haridwar, the area around the caisson at ground level is not sufficiently sealed with a concrete base or clay layer to prevent flood water (or water from an intense precipitation event) seeping down along the outer wall of the caisson to the groundwater table. This hypothesis was tested by simulating the seepage of flood water, using a sodium chloride (NaCl) solution as a tracer, into the ground around the caisson of the wells IW40, 49 and 21 (Figure 2.6). A tap that is supplied by water from the RBF well is also located near the caisson of well IW40. The tap is used by the public for bathing and washing. There is no appropriate drainage for the wastewater into a sewer or drain and consequently it seeps into the ground. Well IW49 is neither protected by a perimeter wall nor by a concrete seal around the periphery of the well.

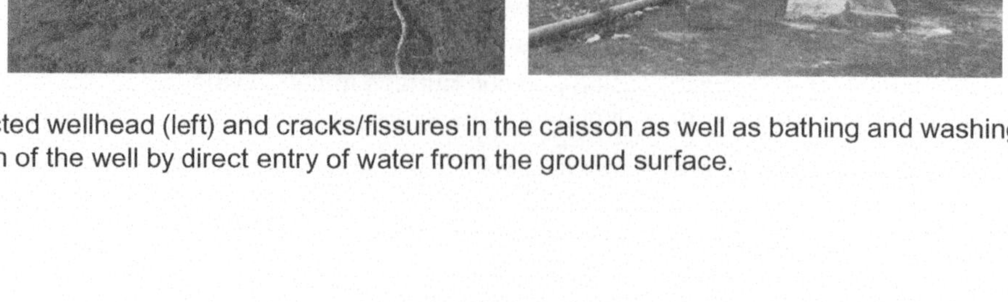

Figure 2.5 Unprotected wellhead (left) and cracks/fissures in the caisson as well as bathing and washing activities (right) can lead to contamination of the well by direct entry of water from the ground surface.

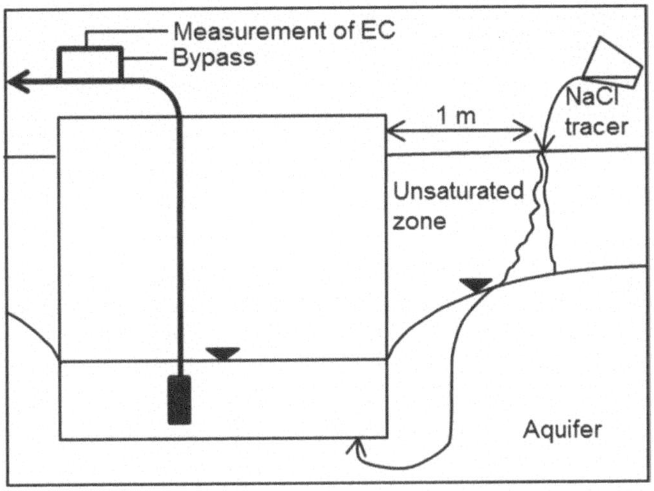

Figure 2.6 Principle of using a NaCl solution as a tracer to illustrate the pathway of contamination to a caisson well due to seepage of surface water.

In comparison, the immediate area within a radius of at least 5 m around well IW21 is protected from encroachment by a perimeter wall and the periphery around the well caisson is protected by a concrete seal (width ~1 m).

Two tests (1 and 2) were performed on well IW40 in December 2013. For the test 1, a NaCl tracer solution having an average electrical conductivity (EC) of 112 mS/cm was poured onto the ground at a distance of around 1 m from the outer edge of the caisson. Test 2 was performed in a slightly different manner by pouring four NaCl tracer solutions, at an interval of 30 minutes, each having an average EC of 260 mS/cm similar to a pulse injection. In both tests the EC of the abstracted water from the well was simultaneously measured using a sensor attached to a potable instant parameter measurement device (WTW Multi 350i). The EC sensor was immersed into a container fed by the abstracted water at a sufficient flow rate from the well via a bypass (Figure 2.6).

In test 1, the EC of the abstracted water from the well IW40 began to increase from its ambient value of 477 μS/cm after ~3 hours after the start of the application of the tracer. The EC then peaked at 489 μS/cm around 4 hours after the start of the experiment. In test 2, an increase in the EC from an ambient value of 473 μS/cm was observed already 10 minutes after the start of the experiment that attained an initial peak at 491 μS/cm after 45 minutes. Thereafter the EC gradually increased marginally and attained a steady value of 496 μS/cm after 4.5 hours of the experiment had elapsed. While a detectable initial peak in the EC is already observed after 45 minutes (test 2), the EC values in both tests 1 and 2 attain a maximum value after 4–4.5 hours elapsed since the start of the experiment. The earlier breakthrough during test 2 could be the result of the tracer flowing into the well along a preferential flow path, e.g. along the caisson of the well. On the other hand, the depth to the groundwater level around the well during test 1 could have been greater (deeper), and thus the tracer arrived later in the well. Thus it may be concluded that for a caisson well, taking well IW40 as an example, water that accumulates on the ground surface within 1 m of the caisson as a result of an intense rainfall event or flood, can take 45 minutes to 4.5 hours to come into contact with the groundwater and eventually flow into the well. Furthermore, the possible seepage of surface water along preferential flow paths (such as the outer wall of the caisson) and shallow depth to the groundwater table may result in shorter travel times for the potentially contaminated water from the surface to arrive in the well.

Two further tests were conducted on wells IW49 and IW21 using a NaCl solution. In case of IW49, having no periphery concrete seal around the caisson, the EC was observed to increase steadily and linearly albeit also marginally by only 10 μS/cm from 300 μS/cm after 100 minutes had elapsed, to 310 μS/cm after 200 minutes had elapsed. But in case of well IW21 that has a concrete seal around the periphery of the caisson, the breakthrough of the tracer was not detected even after 5.5 hours had elapsed since the start of the test. This highlights the importance of sealing the periphery at the base of the caisson wells and also ensuring a well-head protection zone where human encroachment and domestic activities by the public (bathing, washing) are prohibited.

In comparison, the RBF vertical filter well in Srinagar was fitted with a sanitary sealing prior to the flood. The sanitary sealing includes the construction of a concrete and/or clay seal in the immediate vicinity of the well base. Its purpose is to prevent the seepage of water into the ground and along the annulus between the well casing and the aquifer material to the groundwater table ("short circuiting"). In the event that a sanitary sealing is constructed, and as long as the casing pipe and well head remain above the flood level, the risk of direct entry of flood water through the well head or short circuiting along the well casing is lessened but not eliminated. Even if a sanitary seal may exist, the flood water nevertheless comes in direct contact with the casing and thus potential contamination by damage from floating debris or entry of flood water through impervious seals cannot be excluded. Thus, disinfection is a required post-treatment step.

2.4.2 Field investigations on the removal of bacteriological indicators

During the last phase of the monsoon and onset of the post-monsoon period in 2012 (27 September to 7 November), water samples from the production wells PW-DST and PW5 and monitoring wells MW-DST and MW5 as well as the Alaknanda River in Srinagar were taken regularly and analysed for total coliforms, *E. coli* and on one occasion for Enterococci using IDEXX's Colilert-18 Quanti-Tray®/2000 and Enterolert-DW MPN method (Table 2.6). PW5 and MW5 were located between the Alaknanda River and production well PW-DST in Figure 2.4, but both wells were permanently damaged by the flood in June 2013 but were operational till then. It is observed that while the total coliform counts in the Alaknanda River can attain a maximum of nearly 21,000 MPN/100 mL, it is yet considerably lower compared to total coliform counts reported for other RBF sites (e.g. Haridwar, Patna and Mathura) along the Ganga River and its tributaries (Sandhu and Grischek, 2012). The mean total coliform and maximum *E. coli* counts of >7,500 MPN/100 mL and >6,500 MPN/100 mL in the Alaknanda River are however higher than the environmental limit of <5,000 MPN/100 mL specified by the Central Pollution Control Board (CPCB) of India for drinking water sources for which conventional treatment and disinfection is necessary (Saph Pani D1.2, 2013).

Table 2.6 Range and mean total coliform and *E. coli* counts and snap-shot analyses of an Enterococci count in the Alaknanda River and RBF site in Srinagar during September–November 2012 (Saph Pani D1.2, 2013).

Parameter	Sampling Location*				
	Alaknanda River	Production Well PW 5	Monitoring Well MW 5	Production Well PW-DST	Monitoring Well MW-DST
Total coliform counts [MPN/100 mL] (mean)	1,300–20,980 (7,554)	3.1–292 (45)	9.6–770 (229)	1–25 (12)	649–770 (710)
Mean Log removal of TC	–	2.2	1.5	2.8	1.0
E. coli count [MPN/100 mL] (mean)	104–6,570 (1,388)	1–4 (2.2)	2–5.2 (3.6)	<1	<1
Mean Log removal of *E. coli*	–	2.8	2.6	>3.4	>3.4
Enterococci ($n = 1$) [MPN/100 mL]	2	<1	<1	<1	<1

*$n = 5$ samples were taken at each sampling location for the total coliform and *E.coli* counts, $n = 1$ sample was taken at each sampling location for the Enterococci analysis.

On the other hand, the total coliform and *E. coli* counts found in the production wells at the RBF site (PW5 and PW-DST) are significantly lower although not completely absent (Table 2.6). It is observed that the production well PW5 that was located only 5.4 m from the normal monsoon water line of the Alaknanda, had a significantly lower mean total coliform and *E. coli* count of only 45 MPN/100 mL and 2.2 MPN/100 mL compared to the Alaknanda River. These mean values, as also those for the production well PW-DST located around 170 m from the normal monsoon water line and with even lower coliform counts, lie within the environmental limit of <50 MPN/100 mL determined by the CPCB for drinking water sources that do not need conventional treatment but must use disinfection (Saph Pani D1.2, 2013). On one hand this highlights the benefit of RBF as a natural water treatment technology which provides an environmental ecosystem service in terms of a significant natural pre-treatment of raw water. On the other hand, considering the dynamic change of the water line of the river during monsoon and non-monsoon periods, it also highlights the importance of the travel time of the bank filtrate as a critical parameter in order to determine the distance at which the RBF well should be located from the river bank.

In Figure 2.7, the higher end of the range of total coliform counts in the Alaknanda River corresponds to the period when the monsoon begins to retreat (rainfall events lessen in frequency and intensity coupled with receding water line from the riverbank) and passes over into the post-monsoon period. As the post-monsoon period progresses the total coliform counts of the Alaknanda River decrease. However the magnitude of the total coliform counts in the monitoring and production wells remains consistent, especially for PW5 and MW5. Considering that the area around these wells was flooded in

August–September 2012 (up to when the sampling commenced) along with a possible direct entry of flood water into MW5, a breakthrough of coliforms into the wells can be attributed to either or all of the following:

- Seepage of flood water from above ground through the previously upper unsaturated aquifer.
- Short circuit of the flood water along the annulus between the casing of the monitoring well and the aquifer, or
- Breakthrough of coliforms due to increased bank filtrate flow velocity accompanied with very short travel time.

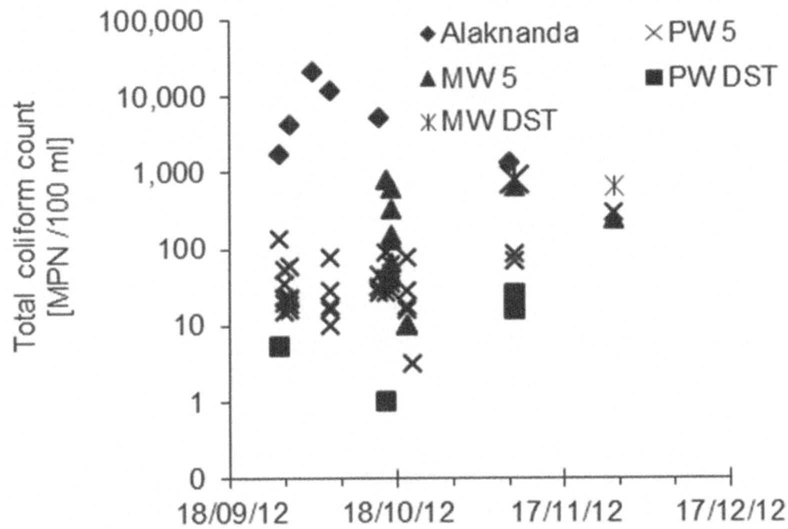

Figure 2.7 Total coliform counts in the Alaknanda River and wells at the Srinagar RBF site (Saph Pani D1.2, 2013).

Considering that the *E. coli* count in PW5 and MW5 is consistently low (<10 MPN/100 mL, Figure 2.8), and comparing the breakthroughs to those of a similar magnitude attained during column experiments conducted in the field and in the laboratory simulating a flood (Saph Pani D1.2, 2013) and from field measurements in well 40 in Haridwar (see modelling of RBF in Chapter 14), it is likely that the breakthrough is due to increased bank filtrate flow velocity (shorter travel time) and seepage through the previously unsaturated aquifer. In case of PW5 and MW5, the possibility of a short circuit of the flood water along the annulus between the casing of the monitoring well and the aquifer is less likely because both wells have a sanitary sealing and an inundation of the area where the wells are located occurred more than one year previously. Furthermore, no anthropogenic activities (personal hygiene) occur at these wells so that the contamination due to seepage of wastewater at the well is also unlikely.

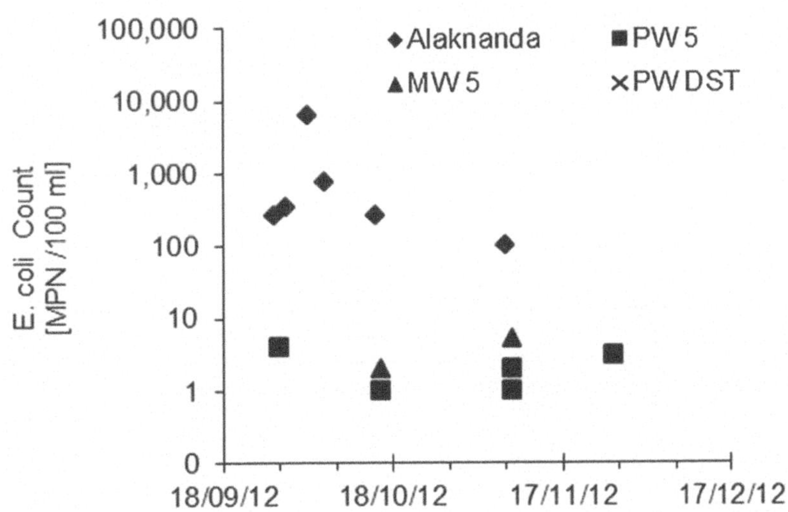

Figure 2.8 *E. coli* counts in the Alaknanda River and wells at the RBF site in Srinagar (Saph Pani D1.2, 2013).

2.4.3 Removal of coliforms under field conditions simulated for the river-aquifer interface

An assessment of the breakthrough of bacteriological indicators was also conducted for the river-aquifer interface by simulating the removal of total coliform and *E. coli* within the first 0.45 cm of flow of surface water through different sub-surface media filled into columns and fed directly with fresh Elbe River in Dresden (Germany) water for a continuous 31 day period at atmospheric temperature (Saph Pani D1.2, 2013). Four columns were filled with glass beads (C1), medium-coarse sand from the artificial recharge basin in the Waterworks Dresden-Hosterwitz (C2), sand and gravel Elbe riverbank material (C3) and a combination of sand and gravel and finer Elbe riverbank material taken adjacent to the Waterworks Dresden-Tolkewitz (C4, Table 2.7). The glass beads (diameter 1.7–2.1 mm) were used to represent well-rounded coarse sand to fine gravel. The discharge ($Q_{outflow}$) was measured once a day on 9–11 days at the outflow of the columns. Accordingly the infiltration rates (I) were calculated as a function of column area for

$$I = \frac{Q_{\text{outflow}}}{A_{\text{column}}} \tag{2.1}$$

The resulting ranges and mean values are summarised in Table 2.7. The residence times (range and median) of the Elbe River water was calculated for each column for the range and median values of their respective effective porosities (Table 2.7).

Table 2.7 Characteristics of materials used to determine coliform removal under field conditions (Saph Pani D1.2, 2013).

Parameter*		Column 1 (C1)	Column 2 (C2)	Column 3 (C3)	Column 4 (C4)
Material filled in column[1]		Glass beads (to represent well-rounded coarse sand- fine gravel)	Medium to coarse sand	Natural sand and gravel riverbed material	Layered combination of finer material (upper layer: 0–10 cm) above natural riverbed material similar to column 3 (10–45 cm)
Grain size distribution range [mm]		1.7–2.1	0.2–2.0	0.06–20	0.06–20
Effective grain size (d_{10}) [mm]		1.74	0.4	0.28	0.26[d]
Effective porosity n_e*		0.30–0.35 (0.325)[a]	0.30–0.35 (0.325)[b]	0.20–0.30 (0.25)[c]	0.25–0.35 (0.29)
Infiltration rate* [m/s]		9.9×10^{-5}–2.5×10^{-3} (6.8×10^{-4})	4.2×10^{-6}–9.3×10^{-5} (3.6×10^{-5})	2.1×10^{-6}–1.6×10^{-5} (6.2×10^{-6})	6.8×10^{-7}–5.2×10^{-5} (1.4×10^{-5})
Travel time*	n_e: 0.20	n.d.	n.d.	2–12 hours (6)	n.d.
	n_e: 0.25	n.d.	n.d.	2–15 hours (7.5)	1–46 hours (17)
	n_e: 0.30	1–82 min (18)	0.4 ≤ 9 hours (3.3)	2–18 hours (9)	1–54 hours (20)**
	n_e: 0.325	2–89 min (22)	0.4 ≤ 10 hours (3.6)	n.d.	1–57 hours (24)
	n_e: 0.35	1–95 min (23)	0.5–10.3 hours (3.9)	n.d.	n.d.

[1]Length of each column: 45 cm; diameter of each column: 10 cm; n_e: effective porosity, n.d.: not determined; [a]Soares (2015); [b,c]: median after Bartak (2011) and Grischek (2003) respectively; [d]: depth-weighted mean of d_{10} = 0.18 mm (for upper layer: 0–10 cm) and d_{10} = 0.28 mm (for lower layer: 10–45 cm); *mean values for infiltration rates are presented in parenthesis; median values for effective porosity and residence time are presented in parenthesis; **an effective porosity of 0.29 is used instead of 0.30.

The log removal rates for total coliforms and *E. coli* for the infiltration rates over a length of 0.45 m of the material used in the columns C1 to C4 (Figure 2.9), together with the travel time of water (Table 2.7), indicates that generally greater average removal is achieved for material having lower effective grain size diameters and consequently lower infiltration rates and longer travel times.

To summarise, for a RBF site where the riverbed material characteristics and travel times are similar, or are expected to be similar to those presented in Table 2.7 and for a range of infiltration rates shown in Figure 2.9, the correlation presented for columns C1 and C2 show that a comparatively lower removal (than C3 and C4) of a maximum of 2 and 3 log orders can be achieved for the glass beads representing well-rounded coarse sand to fine gravel and medium-coarse sand respectively for a travel time ranging from a few minutes up to around 10 hours. On the other hand, for riverbed sediment (C3) with a lower effective grain size diameter (0.28 mm), a greater log removal ranging from a mean of >2.5 (*E. coli*) to >3.5 (total coliforms) to a maximum of >4.2 (*E. coli*) to >3.7 (total coliforms) is achieved within a travel time of 2 to 18 hours, with a median of 6–9 hours (Table 2.7, Figure 2.9). Similarly high removal rates of 2.9 log orders (total coliforms) and greater (*E. coli*) can be observed for the riverbed material covered with finer sediment.

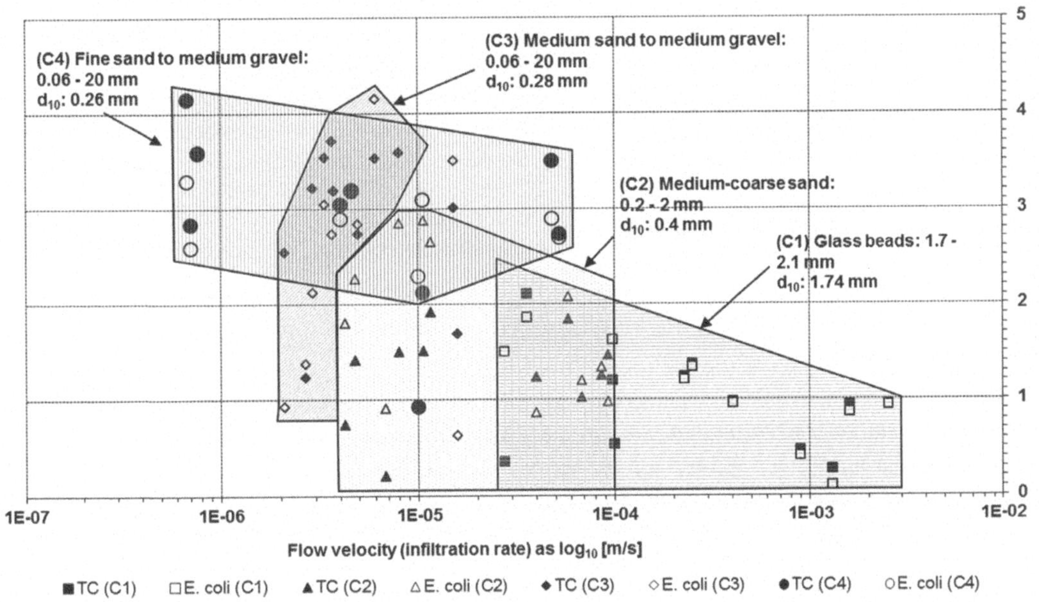

Figure 2.9 Removal of total coliforms and *E. coli* as a function of infiltration rates in different material (Saph Pani D1.2, 2013).

For a range of infiltration rates from 10^{-4} to 10^{-5} m/s, a maximum removal of coliforms of up to around 3 log orders could be expected for medium-coarse sand (artificial recharge basin material, column C2) having a size of 0.2–2 mm (medium to coarse sand) during a travel time ranging from around 0.5 to 10 hours. The infiltration rates of 10^{-5} to 10^{-6} m/s occur for the fine sand to medium gravel column (C3) having a grain size of 0.06–20 mm, whose purpose is to represent a natural river bed immediately after a flood as a consequence of which the overlying clogging layer (finer sediment layer) has been scoured away. A similar range of infiltration rates has also been observed at an RBF site in Austria during a flood event, as a consequence of which the seepage rate in the riverbed changed from around 8.3×10^{-6} to 1.2×10^{-5} m/s (Wett *et al.*, 2002). In contrast, column C4 is intended to represent a natural riverbed immediately before a flood when the naturally formed clogging layer (made-up of finer sediment) is still present. Thus the infiltration rates are also lower in the range of 10^{-7} m/s.

2.5 MITIGATION OF FLOOD-RISKS AT RBF SITES

2.5.1 Risk management plans for RBF sites in Haridwar and Srinagar

Operational and technical aspects

The breakthrough of pathogens in RBF wells has been identified as the most severe probable risk associated with normal monsoon high flow events as well as extreme flood events. A general management plan to address the risk is presented in Table 2.8. The most important aspect is to ensure adequate disinfection at all times. This can only be achieved if a back-up power supply is permanently available. Furthermore, disinfection measures should be installed at certain points along the drinking water supply distribution network in order to guarantee a residual chlorine concentration of 0.2 mg/L. In the event of an extreme flood, like those experienced in September 2010 in Haridwar and in August 2011 and June 2013 in Srinagar, more elaborate long-term measures have to be introduced.

Table 2.8 General flood-risk management plan for RBF wells in Haridwar and Srinagar (Saph Pani D1.2, 2013).

Aspect	Annual Monsoon High Flow Event (Normal Flood)	Extreme Flood
Travel-time of bank filtrate	2–50 days	<1 day to 30 days
Expected risks	Breakthrough of pathogens	– Breakthrough of pathogens and increased turbidity – Failure of power supply – No access to wells – Damage to water supply pipelines and installations
Immediate additional remedy measures	Controlled disinfection	– Back-up power supply – Alternative disinfection measures
Long-term remedy measures	Online-monitoring	– Online-monitoring & inline-electrolyses – Sealing of surface near periphery of wells in Haridwar with clay – Construction of dykes to prevent direct contamination to flood-prone wells – Construction of new flood-proof wells in Srinagar

Health aspects

Additional measures in the formation of a World Health Organisation water safety plan can be implemented prior to engineered post-treatment options that include monitoring and measurement of disinfection residuals throughout the distribution system, and regular sanitary surveys around the well heads, bore holes and well houses, well maintenance, and prohibition of well house housing, public washing, cattle and defecation in or around the wells (Bartak *et al.*, 2015). Watershed protection such as reducing sewer overflow and limiting discharge of untreated wastewater or human excreta into the Ganga River can reduce pathogen numbers by 0.5 to 1 $\log_{10}$ (NHMRC, 2011). Another 1 to 2 $\log_{10}$ unit removal can be achieved by primary and secondary wastewater treatment (NRMMC–EPHC–AHMC, 2006). Currently, around 80% of wastewater upstream from Haridwar is discharged untreated into the Ganga River. During longer religious festivals (e.g. Kumbh Mela), when widespread tented accommodation is provided to pilgrims, temporary sanitation facilities are also constructed at many places (e.g. on Pant Dweep Island). Human excreta are first collected in a pit with a cemented wall, and the overflow is then allowed to seep into the ground through a soak pit. Such soak pits close to RBF wells, especially in areas having a shallow groundwater table (Pant Dweep Island), also pose a high risk. The pathogens can easily be transported into the groundwater and then directly to the well. Optimized well operation during flood events such as increasing abstraction rate of wells with longer travel distance and reduction of abstraction rates at wells along the riverbank is also a potential operation philosophy to minimise risk. Currently when there is contamination with flood water or a loss of disinfection, water supplied in the tap may not be suitable for direct ingestion and needs to be boiled.

2.5.2 Need for construction of flood-proof RBF wells

Criteria for flood protection measures of RBF wells

Taking into account the reoccurrence of the monsoon flood of August 2011 again in the monsoon of 2012 and the highest ever recorded flood of June 2013 in Srinagar, and in order to guarantee high-quality abstracted water by minimising the entry of contaminants, suspended matter and pathogens, the production wells PW5 and PW-DST in Srinagar (Uttarakhand) that were submerged and subsequently buried beneath a 3 m thick layer of fine sand provide a good example of a RBF site requiring flood-proof wells. At the time the need for flood-proofing the wells was identified, the RBF site in Srinagar was attributed with some of the deficiencies and risks from floods listed in Table 2.5. Consequently, considering the availability of local materials and site-specific conditions, the following criteria have been formulated to flood-proof the wells:

- Protection of the well against external factors and trespassing by unauthorised persons.
- Prevention against pollution of groundwater through the well.
- Prevention of rapid seepage of rainfall run-off by providing adequate drainage measures.

- Low maintenance costs and use of non-toxic materials resistant to chemical corrosion and biological degradation.
- Easy access to well for authorised persons.

Sanitary sealing of RBF wells

During the well construction process, especially in case of large-diameter caisson wells, the sub-surface material is always loosened. This favours the infiltration of contaminated surface water. Consequently, it is absolutely necessary to seal all wells around their base to prevent the vertical seepage of water in the immediate vicinity of the well and particularly along the casing. Furthermore, the seal around the well has to protect the aquifer and the water abstracted from the well. Depending on the material used, the thickness and extent of the seal may vary, but must be sufficient.

For large-diameter RBF caisson wells such as those in Haridwar, very large quantities of seal materials are needed. To act economically, it is recommended to place a combination of an impermeable layer of high quality clay and to fill up the remaining space with inferior material like loam. A concrete plate should be placed on top (Figure 2.10).

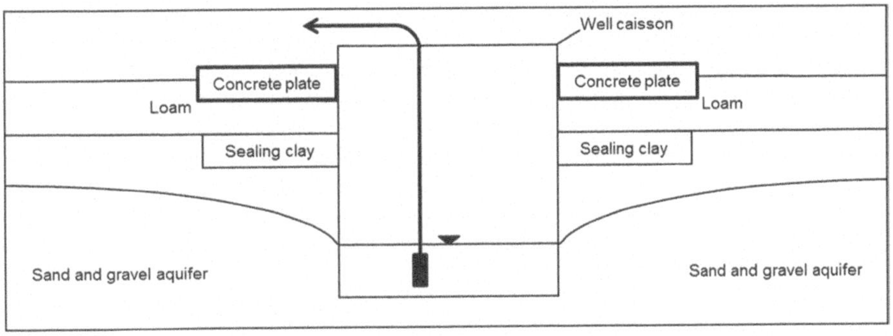

Figure 2.10 Sanitary sealing of a large-diameter well (adapted from Balke, 1999).

The installation of clay above the groundwater table requires the use of dried clay which is broken up and can become compacted through longer periods of compaction. Afterwards the layer has to be soaked very thoroughly, and consequently the clay swells thereby becoming an effective seal after some time.

For vertical filter wells and based on practical experiences in India, it is suggested to excavate an area of at least 1 m^2 (with the well at the centre) to a depth up to the mean groundwater level during the dry pre-monsoon period and fill (seal) the excavation with a commercially available product specifically for well sealing high plasticity, such as bentonite pellets. Thereafter the sealing should be compacted thoroughly. The sealing should subsequently be covered by a concrete plate or a water-tight well-head chamber. This type of sanitary sealing is demonstrated in Saph Pani D1.2 (2013), and was already constructed for the RBF site in Srinagar in November 2011 by UJS. Nevertheless it must be noted that the sanitary sealing, while minimising and in the best case preventing the direct seepage of surface water from above ground near the well (e.g. from an intensive rainfall event), cannot prevent the direct entry of surface water in case of an extreme flood or complete inundation of the well.

ACKNOWLEDGEMENTS

These investigations were conducted for the project 'Saph Pani – Enhancement of natural water systems and treatment methods for safe and sustainable water supply in India', funded by the European Commission (grant agreement no. 282911). The mobility of some authors and persons to conduct field and laboratory investigations, was funded by the projects (no. 27110509 and 56040107) '*Nachhaltige Siedlungswasserwirtschaft* (Sustainable Urban Water Management)' and '*Nachhaltige Trinkwasserversorgung in Uttarakhand* (Sustainable Drinking Water Supply in Uttarakhand)' funded by the German Academic Exchange Service's (DAAD) programme "A New Passage to India" (2009–2011 and 2013–2014). The support of various persons for the field work is gratefully appreciated, and these include W. Macheleidt, F. Musche, J. Lattke, V. D. A. Nguyen, S. Paufler, R. Mink, M. Wolf and R. Schlese of the Division of Water Sciences at the Dresden University of Applied Sciences and the water supply organisations, Uttarakhand Jal Sansthan (UJS) and Drinking Water and Sanitation Department (DWSD) of Jharkhand, especially Mr. D. K. Singh, Mr. R. K. Rohila, Mr. P. K. Saini, Mr. S. Prasad, Mr. P. K. Singh and Mr. S. K. Dawn. The construction of the monitoring well MW5 in Srinagar as part of the Saph Pani Project by Mr. Arun Gulati and his team at the SME well construction company Akshay Jaldhara is gratefully acknowledged.

2.6 REFERENCES

Balke K. D. (1999). Well construction. Lecture notes, Institute for Geology and Paleontology, University of Tübingen, Tübingen, Germany.

Bartak R. (2011). *Untersuchungen zur Rohwasserbeschaffenheit im Wasserwerk Hosterwitz im Hochwasserfall* (Investigations of the raw water quality at the Waterworks Hosterwitz in the event of a flood). Diplom-degree thesis, Institute of Groundwater Management, Department of Hydrosciences, Faculty of Environmental Sciences, Dresden University of Technology, Dresden, Germany.

Bartak R., Page D., Sandhu C., Grischek T., Saini B., Mehrotra I., Jain C. K. and Ghosh N. C. (2015). Application of risk-based assessment and management to riverbank filtration sites in India. *Journal of Water and Health*, **13**(1), 174–189.

Bhunia R., Hutin Y., Ramakrishnan R., Pal N., Sen T. and Murhekar M. (2009). A typhoid fever outbreak in a slum of South Dumdum municipality, West Bengal, India, 2007: Evidence for foodborne and waterborne transmission. *BMC Public Health*, **9**, 115.

Bhunia R. and Ghosh S. (2011). Waterborne cholera outbreak following Cyclone Aila in Sundarban area of West Bengal, India, 2009. *Transactions of the Royal Society of Tropical Medicine and Hygiene*, **105**(4), 214–219.

Bureau of Indian Standards (BIS) 10500 (2012). Drinking Water – Specification. Indian Standard, 2nd Revision. Bureau of Indian Standards, New Delhi, India.

Boving T., Choudri B. S., Cady P., Davis A., Patil K. and Reddy V. (2012). Acceptance of a riverbank filtration system in rural India. *Journal of Indian Water Works Association*, Special Issue on River Bank Filtration (December, 2012), 33–41.

Chakrapani G. J. and Saini R. K. (2009). Temporal and spatial variations in water discharge and sediment load in the Alaknanda and Bhagirathi Rivers in Himalaya, India. *Journal of Asian Earth Sciences*, **35**(6), 545–553.

Central Pollution Control Board (CPCB) (2006). Water quality status of Yamuna river (1999–2005). Report Central Pollution Control Board, New Delhi, India.

Centre for Science and Environment (CSE) (2007). Sewage canal – How to clean the Yamuna. Centre for Science and Environment, New Delhi, ISBN 978-81-86906-40-8.

Central Water Comission (CWC) (2014). River stage gauge and flood forecasting station haridwar. Central Water Commission, Ministry of Water Resources, Government of India, http://www.india-water.gov.in/ffs/data-flow-list-based/flood-forecasted-site (accessed 28 August 2014).

Dash R. R., Mehrotra I. and Kumar P. (2007). Natural Bank/Bed Filtration: Water Supply Schemes in Uttaranchal, India. Proceedings of the World Environmental and Water Resource Congress 2006, Omaha, Nebraska, USA, pp. 1–10, http://dx.doi.org/10.1061/40856(200)98 (accessed 21 August 2014).

Dash R. R., Mehrotra I., Kumar P. and Grischek T. (2008). Lake bank filtration at Nainital, India: Water quality investigations. *Hydrogeology Journal*, **16**(6), 1089–1099.

Dash R. R., Bhanu Prakash E. V. P., Kumar P., Mehrotra I., Sandhu C. and Grischek T. (2010). River bank filtration in Haridwar, India: Removal of turbidity, organics and bacteria. *Hydrogeology Journal*, **18**(4), 973–983.

Derx J., Blaschke A. P., Farnleitner A. H., Pang L., Blöschl G. and Schijven J. F. (2013). Effects of fluctuations in river water level on virus removal by bank filtration and aquifer passage – A scenario analysis. *Journal of Contaminant Hydrology*, **147**(1), 34–44.

Grischek T. (2003). *Zur Bewirtschaftung von Uferfiltratfassungen an der Elbe* (Management of Riverbank Filtration Facilities along the Elbe). PhD thesis, Institute of Groundwater Management, Department of Hydrosciences, Faculty of Environmental Sciences, Dresden University of Technology, Dresden, Germany.

Grischek T., Schoenheinz D. and Ray C. (2002). Siting and design issues for riverbank filtration schemes. In: Riverbank Filtration – Improving source-water quality, Ray C., Melin G. and Linsky R. B. (eds), Water Science and Technology Library, 43, Kluwer Academic Publishers, Dordrecht, The Netherlands, pp. 291–302.

Groeschke M. (2013). Challenges to riverbank filtration in Delhi (India): Elevated ammonium concentrations in the groundwater of an alluvial aquifer. *Zentralblatt für Geologie und Paläontologie*, **1**(1), 1–9.

Hazam R. K., Singla R., Kishore J., Singh S., Gupta R.K. and Kar P. (2010). Surveillance of hepatitis E virus in sewage and drinking water in a resettlement colony of Delhi: What has been the experience? *Archives of Virology*, **155**(8), 1227–1233.

Herrick D. (1996). Caisson well construction. Water Well Journal, National Ground Water Association, Westerville, Ohio, USA, pp. 38–40, http://info.ngwa.org/gwol/pdf/961661990.pdf (accessed 03 July 2014).

Hunt H., Schubert J. and Ray C. (2002). Conceptual design of riverbank filtration systems. In: Riverbank Filtration – Improving source-water quality, Ray C., Melin G. and Linsky R. B. (eds), Water Science and Technology Library, 43, Kluwer Academic Publishers, Dordrecht, The Netherlands, pp. 19–27.

Kale V. S. (2003). Geomorphic effects of monsoon floods on Indian Rivers. *Natural Hazards*, **28**(1), 65–84.

Kale V. S., Hire P. and Baker V. R. (1997). Flood hydrology and geomorphology of monsoon-dominated rivers: The Indian Peninsula. *Water International*, **22**(4), 259–265.

Kimothi P. C., Dimri D. D., Adlakha L. K., Kumar S., Rawat O. P., Patwal P. S., Grischek T., Sandhu C., Ebermann J., Ruppert M., Dobhal R., Ronghang M., Kumar P., Mehrotra I. and Uniyal H. P. (2012). Development of Riverbank Filtration in Uttarakhand. *Journal of Indian Water Works Association*, Special Issue on River Bank Filtration (December, 2012), 13–18.

Kumar P., Mehrotra I., Boernick H., Schmalz V., Worch E., Schmidt W. and Grischek T. (2012). Riverbank filtration: An alternative to pre-chlorination. *Journal of Indian Water Works Association*. Special Issue on River Bank Filtration (December, 2012), 50–58.

Lorenzen G., Sprenger C., Taute T., Pekdeger A., Mittal A. and Massmann G. (2010). Assessment of the potential for bank filtration in a water-stressed megacity (Delhi, India). *Environmental Earth Sciences*, **61**(7), 1419–1434.

Medema G. J., Juhàsz-Holterman M. H. A. and Luijten J. A. (2000). Removal of micro-organisms by bank filtration in a gravel-sand soil. In: Jülich W. and Schubert J. (eds) Proc. Int. Riverbank Filtration Conference, Düsseldorf, 2–4 November 2000, pp. 161–168. International Association of Water Works in the Rhine Basin (IAWR). ISBN 90-707671-27-1.

Murphy J. (1991). Application of ODEX drilling method in a variably fractured volcanic/igneous environment. Proceedings of the 5th National Outdoor Action Conference on Aquifer Restoration, Ground Water Monitoring and Geophysical Methods (5th NOAC), Las Vegas, Nevada, USA, pp. 799–814, http://info.ngwa.org/gwol/pdf/910153152.pdf (accessed 15 July 2014).

Naik S. R., Aggarwal R., Salunke P. N. and Mehrotra N. N. (1992). A large waterborne viral hepatitis E epidemic in Kanpur, India. *Bulletin of the World Health Organization*, **70**(5), 597–604.

NHMRC (2011). Australian drinking water guidelines Paper 6 National Water Quality Management Strategy. National Health and Medical Research Council, National Resource Management Ministerial Council, Canberra, Commonwealth of Australia.

NRMMC–EPHC–AHMC (2006). Australian guidelines for water recycling (Phase 1): Managing Health and Environmental Risks, Natural Resources Ministerial Management Council, Environment Protection and Heritage Council and Australian Health Ministers' Conference, Canberra, Commonwealth of Australia.

Rambags F., Raat K. J., Leunk I. and van den Berg G. A. (2011). Flood Proof wells: Guidelines for the design and operation of water abstraction wells in areas at risk of flooding. PREPARED project report 2011.007, deliverable number 5.2.1. http://www.prepared-fp7.eu/prepared-publications (accessed 28 August 2014).

Rao S. V. N., Kumar S., Shekhar S., Sinha S. K. and Manju S. (2007). Optimal pumping from skimming wells from the Yamuna River flood plain in north India. *Hydrogeology Journal*, **15**(6), 1157–1167.

Ronghang M., Kumar P., Mehrotra I., Kimothi P. C., Adlakha L. K., Sandhu C., Grischek T. and Voltz T. J. (2012). Application of riverbank filtration for year-round drinking water production in a small town in the hills of Uttarakhand. *Journal of Indian Water Works Association*, Special Issue on River Bank Filtration (December, 2012), 19–24.

Sailaja B., Murhekar M. V., Hutin Y. J., Kuruva S., Murthy S. P., Reddy K. S. J., Rao G. M. and Gupte M. D. (2009). Outbreak of waterborne hepatitis E in Hyderabad, India, 2005. *Epidemiology and Infection*, **137**(2), 234–240.

Sandhu C. (2015). A Concept for the Investigation of Riverbank Filtration Sites for Potable Water Supply in India. PhD thesis (submitted), Technische Universität Dresden, Faculty of Environmental Sciences, Institute of Waste Management and Contaminated Site Treatment, Dresden, Germany.

Sandhu C. and Grischek T. (2012). Riverbank filtration in India – using ecosystem services to safeguard human health. *Water Science and Technology: Water Supply*, **12**(6), 783–790.

Sandhu C., Grischek T., Schoenheinz D., Prasad T. and Thakur A. K. (2011a). Evaluation of bank filtration for drinking water supply in Patna by the Ganga River, India. In: Riverbank Filtration for Water Security in Desert Countries, Ray C. and Shamrukh M. (eds), NATO Science for Peace and Security Series, Springer, Dordrecht, The Netherlands, pp. 203–222.

Sandhu C., Grischek T., Kumar P. and Ray C. (2011b). Potential for riverbank filtration in India. *Clean Technologies and Environmental Policy*, **13**(2), 295–316.

Saph Pani D1.1 (2012). Database of relevant pollutants in urban areas and their attenuation at RBF Sites. Saph Pani Project Deliverable. http://www.saphpani.eu/downloads (accessed 19 August 2014).

Saph Pani D1.2 (2013). Guidelines for flood-risk management of bank filtration schemes during monsoon in India. Saph Pani Project Deliverable. http://www.saphpani.eu/downloads (accessed 19 August 2014).

Schubert J. (2000). *Entfernung von Schwebstoffen und Mikroorganismen sowie Verminderung der Mutagenität bei der Uferfiltration* (Removal of particles and microorganisms and reduction of mutagenicity during bank filtration). *gwf Wasser Abwasser*, **141**(4), 218–225.

Shah H. D., Shah V. P. and Desai A. N. (2012). An epidemic outbreak of Vibrio Cholerae El Tor 01 serotype ogawa biotype in a Lalpur town, Jamnagar, India. *Journal of Postgraduate Medicine*, **58**(1), 14–18.

Singh P., Kumar P., Mehrotra I. and Grischek T. (2010). Impact of riverbank filtration on treatment of polluted river water. *Journal of Environmental Management*, **91**(5), 1055–1062.

Soares M. (2011). Recovery of Infiltration rates in columns filled with different media and using Elbe River water after a drought period and type of clogging. Internal report, Division of Water Sciences, University of Applied Sciences Dresden, Germany. In: Saph Pani D1.2 (2013). Guidelines for flood-risk management of bank filtration schemes during monsoon in India, Saph Pani Project Deliverable. http://www.saphpani.eu/downloads.

Sprenger C., Lorenzen G., Grunert A., Ronghang M., Dizer H., Selinka H.-C., Girones R., Lopez-Pila J. M., Mittal A. K. and Szewzyk R. (2014). Removal of indigenous coliphages and enteric viruses during riverbank filtration from highly polluted river water in Delhi (India). *Journal of Water and Health*, **12**(2), 332–342.

Swain S. K., Baral P., Hutin Y. J., Rao T. V., Murhekar M. and Gupte M. D. (2010). A hepatitis E outbreak caused by a temporary interruption in a municipal water treatment system, Baripada, Orissa, India, 2004. *Transactions of the Royal Society of Tropical Medcine and Hygiene*, **104**(1), 66–69.

Thakur A. K., Ojha C. S. P., Grischek T., Ray C. and Sandhu C. (2010). Water quality improvement through river bank filtration in extreme environmental conditions. *Journal of Indian Water Works Association*, **42**(2), 106–115.

Wett B., Jarosch H. and Ingerle K. (2002). Flood induced infiltration affecting a bank filtrate well at the River Enns, Austria. *Journal of Hydrology*, **266**(3–4), 222–234.

Chapter 3

Lake bank filtration for water supply in Nainital

Ankush Gupta, Himanshu Singh, Indu Mehrotra, Pradeep Kumar, Sudhir Kumar, Thomas Grischek and Cornelius Sandhu

3.1 INTRODUCTION

As mentioned in the previous chapters, surface water bodies such as lakes and rivers have been and continue to be major sources of water supply. Alluvial aquifers hydraulically connected to a water course offer a relatively easy option for abstracting surface water with improvement in water quality in terms of turbidity, pathogens, and dissolved organics. This technology called bank filtration makes an important difference in implementations at lake banks and river banks. At lake banks, the colmation layer – the most important purification zone at the soil/water interface – is not disrupted by seasonal changes in water flows as in the case of river banks. This lake bank feature offers the advantage of a consistently higher attenuation of contaminants throughout the year than river banks. On the other hand, there are also potential problems at lake banks that may occur due to clogging.

Lake bank filtration (LBF) is used in many countries around the world for municipal water supply such as Germany, Finland, and the Netherlands. LBF wells at Lake Tegel in Berlin have been used for the city's drinking water supply for more than 100 years (Fritz *et al.*, 2002). In Finland, LBF schemes have been implemented on the island of Hietasalo in Lake Kallavesi and at the banks of lakes Vihnusjärvi and Vesijärvi (Miettinen *et al.*, 1994; Kivimäki *et al.*, 1998). In the Netherlands, an LBF scheme has been implemented along the deep gravel extraction reservoir De Lange Vlieter (Juhász-Holterman *et al.*, 1998), which is recharged with water from River Mass (also known as River Meuse). The travel times for bank filtrates to the LBF wells are estimated to be in the range of one week to a few months based on various tracers such as stable isotopes (δ^2H and $\delta^{18}O$), chloride, boron, and pharmaceutical residues (Massmann *et al.*, 2008; Miettinen *et al.*, 1994). Studies on the colmation layer in the littoral zones of Lake Tegel (Hoffmann & Gunkel, 2011) show that this biological zone in lakes extends to a depth of ~10 cm.

In Nainital town in the Himalayan region of India, an LBF scheme was developed at the bank of Lake Naini starting in 1990 (Figure 3.1). Nainital has one of the oldest piped water supply systems in the Kumaun region implemented in 1898. The development of Nainital's water supply has been described by Dash *et al.* (2008). The water was drawn from springs and pumped to higher elevations using steam engines, which were replaced by diesel engines in 1914. In 1955, increasing water demand led to pumping the lake water in addition to spring water. In 1985, a water treatment plant was installed for purification of the lake water to produce drinking water. To meet the increasing demand, seven tube-wells were installed adjacent to Lake Naini between 1990 and 2007 to abstract lake bank filtrate (Dash *et al.*, 2008). The abstracted filtrate was only chlorinated before supply. A water quality investigation by Dash *et al.* (2008) showed the LBF scheme to be more effective in coliform removal than the sand filtration in the water treatment plant. Subsequently, direct pumping of lake water and its purification was discontinued and in 2008 and 2009, five additional wells were constructed near the lake bank. By November 2011, one of the older wells constructed between 1990 and 2007 was dismantled, and two other tube-wells were shut down because of operational problems; thus the LBF well field comprised nine production wells during the time period of present study. At the time of preparing this manuscript (August, 2014), the two non-functioning wells have been made operational and a total of eleven wells are in operation.

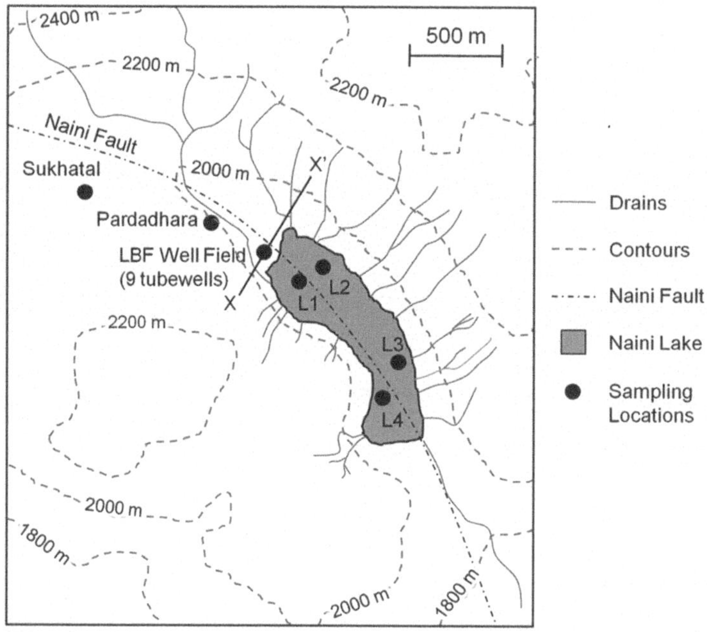

Figure 3.1 Map of the region around Naini Lake (Based on GoogleMaps, 2014 and AHEC, 2002). X–X′ shows the area for which a geological cross-section is shown in Figure 3.5.

Due to being a popular tourist spot, the anthropogenic activities in Nainital have led to severe degradation of the Naini Lake over several decades as it became increasingly eutrophic (Pant *et al.*, 1981). The Macrobenthic community in the lake has almost completely disappeared below a depth of 7 m (Gupta & Pant, 1983). Studies have been conducted on its water quality (Purohit and Singh, 1981), phytoplankton community (Sharma *et al.*, 1982), macrophytes (Purohit *et al.*, 1986), protozoa (Shukla & Gupta, 2001), predacious *Bdellovibrio*-like Organisms (Chauhan *et al.*, 2009), water and sediment geochemistry (Das *et al.*, 1995; Chakrapani, 2002; Purushothaman *et al.*, 2012), morphology and morphometry (Rawat, 1987), sediment accumulation (Das *et al.*, 1994), water balance (Kumar *et al.*, 1999), pesticides (Dua *et al.*, 1998), and heavy metal concentrations (Gupta *et al.*, 2010). Since 2007, several measures have been taken to conserve the lake and control its eutrophication. These measures included upgrading the town's sewer system to prevent any town wastewater from entering the lake, cleaning up and preventing solid waste from being dumped into the lake and, most importantly, introducing hypolimnetic aeration of the lake. After aeration, the algal and diatom population in the lake had decreased, and Cyaneophyceae (blue–green bacteria) had disappeared (Gupta & Gupta, 2012).

This chapter presents studies on water quality of the LBF wells, lake water and groundwater to assess its performance after recent changes in the Naini Lake, its catchment, and the LBF well field. The study was undertaken predominantly over a period of one and a half years during 2012–2013. Stable isotope analysis was done to assess the proportion of groundwater and bank filtrate in each well and the dynamics of the bank filtrate and groundwater in the LBF well field. The results were compared with previous data (Dash *et al.*, 2008) to assess the changes in water quality.

3.2 STUDY SITE

Lake Naini and the LBF system have been described in detail by Dash *et al.* (2008) and AHEC (2002). This section gives a brief overview of the site in light of recent changes in the system. Located in the Kumaun Himalayas in the State of Uttarakhand, India, Lake Naini is a kidney shaped water body. Nainital City, developed around the lake economically as well as socially, has a population of about 41,377 (Census of India, 2011). Additionally, the daily tourist influx in summer months averages around 5,000. The land-use of the lake catchment includes forests and shrubs (42%), buildings (41%), roads (2.1%), water bodies (10.3%), playgrounds (1.1%), and barren lands (3.5%).

Lake Naini is surrounded by steep mountain ranges from three sides and a downhill slope on the south-east side. There are several faults and fractures in the catchment of the lake. A fault called Naini Fault runs midway across the lake as shown in Figure 3.1. Most subsurface inflow to the lake takes place through the faults and fractures (AHEC, 2002). The lake is fed by about twenty water channels, only two of which are perennial open drains. Other water inputs to the lake come from direct precipitation, internal and underwater springs and subsurface groundwater flow from the surrounding mountain ranges. Rainfall is mostly restricted to the monsoon season from June to September during which about 90% of the annual precipitation occurs as shown in Figure 3.2. The year 2013 had unusually high rainfall in February and June, which was an exception in the non-monsoon period.

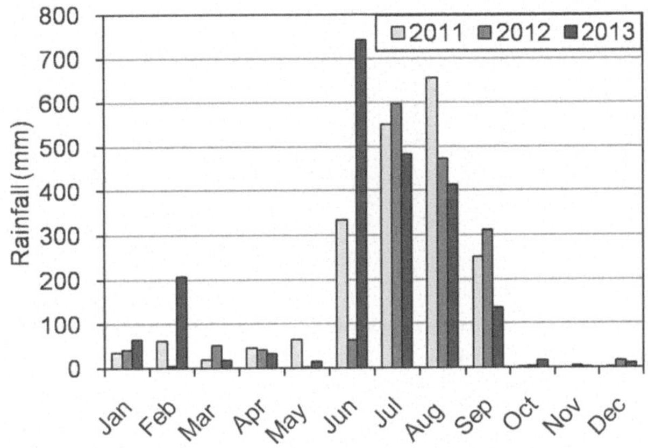

Figure 3.2 Mean monthly rainfall in the Nainital district during 2011–2013 (IMD, 2013).

The water level in the lake is regulated with the help of sluice gates constructed at the south-east end of the lake. Excess water is discharged through sluices and then flows downhill as the Balia River. The capacity of the lake is progressively decreasing. Its volume was estimated to be $7{,}425 \times 10^3$ m^3 in 1899, $6{,}808 \times 10^3$ m^3 in 1969, and $5{,}907 \times 10^3$ m^3 in 1982 (Nachiappan *et al.*, 2002). Based on the sedimentation rates obtained from ^{210}Pb and ^{137}Cs dating techniques, Kumar *et al.* (1999) estimated the lake life as $2{,}480 \pm 310$ years and $2{,}160 \pm 80$ years, respectively. A brief summary of location, morphological, and meteorological data of the lake is given in Table 3.1.

Table 3.1 Location, morphological and meteorological data for Lake Naini. Modified from Dash *et al.* (2008).

Parameter	Values	Parameter	Values
Altitude	1,937 m a.s.l.	Shoreline	3,630 m
Longitude	79°28′ E	Volume of water	5,907,500 m^3
Latitude	29°23′ N	Annual rainfall	2,300 mm
Maximum length	1,432 m	Maximum air temperature	24.6 °C
Maximum breadth	423 m	Minimum air temperature	0.5 °C
Maximum depth	27.3 m	Maximum water temperature	25 °C
Mean depth	16.2 m	Minimum water temperature	10 °C
Surface area	0.48 km^2	Mean water retention time (AHEC, 2002)	1.16 years
Catchment area	3.96 km^2		

For hypolimnetic aeration, two sets of aeration units were installed in the lake in 2007 as shown in Figure 3.3a. Each unit has a set of 15 disk modules that release compressed air at a pressure of ~310 kPa at the bottom of the lake. Ozone is introduced along with the air to prevent clogging of the disk modules (Williams, 2007). The aeration has significantly changed the dissolved oxygen (DO) profile of the lake (Kumar, 2008) as shown in Figure 3.3b. Before aeration, the DO was close to 0 mg/L below a depth of 8 m. After aeration, the lake had a DO of about 4 mg/L down to the bottom. The DO in the bottom zones is maintained around 3–4 mg/L by controlling the aeration rate.

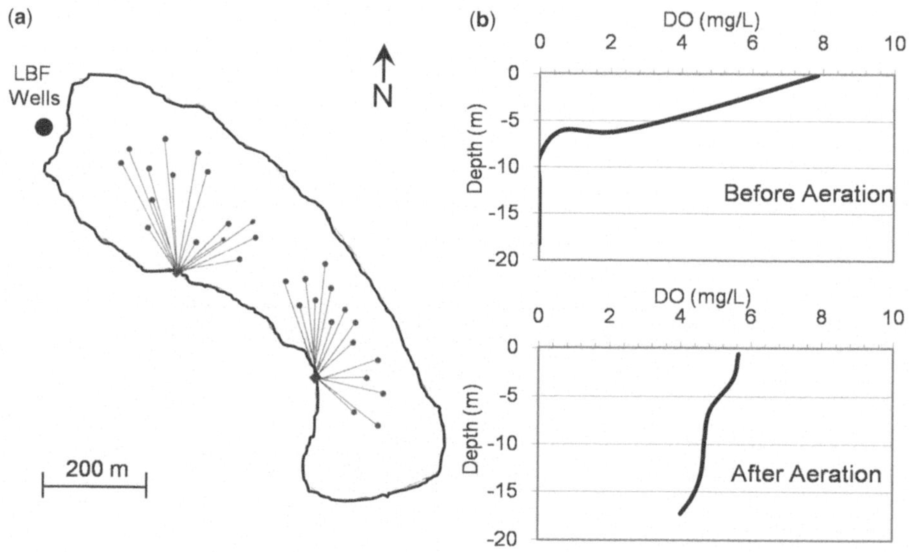

Figure 3.3 (a) Locations of the aeration disks placed in Lake Naini. (b) Representative DO profile of the lake before aeration (on September 6, 2007) and after aeration (on October 22, 2007) (Kumar, 2008).

The LBF wells are located on the north-west bank of the lake as shown in Figure 3.4 and their construction details are given in Table 3.2. As of now (August, 2014), Nainital City has a water demand of about 14 MLD in winter and 18.5 MLD in summer. The LBF wells operate for 14–20 hours per day depending on the demand and are used to abstract about 12 MLD water in winters and 16 MLD in summers. The use of spring water from Pardadhara (Figure 3.1) for water supply continues. There is a seasonal lake called Sukhatal uphill along the fault which fills up with water during the rainy season but drainage through the faults and fractures rapidly dries up this temporary lake. A well has been drilled near the Sukhatal, specified here as Sukhatal tube well (Figure 3.1), to abstract groundwater for local supply.

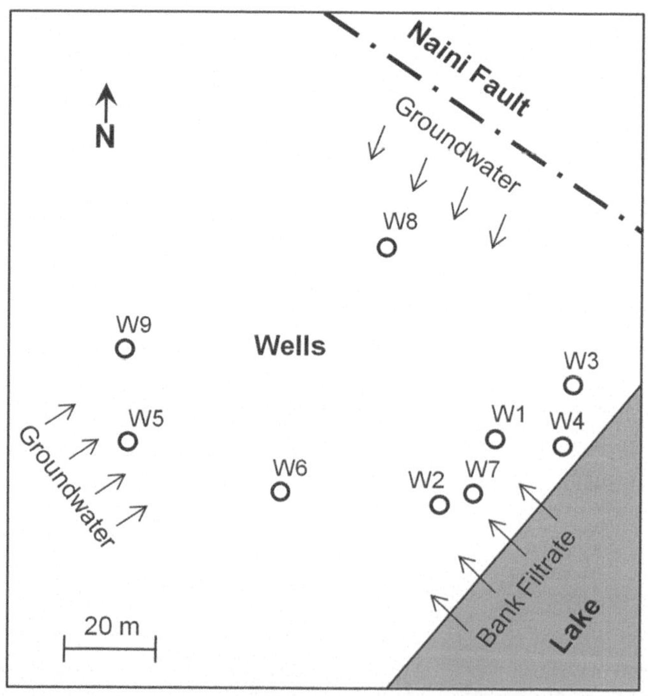

Figure 3.4 Location of wells W1-W9 with respect to the lake bank.

Table 3.2 LBF wells and their parameters in order of increasing distance from the lake.

Well*	Distance from Lake [m]	Year of Installation	Discharge [L/min]	Elevation [m a.s.l.]	Drilling Depth [m]	Filter Screen Depths [m]
W4	4.3	1999	1,600	1,938.5	26.7	10.0–26.0
W7	11.7	2006	1,400	1,938.6	35.9	14.0–34.2
W3	14.5	2009	2,000	1,939.4	36.0	12.7–30.4, 33.2–35.4
W1	14.9	2008	2,000	1,939.0	35.9	13.7–23.6, 25.6–35.2
W2	14.9	2008	2,000	1,939.0	37.2	15.2–24.9, 26.9–36.6
W6	42.9	2006	2,200	1,939.4	36.7	14.0–35.5
W8	51.8	2008	1,800	1,939.7	35.8	15.3–35.1
W5	64.0	2000	2,400	1,939.8	33.4	10.9–33.2
W9	94.1	2008	1,800	1,940.0	37.0	14.1–23.7, 26.6–36.2

*The wells are organized by increasing distance from the lake and well numbers are assigned by the water supply organization: Uttarakhand Jal Sansthan.

3.3 GEOLOGY OF THE TUBE-WELL SITE

A geological profile of the tube-well site is available from an investigatory drilling undertaken near the site to a depth of 132.6 m during 1975–1976 (Ashraf, 1978). The aquifer profile of the well field site has been described by Dash *et al.* (2008). The terrain around the tube-wells gently slopes and consists of debris from recurrent landslides that took place from 1867–1924. The aquifer consists of silty boulders, rock fragments, shale/slate, sand, and clay up to a depth of 50 m, followed by drain deposits, tree trunks, and lake deposits up to 100 m and slide debris for the next 17 m, below which is the bedrock consisting of red and green shale and slate of Middle Krol Formation.

The geology of the area is characterized by a number of folds and faults. Sharma (2001) has mapped the trace of the lake fault shear zone close to the tube-well site to a depth of ~116 m (Figure 3.5). The mountains consist of slates, marls, carbonates, limestones, dolomites, sandstone and conglomerates (Valdiya, 1980, 1988). The aquifer at the LBF well site consisting of irregular collapsed rocks contrasts the weathered sediments and silt making up aquifers in other cases. Based on grain size distribution by sieve analysis of aquifer soil obtained during drilling of well W6, hydraulic conductivities at different depths were estimated by Hazen's method to be 300–460 m/d with an average value of 327.5 m/d (Dash *et al.*, 2008). Pumping tests with Boulton's analysis yielded a hydraulic conductivity of ~275 m/d for wells near the lake and ~430 m/d for well W5 located far from the lake (Sandeep, 2011). This aquifer extends to a depth of about 36 m followed by a clay layer. Based on the hydraulic conductivities, travel times of the bank filtrate to wells W4 and W5 are estimated to be 1–2 days and 11–19 days, respectively.

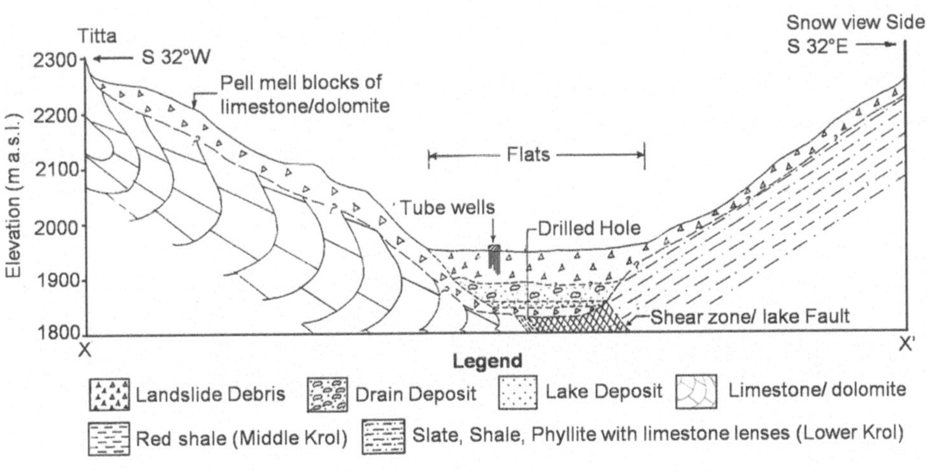

Figure 3.5 Geological cross-section (X–X′ in Figure 3.1) across tube-well/waterworks site near the northern edge of Lake Naini. Modified from Sharma (2001).

3.4 WATER BALANCE

A water balance study of Lake Naini was undertaken by the National Institute of Hydrology from 1994–2001 and a conceptual model was developed (AHEC, 2002). The conceptual model was validated using techniques such as isotope mass balance and chloride mass balance (Nachiappan *et al.*, 2002). Groundwater movement in the lake catchment preferentially takes place along faults and fractures towards the lake. Sukhatal, situated in the catchment of Lake Naini, does not have any surface outflow. Because of the proximity of the lake fault to Sukhatal, most of the water seeps underground and recharges Lake Naini. Sub-surface outflow from the lake mainly takes place through the epilimnion zone.

Average percentages of different components of water inputs and outputs to the lake on an annual basis are shown in Figure 3.6. Average quantities of annual water loss or gain by Lake Naini through different processes have been found to be 7.7×10^6 m^3. The subsurface inflow, pumping from tube-wells, and outflow through sluices are quantitatively prominent processes in water exchange between the lake and its catchment. The evaporation loss, direct rainfall over the lake surface area, outflow through springs and inflow through the drains are minor components.

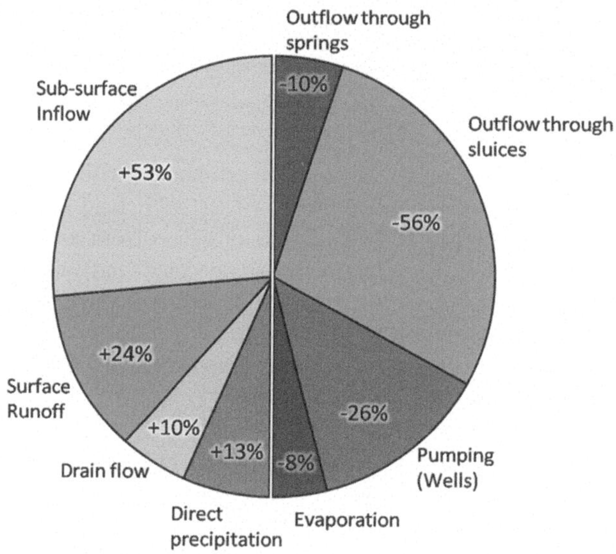

Figure 3.6 Water balance components and their average annual percentages for Lake Naini (based on Nachippan *et al.*, 2002 and AHEC, 2002). Positive and negative values represent inflows to and outflow from the lake, respectively.

3.5 METHODOLOGY

3.5.1 Sample collection

Water samples were collected from the lake and the LBF wells almost monthly from April 2012 to November 2013. Groundwater samples were collected from Pardadhara from Jan 2013 to Nov 2013 and from Sukhatal from February 2013 to November 2013. A few samples from the lake and LBF wells for coliform studies were also collected from January to May, 2011. Lake water samples were collected from ~0.5 m below the lake surface. Electrical conductivity (EC), pH, temperature, and dissolved oxygen (DO) were measured on-site using a portable multi-parameter probe (HQ40d, Hach, Loveland, USA). Well water samples were collected while the pumps were in operation for supply.

For the analysis of dissolved ions, water samples were collected in 1,000 mL polyethene and polypropylene bottles. Samples for ^{18}O isotope analyses were collected in 15 mL polypropylene bottles ensuring that no air bubbles were trapped in the bottle during collection. Samples for bacteriological analysis were collected in sterilized glass bottles. Samples were stored cool (4°C), in the dark and transported to the laboratory at IIT Roorkee for further analysis within 24 hours.

3.5.2 Sample analysis

Total and faecal coliform counts in the samples were determined by the multiple tube fermentation technique. For the analysis of dissolved ions and dissolved organic carbon (DOC), the samples were filtered with 0.22 μm size filter (Millipore,

GVWP). The ions (sodium, potassium, ammonium, calcium, magnesium, chloride, fluoride, nitrate, nitrite, sulphate, and total phosphate) were determined by ion chromatography (Metrohm, AG-861). DOC was measured using TOC-V_{CSN} Total Organic Carbon Analyser (Shimadzu). UV absorbance (UV-A) at 254 nm was measured using DR5000 spectrophotometer (HACH) using a 10 mm quartz cell. Alkalinity was determined by titration with N/50 H_2SO_4 (aq.) with bromocresol green as the indicator. All procedures for sampling, transportation, storage, and analyses were in accordance with the procedures given in Standard Methods (APHA *et al.*, 2005).

Isotopic analysis for $\delta^{18}O$ in H_2O in the samples was done at National Institute of Hydrology, Roorkee (India) using GV Isoprime Dual Inlet Isotope Ratio Mass Spectrometer. For $\delta^{18}O$ analysis, 400 µL of water samples were equilibrated for 7 hours with CO_2 reference gas. The measured delta (δ) values are given with respect to Vienna Standard Mean Ocean Water. The precision of measurement for $\delta^{18}O$ is ± 0.1‰.

3.6 RESULTS AND DISCUSSION

The performance of the LBF system at Lake Naini was evaluated in terms of two criterions: the quality of water abstracted in terms of drinking water standards and the fraction of lake water abstracted in the wells. Mixing of bank filtrate with ground water in the LBF wells was assessed by stable isotope analysis. This section first briefly discusses the spatial and seasonal variations in the lake water quality during the study period and then presents the results of isotopic investigations and water quality analysis.

3.6.1 Spatio-temporal variation in lake water quality

The most important parameter of lake water quality is DO. During the study period, the surface DO in Lake Naini underwent seasonal cycles as shown in Figure 3.7. From June to November, it dropped to 4–6 mg/L. From December to May, it reached 7–10 mg/L, close to its saturation value. The temperature of the lake surface water varied between 10°C and 22°C. EC of the lake water also showed annual variation between 570–630 µS/cm with a slight decrease in the monsoon season and an increase in winter. The variations in EC between various points (L1-L4, Figure 3.1) on the lake were up to ~60 µS/cm. Point L1, shown in Figure 3.1, was chosen as a representative point for infiltrating lake water in LBF wells because of its proximity to the wells.

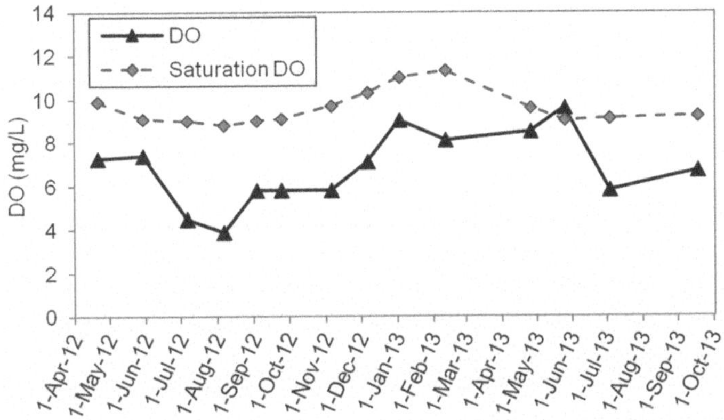

Figure 3.7 Dissolved oxygen (DO) concentration of the lake at its surface and its saturation concentration at the water temperature as a function of time.

3.6.2 Proportion of bank filtrate and groundwater in the wells

The $\delta^{18}O$ values for the waters from various sites are presented in Figure 3.8. The lake was predominantly fed by groundwater but because of a long retention time of ~1 year and hence prolonged evaporative enrichment, the lake water was isotopically richer in ^{18}O than the groundwater. Among the well waters, the $\delta^{18}O$ values of W9 were the lowest throughout the year and were comparable with the spring water samples from Pardadhara. Due to lack of groundwater data for the complete study period, W9 values were considered as indicative of the groundwater isotopic values (Figure 3.8). It is consistent with the expectation that W9 would abstract the highest portion of ambient land-side groundwater as W9 is located farthest from the lake at a distance of 94 m (Figure 3.4, Table 3.2).

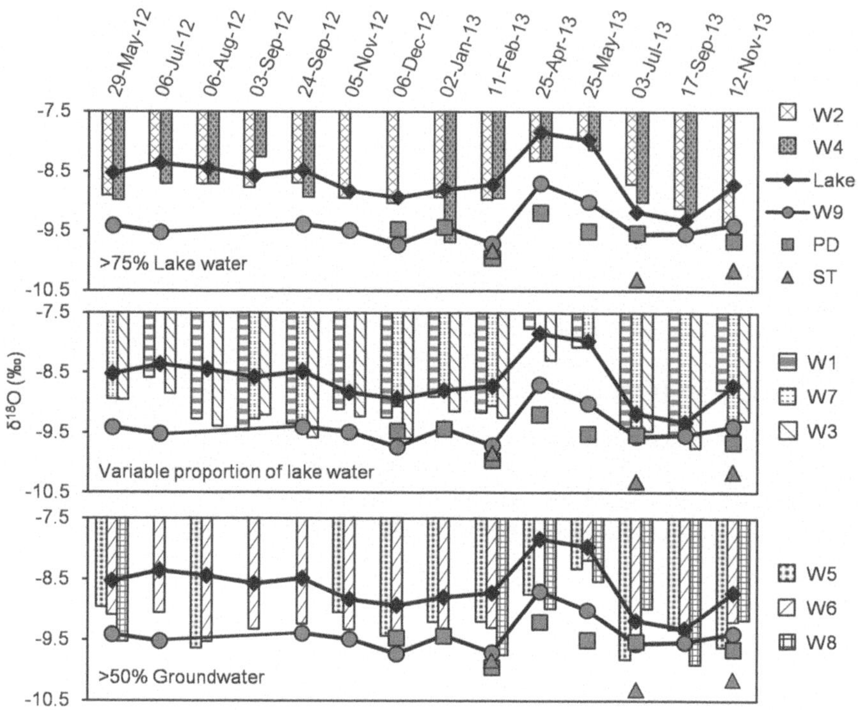

Figure 3.8 Plots of $\delta^{18}O$ values for the lake water, various wells, and the groundwater sources.

The $\delta^{18}O$ values of the well waters and their temporal variations could be categorized into the following three groups, represented by the three graphs in Figure 3.8:

(a) The first group consisting of wells W2 and W4 had $\delta^{18}O$ values similar to the lake water almost throughout the year, suggesting that these wells continuously drew bank filtrate. Average $\delta^{18}O$ values for each well calculated using the two-component mixing model, shown in Table 3.3, suggest that the two wells abstracted ~80% bank filtrate.

(b) The second group consisting of wells W1, W3, and W7 had waters isotopically similar to the groundwater during the monsoon period and similar to the lake water during the non-monsoon period. Average values (Table 3.3) indicate that W1 and W7 abstracted 50–75% bank filtrate annually while well W3 abstracted only ~40% bank filtrate. These results are surprising particularly because wells W1, W2, and W3 are located at about the same distance from the lake bank but abstracted such different proportions of bank filtrate. Such anomalous observations suggest the presence of subsurface seepage from the Naini Fault or highly irregular flow of groundwater in the lake bank aquifer. The proposed groundwater flow from the fault seepage is indicated in Figure 3.4, where higher groundwater proportion in W3 can be explained by its proximity to the fault.

(c) The third group consists of wells W5, W6, W8, and W9, which all had isotopic signatures consistently similar to the groundwater, showing that these wells predominantly abstracted groundwater throughout the year. Average values (Table 3.3) indicate that W5, W6, and W8 abstracted 25–50% bank filtrate while W9 abstracted <25% bank filtrate. These conclusions about the bank filtrate proportion are also supported by the water quality results (*vide infra*).

Temperature profiles of water in the wells and their correlation with the lake temperature (Figure 3.9) were also different for the three groups of wells. Lake surface temperature showed a smooth temperature variation corresponding to the seasonal changes. Temperatures of the wells W2 and W4 varied similar to the lake with a slight delay in the maxima of about 1 month. This delay might be less than one month, but it cannot be determined exactly because no sample was taken during this period. This delay – much longer than the travel time of water – was likely caused by temperature retardation by the aquifer material. Temperature profiles of the wells W1, W3, and W7 were also similar to the lake but had a slightly lower correlation with the lake as compared to wells W2 and W4. The temperature equilibration effect of the aquifer is likely to reduce the temperature difference between the first and second groups of wells. Wells W5, W6, W8, and W9 had poor correlation with the lake water. These wells also had a very narrow temperature range, characteristic of groundwater – consistent with the high groundwater proportion in these wells.

Table 3.3 Percent bank filtrate in various wells based on average $\delta^{18}O$ values for the lake water, various wells and groundwater sources.

Water Source	$\delta^{18}O$ [‰]	Percent Bank Filtrate [%]
Lake	–8.60	–
W2	–8.78	83
W4	–8.81	80
W1	–8.97	65
W7	–9.08	54
W3	–9.23	40
W5	–9.21	42
W6	–9.24	39
W8	–9.26	37
W9	–9.40	24
Groundwater (Pardadhara & Sukhatal)	–9.65	–

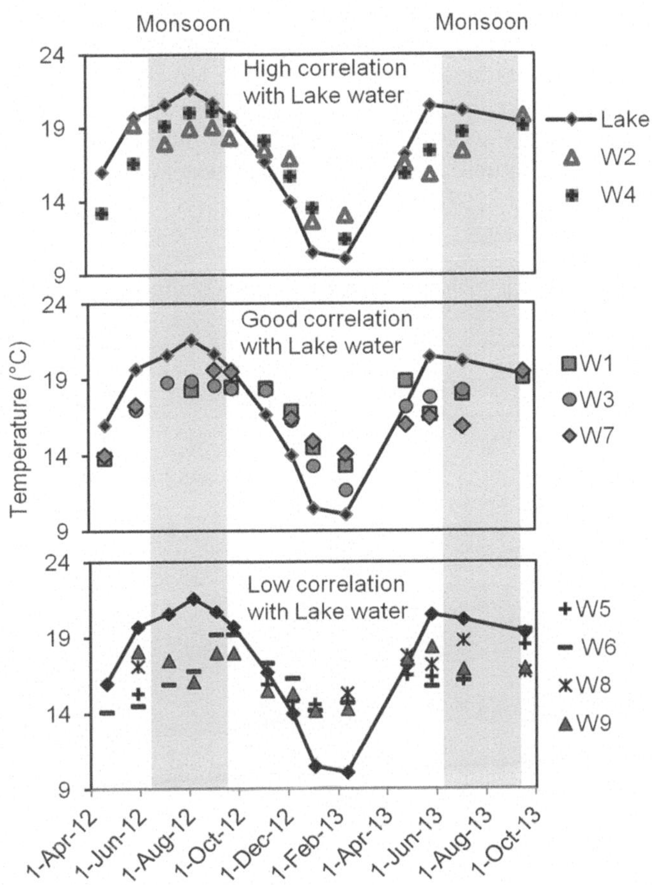

Figure 3.9 Temperature profiles of the lake and well waters.

3.6.3 Attenuation of coliforms, turbidity and dissolved organics

After aeration, the coliform counts, the turbidity and the concentrations of organics reduced significantly in the lake. Figure 3.10 shows the total and faecal coliform counts in the lake water before and after aeration. The coliform MPN counts in the lake reduced by two orders of magnitude to low levels of ~1,000 MPN/100 mL. This reduction in the coliform counts is also likely to be due to disinfection by ozone used in the aerators.

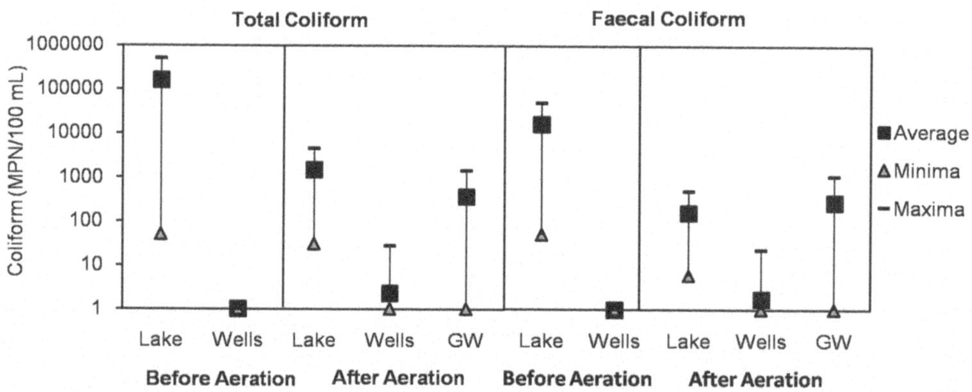

Figure 3.10 Total and faecal coliform in the lake, the wells, and the groundwater sources (includes Pardadhara and Sukhatal) before (Dash *et al.*, 2008) and after aeration (2011, 2012–2013). The five sets of samples for each parameter have *n* values 72, 30 (5 wells), 18, 116 (9 wells), and 8, respectively.

As mentioned before, the LBF wells had delivered coliform-free water even before aeration (Dash *et al.*, 2008). The median and range of the coliform values for different sources during the recent sampling period (2011 to 2013) is given in Table 3.4. Pardadhara water often showed coliform contamination. The wells were largely free of coliform except for a breakthrough that was observed in summer months. During this period, coliform level was at its highest in the lake, a few wells, and Pardadhara. The highest level of contamination was observed in wells W5, W9, and W2. Contamination in wells W1 and W4 was much lower, indicating that the contamination came from the groundwater or localized infiltration of contaminated water and not from the bank filtrate.

Table 3.4 Variation of total and faecal coliform counts in various sources. The entries represent median (range).

Sample (n)	Total Coliforms [MPN/100 mL]	Faecal Coliforms [MPN/100 mL]	No. of Samples with Total Coliforms >2 MPN/100 mL
Lake (18)	1,600 (30–4,500)	141 (6–500)	18
W4 (13)	<2 (<2–17)	<2	2 (Jun. 2012, Sep. 2012)
W2 (17)	<2 (<2–1,553)	<2 (<2–1,553)	1 (Jul. 2013)
W1 (18)	<2 (<2–17)	<2 (<2–3)	2 (Jun. 2012; Jul. 2013)
W7 (15)	<2 (<2–36)	<2 (<2–23)	3 (Jan. May, 2011; Jul. 2013)
W3 (18)	<2 (<2–70)	<2 (<2–16)	2 (Jun. 2012; Jul. 2013)
W6 (17)	<2	<2	–
W8 (7)	<2 (<2–27)	<2 (<2–14)	3 (Feb. 2011; Jul.–Sep. 2013)
W5 (6)	<2 (<2–2,420)	<2 (<2–2,420)	1 (Jul. 13)
W9 (17)	<2 (<2–2,420)	<2 (<2–2,420)	5 (Feb.–Mar. 2011; Jul.–Nov. 2013)
Pardadhara (4)	709 (5–1,414)	561 (<2–1,120)	4 (May–Nov. 2013)
Sukhatal (4)	<2	<2	–

It is important to note that the well W6 still was not coliform contaminated while other wells on both sides occasionally were. This observation suggests that W6 gets a stream of groundwater that did not pass through wells W5 and W9 (Figure 3.4). Apparently, the groundwater flow direction from the hills on the west is towards the northeast as shown in Figure 3.4.

The turbidity of the lake decreased from 3.7–7.3 NTU before aeration (Dash *et al.*, 2008) to 1.7–5.5 NTU post aeration as shown in Figure 3.11. The turbidity in the wells monitored earlier had been consistently below 1 NTU. During the present sampling, the turbidity of the well water was 0.2 NTU most of the time, but occasionally it rose above 1 NTU. The mean turbidity and its range for various waters are given in Table 3.5. The turbidity in the wells, however, was within the limits for drinking water of 5 NTU according to the Indian standards BIS-10500 (2012).

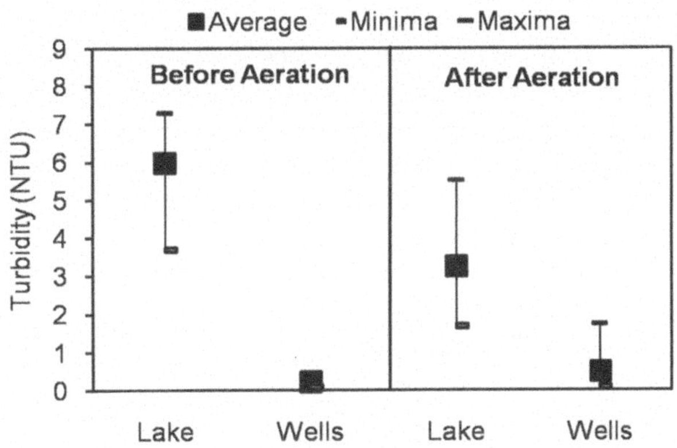

Figure 3.11 Turbidity in the lake and wells before (Dash *et al.*, 2008) and after aeration (2012–2013).

Table 3.5 Mean (range) of turbidity, DOC, and SUVA values for various water sources. Single observations with values outside the normal range are mentioned in remarks column.

Sample	Turbidity [NTU]	DOC [mg/L]	SUVA [L/(mg m)]	Remarks
Lake	3.3 (1.7–5.5)	3.1 (1.2–4.7)	1.3 (0.7–2.5)	
W4	0.5 (0.2–1.3)	1.7 (0.8–3.6)	0.8 (0.3–1.7)	
W2	0.6 (0.2–1.2)	1.7 (0.6–3.7)	1.2 (0.3–1.4)	
W1	0.5 (0.1–1.2)	1.4 (0.6–3.8)	0.8 (0.3–1.4)	
W7	0.6 (0.2–1.8)	1.6 (0.6–3.2)	1.0 (0.2–2.2)	
W3	0.5 (0.2–1.0)	1.7 (0.5–3.6)	0.9 (0.3–1.3)	
W6	0.4 (0.2–1.2)	1.4 (0.4–3.3)	1.1 (0.1–2.3)	
W8	0.6 (0.2–0.9)	1.4 (0.4–3.2)	0.6 (0.2–1.9)	
W5	0.4 (0.2–1.1)	1.5 (0.6–3.5)	0.8 (0.2–2.4)	High turbidity (3.1 NTU) in Nov. 2013
W9	0.6 (0.1–1.6)	1.4 (0.4–3.1)	0.5 (0.2–1.1)	High SUVA (4.4 L/(mg m)) in Jul. 2013
Pardadhara	0.7 (0.5–1.0)	1.1 (0.4–3.2)	1.4 (0.1–3.0)	High turbidity (5.7 NTU) in Apr. 2013
Sukhatal	1.4 (1.2–1.7)	1.0 (0.3–1.6)	1.2 (0.3–1.7)	

The organics concentration was low in both the lake and the wells. The mean UV-A of the lake had been 0.059/cm before aeration, which was now reduced to 0.035/cm. Wells had UV-A consistently below 0.020/cm, except for W5 and W9 that had an increase in UV-A during monsoon of 2013. The LBF wells also had much lower average DOC of ~1.6 mg/L than the lake (Table 3.5). Specific ultraviolet absorbance (SUVA) values of well waters (Table 3.5) were slightly lower than the lake water, suggesting a slight decrease in aromatic compounds (Weishaar *et al.*, 2003) in the wells compared to the lake. Since wells W2 and W4 were receiving almost completely bank filtrate, we can estimate that bank filtration led to DOC reduction of 19% to 76% with a mean reduction of 52%.

3.6.4 Ionic composition of waters

The lake water had an average EC of 598 μS/cm – lower than the groundwater (Table 3.6), likely because of mixing of groundwater with low conductivity rainwater and surface run-off. The EC of the wells varied from 630 μS/cm to 900 μS/cm, with significant differences between the three groups of wells. The first group of wells, W2 and W4, had an EC ~20 μS/cm higher than that of lake water with much lower seasonal variation than the lake. The lower seasonal variation may be due to averaging effect of bank filtrate stored in the aquifer. The second group of wells, W1, W3, and W7 showed higher conductivity than the lake water by about 100–200 μS/cm. Their EC values were also comparable to the groundwater samples from Sukhatal and Pardadhara. The third group of wells, W5, W6, W8, and W9, had EC values that were consistently 250–300 μS/cm higher than the lake water and 100–150 μS/cm higher than the groundwater samples. Such high differences observed consistently in EC values indicate that either there is high mineralization of groundwater in the aquifer around these wells or that the groundwater coming to these wells is different than the groundwater at Pardadhara and Sukhatal. The range of seasonal variation in EC was also different for wells W9, W5, and W6, which supports the latter possibility and is consistent with the proposed groundwater flow direction shown in Figure 3.4. The differences in conductivities in the first group of wells, W2 and W4, and second group of wells, W1, W3, and W7, located very close to each other further supports the hypothesis of groundwater mixing rather than mineralization from the surrounding aquifer.

Table 3.6 Mean (range) concentrations of major ions in various water sources.

Source	EC [μS/cm]	Na^+ (mg/L)	K^+ [mg/L]	Ca^{2+} [mg/L]	Mg^{2+} [mg/L]
Lake	598 (498–698)	11 (7–14)	6 (3–9)	51 (38–80)	42 (37–50)
W4	633 (588–660)	13 (11–18)	6 (3–7)	55 (42–82)	47 (40–54)
W2	647 (552–705)	11 (9–14)	6 (4–9)	58 (46–84)	45 (36–53)
W1	697 (617–817)	15 (11–22)	6 (5–8)	60 (48–83)	49 (37–56)
W7	685 (610–862)	14 (10–19)	6 (4–8)	62 (47–84)	49 (36–57)
W3	708 (643–832)	16 (12–24)	6 (4–7)	63 (43–92)	55 (43–62)
W6	795 (688–894)	15 (11–19)	6 (4–8)	72 (43–95)	60 (44–68)
W8	823 (749–893)	20 (11–29)	6 (5–7)	75 (63–105)	56 (46–72)
W5	794 (693–915)	14 (10–22)	5 (4–7)	74 (62–99)	57 (42–66)
W9	844 (788–882)	15 (10–22)	6 (5–7)	75 (63–105)	63 (50–71)
Pardadhara	724 (683–766)	10 (7–13)	6 (4–13)	73 (62–110)	50 (46–58)
Sukhatal	693 (675–702)	9 (7–11)	6 (4–7)	73 (56–89)	50 (46–53)
Source	**Cl^- [mg/L]**	**NO_3^- [mg/L]**	**SO_4^{2-} [mg/L]**	**Alkalinity [mg/L]**	**Hardness [mg/L of $CaCO_3$]**
Lake	8 (6–9)	6 (2–16)	91 (74–109)	213 (189–319)	302 (248–361)
W4	10 (8–18)	4 (0.4–11)	95 (72–109)	226 (196–239)	333 (273–378)
W2	8 (7–10)	4 (ND–6)	94 (68–147)	233 (186–276)	334 (273–370)
W1	12 (9–18)	13 (1–22)	100 (66–135)	255 (232–310)	356 (316–398)
W7	10 (8–13)	6 (2–15)	103 (98–165)	269 (240–318)	363 (301–408)
W3	11 (9–19)	13 (1–32)	105 (75–121)	251 (214–288)	385 (317–488)
W6	12 (9–16)	14 (2–31)	133 (98–165)	273 (234–324)	435 (316–478)
W8	14 (10–21)	11 (<0.5–34)	121 (107–137)	287 (252–333)	424 (362–502)
W5	10 (8–19)	10 (<0.5–22)	129 (96–176)	264 (260–268)	420 (362–463)
W9	12 (9–13)	14 (<0.5–28)	142 (121–158)	281 (240–301)	448 (384–495)
Pardadhara	9 (7–12)	13 (12–14)	136 (111–156)	259 (246–265)	403 (362–467)
Sukhatal	11 (9–14)	11 (9–14)	127 (110–141)	265 (238–312)	391 (357–442)

The lake and Sukhatal waters were predominantly alkaline with average pH values of 8.0 and above, while the well waters and Pardadhara were less alkaline with pH between 7.0 and 8.0. There were no significant differences among pH values of various wells, suggesting that the pH was largely determined by the aquifer and not the proportion of bank filtrate.

In terms of ionic composition, the lake, wells, and ground waters contained predominantly calcium, magnesium, sulphate and hydrogen carbonate in major amounts and sodium, potassium, chloride and nitrate in smaller amounts (Table 3.6). This is consistent with the findings of previous researchers (Nachiappan & Kumar, 1999; Chakrapani *et al.*, 2002), who ascribed the presence of limestone, dolomite, and other carbonates as sources for calcium, magnesium, and hydrogen carbonate ions; silicate minerals as source for sodium and potassium ions; and oxidation of pyrite minerals as a source for sulphate ions. Small amounts of chloride and nitrate are ascribed to sewage, and forest and small agricultural run-off. The hardness (Table 3.6) followed the same pattern as EC.

The concentrations of calcium and magnesium often exceed the desirable limits of 75 mg/L and 30 mg/L, respectively, for drinking water as per Indian drinking water standards BIS-10500 (2012). However, the concentration values remained within the maximum permissible limits of 200 and 100 mg/L, respectively, prescribed in BIS-10500 (2012) for the situations where an alternative source of water supply is not available. Consequently, the hardness values in the wells were also above the desirable limits for drinking water (Bureau of Indian Standards, 2012), but were lower than the maximum permissible limit prescribed when an alternative source is not available.

Monthly variations in the concentrations of calcium, magnesium, and alkalinity, shown in Figures 3.12–3.14, had different patterns for the three groups of wells. These differences were not so pronounced for other ions. Wells W2 and W4 had calcium and magnesium concentrations similar to and alkalinity slightly higher than the lake water. In the second group of wells W1, W3, and W7, the values of the three parameters were higher than in the lake during the summer and monsoon months but comparable to those of the lake during other times. These observations were consistent with the variations in bank filtrate proportions in these wells. The concentrations in the third group of wells, W5, W6, W8, and W9, were much higher than the lake and comparable to the groundwaters. In general, calcium and magnesium concentrations in the lake and the bank filtrate decreased during the summer and monsoon periods and remained roughly constant in the groundwater.

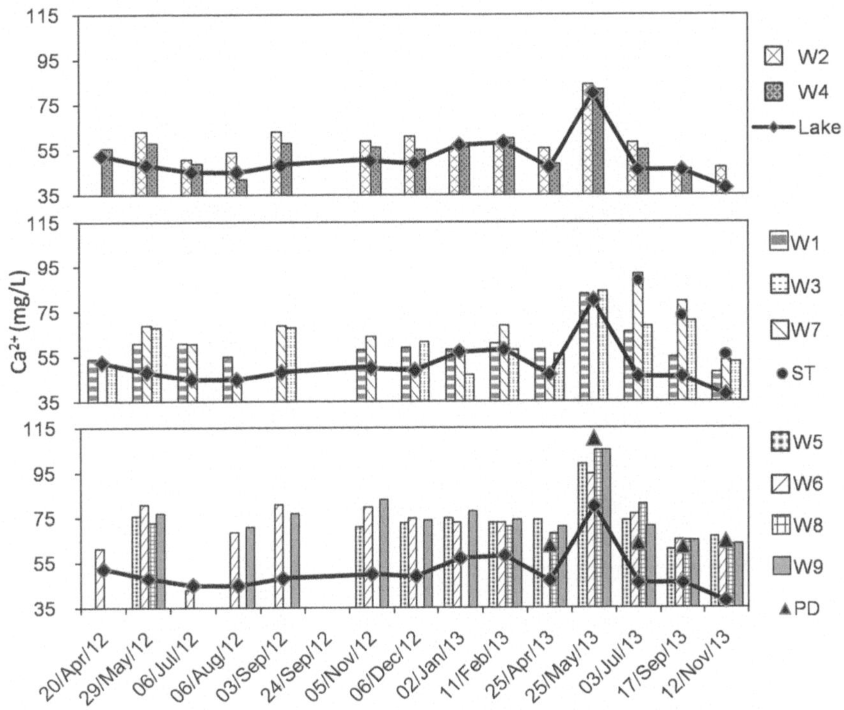

Figure 3.12 Monthly variations of calcium (Ca^{2+}) concentration in various waters.

In spite of different values among various water sources, the calcium:magnesium molar ratio for all the sources studied lied around 0.75. This ratio is indicative of predominantly magnesium-containing minerals such as dolomite contributing to the mineralization of the water. Compared to the water quality of these sources before aeration (Dash *et al.*, 2008), this ratio has increased after aeration. Calcium concentrations in the waters have increased by about 10–15 mg/L, and magnesium concentrations have decreased by 5–10 mg/L. There was no significant difference in the concentration values of other ions before aeration and the values measured during the sampling campaign.

In general, the observations show that bank filtrate had low mineral content (comparable to the lake water) and better water quality in terms of organics, turbidity, and coliforms. The groundwater showed occasional contaminations. Therefore, wells W2 and W4 should be pumped more to obtain more of (better quality) bank filtrate. Abstraction from other wells can be regulated with seasonal changes to obtain larger amount of bank filtrate. Compared to the LBF well field in 2001 when water balance study was undertaken (Figure 3.6), there are more wells now, but the additional wells abstract mostly groundwater. As mentioned in Section 3.4, LBF wells in 2001 pumped 26% of the lake water output and 56% of the lake outflow was through sluices. Since evaporation and subsurface drainage cannot be prevented, LBF wells can extract a maximum of 82% of the lake outflow. Beyond this limit, water abstraction may not remain sustainable and would reduce the water quantity in the lake and groundwater level in the region. Based on a lake volume of 5,907 ML (5,907,000 m^3) and a mean retention time of 1.16 years (Table 3.1), this quantity turns out to be ~13.9 MLD. Of the current water abstraction of 12–16 MLD (Section 3.2), since LBF wells abstract ~40–50% bank filtrate on average, the quantity of bank filtrate abstraction can be increased only by up to a factor of 2.

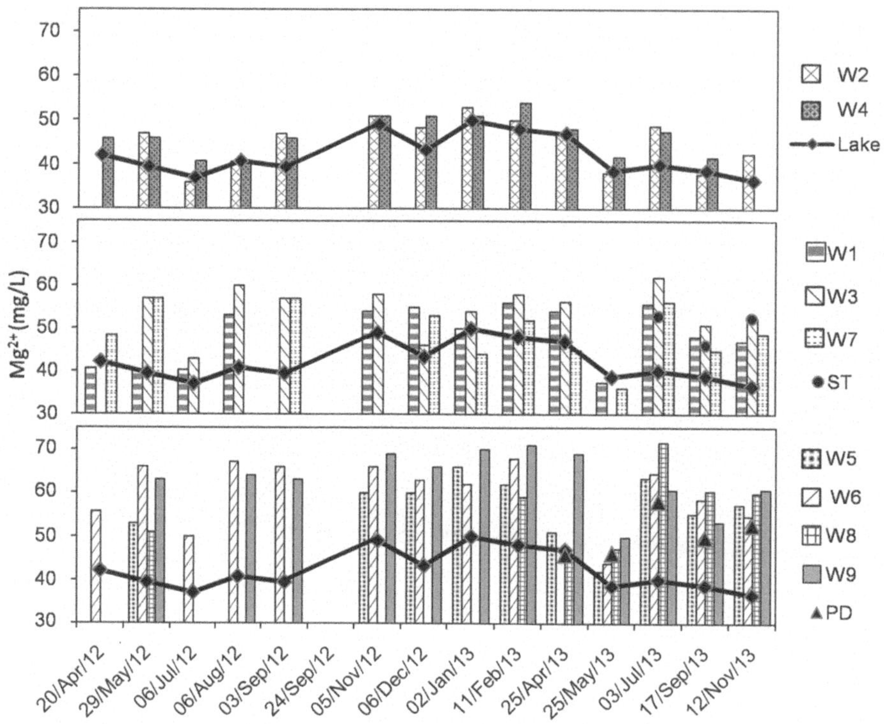

Figure 3.13 Monthly variations of magnesium (Mg^{2+}) concentration in various waters.

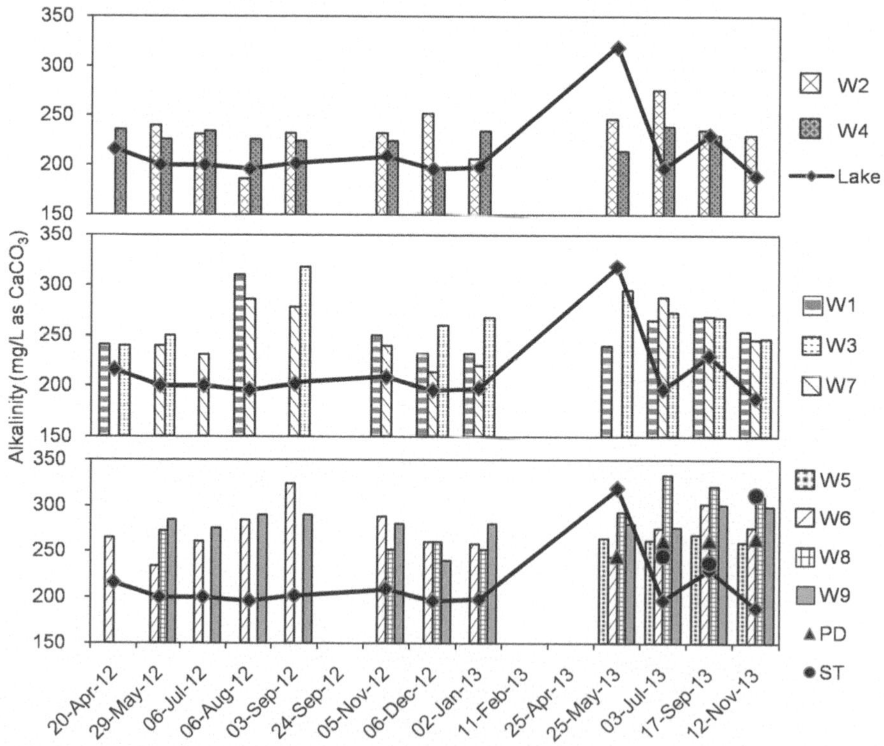

Figure 3.14 Monthly variations of alkalinity in various waters.

This study also showed that even though LBF systems are highly effective, groundwater dynamics can have a strong impact on the quality of water that is abstracted from the LBF wells. Local geology, hydrology, seasonal variations in groundwater dynamics, and potential impact of anthropogenic activities in the catchment need to be studied for effective decision making on LBF abstraction systems.

3.6.5 Comparison with previous literature

These conclusions about the proportions of bank filtrate in the wells are quite different from that of Nachiappan *et al.* (2002) made based on isotopic studies during 1994–1995. For the wells W1-W5, Nachiappan *et al.* (2002) estimated higher proportion of bank filtrate (80%) during monsoon and a lower proportion (20%) during non-monsoon. Although this difference in conclusions could be reflecting actual changes in the local hydrology in the lake bank aquifers over the years, a few important differences also exist between present data and analysis and that of Nachiappan *et al.* (2002). The first and foremost difference is that both the lake and the groundwater were isotopically lighter in the present study than what was observed during 1994–1995, either because of change in rainfall pattern in the region or because of a change in the origin of groundwater coming to the lake catchment region. Secondly, Nachiappan *et al.* (2002) sampled at multiple locations in the lake at different depths and took the average of the isotope values. Investigations at Lake Tegel in Berlin (Fritz *et al.*, 2002) show that the majority of the infiltration at lake banks is likely to be from the top epilimnion zone of the lake. Therefore, water near the lake surface close to the infiltration wells is more likely to be representative of infiltrating water rather than average values of samples across the lake. We tested water samples from various points across the lake and observed differences in $\delta^{18}O$ value of up to 0.4‰ between various points. Therefore, our analysis based on the lake surface water close to the wells would lead to different results than the average derived from sampling various points in the lake as adopted by Nachiappan *et al.* (2002). Thirdly, groundwater samples in the present study were collected from sources close to the LBF wells, whereas previous groundwater data was collected from the springs in the whole region, many of which dried up in recent years.

In terms of organic content, the lake has become one of the cleaner lakes in the world. The average DOC of the lake – 3.1 mg/L (Table 3.5) was much lower than typical DOC concentrations in most lakes in the world. Median DOC of ~7,541 lakes in the world is 5.7 mg/L (Sobek *et al.*, 2007). The SUVA values observed in this present study were also lower than the values observed for lake water and bank filtrate in Lake Tegel (Grünheid *et al.*, 2005), and were also lower than the values observed in deep groundwaters (Inamdar *et al.*, 2012) suggesting the dominance of low molecular weight aliphatic compounds in Nainital waters.

3.7 CONCLUSIONS

Recent isotope studies showed that during the present study period, the LBF wells at Naini lake bank abstracted different proportions of bank filtrate and these proportions were not necessarily determined by the distance from the lake. Wells W2 and W4 abstracted ~80% bank filtrate almost throughout the year. Wells W1, W3, and W7, although located at similar distances (12–15 m) as well W2 (14.9 m), abstracted >50% bank filtrate during non-monsoon season but mostly groundwater during monsoon seasons. Wells W5, W6, W8, and W9, located >40 m from the lake, abstracted >75% groundwater almost throughout the year. One possible reason for such unusual hydrology is the presence of the Naini fault and other sub-faults very close to the well field. Groundwater seepage from these faults can affect the flow pattern of bank filtrate in the lake bank aquifer. In addition, steep slopes on the north, west, and south sides of the well field also complicate the flow patterns.

This study showed that the water quality of Lake Naini considerably improved with lower coliforms, turbidity, and organics and higher DO as a result of various cleaning efforts including hypolimnetic aeration. The water quality of the wells continues to be good, but a few occasional contaminations are observed for brief periods.

The waters in the region had calcium, magnesium, hydrogen carbonate and sulphate as the predominant ions with a calcium:magnesium ratio of ~0.75 in all the waters. The data suggests the predominant mineral responsible for this mineralization of water to be magnesium and carbonate based.

The bank filtrate had much better water quality than the groundwater in terms of inorganic ion concentrations throughout the year. The wells W2 and W4 should be pumped more to obtain a better quality bank filtrate. Abstraction from other wells can be regulated with seasonal changes to obtain the maximum amount of bank filtrate. Net abstraction of bank filtrate, however, should not increase by more than two times, beyond which water level of the lake and groundwater may go down making the system unsustainable.

Overall the lake bank filtration system in Nainital is highly effective in providing drinking quality water, provided contamination through groundwater can be kept in check.

ACKNOWLEDGEMENTS

This study was conducted as a part of the project 'Saph Pani – Enhancement of natural water systems and treatment methods for safe and sustainable water supply in India', funded by the European Commission (grant agreement no. 282911). The authors are grateful to Uttarakhand Jal Sansthan for their co-operation in this study. Laxmi Das, Anand Bharti, Fuzail Ahmed, Sandeep C and Pooran Patwal at IIT Roorkee and F. Mütze and S. Zschau at HTW Dresden are thanked for help in the investigations.

3.8 REFERENCES

AHEC (2002). Detailed project report on conservation and management plan for Nainital Lake. AHEC/2002, Alternate Hydro Energy Centre, Indian Institute of Technology Roorkee, Uttaranchal (now Uttarakhand), India.

APHA, AWWA and WPCF (1995, 1998, 2005). Standard methods for the examination of water and wastewater, 19th, 20th and 21st edn., American Public Health Association, Washington DC, United States.

Ashraf Z. (1978). A geological report on the drilling explorations carried out for the study of Nainital hill slope stability. Nainital district, U.P. Report of Geological Survey of India, Northern Region, Lucknow, India.

Bureau of Indian Standards (BIS) 10500 (2012). Drinking Water – Specification. Indian Standard, 2nd Revision. Bureau of Indian Standards, New Delhi, India.

Census of India (2011). Office of the Registrar General and Census Commissioner, India. http://www.censusindia.gov.in (accessed 28 August 2014).

Chakrapani G. J. (2002). Water and sediment geochemistry of major Kumaun Himalayan lakes, India. *Environmental Geology,* **43**(1–2), 99–107.

Chauhan A., Fortenberry G. Z., Lewis D. E. and Williams H. N. (2009). Increased diversity of predacious *Bdellovibrio*-Like Organisms (BLO) as a function of eutrophication in Kumaon lakes of India. *Current Microbiology,* **59**(1), 1–8.

Das B. K., Singh M. and Borkar M. D. (1994). Sediment accumulation rate in lakes of Kumaun Himalaya, India using ^{210}Pb and ^{226}Ra. *Environmental Geology*, **23**(2), 114–118.

Das B. K., Singh M. and Van Grieken R. (1995). The elemental chemistry of sediments in the Nainital lake, Kumaun Himalaya, India. *Science of the Total. Environment*, **168**(1), 85–90.

Dash R. R., Mehrotra I., Kumar P. and Grischek T. (2008). Lake bank filtration at Nainital, India: Water quality investigations. *Hydrogeology Journal*, **16**(6), 1089–1099.

Dua V. K., Kumari R., Johri R. K., Ojha V. P., Shukla R. P. and Shukla V. P. (1998). Organochlorine insecticide residues in water from five lakes of Nainital (U.P.), India. *Bulletin of Environmental Contamination and Toxicology*, **60**(2), 209–215.

Fritz B., Sievers J., Eichhorn S. and Pekdeger A. (2002). Geochemical and hydraulic investigations of river sediments in a bank filtration system. In: Management of aquifer recharge for sustainability, P. Dillon (ed.), Swets and Zeitlinger, Lisse, The Netherlands, pp. 95–100.

GoogleMaps (2013). http://maps.google.com/ (accessed 1 October 2014).

Grünheid S., Amy G. and Jekel M. (2005). Removal of bulk dissolved organic carbon (DOC) and trace organic compounds by bank filtration and artificial recharge. *Water Research*, **39**(14), 3219–3228.

Gupta P. K. and Pant M. C. (1983). Microbenthos of lake Naini Tal (U.P., India) with particular reference to pollution. *Water, Air & Soil Pollution*, **19**(4), 397–405.

Gupta R., Bhagat P., Joshi M., Inaotombi S. and Gupta P. K. (2010). Heavy metal pollution status of Lake Nainital, Uttarakhand. *Indian Journal of Scientific Research*, **1**(1), 15–19.

Gupta R. and Gupta P. K. (2012). The impact of artificial circulation on phytoplankton community on a fresh water eutrophic lake. *Journal of Ecophysiology and Occupational Health*, **12**(3–4), 1–20.

Hoffmann A. and Gunkel G. (2011). Bank filtration in the sandy littoral zone of Lake Tegel (Berlin): Structure and dynamics of the biological active filter zone and clogging processes. *Limnologica*, **41**(1), 10–19.

Inamdar S., Finger N., Singh S., Mitchell M., Levia D., Bais H., Scott D. and McHale P. (2012). Dissolved organic matter (DOM) concentration and quality in a forested mid-Atlantic watershed, USA. *Biogeochemistry*, **108**(1–3), 55–76.

IMD (2013). Mean rainfall data of Nainital District. HYDROMET Division, Indian Metrological Department, New Delhi. http://www.imd.gov.in/section/hydro/distrainfall/webrain/uttarakhand/nainital.txt (accessed 20 August 2014).

Juhász-Holterman M. H. A., Peters J. H. and Geets J. J. G. M. (1998). Artificial recharge of a lake excavated for gravel extraction. In: Artificial recharge of groundwater, J. H. Peters *et al.* (eds), Balkema, Rotterdam, pp. 237–242.

Kumar A. (2008). Hypolimnetic withdrawal for lake conservation. In: Proceedings of Taal2007: The 12th World Lake Conference, M. Sengupta and R. Dalwani (eds), Ministry of Environment and Forests, New Delhi, India, pp. 812–818.

Kumar B., Nachiappan R. P., Rai S. P., Sarvanakumar U. and Navada S. V. (1999). Improved prediction of life expectancy for a Himalayan lake: Nainital, U.P., India. *Mountain Research and Development*, **19**(2), 113–121.

Kivimäki A. L., Lahti K., Hatva T., Tuominen S. M. and Miettinen I. T. (1998). Removal of organic matter during bank filtration. In: Artificial recharge of groundwater, J. H. Peters *et al.* (eds), Balkema, Rotterdam, The Netherlands, pp. 107–112.

Massmann G., Sültenfuss J., Dünnbier U., Knappe A., Taute T. and Pekdeger A. (2008). Investigation of groundwater residence times during bank filtration in Berlin: A multi-tracer approach. *Hydrological Processes*, **22**(6), 788–801.

Miettinen I. T., Martikainen P. J. and Vartiainen T. (1994). Humus transformation at the bank filtration water plant. *Water Science and Technology*, **30**(10), 179–187.

Nachiappan R. P. and Kumar B. (1999). Study of the interconnection between a lake and surrounding springs using environmental tracers. Proceedings IUGG 99 Symposium HS4, Birmingham, IAHS (258), pp. 215–219.

Nachiappan R. P., Kumar B. and Manickavasagam R. (2002). Estimation of sub-surface components in the water balance of Lake Nainital (Kumaun Himalaya, India) using environmental isotopes. *Hydrological Sciences Journal*, **47**, 41–54.

Pant M. C., Gupta P. K., Pande J., Sharma P. C. and Sharma A. P. (1981). Aspects of water pollution in Lake Nainital, U.P., India. *Environmental Conservation*, **8**(2), 113–117.

Purohit R. and Singh S. P (1981). Seasonal variation in physico-chemical limnology of shallow zones in Nainital Lake, Western Himalaya (India). *Proc. Indian Natn. Sci. Acad.*, **B47**(2), 194–203.

Purohit R., Singh S. P. and Upreti N. (1986). Effect of non-removal of the macrophytic biomass on the characteristics of water and plant community in Lake Naini Tal, U.P., India. *Internationale Revue Der Gesamten Hydrobiologie*, **71** (2), 245–257.

Purushothaman P., Mishra S., Das A. and Chakrapani G. J. (2012). Sediment and hydro biogeochemistry of Lake Nainital, Kumaun Himalaya, India. *Environmental Earth Sciences*, **65**(3), 775–788.

Rawat J. S. (1987). Morphology and morphometry of the Naini Lake, Kumaun, Lesser Himalaya. *Journal of the Geological Society of India*, **30**(6), 493–498.

Sandeep C. (2011). Lake bank filtration induced transport of microorganisms. M. Tech. Dissertation, Department of Civil Engineering, Indian Institute of Technology Roorkee, Uttarakhand, India.

Sharma A. P., Jaiswal S., Negi V. and Pant M. C. (1982). Phytoplankton community analysis in lakes of Kumaun Himalaya. *Archiv für Hydrobiologie*, **93**, 173–193.

Sharma V. K. (2001). Geological investigations at the proposed water treatment plant (WTP) site at Children Park (Flats), Nainital Township, Kumaun Himalaya. Report of Geological Survey of India, Northern Region, Lucknow, India.

Shukla U. and Gupta P. K. (2001). Assemblage of ciliated protozoan community in a polluted and non-polluted environment in a tropical lake of central Himalaya: Lake Naini Tal India. *Journal of Plankton Research*, **23**(6), 571–584.

Sobek S., Tranvik L. J., Prairie Y. T., Kortelainen P. and Cole J. J. (2007). Patterns and distribution of dissolved organic carbon: An analysis of 7500 widely distributed lakes. *Limnology and Oceanography*, **52**(3), 1208–1219.

Valdiya K. S. (1980). Geology of Kumaun Lesser Himalaya. Gyānodaya Prakāshan, Nainital, India.

Valdiya K. S. (1988). Geology and Natural Environment of Nainital Hills, Kumaun Himalaya. Gyānodaya Prakāshan, Nainital, India.

Weishaar J. L., Aiken G. R., Bergamaschi B. A., Fram M. S., Fujii R. and Mopper K. (2003). Evaluation of specific ultraviolet absorbance as an indicator of the chemical composition and reactivity of dissolved organic carbon. *Environmental Science and Technology*, **37**(20), 4702–4708.

Williams D. (2007). Lake aeration system of Nainital, India. http://www.paradisenw.com/cases/lake_nainital_in.php (accessed 24 August, 2014).

Chapter 4

Application of bank filtration in aquifers affected by ammonium – The Delhi example

Maike Groeschke, Theresa Frommen, Gesche Grützmacher, Michael Schneider and Dhruv Sehgal

4.1 INTRODUCTION

Bank filtration (BF), sometimes also called riverbank filtration (RBF), is used worldwide for drinking water production (Tufenkji *et al.*, 2002; Doussan *et al.*, 1997; Grünheid *et al.*, 2005). It has two main advantages: (1) Sufficient quantity of water can be produced independent of the usable groundwater capacity as BF is a form of artificial groundwater recharge (Bouwer, 2002; Dillon, 2005). (2) Low cost post-treatment is often sufficient for the raw water as the process of bank filtration takes advantage of the natural filter capacity of the sediments during the soil passage (Kuehn & Mueller, 2000). Usually, there is a significant increase in water quality for the bank filtrate compared to the surface water regarding organic substances, colour, coliform bacteria and faecal contaminants (Singh *et al.*, 2010; Weiss *et al.*, 2005). Additionally, the water quality at bank filtration wells is relatively constant, and, therefore, it is easier to treat than surface water, which often shows high variation in many quality parameters (Ray, 2004; Tufenkji *et al.*, 2002).

Typically, bank filtration sites are planned according to the local hydrogeological conditions and the wells are constructed to achieve both of the above mentioned effects (Ray *et al.*, 2002). In developing countries there is usually an emphasis on securing sufficient water quantity for drinking. Wells are often constructed along rivers and lakes because alluvial or riparian aquifers generally have good hydraulic properties (Rosenshein, 1988). The shallow depths of the sediments make them easy to exploit and help to reduce drilling costs (Doussan *et al.*, 1997). However, when bank filtration is applied at sewage contaminated surface waters, which is often the case in developing countries (Ray, 2008), a range of problems can arise as contaminated water infiltrates into the aquifer in large quantities and the capacity of the soil to filter the contaminants is often exceeded (Heberer, 2002). The resulting contamination of the aquifer can prevail for many decades, making post-treatment and/or remediation measures necessary. Such contamination is often caused by nitrogen, especially the species ammonium (Hiscock & Grischek, 2002).

Increasing ammonium concentrations at one specific well (P3) in a well field located at the Yamuna River in East Delhi are a cause for concern as the well field is used for drinking water production. Elevated ammonium concentrations at the well were reported since 2006 (Sprenger & Lorenzen, 2014) and Groeschke (2013) identified the sewage contaminated river water as the main source of ammonium in the raw water. Ammonium concentrations are already about ten times higher (5.5–8 mg/L) than the Indian guideline value of 0.6 mg/L total ammonia [0.5 mg/L total ammonia-N] as specified in BIS 10500:2012 and are expected to increase further. As water suppliers have to plan several decades ahead in order to be able to develop appropriate water management concepts, it is important to know about the development of future ammonium concentrations at that well field and other well fields along the Yamuna River to be able to choose appropriate remediation and treatment options.

Numerous studies of ammonium contaminations in groundwater have been conducted, mostly focusing on contamination from point sources - such as septic tank effluents (Hinkle *et al.*, 2007), leachate from sewage farms (Hamann, 2009), leachate from coking plants (Haerens *et al.*, 2002a) or chemical companies (Clark *et al.*, 2008) – and from contaminations resulting from the infiltration of treated sewage water (LeBlanc, 1984; Ceazan *et al.*, 1989; Böhlke *et al.*, 2006; DeSimone & Howes, 1996, 1998). A comprehensive review of published literature on ammonium retardation is given by Buss *et al.* (2004). Doussan *et al.* (1997, 1998) studied the transport of nitrogen species at a RBF site at the Seine (France), where the river water was a main source of nitrogen – in the form of nitrate. Reducing conditions prevailed in the aquifer owing to the decay of organic matter and the nitrate was reduced to ammonium during the soil passage, while the mineralization of organic matter was an additional source of ammonium. In central Delhi, reducing conditions in the aquifer are caused by the infiltration of reducing surface water. It is expected that without infiltration of reducing surface water, a redox sequence from oxidizing conditions to iron-reducing conditions further away from the river would prevail – as reported by Lorenzen *et al.* (2010a) for a field site upstream Delhi.

In order to understand the behaviour of ammonium in aquifers at BF sites at surface waters highly polluted by untreated sewage, field data have been collected and laboratory column studies have been conducted with aquifer material from Delhi. Results of the analyses and experiments are summarized in section 4.3 and were used as the basis for recommendations about the application of BF in nitrogen contaminated aquifers.

4.2 NITROGEN

4.2.1 Occurrence and effects

Nitrogen is a redox-sensitive parameter which can occur in different species. The most common forms of nitrogen in the water-soil environment are, in order of decreasing oxidation state (Metcalf & Eddy Inc, 2014; Stumm & Morgan, 1996):

- Nitrate (NO_3^-, +V)
- Nitrite (NO_2^-, +III)
- Nitrogen gas (N_2, 0)
- Ammonia and ammonium (NH_3, and NH_4^+, both –III)
- Organic nitrogen (OrgN, mostly –III)

Whether the reduced form of nitrogen occurs as un-ionized ammonia (NH_3) or in the form of ammonium ions (NH_4^+) depends on the temperature and, to a stronger extent, the pH of the solution (Table 4.1).

Table 4.1 Proportions of ammonium (NH_4^+) and ammonia (NH_3) at different pH values (Metcalf & Eddy Inc, 2014, p.94).

$NH_3 + H_2O$	$\longleftrightarrow$	$NH_4^+ + HO^-$	pH	Temperature
10%		90%	8.3	20°C
50%		50%	9.25	20°C

At pH and temperature conditions commonly found in natural waters, ammonium is the principal species (Hem, 2005). Sometimes the term "total ammonia" is being used referring to the sum of ionized and un-ionized ammonia.

Nitrogen pollution can cause problems such as eutrophication of surface water bodies (Howarth & Marino, 2006), which can lead to toxic algal blooms or decreasing dissolved oxygen concentrations and related issues such as a decrease in animal and plant diversity. Furthermore, NH_3 is toxic for aquatic species (Randall & Tsui, 2002), but not for humans at low concentrations (GESTIS Substance database, 2014). Nitrite (NO_2^-) is also extremely toxic to fish or other aquatic species (Metcalf & Eddy Inc, 2014). For humans, excessive nitrogen intake in the form of nitrate (NO_3^-) or nitrite through water can result in diarrhoea or methaemoglobinemia (blue-baby syndrome) in infants (Ward *et al.*, 2005). When chlorination is used for the disinfection of drinking water, the presence of ammonium in raw water, even at low concentrations, causes the formation of chloramines (Weil & Morris, 1949). Higher chlorine doses are necessary to achieve required minimum residual chlorine concentration at the outlet of the water treatment plants (WTPs) and in the distribution system (Duong *et al.*, 2003).

4.2.2 Guideline values

Guideline values for nitrogen species given in the Indian standard (BIS, 10500, 2012) and the WHO drinking water quality guidelines (WHO, 2011) are compared in Table 4.2. The WHO did not establish a guideline value for total ammonia because it usually occurs in drinking water at concentrations well below those of health concern. Because the WHO includes the non-ionized form NH_3 and the ionized form NH_4^+ in their definition of ammonia, it is assumed that this is also the case in the Indian Standard BIS 10500, although it is not further defined.

Table 4.2 Guideline values for nitrogen species in drinking water.

Parameter	Unit	BIS 10500:2012 (India)	WHO (2011)
Nitrate (as NO_3^-)	mg/L	45	50
Nitrite (as NO_2^-)	mg/L	No guideline value	3
Ammonia (as NH_3-N and NH_4^+-N)*	mg/L	0.05	No guideline value

*Definition of WHO, not specified in BIS 10500.

4.2.3 Nitrogen in surface water bodies

In surface water bodies, nitrogen concentrations depend on several factors, mainly land use, sewage disposal and the water balance of the water body, as dilution strongly affects the pollutant concentrations. In European rivers agricultural fertilizers are the main source of nitrogen. NH_4^+-N concentrations range between 0.1 and 0.3 mg/L and median NO_3-N concentrations are around 3 mg/L in large rivers (EEA, 2001). In urban agglomerations in developing or newly industrialized countries nitrogen loads in surface water bodies are expected to be substantially higher: Nitrogen contamination of surface water through untreated or partially treated domestic sewage water is a concern in many of these countries. According to Corcoran *et al.* (2010), up to 90% of sewage water in developing or newly industrialized countries is not collected or treated but discharged directly into rivers, lakes and coastal areas or leached into the subsoil. In the Asia-Pacific region alone, this amounts to approximately 150–250 million m^3 per day of untreated (domestic) wastewater from urban areas released to the environment (WWAP, 2012). However, data on total inorganic nitrogen concentrations (NH_4^+, NO_2^-, NO_3^-) is scarce, as most studies on river water quality in Asian megacities only report concentrations for nitrate and nitrite, e.g. Sikder *et al.* (2013) and Kido *et al.* (2009). But because of the high chemical oxygen demand in sewage contaminated rivers, the prevailing form of inorganic nitrogen is expected to be ammonium.

4.2.4 Nitrogen in sewage water

In sewage water about 60–70% of the nitrogen is present in the form of ammonia/ammonium, depending on the pH of the solution, while the remaining 30–40% is mostly found as biodegradable or non-biodegradable organic nitrogen (Metcalf & Eddy Inc, 2014, p. 712). Fresh domestic wastewater usually contains no more than 1% nitrates and nitrites (Eckenfelder & Argaman, 1991) before it is aerated in the nitrification step at wastewater treatment plants (WWTPs). Total nitrogen concentrations in wastewater are highly variable depending on the diet of the population (Pescod, 1992; Patterson, 2003) and the per capita wastewater flow rate (Eckenfelder & Argaman, 1991). Concentrations between 20 and 85 mg/L are reported as an average for typical domestic wastewater in Metcalf & Eddy Inc (2014), and maximum concentrations can be much higher (sometimes above 150 mg/L – e.g. Ammary, 2007). The main source of nitrogen in wastewater is urea, which is contained in urine and is degraded by biological hydrolysis (Mobley & Hausinger, 1989, Table 4.3).

Table 4.3 Reactions of urea in the environment (Mobley and Hausinger, 1989).

Reactant		Product	Explanation
urea + water		*ammonia + carbamate*	Hydrolysis of urea
$(NH_2)_2CO + H_2O$	*→	$NH_3 + H_2NCOOH$	*Naturally occurring enzyme urease catalyzes reaction
carbamate + water		*ammonia + carbonic acid*	Hydrolysis of carbamate
$H_2NCOOH + H_2O$	→	$NH_3 + H_2CO_3$	
carbonic acid		*hydrogen ion + bicarbonate*	Dissociation of carbonic acid
H_2CO_3	→	$H^+ + HCO_3^-$	Increase in pH
ammonia + water		*ammonium + hydroxide*	
$2NH_3 + 2H_2O$	→	$2NH_4^+ + 2OH^-$	Ammonia molecules equilibrate with water

Other sources are faeces (a source of organic nitrogen) and grey water from laundry and personal washing (a source of NH_3/NH_4^+) (Patterson, 2003). Products like toilet paper had the lowest nitrogen load contribution (Tjandraatmadja *et al.*, 2010). In countries without a regular waste disposal system, garbage disposal in wastewater is another large source of organic nitrogen.

4.3 THE DELHI CASE STUDY

4.3.1 Overview

The well fields in East Delhi are an example of unintended contamination by RBF use by placing wells along a sewage contaminated river. Delhi is a megacity with a population of currently 16 million people (Census of India, 2011) located in the centre of the Indo-Gangetic Plain (Figure 4.1). The Yamuna River, the largest tributary of the Ganges River, flows through Delhi in north-southerly direction. As many rivers in Asia, the Yamuna is characterized by high sediment loads during monsoon season (Jha *et al.*, 1988) and a constantly changing riverbed (Khan & Bajpai, 2014). Although numerous dykes and embankments were constructed within the city to control the flow, the river still has the opportunity to shift within certain limits between the embankments (Figure 4.2). Thus, the location of the riverbank frequently changes. The river is dammed up by two barrages within the city area, Wazirabad barrage in the North and Okhla barrage in the South, and the 22 km river stretch between the two barrages is highly polluted by sewage water.

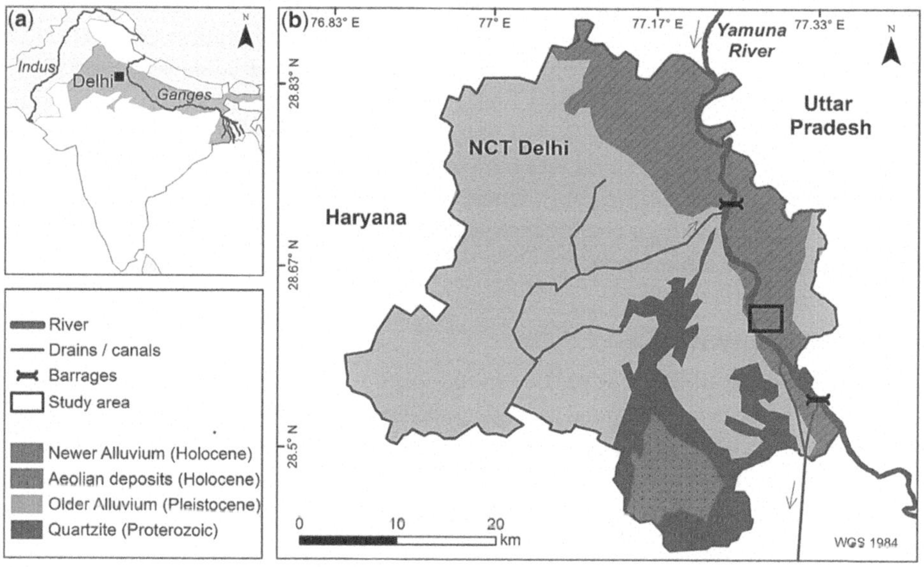

Figure 4.1 (a) Delhi is located in the Indo-Gangetic plain (Indian part marked in medium grey). Data source: Natural Earth (2011) (b) The study area is located on the East bank of the Yamuna River, where sewage influenced river water infiltrates into the sediments of the Newer Alluvium. Geological Map modified after Geological Survey of India (2006).

The floodplain covers the eastern and western bank along the entire stretch of the river in Delhi. It is mostly undeveloped and mainly used for agriculture. The floodplain sediments are mostly medium grained sands with a layer thickness of up to 70 m in the north of Delhi (Shekhar & Prasad, 2009) and about 20 m in the southern part of the city. They constitute what is known as the floodplain aquifer or Newer Alluvium. Compared to other groundwater sources in Delhi, the groundwater of the Newer Alluvium is found at shallow depth and is characterized by only minor water table fluctuations of about 0–2 m throughout the year and over decades – as opposed to about 4–20 m in other aquifers in Delhi (CGWB, 2012; Shekhar *et al.*, 2009).

Numerous tube wells and about 20 Ranney wells (radial collector wells) were constructed on the floodplain in Delhi, tapping the Newer Alluvium. The wells are not arranged parallel to the riverbank but were constructed across the complete width of the upper floodplain. Owing to losing stream conditions (Lorenzen *et al.*, 2010a) it can be assumed that the wells situated along the river draw a high share of bank filtrate, although they have not been specifically designed for bank filtration.

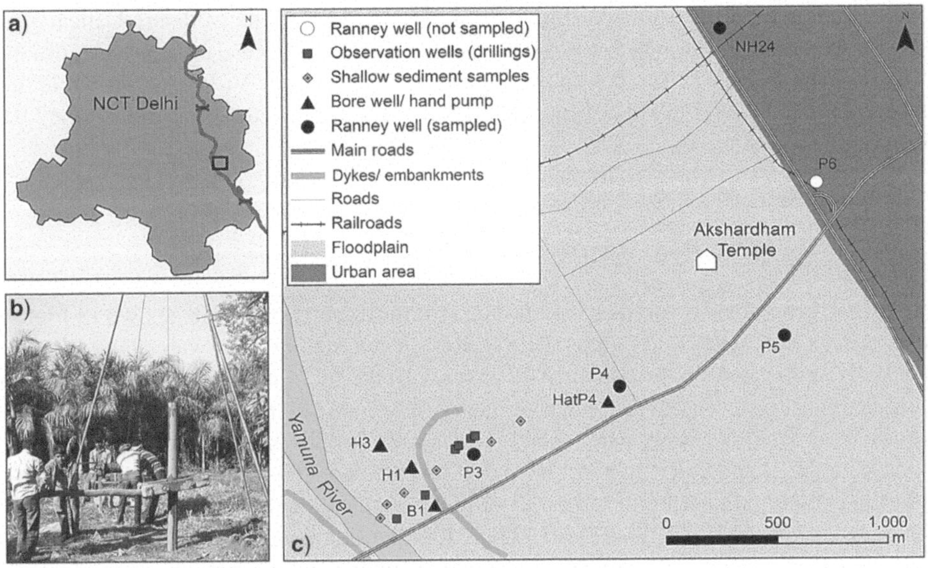

Figure 4.2 (a) Location of the study area (b) Drilling of observation well (c) Location of the hand pumps, Ranney wells and of the shallow and deep drillings conducted during the field work (modified after Groeschke, 2013). The river is able to shift between the dykes on the eastern and the western bank.

4.3.2 Study area

The study area covers an area of about 2.5 km^2 on the east bank of the Yamuna River in East Delhi near the Akshardham Temple between 720800 and 722800 m E and 3165900 and 3168200 m N (UTM, WGS84 Zone 43 N). Here the undeveloped floodplain is about 2.4 km wide. Four Ranney wells operated by the Delhi water company Delhi Jal Board (DJB – P3, P4, P5, P6) and one Ranney well operated by the Indian Railways (NH24) as well as numerous hand pumps and bore wells used by the local population are located within the area and some of them were used for water sampling. Additionally, four observations wells and two hand pumps were installed in the frame of Saph Pani (see section 4.3.3). The study site with the location and type of the sampling points are shown in Figure 4.2.

Previous research

Previous research at this location was conducted 2006 in the frame of the feasibility study IDB India (International Development of Bank Filtration: Case Study India) and 2007–10 in the frame of the TECHNEAU project (Lorenzen *et al.*, 2007; Pekdeger *et al.*, 2008; Sprenger *et al.*, 2008; Lorenzen *et al.*, 2010a; Lorenzen *et al.*, 2010b; Lorenzen, 2011; Sprenger, 2011; Sprenger & Lorenzen, 2014).

Geology and hydrogeology

In the study area, like in the entire flood plain in Delhi, the Holocene alluvial sands (Newer Alluvium) are underlain by finer grained Pleistocene sediments of the Older Alluvium. At 38 mbgl Precambrian bedrock was encountered (Sprenger, 2011, p.70). Hydraulic conductivities of the Newer Alluvium are in the range of 2×10^{-4} to 7×10^{-4} m/s (Chatterjee *et al.*, 2009) and Sprenger (2011, p.66) reported an average pore water velocity of 0.9 m/d for this unit at the field site. The hydraulic conductivities of the Old Alluvium are between 3×10^{-5} and 5×10^{-5} m/s (Chatterjee *et al.*, 2009). According to Lorenzen *et al.* (2010a) losing stream conditions prevail on the east bank of the river. Sprenger (2011, p.66) reported infiltration rates of 6.4×10^{-7} m^3/m^2/s for monsoon times and 4.2×10^{-7} m^3/m^2/s for non-monsoon times. No information is available about the west bank of the river at this location.

Description of the production wells

The Ranney wells of the Delhi Jal Board in the study area were constructed in 1973 and commenced operation in 1975. They are about 15 m deep and thus tap the floodplain aquifer. Each well has ten laterals which are each about 30 m long. The recorded discharge of the wells is about 150–300 m³/h (Chatterjee *et al.*, 2009). The wells are typically operated about eight hours every day, except for well number P3, which is sometimes not operating because of elevated ammonium concentrations in the groundwater. The water of well P4 is directed to the Commonwealth Games Village WTP while the water of wells P3, P5, P6 is supplied to the Okhla WTP. Well NH24 of the Indian Railways is constructed similar to the DJB Ranney wells but the water is not used for public water supply.

4.3.3 Field studies

Water and sediment sampling

In the frame of Saph Pani, 72 groundwater samples were taken from the sampling points shown in Figure 4.2, including the newly constructed observations wells and hand pumps. In addition, eleven regular river water samples and two samples of the flood event in July 2013 were taken and analyzed (for NO_3^-, NO_2^-, NH_4^+, main cations and anions, pH, oxidation reduction potential, electrical conductivity, dissolved oxygen, trace elements). The four new observation wells were installed at distances between 500–550 m to the riverbank and the two new hand pumps were installed at distances of 35 m and 250 m to the river. The drillings were done by manual auger drilling and have depths between 7.6 and 28 m. Sediment samples were collected of the encountered lithological units. In addition, sediment samples were collected from seven shallow drillings with depths between 2.5 and 4.3 m at distances of 5 m, 75 m, 200 m, 375 m, 500 m, 600 m, and 775 m to the river. This was conducted by using an Eijkelkamp hand drilling device. To sample river bottom sediments and to measure water depths, three profiles were taken across the Yamuna River in December 2013, using a Van-Veen grab sampler.

Results: Ammonium concentrations at the field site

Varying ammonium concentrations were found in the aquifer close to the river (Figure 4.3). In 2012, a similar trend was observed in ammonium concentrations at the three sampling points B1, H1, and H3, with values between 4.5 mg/L in June 2012 and 26 mg/L in December 2012 (Groeschke, 2013). In 2013, ammonium concentrations still fluctuated (between 6.4 and 35 mg/L), but no trend could be discerned. In the Ranney well P3 at a distance of 500 m from the river, ammonium concentrations varied between 5.5 and 8 mg/L in 2012 and 2013. In wells farther away from the riverbank, ammonium concentrations remained below 1.7 mg/L in both years. In the river water, ammonium concentrations up to 20 mg/L were measured in 2012 and up to 16 mg/L during the field campaigns in 2013. Maximum and minimum concentrations at the field site are summarized in Figure 4.3. Data obtained from water samples were further used to set-up the 1D model (section 4.3.5). A detailed description of the ammonium plume is given in Groeschke *et al.* (2015a).

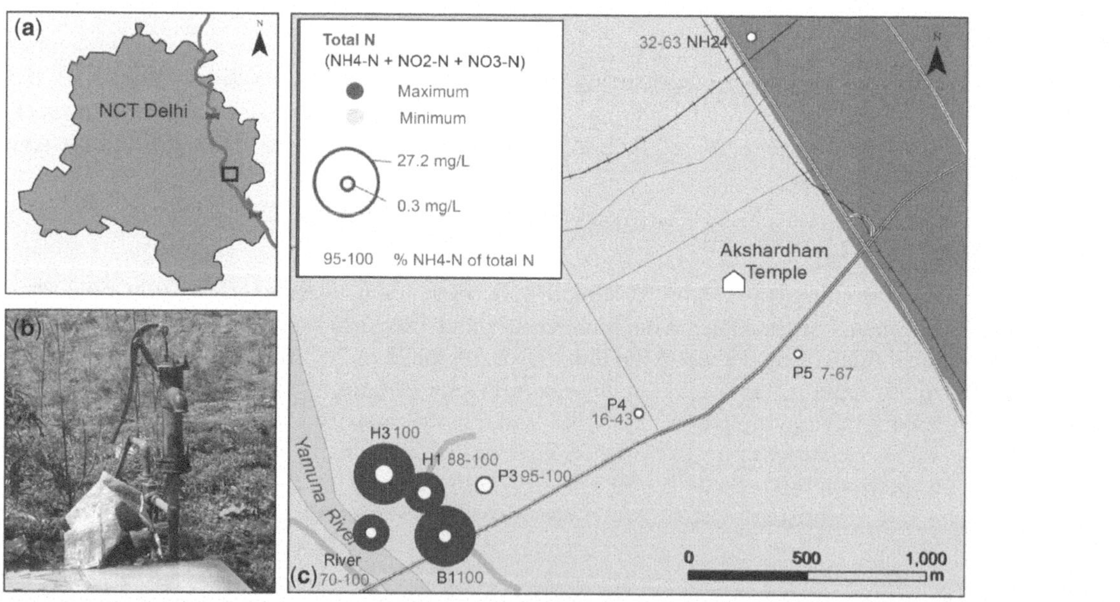

Figure 4.3 (a) Location of the study area (b) Hand pump H1 (c) Minimum and maximum total nitrogen concentrations in the water samples taken between March 2012 and December 2013 work (modified after Groeschke, 2013).

4.3.4 Laboratory studies

Sediment analyses

To characterize the alluvial aquifer a total of 25 sediment samples from the unsaturated zone were studied (Groeschke *et al.*, 2015b) and 14 samples from the saturated zone as well as one sediment sample from the river bed were analysed (Groeschke *et al.*, 2015a). Grain size distribution was determined by sieve test and hydrometer method, organic content and carbonate content were measured through loss on ignition. The cation-exchange capacity (CEC) was determined using a barium chloride ($BaCl_2$) percolation method for sands and gravels and a $BaCl_2$ batch method for fine grained sediments.

The sediment's main components range between silt and gravel. The particle sizes increase with increasing sampling depths. Silts and silty fine sands are predominant in the unsaturated zone, while the saturated zone is dominated by well sorted medium sands. Below is a gravelly layer at a depth of about 13.5 m, which consists of real gravel particles (grain size of >2 mm) and of concretions >2 mm made up of clay and silt. It is underlain by silty clays, presumably of the Old Alluvium, at a depth of about 16 m. In the New Alluvium, the hydraulic conductivity (k value) ranges between 1.0×10^{-7} m/s in the unsaturated zone and 4.2×10^{-3} m/s in the saturated zone if calculated according to Beyer (1964) or – in case of the fine grained sediments – to the U.S. Bureau of Soil Classification (Richter, 1966). The hydraulic conductivities increase with increasing particle size (and thus with depth). The organic content of the sediments ranges between 0.5% and 14.9%. In the unsaturated zone, the organic content was generally higher than in the saturated zone. The carbonate content of the sediments ranges between 0.9% and 18%. The highest carbonate content was found in the gravel layer, which contains the concretions. The CEC ranges between 1.2 meq/100 g sediment in the saturated zone and 37.2 meq/100 g sediment in the unsaturated zone. In the saturated zone, the CEC is slightly higher in the gravelly layer (2.1 meq/100 g sediment) than in the sand (1.6 meq/100 g sediment, Groeschke *et al.*, 2015c). Calcium has the highest share in the CEC, leading to the conclusion that the carbonates consist mainly of calcium carbonate and that the concretions are probably the typical calcite concretions locally known as kankar (Eybing, 2014). The gravel layer will be referred to as kankar in the following text.

A detailed description of the unsaturated zone and an evaluation of its significance for the ammonium contamination is given by Groeschke *et al.* (2015b). The saturated zone, and especially the kankar layer and its significance for flow and transport are described in detail in Groeschke *et al.* (2015a; and 2015c).

Column experiments

The transport and fate of ammonium in the sand and kankar aquifer materials from the Yamuna floodplain was further investigated in laboratory column experiments at Freie Universität Berlin (Groeschke *et al.*, 2015b; and 2015c). Such experiments are a common method in hydrogeology to determine specific sediment parameters. The goal of these series of experiments was to provide data regarding sorption, degradation and fixation of ammonium under field site conditions. The data was later used to set up a reactive transport model of the field site to predict the future development of the ammonium plume. The set-up of the experiments is shown in Figure 4.4.

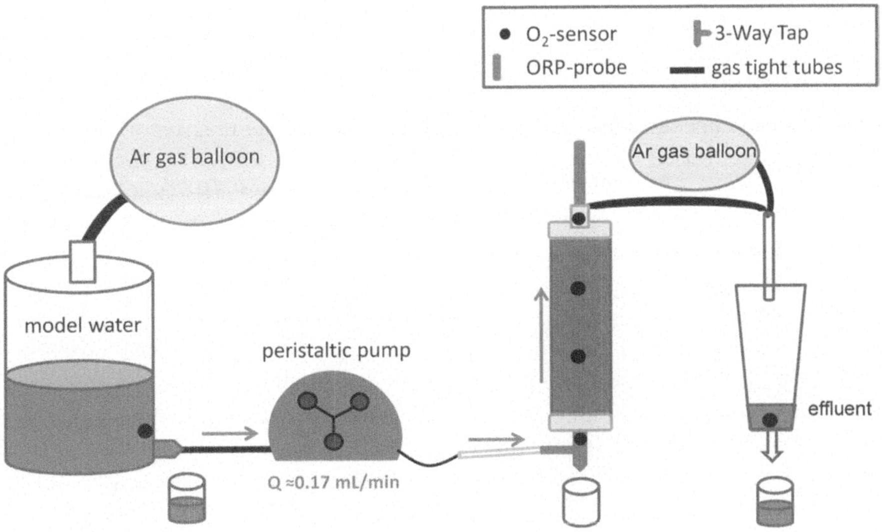

Figure 4.4 Set-up of the column experiments.

The experiments were conducted under suboxic or anoxic conditions, whereby the latter prevail in the aquifer. To achieve laboratory conditions similar to those at the study site, the following parameters were adapted:

- The model water was adjusted for the main cations to be comparable to the concentrations in the groundwater.
- Anoxic conditions were established by using argon balloons to create an oxygen- (and nitrogen-) free atmosphere above the model water container and the effluent sampling flasks. Glass and gas-tight tubing materials were used to inhibit gas exchange. The oxygen concentration was monitored with chemical optical oxygen sensor spots (PreSens) at six points during the entire duration of the experiments. Redox potential was measured using oxidation reduction potential -probes at the outlet of the columns.
- A peristaltic pump maintained a flow rate of ~0.17 mL/min, which correlates to a flow velocity at the field site of about 0.9 m/d (Sprenger, 2011, p.66).

The glass columns had an inner diameter of 45 mm and a sediment filled length of 146 mm and were flushed upflow with the model water. Three sets of experiments were conducted with this set-up: pre-tests, adsorption experiments and desorption experiments.

- During pre-tests, the freshly filled columns were flushed with nitrogen-free model water until nitrogen concentrations in the effluent were sufficiently low.
- In the adsorption experiments, the columns were flushed with model water with ammonium concentrations of either 20 mg/L or 10 mg/L until the ammonium concentrations in the effluent were equal to the concentrations in the model water.
- During the subsequent desorption experiments, the columns were again flushed with nitrogen-free model water until nitrogen concentrations in the effluent were low and did not decrease any further.

Each experiment was conducted with two or three columns filled with the same sediment (doubles or triplets). To check for reproducibility, most experiments except for the pre-tests were repeated one or two times.

Results of the column experiments

The column experiments indicate that there is some degradation or fixation of ammonium in the sediments of the unsaturated zone and no or very little natural degradation potential in the sediments of the saturated zone. The transport of ammonium is therefore mainly controlled by cation exchange. In the sand columns, 10–12 pore volumes were necessary to observe ammonium saturation in the sediment and subsequently same ammonium concentrations in the column effluent as in the feed water (100% breakthrough of ammonium) and about 15 pore volumes to flush the ammonium out of the sediment. In the kankar, 30–35 pore volumes were necessary to observe the 100% breakthrough in the adsorption experiments and the flushing of the ammonium in the desorption experiments took about 40 pore volumes (Groeschke *et al.*, 2015c).

4.3.5 1D Transport modelling

With the aim to predict the future concentrations of ammonium at well P3, 1D reactive transport models (Haerens *et al.*, 2002b) were set up for the field site. Based on the results of the 1D reactive transport column models, of Groeschke *et al.* (2015c), two flow paths in representative aquifer sediments were modelled with PHREEQC v3 (Parkhurst & Appelo, 2013; Figure 4.5). One flow path comprises 500 m distance from the riverbank to Ranney well P3. These 500 m were set up as a column divided into 139 cells with a cell length of 3.6 m each. The time step was set to 4 d, resulting in the average linear velocity of 0.9 m/d as determined by Sprenger (2011). Transport parameters (effective porosities, number of exchange sites, and selectivity coefficients for the cation exchange) were taken from the 1D column modelling without any further adjustments (Table 4.4). Dispersivities were adjusted to the model length. Although dispersion is generally higher at the field scale than at the laboratory scale because of sediment inhomogeneities which are not present in laboratory columns (Gelhar *et al.*, 1992), the dispersivities of the field model were adjusted to represent the magnitude measured in the column experiments: the longitudinal dispersivity was set to 5 m (1/100 of the flow path) in the sand and to 50 m (1/10 of the flow path) in the kankar. Because the sediment is carbonatic (Eybing, 2014) and most water samples at the field site are slightly oversaturated with calcite, calcite was included as an equilibrium phase in the model. To check for numerical errors, the models were also run with 278 cells (1.8 m cell lengths) and 2 d time steps and with 556 cells (0.9 m cell lengths) and 1 d time steps.

Table 4.4 Transport parameters used in the simplified 1D model.

Parameter	Unit	Sand	Kankar
Effective Porosity (n_e)*	—	0.24	0.175
Number of exchange sites	meq/1L water	0.054	0.21
$log\_k_{Na\backslash K}$	—	0.67	0.98
$log\_k_{Na\backslash Ca}$	—	0.1	0.18
$log\_k_{Na\backslash Mg}$	—	−0.28	−0.09
$log\_k_{Na\backslash NH4}$	—	0.55	0.81

*Effective porosities are not explicitly included in PHREEQC models. They are incorporated through the number of exchange sites.

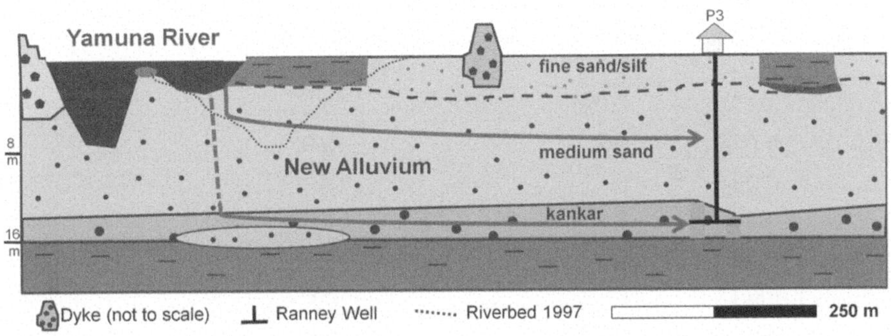

Figure 4.5 Flow paths from the river to well P3 in the 1D reactive transport models. The vertical flow from the river to the kankar layer was not considered and modelled. Cross section after geological information given in Groeschke *et al.* (2015a).

To keep the model minimal and straightforward, the following assumptions and simplifications were applied:

- Source water composition (displacing solution) was kept constant, although in reality there is a seasonal variability in the river water due to monsoon–non monsoon compositions.
- Ammonium was decoupled from the nitrogen cycle, meaning it cannot be oxidized to nitrate in the model. This would be representative of anoxic conditions in the aquifer, which by no means must prevail after an improvement of source water quality.
- An average linear flow velocity of 0.9 m/d (Sprenger, 2011) was assumed for both flow paths. It is very likely that flow velocities are much higher in the kankar layer, but real data for this layer are not available.

Adsorption modelling

To estimate the increase of ammonium concentrations at well P3, the infiltration of sewage influenced river water into the aquifer was modelled. The cells were equilibrated with water samples taken at sampling points still uninfluenced by the ammonium plume. A sample taken from HatP4 in December was used for equilibrating the sand layer and a sample taken at P4 in December 2013 was used to equilibrate the kankar layer. The cells were then flushed with a displacing solution with the composition of a sewage influenced river water sample taken at the field site in December 2012 with an ammonium concentration of 20 mg/L. The compositions of the water samples are summarized in Table 4.5. In the models, it took about 15 years to reach the 100% ammonium breakthrough in the sand layer and 62 years to reach the 100% ammonium breakthrough in the kankar layer (Figure 4.6).

Desorption modelling

How long the ammonium contamination will prevail in the aquifer after source water quality improves depends not only on the sediment properties, but also on the extent of the ammonium plume. To model ammonium desorption, it was assumed that

- the ammonium plume has completely reached the well P3 and
- the ammonium distribution within the plume is homogenous at 35 mg/L NH_4^+ in the groundwater in the sand layer and 26 mg/L in the kankar layer.

Table 4.5 Composition of the equilibrating and displacing solutions. Water samples were charge-balanced with alkalinity as HCO_3^-. Groundwater samples were taken in December 2013. River water sample was taken in December 2012.

Parameter	Unit	Equilibrating Solution Sand (HatP4)	Equilibrating Solution Kankar (P4)	Displacing Solution (River Water Field Site)
Temperature (T)	°C	26.2	26.4	20.5
pH	pH	7.58	7.4	7.6
Reduction potential (E_h)	mV	160	175	82
Electrical conductivity (EC)	µS/cm	495	893	1588
Sodium (Na)	mg/L	19.9	67.5	171
Potassium (K)	mg/L	5.4	6.8	15.4
Magnesium (Mg)	mg/L	14	23	33.7
Calcium (Ca)	mg/L	63.7	80	65.4
Iron (Fe)	mg/L	0.09	0.1	0.07
Manganese (Mn)	mg/L	0.09	0.3	0.3
Hydrogencarbonate (HCO_3^-)	mmol/L	5.2	5.9	6.5
Chloride (Cl)	mg/L	6	78	218
Sulphate (SO_4^{2-})	mg/L	2	53	125
Sulfide (S^{2-})	mg/L	0	0	0
Ammonium (NH_4^+)	mg/L	0	0.6	20
Nitrite (NO_2^-)	mg/L	0.005	0.03	0.02
Nitrate (NO_3^-)	mg/L	0	3.5	0

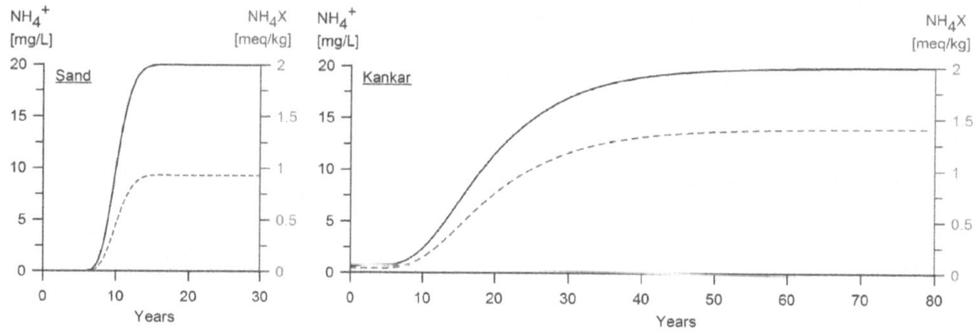

Figure 4.6 Results of adsorption modelling. Solid black line: Ammonium concentration in the water in mg/L (shown on the primary y-axis), dashed grey line: Ammonium concentrations on the exchanger in meq/kg sediment (shown on the secondary y-axis). The amount of ammonium sorbed on the sediment depends on the number of exchange sites available, on the solute composition and on the selectivity coefficients, which are sediment-dependent. It can be clearly seen that more ammonium can be adsorbed on the kankar material than on the sand.

The cells of the sand flow path were equilibrated with a water sample from hand pump B1 taken in December 2013 and the cells of the kankar flow path were equilibrated with water composition of sample H250 taken in December 2013 (Table 4.6). After equilibration, the column was flushed with a displacing solution with the composition of the river water upstream Delhi at Palla, where the Yamuna is still uninfluenced by sewage water. Assuming the same average linear velocity of 0.9 m/s in the sand and the kankar, Ammonium concentrations were below the drinking water limit value of 0.5 mg/L after about 19 years in the sand layer and after about 61 years in the kankar layer (Figure 4.7). This is due to the higher number of exchange sites in the kankar and to different selectivity coefficients in both materials. Because degradation of ammonium was not implemented in the models, the results can only be seen as conservative estimates. Furthermore, the average linear velocity in the kankar is probably higher than in the sand and flushing the ammonium out of the kankar layer might therefore be faster than 61 years.

Table 4.6 Composition of the equilibrating and displacing solutions. Water samples were charge-balanced with alkalinity as HCO_3^-. Groundwater samples were taken in December 2013. River water sample was taken in March 2007 in the frame of the TECHNEAU project.

Parameter	Unit	Equilibrating Solution Sand (B1)	Equilibrating Solution Kankar (H250)	Displacing Solution (River Water Upstream)
Temperature (T)	°C	25.2	24.3	22.3
pH	pH	6.93	7.23	8.56
Reduction potential (E_h)	mV	105	84	268
Electrical conductivity (EC)	µS/cm	1615	1153	457
Sodium (Na)	mg/L	97	79.7	35
Potassium (K)	mg/L	17.3	13.2	9
Magnesium (Mg)	mg/L	38.7	24.8	14
Calcium (Ca)	mg/L	126.5	89.1	44
Iron (Fe)	mg/L	16.9	5.2	0.62
Manganese (Mn)	mg/L	0.42	0.27	0.05
Hydrogencarbonate (HCO_3^-)	mmol/L	11.9	8.3	2.7
Chloride (Cl)	mg/L	141	115	38
Sulphate (SO_4^{2-})	mg/L	5	4	46
Sulfide (S^{2-})	mg/L	0.04	0	0
Ammonium (NH_4^+)	mg/L	35	26	0.1
Nitrite (NO_2^-)	mg/L	0.005	0.005	0.2
Nitrate (NO_3^-)	mg/L	0	0.05	6

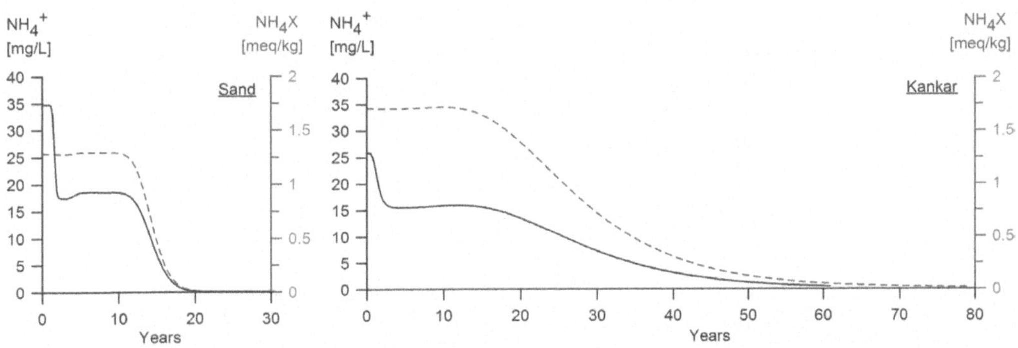

Figure 4.7 Results of the desorption modelling. Solid black line: Ammonium concentration in water in mg/L (shown on the primary y-axis), dashed grey line: Ammonium concentrations on the exchanger in meq/kg sediment (shown on the secondary y-axis). Like in the adsorption-model runs, more ammonium is adsorbed on the kankar material than on the sand. However, the actual amounts adsorbed in the beginning of the desorption-model run are higher than the amounts adsorbed in the adsorption-models (Figure 4.6). The reason for this discrepancy is that the desorption-model was equilibrated with higher concentrations of ammonium (see Table 4.6).

4.4 OVERVIEW OF REMEDIATION AND POST-TREATMENT OPTIONS

To make ammonium-contaminated groundwater suitable for drinking purposes, either remediation measures have to be applied to lower ammonium concentrations in the groundwater, or the raw water has to be treated before distribution. For remediation, there are generally three different approaches: (1) source control, meaning the removal or control of all known and suspected sources of contamination, (2) in-situ treatment methods, where the remediation takes place in the aquifer itself, or (3) pump and treat measures, which involve treating the water above-ground and re-injecting it into the aquifer. In Table 4.7 a short overview of possible remediation measures is given and the use of the measures in the Indian context is briefly discussed. Post-treatment options, which include physico-chemical treatment options (ammonia stripping, ion exchange, breakpoint chlorination and reverse osmosis) and biological filters, are summarized in Table 4.8 together with a brief assessment of the advantages and disadvantages regarding the application in India.

Table 4.7 Overview of remediation options for aquifers contaminated with ammonium (NH_4^+).

Method	Principle	Advantages	Disadvantages	Use in Delhi
Source Control				
Improvement of source water quality[1]	Stopping discharge of untreated sewage into rivers would allow for infiltration of aerated, NH_4^+-free water into the aquifer.	– Might decrease arsenic concentrations in bank filtrate – Decrease in odour pollution, improvement of environment	– Long-term solution: WWTPs need to be planned and built – NH_4^+ contamination will prevail decades after river quality improves	– Likely to change redox conditions in aquifer (arsenic demobilization) – Indian guideline limits for WWTP effluents[1] permit high nitrogen discharge into rivers
Aeration basins	Construction of aeration basins along riverbanks would lead to infiltration of NH_4^+-free water.	– Short- term improvement of river water quality – Low-tech	– Sufficient distribution of O_2 is difficult – Might not be suitable for variable flow – Clogging of diffusers might occur	– High COD and BOD in Yamuna in Delhi – Shifting riverbed: adjustment of facilities – Bacterial contamination: safety measures
In-Situ				
Oxygen injection – Bioxwand*[2]	In-situ bioremediation through oxygen gas injections into the aquifer using lances.	– No addition of further chemicals. Suitable for drinking water protection zones.	– O_2 consumed by oxidation of Fe^{2+}, Mn^{2+} and organic material present in the aquifer. – Increase in SO_4 and hardness	– High GW temperatures: fast nitrification process – Post-treatment for increased hardness of water at the WTPs might be necessary
Sequential permeable reactive barrier using polymer mats[3]	Use of in-situ polymer mats. Up-gradient mat delivers O_2 to induce nitrification as groundwater moves past. Down-gradient mat delivers ethanol to induce denitrification.	– 90% reduction in total N in field experiments – In-situ remediation over short time frame or GW flow distance. – Cost effective	– Measure changes local groundwater flow regimes. – Competing chemicals might exist in the aquifer. – Limited studies available.	– Thinkable to construct shield around wells with high NH_4^+ concentrations – Mats sensitive to fluctuating GW tables (private wells cause drawdown zones)

Groundwater circulation wells (GCW) and virtually permeable reactive barrier*[4]	GCWs induce circulating flow. Increasing pH in flow cell converts NH_4^+ to NH_3; removal by negative pressure air stripping. Down gradient further GCWs induce aerobic and anaerobic flow cells.	– Less negative impact on land use compared with pump and treat methods[5] – Simultaneous treatment of unsaturated zone and capillary fringe by vapour extraction	– Anisotropy of aquifer must be within range that allows circulation cells to develop[6] – Limited effectiveness in shallow aquifers[6] – Wells may become clogged[6].	– 20 mg/L is target concentration for the negative pressure air stripping. Peak NH_4^+ concentrations in plume in Delhi are 20–35 mg/L: Sufficient to install the VPR barrier only. This requires the installations of numerous GCWs.
Pump and Treat				
IVEY-sol aided sorbitall filtration method*[7]	Installation of injection and abstraction wells across plume. Injection of Ivey-sol surfactant to desorb NH_4^+ from soil. Water is abstracted and treated above ground using ion exchange. Treated water is re-injected.	– >96% reduction in the dissolved NH_4^+at the subject site. – Addition of Ivey sol enhances bioremediation by 40 to 60%. Ivey sol does not negatively affect water treatment systems. – Works well in fine-grained soils	– High operation and maintenance costs – Complex operation and maintenance – Injection of additives into the aquifer	– Land use would have to be changed in order to install 5-spot-pattern of wells across the ammonium plume. – Injection of chemicals problematic in drinking water protection zone
Nitrification and denitrification remediation*[8]	To trigger in-situ nitrification–denitrification reactions, GW is extracted, mixed with O_2 + nutrients or carbon + nutrients and re-injected. Separate injection and abstraction wells needed for nitrification and denitrification.	– Especially applicable in sand and sandy gravel aquifers – In aquifers with high hydraulic conductivities reaction cell size of injection wells can be maximized and number of injection wells minimized	– Not suitable for low hydraulic conductivities – Four drillings required in one line parallel to GW flow to complete nitrification and denitrification – High maintenance costs and space requirement	– Method is costly and requires much space and thus might conflict with agricultural land use – Method would involve two extraction-injection cycles

*Patented technology; COD = chemical oxygen demand; BOD = biological oxygen demand.
[1] Government of India, 1986; [2]BWB, 2007; Horner *et al.*, 2009; [3]Patterson *et al.*, 2002, 2004; [4]IEG Technology, 2008; [5]Elmore & Graff, 2002; [6]OST, 2002; [7]IVEY, 2012a, 2012b; [8]Mailath & Chu, 2005

Table 4.8 Overview of post-treatment options for raw water with elevated ammonium (NH_4^+) concentrations.

Method	Principle	Advantages	Disadvantages	Use in Delhi
Physico-Chemical				
Air stripping of ammonia	Adding lime increases pH of water to ~11; NH_4^+ converts to NH_3^1; in contact with air, a gradient exists across the gas/liquid interface and NH_3 will be stripped to the air in a stripping tower.[2] Applied mostly in wastewater treatment.[3]	– Increase in pH leads to killing of some pathogens and micro-organisms.[4]	– To make water suitable for drinking purposes re-carbonation would have to be adopted.[4] – A volumetric air: water ratio of about 3000:1 is required to achieve effective NH_3 removal.[1] – Space for stripping towers.	– Suitable for Indian climate: high temperatures increase efficiency.[2] – Neutralization before distribution is necessary and WTP effluent has to be tightly monitored.
Ion exchange	Zeolites (e.g. Clinoptilolite and Chabazite) are selective for NH_4^+ ions. Water is passed through a bed of zeolites to achieve 86–99% ammonium removal, depending on concentration and water composition. Method successful for up to 200 mg/L NH_4^+.[3]	– Zeolites can be regenerated using NaCl[5] or biological regeneration.[6] – Method has no sensitivity to fluctuation in NH_4^+ influent concentration.[6]	– Inconvenient for WTPs with a capacity >80,000 m^3/d because of space required for ion exchange columns.[7] – Presence of Ca reduces NH_4 adsorption onto the zeolite.[8] – High input concentrations require large volumes of zeolites.[9]	– WTPs in Delhi have a capacity >180,000 m^3/d[10]. Therefore the method is not suitable to retrofit into existing WTPs. Could be used for treating NH_4^+ contaminated GW from Ranney wells by clustering a few wells and installing a small treatment facility for the NH_4^+ treatment prior to conventional treatment.
Breakpoint chlorination	By adding chlorine to water, a stepwise reaction takes place with the NH_4^+-N, first forming mono- and di-chloramines and then, at the breakpoint, N_2, NO_3^- and free residual chlorine. Method best suitable for NH_4^+ << 1 mg/L[1].	– The mono-chlor-amine and di-chloramine formed, act as a potential disinfectant.[11] – Can be combined with As removal.[12]	– Cl reacts with organic material and by-products are formed. Activated carbon adsorber needs to be installed too, making treatment expensive.[12,13] – Requires frequent monitoring of NH_4^+ and chlorine concentrations.[3] – High ammonia-N:chlorine ratios.[14]	– Frequent variations in raw water quality: chlorine dosage has to be continuously adjusted to reach breakpoint. – Treatment presumably has to be followed by carbon adsorbers because of organic compounds in the raw water. – NH_4^+ in raw water often >1 mg/L.

Reverse Osmosis (RO)	Water is forced across a semi permeable membrane and molecules and ions, (NH_4^+), are retained.[3] 94% NH_3 removal in full scale tests with feed concentrations of 33 mg/L.[15] In other studies >98%[16] and >96%.[7]	– Small space required. – Low start-up time and continuous operation.[7] Initial water parameters don't have major effect on treatment process.	– Often pre-filtration for particle removal + other pre-treatment steps.[3] – High investment costs, but comparatively low operating costs.[7] – Mineral imbalances can increase corrosive nature of the effluent and post-treatment might be necessary.[3]	– Membrane technique would be best to use as an intermediate solution to treat peak concentrations. Too expensive for regular use.
Biological				
Biological filters	Biofilms form through accumulation of nitrifying bacteria on filter material. NH_4^+ is oxidized to NO_2^- and then to NO_3^- by different bacteria.[17,18] Sufficient O_2 has to be supplied through aeration step prior to filtration.[19] Removal rates <95% with feed concentrations below 4.5 mg NH_4-N/L[3]	– Low construction and maintenance costs.[17] – Simplicity in operation.[17] – Biodegradable matter (BOM) removed simultaneously.[17]	– Colonization takes 2–3 months.[19] – Increases nitrate levels, may release bacteria into the treated water.[3] – Incomplete nitrification can occur due to: occurrence of elevated BOM concentrations,[20] competition for phosphate with other bacteria,[21] fluctuations in feed concentration,[22] if O_2 concentrations are low[19]	– Filters are sensitive to changes in raw water concentrations, which are common in Delhi due to mixing of different groundwater sources and surface water. Microbiology is sensitive to this and reliable functioning is not guaranteed. – Malfunctions detected only through regular, very accurate sampling of influents and effluents.

[1]Gauntlett, 1980; [2]Huang & Shang, 2006; [3]Health Canada, 2013; [4]Jones *et al.*, 2005; [5]Abd El-Hady *et al.*, 2001; [6]Rahmani & Mahvi, 2006; [7]Kurama *et al.*, 2002; [8]Weatherley & Miladinovic, 2004; [9]Li *et al.*, 2011; [10]Govt. of Delhi, 2011; [11]Donnermair & Blatchley III, 2003; [12]Takó and Laky, 2012; [13]Janda & Rudovský, 1994; [14]Griffin & Chamberlin, 1941; [15]Bellona *et al.*, 2008; [16]Bodalo *et al.*, 2005; [17]Yu *et al.*, 2007; [18]Andersson *et al.*, 2001; [19] Lytle *et al.*, 2013; [20]Manem & Rittmann, 1992; [21]De Vet *et al.*, 2010; [22]Rittmann, 1990.

4.5 CONCLUSION AND RECOMMENDATIONS

The use of BF in the Yamuna floodplain in Delhi and in similar hydrogeological settings is basically recommended. However, in these locations, bank filtration should not be seen as a treatment option, but as an option to adapt and improve water management measures. The two main advantages are (1) temporary water storage in the aquifer and (2) a relatively constant raw water composition, improving operating conditions for WTPs. In addition, it is advised to set up a post-treatment concept that is designed specifically for the groundwater parameters at the particular BF location. Such an adapted site-specific post-treatment concept would have the advantage that it would not only reduce elevated ammonium concentrations caused by the infiltration of sewage water, but it would also allow treating other (geogenic) parameters of concern, for example arsenic and fluoride.

In the long term, it is essential to improve the river water quality by implementing sufficient sewage treatment capacity. As this has been widely recognized several new wastewater treatment plants are planned or under construction, e.g. five WWTPs with a designed capacity of 360,000 m^3/d are likely to be commissioned in 2014–5 (Government of India, 2014). However, elevated ammonium concentrations will prevail long after source water quality has improved.

According to the laboratory column experiments (section 4.3.4) and a simplified 1D reactive transport model that was set up for the field site (section 4.3.5), ammonium desorption in the kankar layer – where the laterals of the contaminated well (P3) are presumably located – will take decades. With the assumptions described in section 4.3, the period of ammonium desorption to concentrations <0.5 mg/L will last for about 61 years in the 500 m strip along the river. This result of the simple 1D model is a conservative estimate, as the average linear flow velocity in the kankar layer is probably much higher than the literature value (Sprenger, 2011) suggests. In general, due to the continuing accumulation of ammonium on the aquifer matrix, desorption times will increase with prolonged infiltration of contaminated surface water. Thus, a short and medium term solution such as post-treatment remains a necessity for the investigated site, even if river water quality improves fast. More detailed 2D models are strongly recommended to make more precise and realistic predictions.

4.5.1 Recommended remediation

In general, ammonium remediation options are complex and expensive. At BF sites such as Delhi remediation would often be further complicated by the fact that wells are spread along a long stretch of the river. In-situ or pump and treat remediation measures would, therefore, have to be installed over large areas making remediation extremely costly. If it is decided to develop such a remediation concept, it is of utmost importance to implement the following recommendations:

- *Installation of multi-level observation wells at the well field including levelling survey*: Regular measurements of the water level when wells are operating and when they are switched off, and creation of groundwater contour maps for the different seasons.
- Development of a groundwater flow model based on the water level data. The hydraulic conditions at the well field have to be well known in order to be able to decide on a concept; it is especially important to understand, how the wells influence the flow regime.
- Modelling of scenarios of possible remediation concepts.
- Implementation of groundwater monitoring by regularly taking groundwater samples from the observation wells. To be able to evaluate remediation measures, groundwater quality has to be known and documented before the start of remediation measures.
- Implementation of accompanying groundwater monitoring during remediation.

In case a remediation option is wanted, it is not recommended to use any option involving the injection of chemicals or additives other than oxygen into the aquifer. Although those methods usually are characterized by a faster removal of ammonium, there is a risk of unwanted secondary reactions and formation of by-products, which might not get degraded on the short flow paths to the production wells. BF sites are always water protection zones and therefore special precaution should be taken.

4.5.2 Recommended post-treatment

Two options were identified as the most applicable for the given context of high ammonium concentrations in raw water: (1) raw water from affected wells can be mixed with raw water from other sources before treatment or (2) raw water from wells can be treated separately in independent WTPs.

In Delhi, the first option is generally chosen. Raw water from the Ranney wells is usually mixed with surface water before treatment. This has the advantage that the groundwater from the Ranney wells is diluted and parameters such as arsenic remain below the guideline values and do not need to be treated. A major disadvantage of this method is that water quality is not constant. Nitrogen concentrations in the Yamuna River upstream Delhi show high variations. As the treatment plants are not designed to cope with peak concentrations, it is not always possible to remain below the guideline values for ammonium and/or nitrate. Furthermore, the mixing of ammonium contaminated groundwater with surface water might increase ammonium concentrations in the WTP influents to a level where chlorination is negatively affected.

The second option is therefore recommended. It is better to treat the raw water from the floodplain aquifer separately, e.g. by further pursuing the concept started with the Commonwealth Games Village WTP, a 4.5 million litres per day (MLD) WTP for the water from P4 and nearby bore wells, and the 27 MLD nitrification plant in Okhla.

Although microbial filters (nitrification filters, Table 4.5), as for example used in the Okhla nitrification plant, are a common and cost-efficient option to treat ammonium in raw water, most studies about biological filters for drinking water treatment were not conducted under conditions met in India and the results cannot directly be transferred. Challenges to be met with this technique in locations like Delhi include:

- Supply enough oxygen to cope with the high ammonium concentrations.
- Monitor both inlet and outlet concentrations closely and adapt the hydraulic loading to stabilize nitrogen loading and thus achieve complete nitrification.

Lee *et al.* (2014) reported stable ammonium removal in new, less concentration-sensitive biological filters. However, the reported range of ammonium concentrations is an order of magnitude lower than in the raw water of the Ranney wells in Delhi. Therefore, pilot and full scale studies to find optimum operating conditions for the specific local requirements are recommended if the application of this technique should be further enhanced in India. As an alternative to biological filters, a method with more robustness towards fluctuating input parameters and less downtime in case of failures should be considered, such as ion exchange using zeolites.

ACKNOWLEDGMENTS

This research was conducted in the framework of Saph Pani that was co-financed by the European Commission within the Seventh Framework Programme Grant agreement No. 282911 and Veolia Eau. Vikas Gupta, Shams Ul Haq and Sajil Kumar were part of the team conducting the field work. We thank Dietrich Lange for constructing the laboratory columns. Laura Krömer conducted some of the ammonium experiments and Mario Eybing did sediment analyses with some of the sediment samples. We thank Andreas Winkler, Enrico Hamann and Thomas Taute for the helpful discussions about the column experiments and modelling and Gunnar Lorenzen and Christoph Sprenger for letting us use data from their field work in the frame of the TECHNEAU project and for the discussions about the field site. We thank Hella Schwarzmueller for her comments and suggestions.

4.6 REFERENCES

Abd El-Hady H. M., Grünwald A., Vlčková K. and Zeithammerová J. (2001). Clinoptilolite in drinking water treatment for ammonia removal. *Acta Polytechnica*, **41**(1), 41–45.

Ammary B. Y. (2007). Wastewater reuse in Jordan: Present status and future plans. *Desalination*, **211**(1–3), 164–176.

Andersson A., Laurent P., Kihn A., Prévost M. and Servais P. (2001). Impact of temperature on nitrification in biological activated carbon (BAC) filters used for drinking water treatment. *Water Research*, **35**(12), 2923–2934.

Bellona C., Drewes J. E., Oelker G., Luna J., Filteau G. and Amy G. (2008). Comparing nanofiltration and reverse osmosis for drinking water augmentation. *Journal American Water Works Association*, **100**(9), 102–116.

Beyer, W. (1964). Zur Bestimmung der Wasserdurchlässigkeit von Kiesen und Sanden aus der Kornverteilung (Determination of the water permeability of gravels and sands by grain size distribution). *Wasserwirtschaft-Wassertechnik*, **14**(6), 165–169.

BIS 10500 (2012). Drinking Water – Specification. Indian Standard, 2nd Revision. Bureau of Indian Standards, New Delhi, India.

Bodalo A., Gomez J. L., Gomez E., Leon G. and Tejera M. (2005). Ammonium removal from aqueous solutions by reverse osmosis using cellulose acetate membranes. *Desalination*, **184**(1), 149–155.

Böhlke J. K., Smith R. L. and Miller D. N. (2006). Ammonium transport and reaction in contaminated groundwater: Application of isotope tracers and isotope fractionation studies. *Water Resources Research*, **42**(5), W05411.

Bouwer H. (2002). Artificial recharge of groundwater: Hydrology and engineering. *Hydrogeology Journal*, **10**, 121–142.

Buss S. R., Herbert A. W., Morgan P. and Thomton S. F. (2004). A review of ammonium attenuation in soil and groundwater. *Quarterly Journal of Engineering Geology and Hydrogeology*, **37**(4), 347–359.

Berliner Wasserbetriebe (BWB) (2007). Annual Report 2007, Berlin, Germany.

Ceazan M. L., Thurman E. M. and Smith R. L. (1989). Retardation of ammonium and potassium transport through a contaminated sand and gravel aquifer: The role of cation exchange. *Environmental Science & Technology*, **23**(11), 1402–1408.

Census of India (2011). Provisional Population Totals: Cities having Population 1 lakh and Above, Census of India 2011, Office of the Registrar General and Census Commissioner, India, Ministry of Home Affairs, Government of India. http://www.censusindia.gov.in/2011-prov-results/paper2/data_files/India2/Table_2_PR_Cities_1Lakh_and_Above.pdf (accessed 24 September 2012).

Central Groundwater Board (CGWB) (2012). Groundwater Year Book 2011–12. Central Ground Water Board, Ministry of Water Resources, Government of India. http://cgwb.gov.in/documents/Ground%20Water%20Year%20Book%20-%202011–12.pdf (accessed 7 September 2012).

Chatterjee R., Gupta B. K., Mohiddin S. K., Singh P. N., Shekhar S. and Purohit R. (2009). Dynamic groundwater resources of national capital territory, Delhi: Assessment, development and management options. *Environmental Earth Sciences*, **59**(3), 669–686.

Clark I., Timlin R., Bourbonnais A., Jones K., Lafleur D. and Wickens K. (2008). Origin and fate of industrial ammonium in anoxic ground water—15N. evidence for anaerobic oxidation (Anammox). *Ground Water Monitoring & Remediation*, **28**(3), 73–82.

Corcoran E., Nellemann C., Baker E., Bos R., Osborn D. and Savelli H. (2010). Sick Water. The Central Role of Wastewater Management in Sustainable Development. A Rapid Response Assessment. United Nations Environment Programme, UN-HABITAT, GRID-Arendal, Norway.

De Vet W. W. J. M., Van Genuchten C. C. A., Van Loosdrecht M. C. M. and Van Dijk J. C. (2010). Water quality and treatment of river bank filtrate. *Drinking Water Engineering and Science*, **3**(1), 79–90.

DeSimone L. A. and Howes B. L. (1996). Denitrification and nitrogen transport in a coastal aquifer receiving wastewater discharge. *Environmental Science & Technology*, **30**(4), 1152–1162.

DeSimone L. A. and Howes B. L. (1998). Nitrogen transport and transformations in a shallow aquifer receiving wastewater discharge: A mass balance approach. *Water Resources Research*, **34**(2), 271–285.

Dillon P. (2005). Future management of aquifer recharge. *Hydrogeology Journal*, **13**, 313–316.

Donnermair M. M. and Blatchley III E. R. (2003). Disinfection efficacy of organic chloramines. *Water Research*, **37**(7), 1557–1570.

Doussan C., Poitevin G., Ledoux E. and Detay M. (1997). River bank filtration: Modelling of the changes in water chemistry with emphasis on nitrogen species. *Journal of Contaminant Hydrology*, **25**, 129–156.

Doussan C., Ledoux E. and Detay M. (1998). River-groundwater exchanges, bank filtration, and groundwater quality: Ammonium behavior. *Journal of Environmental Quality*, **27**(6), 1418–1427.

Duong H. A., Berg M., Hoang M. H., Pham H. V., Gallard H., Giger W. and von Gunten U. (2003). Trihalomethane formation by chlorination of ammonium- and bromide-containing groundwater in water supplies of Hanoi, Vietnam. *Water Research*, **37**(13), 3242–3252.

Eckenfelder W. W. and Argaman Y. (1991). Principles of biological and physical/chemical nitrogen removal. In: Sedlak, R. I. Phosphorous and Nitrogen Removal from Municipal Wastewater. Principles and Practice. 2nd edn, Lewis Publishers, United States.

European Environment Agency (EEA) (2001). Nitrogen and Phosphorus in River Stations by River Size and Catchment Type. Core Set of Indicators, EEA, Copenhagen, Denmark. http://www.eea.europa.eu/data-and-maps/indicators/nitrogen-concentrations-in-rivers/nitrogen-and-phosphorus-in-river-stations (accessed 14 June 2014).

Elmore C. A. and Graff T. (2002). Best available treatment technologies applied to groundwater circulation wells. *Remediation Journal*, **12**(3), 63–80.

Eybing M. (2014). *Charakterisierung eines alluvialen Grundwasserleiters in Delhi (Indien): Sedimentanalysen von Bohrungen an einem Uferfiltrationsstandort am Yamuna* (Characterization of an Alluvial Aquifer in Delhi (India): Sediment Analysis from Drillings at a Riverbank Filtration Site at the Yamuna). Bachelor thesis, Hydrogeology Group, Freie Universität Berlin, Berlin, Germany.

Gauntlett R. B. (1980). Removal of nitrogen compounds. In: Developments in Water Treatment – 2, Lewis W. M. (eds), Applied Science Publisher, United Kingdom.

Gelhar L. W., Welty C. and Rehfeldt K. R. (1992). A critical review of data on field-scale dispersion in aquifers. *Water Resources Research*, **28**(7), 1955–1974.

Genske D. (2006). *Ingenieurgeologie Grundlagen und Anwendungen* (Engineering Geology Basics and Applications). Springer Verlag, Heidelberg.

Geological Survey of India (2006). Geological, Geomorphologic and Mineral Map of Delhi. Geological Survey of India Misc. Publications No. 30 Pt. XVI. http://www.portal.gsi.gov.in/gsiDoc/pub/MP30_GM_Delhi.pdf (accessed 23 January 2013).

GESTIS Substance Database (2014). Nitrogen. Institute for Occupational Safety and Health of the German Social Accident Insurance (IFA). http://www.gestis-en.itrust.de/ (accessed 01 July 2014).

Government of Delhi (2011). List of Water Treatment Plants in Delhi. http://www.delhi.gov.in/wps/wcm/connect/056b83804a10d627b181fbe4899821f2/WTP+in+Delhi.pdf?MOD=AJPERES&lmod=1170192194&CACHEID=056b83804a10d627b181fbe4899821f2 (accessed 10 July 2014).

Government of India (1986). The Environment (Protection) Rules, 1986 Schedule–VI General Standards for Discharge of Environmental Pollutants Part-A: Effluents. Central Pollution Control Board, Government of India, New Delhi, India.

Government of India (2014). Pollution in Yamuna. Ministry of Water Resources, River Development and Ganga Rejuvenation. http://admin.indiaenvironmentportal.org.in/files/file/Yamuna%20pollution%20RS%2021%20July%202014.doc (accessed 10 August 2014).

Griffin A. E. and Chamberlin N. S. (1941). Relation of ammonia-nitrogen to break-point chlorination. *American Journal of Public Health Nations Health*, **31**(8), 803–808.

Groeschke M. (2013). Challenges to riverbank filtration in Delhi (India): Elevated ammonium concentrations in the groundwater of an alluvial aquifer. *Zbl. Geol. Paläont.* **1**(1/2), 1–9.

Groeschke M., Taute T., Eybing M. and Schneider M. (2015a). Assessment of an ammonium contamination at a riverbank filtration site in central Delhi, India. *In Preparation.*

Groeschke M., Kumar P., Winkler A., Grützmacher G. and Schneider M. (2015b). The role of agricultural activity for ammonium contamination at a riverbank filtration site in central Delhi (India). *In Preparation.*

Groeschke M., Hamann, E., Frommen T., Grützmacher G. and Schneider M. (2015c). Transport of ammonium in alluvial sediments – laboratory column experiments and reactive transport modelling. *In Preparation.*

Grünheid S., Amy G. and Jekel M. (2005). Removal of bulk dissolved organic carbon (DOC) and trace organic compounds by bank filtration and artificial recharge. *Water Research*, **39**(14), 3219–3228.

Haerens B., Dassargues A., Prommer H. and Lerner D. (2002a). Multicomponent reactive modelling of ammonium contamination at a former coal carbonisation plant. *Aardkundige Mededelingen*, **12**, 211–214.

Haerens B., Dassargues A. and Lerner D. N. (2002b). Reactive transport modelling of ammonium: 1D conceptual modelling and comparison of reactive transport codes. *Acta Universitatis Carolinae Geologica,* **46**(2–3), 27–31.

Hamann E. (2009). *Reaktive Stofftransportmodellierung einer urbanen Grundwasserkontamination aus einem ehemaligen Rieselfeld* (Reactive Transport Modeling of an Urban Groundwater Contamination Originating from a Former Sewage Farm). PhD thesis, Faculty of Mathematics and Science II, Humboldt Universität zu Berlin, Berlin, Germany.

Hazen A. (1893). Some physical properties of sand and gravel with special reference to their use infiltration, annual report of mass. *State board of health,* **24**, 541–556.

Health Canada (2013). Guidelines for Canadian Drinking Water Quality: Guideline Technical Document - Ammonia. Federal-Provincial-Territorial Committee on Drinking Water, Canada.

Heberer T. (2002). Tracking persistent pharmaceutical residues from municipal sewage to drinking water. *Journal of Hydrology*, **266**, 175–189.

Hem J. D. (2005). Study and Interpretation of the Chemical Characteristics of Natural Water. Geological Survey Water-Supply Paper, University Press of the Pacific, Honululu, Hawai.

Hinkle S. R., Boehlke J. K., Duff J. H., Morgan D. S. and Weick R. J. (2007). Aquifer-scale controls on the distribution of nitrate and ammonium in ground water near La Pine, Oregon, USA. *Journal of Hydrology*, **333**(2–4), 486–503.

Hiscock K. M. and Grischek T. (2002). Attenuation of groundwater pollution by bank filtration. *Journal of Hydrology*, **266**(3–4), 139–144.

Horner C., Engelmann F. and Nützmann G. (2009). Model based verification and prognosis of acidification and sulphate releasing processes downstream of a former sewage field in Berlin (Germany). *Journal of Contaminant Hydrology*, **106**, 83–98.

Howarth R. W. and Marino R. (2006). Nitrogen as the limiting nutrient for eutrophication in coastal marine ecosystems: Evolving views over three decades. *Limnology and Oceanography*, **51**(1), 364–376.

Huang J.-C. and Shang C. (2006). Air stripping. In: Handbook of Environmental Engineering. Vol. 4: Advanced Physicochemical Treatment Processes, L. K. Wang, Y.-T. Hung and N. K. Shammas (eds.), Humana Press, Totowa, New Jersey, United States, pp. 47–79.

IEG Technology (2008). IEG Technical Briefing Note No. 12: In Situ Remediation of Ammonia and Nitrate Impacted Groundwater. http://iegtechnologies.co.uk/files/pdfs/IEG%20Ammonia%20&%20Nitrate%20-%20TB12.pdf (accessed 24 July 2014).

IVEY International Inc. (2012a). CASE STUDY: In-situ Ivey-sol TM Surfactant Enhanced SORBITALLTM Ammonia Remediation of Ammonia Nitrogen Groundwater Plume at a Fertilizer Facility.

IVEY International Inc. (2012b). SorbitAll: Material Safety Data Sheet Version No: 120808–03.

Janda V. and Rudovský J. (1994). Removal of ammonia from drinking water by biological nitrification. *Journal of Water Supply: Research and Treatment-Aqua*, **43**(3), 120–125.

Jha P. K., Subramanian V. and Sitasawad R. (1988). Chemical and sediment mass transfer in the Yamuna river—a tributary of the ganges system. *Journal of Hydrology*, **104**, 237–246.

Jones C. S., Corrigan T. P., Graham D. P. and McMullen L. D. (2005). Reduced lime feeds: Effects on operational costs and water quality, Report Des Moines Water Works, Des Moines, Iowa, United States.

Khan A. and Bajpai V. N. (2014). Constriction of the Yamuna river floodplain within Delhi region since 19th century: A serious concern. *International Journal of Research in Engineering and Technology*, **3**, 430–436.

Kido M., Yustiawati, Syawal M. S., Sulastri, Hosokawa T., Tanaka S., Saito T., Iwakuma T. and Kurasaki M. (2009). Comparison of general water quality of rivers in Indonesia and Japan. *Environmental Monitoring and Assessment*, **156**(1–4), 317–329.

Kuehn W. and Mueller U. (2000). Riverbank filtration: An overview. *Journal American Water Works Association*, **92**(12), 60–69.

Kurama H., Poetzschke J. and Haseneder R. (2002). The application of membrane filtration for the removal of ammonium ions from potable water. *Water Research*, **36**(11), 2905–2909.

LeBlanc D. R. (1984). Sewage Plume in a Sand and Gravel Aquifer, Cape Cod, Massachusetts. United States Geological Survey, Water-Supply Paper, United States.

Lee C. O., Boe-Hansen R., Musovic S., Smets B., Albrechtsen H. J. and Binning P. (2014). Effects of dynamic operating conditions on nitrification in biological rapid sand filters for drinking water treatment. *Water Research*, **64**, 226–236.

Li M., Zhu X., Zhu F., Ren G., Cao G. and Song L. (2011). Application of modified zeolite for ammonium removal from drinking water. *Desalination*, **271**(1–3), 295–300.

Lorenzen G. (2011). Riverbank Filtration in Delhi, India: Insights from Hydrogeological Field Investigations and Perspectives for a Water-Stressed Megacity. PhD thesis, Hydrogeology Group, Freie Universität Berlin, Berlin, Germany.

Lorenzen G., Sprenger C. and Greskowiak J. (2007). Results of Background Work and Data Integration of MAR Systems for an Integrated Water Resources Management. Report TECHNEAU, D 5.2.1, European Commission.

Lorenzen G., Sprenger C., Taute T., Pekdeger A., Mittal A. and Massmann G. (2010a). Assessment of the potential for bank filtration in a water-stressed megacity (Delhi, India). *Environmental Earth Sciences*, **61**(7), 1419–1434.

Lorenzen G., Sprenger C. and Pekdeger A. (2010b). Investigation of RBF Potential in Developing and Newly Industrializing Countries: Lessons Learnt from the Delhi Case Studies. Report TECHNEAU, D 5.2.8., European Commission.

Lytle D. A., White C., Williams D., Koch L. and Nauman E. (2013). Innovative biological water treatment for the removal of elevated ammonia. *Journal – American Water Works Association*, **105**(9), 87–88.

Mailath S. B. and Chu A. C. (2005). Patent Application Title: In-Situ Groundwater Nitrification and De-Nitrification Remediation System. Patent: WO 2005123610 A1, CIPO, Canada.

Manem J. A. and Rittmann B. E. (1992). Removing trace-level organic pollutants in a biological filter. *Journal American Water Works Association*, **84**(4), 152–157.

Metcalf & Eddy Inc (2014). Wastewater Engineering: Treatment and Resource Recovery. 5th International edn, McGraw-Hill Book Co., Singapore.

Mobley H. L. T. and Hausinger R. P. (1989). Microbial ureases: Significance, regulation, and molecular characterization. *Microbiological Reviews*, **53**(1), 85–108.

Natural Earth (2011). Free Vector and Raster Map Data. http://www.naturalearthdata.com/ (accessed 01 August 2011).

Office of Science and Technology (OST) (2002). In-Well Vapor Stripping Technology. Department of Energy, Office of Environmental Management, United States.

Parkhurst D. L. and Appelo C. A. J. (2013). Description of Input and Examples for PHREEQC Version 3-a Computer Program for Speciation, Batch-reaction, One-dimensional Transport and Inverse Geochemical Calculations.

Patterson B. M., Grassi M. E., Davis G. B., Robertson B. S. and McKinley A. J. (2002). Use of polymer mats in series for sequential reactive barrier remediation of ammonium-contaminated groundwater: Laboratory column evaluation. *Environmental Science & Technology*, **36**(15), 3439–3445.

Patterson B. M., Grassi M. E., Robertson B. S., Davis G. B., Smith A. J. and McKinley A. J. (2004). Use of polymer mats in series for sequential reactive barrier remediation of ammonium-contaminated groundwater: Field evaluation. *Environmental Science & Technology*, **38**(24), 6846–6854.

Patterson R. A. (2003). Nitrogen in wastewater and its role in constraining on-site planning in future directions for on-site systems: Best management practice. In: *Proceedings of On-site '03 Conference*, Patterson R. A. and Jones M. J. (eds), Lanfax Laboratories Armidale, Armidale, Australia, pp. 313–320. ISBN 0-9579438-1-4.

Pekdeger A., Lorenzen G. and Sprenger C. (2008). Preliminary report on Data of All Inorganic Substances and Physicochemical Parameters Listed in the Indian and German Drinking Water Standards from Surface Water and Groundwater at the 3 (+1) Field Sites. Report TECHNEAU, D 5.2.2, European Commission.

Pescod M. B. (1992). Wastewater Treatment and Use in Agriculture. FAO Irrigation and Drainage Paper 47, Food and Agriculture Organization of the United Nations, Rome.

Rahmani A. R. and Mahvi A. H. (2006). Use of ion exchange for removal of ammonium: A biological regeneration of zeolite. *Global NEST Journal*, **8**(2), 146–150.

Randall D. J. and Tsui T. K. N. (2002). Ammonia toxicity in fish. *Marine Pollution Bulletin*, **45**(1–12), 17–23.

Ray C., Grischek T., Schubert J., Wang J. and Speth T. (2002). A perspective of riverbank filtration. *Journal American Water Works Association*, **94**(4), 149–160.

Ray C. (2004). Modeling RBF efficacy for mitigating chemical shock loads. *Journal American Water Works Association,* **96**(5), 114–128.

Ray C. (2008). Worldwide potential of riverbank filtration. *Clean Technologies and Environmental Policy*, **10**(3), 223–225.

Richter, L (1966). Aufgaben der hydrogeologischen Erkundung und Nacherkundung in der Braunkohlenindustrie. Zeitschrift für Angewandte Geologie **12**(6), 288–297.

Rittmann B. E. (1990). Analyzing biofilm processes used in biological filtration. *Journal American Water Works Association*, **82**(12), 62–66.

Rosenshein J. S. (1988). Region 18, Alluvial Valleys. In: Hydrogeology – The Geology of North America, Back W., Rosenshein J. S. and Seaber P. R. (eds), The Geological Society of America, Vol. 0–2, Boulder, Colorado, United States, pp. 165–175.

Schultze E. and Muhs H. (1967). *Bodenuntersuchungen für Ingenieurbauten* (Soil Analyses for Civil Engineering). Springer Verlag, Heidelberg.

Shekhar S. and Prasad R. K. (2009). The groundwater in the Yamuna flood plain of Delhi (India) and the management options. *Hydrogeology Journal*, **17**(7), 1557–1560.

Shekhar S., Purohit R. R. and Kaushik Y. B. (2009). Groundwater management in NCT Delhi. In: Technical Paper Included in the Special Session on Groundwater in the 5th Asian Regional Conference of INCID, New Delhi, India.

Sikder M. T., Kihara Y., Yasuda M., Yustiawati, Mihara Y., Tanaka S., Odgerel D., Mijiddorj B., Syawal S. M., Hosokawa T., Saito T. and Kurasaki M. (2013). River water pollution in developed and developing countries: Judge and assessment of physicochemical characteristics and selected dissolved metal concentration. *Clean – Soil, Air, Water*, **41**(1), 60–68.

Singh P., Kumar P., Mehrotra I. and Grischek T. (2010). Impact of riverbank filtration on treatment of polluted river water. *Journal of Environmental Management,* **91**(5), 1055–1062.

Sprenger C., Lorenzen G. and Pekdeger A. (2008). Occurrence and Fate of Microbial Pathogens and Organic Trace Compounds at Riverbank Filtration Sites in Delhi, India. Report TECHNEAU, D 2.5.6, European Commission.

Sprenger C. (2011). Surface-/Groundwater Interactions Associated with River Bank Filtration in Delhi (India) – Investigation and Modelling of Hydraulic and Hydrochemical Processes. PhD thesis, Hydrogeology Group, Freie Universität Berlin, Berlin, Germany.

Sprenger C. and Lorenzen G. (2014). Hydrogeochemistry of urban floodplain aquifer under the influence of contaminated river seepage in Delhi (India). *Aquatic Geochemistry*, **20**(5), 519–543.

Stumm W. and Morgan J. J. (1996). Aquatic Chemistry: An Introduction Emphasizing Chemical Equilibria in Natural Waters. 3rd edn, John Wiley, New York, United States.

Takó S. and Laky D. (2012). Laboratory experiments for arsenic and ammonium removal—the combination of breakpoint chlorination and iron(III)-coagulation. *Journal of Environmental Science and Engineering*, **A1**, 1165–1172.

Tjandraatmadja G., Pollard C., Sheedy C. and Gozukara Y. (2010). Sources of Contaminants in Domestic Wastewater: Nutrients and Additional Elements from Household Products. Water for a Healthy Country, Flagship Report Series ISSN: 1835–095X, CSIRO, Australia.

Tufenkji N., Ryan J. N. and Elimelech M. (2002). The promise of bank filtration. *Environmental Science & Technology*, **36**(21), 422A–428A.

Ward M. H., deKok T. M., Levallois P., Brender J., Gulis G., Nolan B. T. and VanDerslice J. (2005). Workgroup report: Drinking-water nitrate and health – recent findings and research needs. *Environmental Health Perspectives*, **113**(11), 1607–1614.

Weatherley L. R. and Miladinovic N. D. (2004). Comparison of the ion exchange uptake of ammonium ion onto New Zealand clinoptilolite and mordenite. *Water Research*, **38**(20), 4305–4312.

Weil I. and Morris J. C. (1949). Kinetic studies on the chloramines. I. The rates of formation of monochloramine, N-chlormethylamine and N-chlordimethylamine. *Journal of the American Chemical Society*, **71**(5), 1664–1671.

Weiss W. J., Bouwer E. J., Aboytes R., LeChevallier M. W., O'Melia C. R., Le B. T. and Schwab K. J. (2005). Riverbank filtration for control of microorganisms: Results from field monitoring. *Water Research*, **39**(10), 1990–2001.

World Health Organization (WHO)(2011). Guidelines for Drinking-Eater Quality – 4th edn, WHO, Geneva, Switzerland.

United Nations World Water Assessment Programme (WWAP) (2012). The United Nations World Water Development Report 4. Managing Water under Uncertainty and Risk. UNESCO, Paris, France.

Yu X., Qi Z., Zhang X., Yu P., Liu B., Zhang L. and Fu L. (2007). Nitrogen loss and oxygen paradox in full-scale biofiltration for drinking water treatment. *Water Research*, **41**(7), 1455–1464.

Chapter 5

Overview of Managed Aquifer Recharge in India

Anders Nättorp, Jeremias Brand, Devinder Kumar Chadha, Lakshmanan Elango, Narayan C. Ghosh, Gesche Grützmacher, Christoph Sprenger and Sumant Kumar

5.1 INTRODUCTION

5.1.1 Scope

Groundwater exploitation in India has increased rapidly over the last 50 years as reflected by the growth of the number of groundwater abstraction structures (from 3.9 million in 1951 to 18.5 million in 2001) and shallow tube wells (from 3,000 in 1951 to 8.5 million in 2001) (Singh & Singh, 2002). Today, groundwater is the source for more than 85% of India's rural domestic water requirements, 50% of urban water and more than 50% of irrigation demand. The increase in demand in the last 50 years has led to declining water tables in many parts of the country. For example, 15% of the assessment units (Blocks/Mandals/Talukas) have groundwater extraction in excess of the net annual recharge (Central Ground Water Board (CGWB), 2007). According to Rodell *et al.* (2009), the extent of groundwater depletion between 2002 and 2008 was 109 km^3, which is about half the capacity of India's surface-water reservoirs.

One way to address the dwindling groundwater resources is through the use of Managed Aquifer Recharge (MAR). It is estimated that about 29% of the total land area in India is suitable for additional MAR and that a volume of 86 km^3 is available for additional groundwater recharge annually. This is equivalent to an average depth of 90 mm over the suitable area and the volume that could be recharged equates 27% of the 231 km^3 of groundwater that is currently utilized annually (CGWB, 2013).

This chapter presents an inventory of the MAR applications in India, performed as a starting point for the work in Saph Pani. It focuses on the technical aspects associated with MAR, discussing the state-of-the art with respect to the techniques used and to the amount of water being artificially recharged. This chapter does not consider, or only tangentially touches, on the socio-economic impacts of MAR. These aspects will be treated in chapter 17: Rapid assessment and SWOT analysis of non-technical aspects of natural wastewater treatment systems. The goal of this chapter is to give a comprehensive overview of the potentials and limitations of MAR techniques for natural water treatment in India and derive ideas for action.

5.1.2 Definition of Managed Aquifer Recharge (MAR)

MAR has been defined as intentional storage and treatment of water in aquifers (Dillon *et al.* 2009; Sharma *et al.* 2011). Dillon *et al.* (2009) and Sharma *et al.* (2011) included the techniques Soil Aquifer Treatment (SAT), Aquifer Storage and Recovery (ASR), Aquifer Storage Transfer and Recovery (ASTR), Subsurface groundwater treatment (SGT) and Bank filtration (BF) in the wider frame of MAR. The term Artificial Recharge commonly used in India denotes recharge of the aquifer for later use. Artificial Recharge is practiced in order to increase water quantity without reference to recharged water quality. It is similar to the term Aquifer Recharge and Recovery used by Sharma *et al.* (2011) and encloses ASR and ASTR.

In the Saph Pani project, MAR denotes the replenishment of the aquifer with the intention to compensate for prior use and/ or to store for future use. ASR, ASTR and SAT all fall under this definition of MAR (Table 5.1)

Table 5.1 Characterization of techniques for MAR with respect to the intention and the water flow.

Technique	ASR Aquifer Storage and Recovery	ASTR Aquifer Storage Transfer and Recovery	SAT Soil Aquifer Treatment	SGT Subsurface Groundwater Treatment	BF Bank Filtration
Intention	Mainly storage	Storage and treatment	Storage and treatment	Treatment	Treatment
Water flow	Infiltration, subsequent abstraction	Infiltration, subsequent abstraction	Infiltration, subsequent abstraction	Small quantity infiltrated in order to cause treatment	Abstraction leading to infiltration

Table 5.1 ASR being practiced mainly for storage of water, whereas ASTR and SAT also have the intention to improve quality by controlled underground treatment. Subsurface groundwater treatment and bank filtration are exclusively intended for treatment and consequently do not fall under the definition of MAR in Saph Pani. BF is the subject of a separate work package in Saph Pani.

5.1.3 Structures for MAR

MAR can be performed by a multitude of structures (Figure 5.1) which capture and store the water and enable it to infiltrate into the underground. Thereby the flow of water is better controlled, limiting flooding and erosion. Both new structures and modification of existing structures (e.g. rooftop rainwater harvesting and dugwell recharge) are possible. Because of India's long tradition of water harvesting and its many languages, there are many traditional structures and also different names for similar structures. The following list of structures is thus not exhaustive, but covers the main types. More detailed information is given by CGWB (2007) and Dilllon *et al.* (2009).

Surface spreading

Surface spreading structures (Figure 5.1) aim to increase the area which is in contact with surface water and also the time over which this contact takes place. In this way infiltration is improved and evaporation decreases. This can be achieved through managed flooding between constructed canals or streambeds or by constructing a system of ditches and furrows.

Contour bund and contour trench

A bund is an embankment of earth. Contour bunds and trenches (Figure 5.1) break the flow of water and thus increase infiltration and limit erosion. They are constructed along contours of equal land elevation. Between two contours, agriculture can be practiced and tree plantation on the bund is possible. Bunds trees/plants can help to fix nitrogen in the soil for the crop plants. During rainfall the contour bund acts as a barrier to the water flow, reducing the speed of run-off water and thus also the washing out of nutrients.

Bench terracing

Bench terracing (Figure 5.1) is practiced in hilly areas where the original slope is levelled stepwise by cutting and filling. Under suitable conditions the structure helps to reduce surface run-off and enhances soil moisture conservation, crop production and aquifer recharge.

Percolation or infiltration pond or tank and recharge basin

Percolation tank or pond (Figure 5.1) is a term used in India to describe harvesting of water in storages built in ephemeral streams or off-stream where water is detained and infiltrates through the permeable base to enhance storage in unconfined aquifers. Recharge basins differ from percolation ponds in that they are designed to accommodate a flow through a series of basins not retaining the whole amount of water in a single basin like in a percolation pond. For both types of structures the water is usually desilted to prevent clogging.

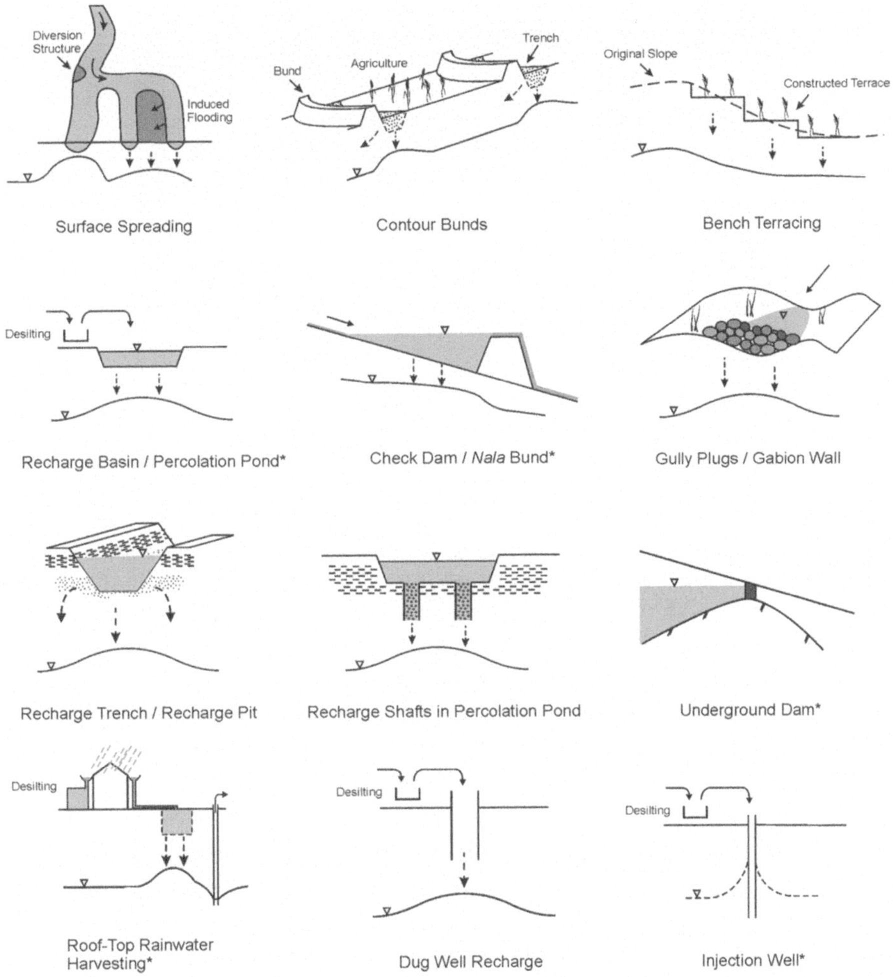

Figure 5.1 Sketches of managed aquifer recharge structures commonly used in India. *Modified from Gale (2005).

Check dam, nala bund

Check dams (Figure 5.2) or Nala bunds (Figure 5.3) are barriers built across the direction of water flow in rivers. These dams retain part of the water flow during monsoon rains in the area upstream of the structure. The increased accumulation of water in the reservoir area increases the infiltration rate.

Figure 5.2 Check dam in Araniyar River in Tamil Nadu (*Source*: Christoph Sprenger).

Figure 5.3 Nala bund (*Source*: Lakshmanan Elango).

Gully plug and gabion wall

Gullies are formed due to erosion of top soil by the flow of rain water. Gully plugs are built with local stones, sand, clay and plants. It is a simple technique for conservation of soil and moisture by reducing the speed of run-off water during floods. Gabions are wire mesh baskets filled with rocks and have a permeable, flexible structure (Figure 5.1). Gabions walls are used often for erosion control, bank stabilization, channel linings and weirs. They are also constructed to protect the bank of lakes and rivers against erosion due to water and waves. Sludge and small stones deposit in the interstices, leading to growth of vegetation and ultimately a natural reservoir is formed. It retains water for dry periods to serve agriculture and also replenishes groundwater.

Recharge pit

Recharge pits (Figure 5.4) are dug out pits and trenches, which have been dug through a layer of low permeability to improve infiltration to a shallow phreatic (unconfined) aquifer (Figure 5.1). They differ from percolation ponds and recharge basin in that they are deeper and frequently recharge takes place through the sides of the pit as well. Abandoned mine shafts and quarries are often converted to recharge pits if they are in contact with an underlying aquifer.

Figure 5.4 Recharge pit at Raipur Municipal Corporation headquarters (*Source*: Raipur Municipal Corporation).

5.1.3.8 Recharge shaft

Recharge shafts like recharge pits (Figure 5.1) are recharge structures which penetrate an upper layer of low permeability into the underlying phreatic aquifer. They are constructed at the bottom of surface structures (ponds/tanks/channels) which do not connect to the permeable layer. In contrast to injection or recharge wells they are backfilled with coarse sand and stones thereby creating columns of porous, permeable soil which connect the recharge pit to the aquifer.

Injection well or recharge well

Injection wells (Figure 5.1) are tube wells constructed for the purpose of recharge. Injection wells are primarily used to recharge deep lying aquifers and the water is injected under pressure or using gravity alone. Many of them are constructed with slotted PVC pipe and surrounded with some kind of clogging protection.

Underground dam

Underground dams (Figure 5.1) are built in ephemeral streams where basement ridges constrict flows. A trench is dug across the streambed keyed to the basement and backfilled with low permeability material to help retain flood flows in saturated alluvium for stock and domestic use.

Rooftop rainwater harvesting structure

Rooftop Rainwater harvesting (Figure 5.1) collects and infiltrates the roof run-off from buildings. Most commonly injection takes place through dug or bore wells, but recharge through percolation ponds is also possible.

Dug well recharge

Dug wells (Figure 5.1) which have run dry can be adapted for use as recharge structures. This is done by diverting surface water into the well. It is normally desilted before infiltration to avoid clogging.

5.2 HYDROLOGIC CYCLE OF INDIA

5.2.1 Current overall situation

The main features of India's hydrologic cycle are shown in Figure 5.5 and given in Tables 5.2 and 5.3. The availability and use of water and the interactions between surface and groundwater are shown, allowing an appraisal of the role of MAR in the Indian water supply.

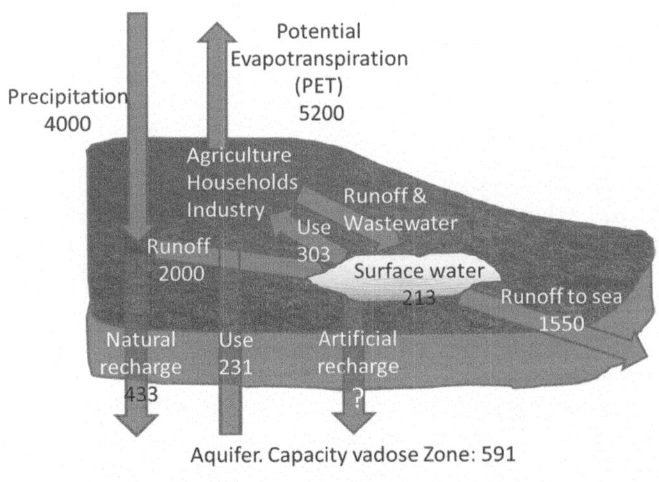

Figure 5.5 Water cycle of India with flows in km^3/year and capacities in km^3. Values and literature references in Tables 5.2 and 5.3.

Potential Evapotranspiration is the maximum amount of water evaporated and lost by transpiration from vegetation (Allen, 1998). India's potential evapotranspiration (5,200 km^3) is higher than its rainfall (4,000 km^3). In other words, the amount of irrigation needed to keep the Indian landmass moist all year, exceeds the water available through rainfall. Today approximately half of the rainwater (2,000 km^3) flows as run-off into natural or manmade surface water bodies and 39% of the rainwater (1,550 km^3) flows into the sea, mainly during the monsoon period.

Table 5.2 Water balance of India.

	[km^3]	[mm][a]
Annual precipitation	4,000[b]	1,200
Annual potential evapotranspiration	5,200	1,580[c]
Annual run-off to surface water	1,890[d]–2,440[e]	570–740
Annual run-off to sea	1,548[f]	470
Annual surface water use	Irrigation: 228[g] Other: 75	Irrigation: 69 Other: 23
Surface storage	213[h]	65

[a]In general the average number of mm was recalculated from the volume using the area of India (3,288,000 km^2); [b]Central Water Commission (2005), Water Resources Information System Data Book-2005; [c]India Meteorological Department (1971); [d]Chaturvedi(1976); [e]Zade *et al.* (2005); [f]Parikh *et al.* (2007); [g]Central Water Commission (2005), Water Resources Information System Data Book-2005: Data for year 2000; Central Ground Water Board (2006), Dynamic Ground Water Resources of India: Difference between total and groundwater use; [h]Central Water Commission (2005), Water Resources Information System Data Book-2005: Data for 2002.

Table 5.3 Indian groundwater balance. Values for natural recharge, groundwater use and annual balance are largely confirmed in the revised master plan for artificial recharge to groundwater of CGWB (2013).

Groundwater Balance	[km^3]	[mm]
Annual natural recharge	433[a]	132
Annual groundwater use	Irrigation: 213[a] Other: 18	Irrigation: 65 Other: 6
Annual natural discharge non-monsoon	34[a]	10
Annual balance	168[a]	51
Unsaturated aquifer deeper than 3 m below ground	591[b]	180

[a]Central Ground Water Board (2006), Dynamic Ground Water Resources of India; [b]Central Ground Water Board (1996), National Perspective Plan for utilization of surplus run-off for augmentation of groundwater resources in India.

Approximately 11% of the rainwater (433 km^3) is naturally recharged to the groundwater, either directly in the rainfall area or from the surface water bodies, whereas abstraction for irrigation and other uses from groundwater represents 6% and abstraction from surface water bodies represents 8%. Part of the water evaporates during use and part of it returns to the groundwater table and surface water bodies.

There is no official figure for the total volume of water harvested by MAR in India (see section 5.3.2). The statistical data on surface water bodies also includes the volumes of MAR structures. Often structures are used conjunctively for irrigation and infiltration. The extent of infiltration in that case depends on whether a passage to the vadose zone has been freed and whether silt has accumulated since then. The total volume (83km^3) to be recharged in the structures suggested in the Master Plan of the CGWB (2013) is 2% of the total rainfall. This gives an idea of the potential importance of MAR in India's water cycle. It is a small fraction of the total rainfall, but could make a sizeable contribution (62km^3, 27%[1]) to the amount of groundwater used (231 km^3). Large volumes of water (213 km^3) are used for irrigation and consequently demand side management (Dillon *et al.* 2009) influencing the type and the number of crops and the irrigation methods also has a large potential for attenuating the groundwater scarcity.

[1]The amount of infiltrated water is calculated from the amount of recharge water (85 km^3) and the efficiency of the structure (75% was assumed).

5.2.2 Spatial and seasonal variation

The parameters of the Indian water cycle (Figure 5.5, Table 5.2 and 5.3) are average values. India has high spatial variability of rainfall across the country, ranging from 150 mm in the west to 11,690 mm in the northeast (Indian Meterological Department, 2004). Thus the water availability and the possibilities for MAR are very different in different parts of the country.

India has an average precipitation comparable to many European countries (European Environment Agency, 2012). However, the seasonal variation is much more pronounced in India, which makes MAR and water storage in general more important.

The period of rainfall in India, the monsoon, comes either from the southwest or the northeast. All states are subject to the southwest monsoon that accounts for about 74% of the annual rainfall in the country (Guhathakurta & Rajeevan, 2006), while some stretches in the peninsular India are also subject to the north-east monsoon which accounts for about 11% of the annual rainfall. Whereas the eastern part of the peninsular India receives most of the rainfall (over 60%) during the north east monsoon.

The rain period can be characterized by recording the shortest period in which 10% and 90%, respectively, of the annual rain falls. "The 10% wet period occurs in the months of July/August with an average duration of 1–3 days and rainfall intensity varying from 44 to 89 mm/d. The duration of the 90% wet period varies from 112 days in the central part of the country to 186 days in the north." (Deshpande & Singh, 2010, p. 561).

The rivers are fed by the monsoon and to some extent by snow melt and experience high seasonal variations. The Ganges peak flow during monsoon in the Himalayan foreland was measured to be 17 times higher than during non-monsoon (Chakrapani & Saini, 2009). In the Indo Gangetic plains, the average dry season to monsoon discharge ratio is about 1 to 6 (Qader, 2005). In the southern part of the country, streams dry out during non-monsoon season.

5.2.3 Future water demand

The water demand in India is expected to increase by some 15% between 2010 and 2025 (0.9%/year) (Kumar *et al.* 2005) (Table 5.4). India Infrastructure Research (2012) predicts an increase in yearly demand of 68 km^3 for irrigation and 28 km^3 for domestic purposes between 2000 and 2025. This would correspond to an increase of 0.6% per year for these two major sectors. The total Indian consumption reported by Kumar *et al.* (2005) and India Infrastructure Research (2012) is above 600 km^3/year in 2010, which is somewhat higher than the values given by CGWB (2006) and CWC (2005) (total of 534 km^3 in 2005 in Table 5.2). Thus, although absolute values and growth rates vary considerably, the sources indicate an increasing water demand. This is attributed among other things to rising population and living standards.

Table 5.4 Expected increase in total water consumption (Kumar *et al.* 2005).

	1997/1998	2010	2025	2050
Total consumption [km^3]	629	694–10	784–843	973–1180
Increase [%]		12	29	71

5.3 COORDINATED ACTIONS FOR PROMOTING ARTIFICIAL RECHARGE

5.3.1 Pilot schemes of the Central Ground Water Board (CGWB)

The CGWB, a subordinate office of the Ministry of Water Resources, Government of India (GoI), is entrusted with the responsibilities of providing scientific inputs for management, exploration, monitoring, assessment, augmentation and regulation of groundwater resources of the country (CGWB, 2012).

The Central Ground Water Board undertakes macro/micro-level groundwater management studies, exploratory drilling programs and also monitoring of groundwater levels and water quality through a network of groundwater observation wells. Periodic assessment of replenishable groundwater resources of the country is carried out by the Board jointly with the concerned State Government agencies. Geophysical studies, remote sensing and GIS studies and groundwater modelling as well as special studies on groundwater sector such as aquifer mapping, groundwater depletion, seawater ingress, groundwater contamination, conjunctive use of surface and groundwater and water balance are also part of the CGWB activities. In addition, the CGWB organizes internal and external capacity building activities as well as mass awareness campaigns on the importance of water conservation and judicious groundwater management (CGWB, 2012).

In the post-independence period, the CGWB first initiated the pilot programme for water harvesting and water conservation during the period 1972 to 1984 with the UN Department of Technical Cooperation for Development collaboration (Table 5.5).

After an inactive period, pilot projects were taken up again in 1992 to demonstrate the technology for different types of recharge structures. Up to 1997 a total of over 700 pilot recharge structures were constructed.

Table 5.5 Artificial recharge studies undertaken by the CGWB during different five year plans (Chadha, 2012; CGWB, 2012).

Period and Plan	Status	Cost* [Million INR]
1972–1984	Haryana, Kerala, Gujarat	NA
1984–1992	No rainwater harvesting or groundwater development programs	0
1992–1997, VIII	Maharashtra, Karnataka, Andhra Pradesh, Delhi, Kerala, Madhya Pradesh, Tamil Nadu, West Bengal & Chandigarh (Total States/UT – 9)	32.3
1997–2002, IX	Andhra Pradesh, Arunachal Pradesh, Assam, Bihar, Chandigarh, Gujarat, Haryana, Himachal Pradesh, Jammu & Kashmir, Jharkand, Kerala, Lakshdweep, Madhya Pradesh, Maharashtra, Meghalaya, Mizoram, Nagaland, NCT Delhi, Orissa, Punjab, Rajasthan, Tamil Nadu, Uttar Pradesh and West Bengal (Total States/UT – 25)	331
2002–2007, X	Andhra Pradesh, Karnataka, Madhya Pradesh and Tamil Nadu (Total States – 4); Pilot projects 18; 197 structures	56
2007–2012, XI	Arunachal Pradesh, Punjab, Tamil Nadu, Kerala, Karnataka, West Bengal, Andhra Pradesh, Uttar Pradesh, Madhya Pradesh, Delhi, Chandigarh, Gujarat, Maharashtra, Jharkhand, Himachal Pradesh, Jammu & Kashmir, Orissa, Rajasthan and Bihar (Total States/UT – 19) Pilot projects 82; 1475 structures	1,000

*Costs in [Million EUR]: VIII Plan = 0.39; IX Plan = 3.97; X Plan = 0.67 and XI Plan = 12.30. (Average currency exchange rate of year 2014: INR EUR = 0.0123; Online Currency Converter, 2015).

During the plan period 2007–2012, 82 pilot projects with a total of 1,475 structures were to be constructed in areas which are marked by declining groundwater level, in coastal areas and on islands affected by saline water ingress, in areas of inland salinity, in urban areas showing steep decline in groundwater levels and in sub-mountainous/hilly areas of the country. Since 1972 and increasingly since 1997 (Table 5.5) all common types of structures such as check dams, percolation ponds/tanks, subsurface dykes, rooftop rainwater harvesting, recharge wells and shafts and others were financed, documented and evaluated by the CGWB. In the last five years the structures financed by the CGWB with the purpose of "demonstrating artificial recharge and rain water harvesting techniques in overexploited and critical areas, urban areas and areas affected by water quality" (CGWB, 2012).

Based on the experience acquired in the pilot programs the CGWB strives to contribute to large scale implementation of MAR. The CGWB Perspective Plan for Artificial Recharge (1996) estimated the non-committed surplus monsoon run-off available for recharge in India by adding data from different basins (872 km^3). Furthermore the sub-surface storage potential available on saturation of the vadose zone up to 3 meters below ground level was calculated (590 km^3). By selecting the lowest of those two values for each basin the "feasible groundwater storage" was calculated (234 km^3). This is the amount of water which is available in the basin and for which there is also storage potential in the basin.

A Master Plan for Artificial Recharge to groundwater (CGWB, 2002) was prepared and approved by the Ministry of Water Resource on the basis of hydrogeological parameters and hydrological data available for each state. The identification of feasible areas for artificial recharge to groundwater was made on the basis of depth and declining trend of groundwater levels. The plan provides information about area specific artificial recharge techniques to augment the ground water storages based on the availability of source water and the capability of subsurface formations to accommodate it. As a part of the Master Plan of 2002, a number of demonstration projects were implemented between 2007 and 2012 as mentioned above. In 2013 a revised Master Plan was published (CGWB, 2013). The revised master plan of 2013 identifies more than twice as much water to be recharged as the original Master plan of 2002 (Table 5.6). It remains to be seen if the huge amounts of funds needed for this plan (see section 5.3.2) can be raised and whether the volumes of the revised plan of 2013 or only those of the Master plan of 2002 can be recharged.

Table 5.6 List of structures proposed under the master plan of 2013 (Values of 2002 in brackets) (CGWB, 2002; and 2013).

Structures Proposed under the Master Plan of 2013	Values
Area Identified for Artificial Recharge [km^2]	942,000 (449,000)
Volume of water to be recharged [km^3]	85.6 (36.5)
Number of structures in rural areas	2,283,000 (225,000)
Number of structures in urban areas (rooftop rainwater harvesting)	8,799,000 (3,700,000)
Total number of structures proposed	11,082,000 (3,925,000)
Total cost of structures proposed [Million INR]	792,000 (245,000)*

*Costs in [Million EUR]: 9,741.6 (Master Plan 2013) and 3,013.5 (Master Plan 2002). (Average currency exchange rate of year 2014: INR EUR = 0.0123; Online Currency Converter, 2015).

5.3.2 Implementation schemes

Right from the ancient days, canals, ponds, anicuts and reservoirs have been dug and constructed in India to improve the water availability. There are numerous examples and stone inscriptions from as early as 600 A.D. citing that ancient kings and other benevolent persons considered construction of small ponds to collect rainwater which also assisted increasing groundwater recharge. Traditionally each village had a pond to store surface run off and to augment groundwater recharge. Most of the temples had a tank which also serves as a structure for groundwater recharge.

Over the last few decades several initiatives have been taken to improve the groundwater potential by increasing the rainfall recharge. By now India counts innumerable structures mainly in peninsular India (0.5 Million) (Sakthivadivel, 2007) or even 1.5 Million (Pandey *et al.* 2003). Several agencies in India provide financial support for constructions which facilitate improvement of groundwater conditions. These agencies are from both government and non-governmental sectors. Several Departments/Boards under the Ministry of Water Resources and Ministry of Rural Development fund groundwater recharge related projects (Table 5.7).

Table 5.7 Main features of some important programs of the Government of India (GoI) involving MAR.

Year	Name of the Program	Financing Organization	Budget*	Additional Info
1995–	Integrated Watershed Management Program (IWMP)	Ministry of Rural Development, GoI	INR 43,616 million released until 2012 (EUR 536.4 million)	All states. 1,900 projects covering 107,000 km^2 were financed until 2012.
2007–2012	Repair, Renovation and Restoration (RRR) scheme	Ministry of Water Resources, GoI	INR 60,000 million (partly local government; EUR 738 Million)	Planned were 23,000 water bodies for irrigation of 17,000 km^2. One of ten goals was MAR.
2005–2009	Bharat Nirman	Ministry of Rural Development, GoI	INR 223,992 million (EUR 2,755.1 million)	Only a minor part is related to water. 28% of irrigation capacity shall be crated from groundwater and 10% from the RRR scheme mentioned above (out of total of 100,000 km^2). Two investment areas (irrigation and drinking water) out of six are related to Groundwater/MAR.
2008–	Artificial Recharge of Groundwater through Dugwells	MGNREGA Ministry of Rural Development, GoI	INR 17,987 million (EUR 221.2 million)	Seven states are involved. 4.5 million dug-wells proposed.

*(Average exchange rate of year 2014: INR EUR = 0.0123; Online Currency Converter, 2015).

The Department of Land Resources have integrated and consolidated three programmes namely, Drought Prone Areas Programme, Desert Development Programme and Integrated Wastelands Development Programme (IWDP) into a single modified programme called Integrated Watershed Management Programme (IWMP). The major activities of this project include rainwater harvesting activities like farm ponds, percolation tanks, check dams etc. The projects under the programme are being implemented in 470 districts in all 28 states of the country. From 1995 to 2007, 1,877 IWDP projects covering an area of 107,000km^2 have been sanctioned. A total number of 770 projects covering an area of 49,000 km^2 were completed by 2011. Other projects are at various stages of implementation in different States. Central funds to the tune of INR 43,616 million (EUR 536.4 million[2]) have been released by December 31, 2012 (Ministry of Rural Development, 2012).

The Ministry of Water Resources (2009) writes "In India, tanks/ponds and lakes have traditionally played an important role in irrigation, drinking water supply, hydropower, ecology, tourism/culture and domestic use. Relative importance of some of these water bodies has waned due to a number of reasons such as shifting away from community based tank system to individual beneficiary oriented ground water dependent system, encroachments, silting, population pressure, multiplicity of agencies responsible for their upkeep, etc."

The Repair, Renovation and Restoration (RRR) scheme was introduced in 2005 in order to restore these bodies, and one of the ten goals is groundwater recharge. The scheme is financed partly by the central government (in most states 25%; in some states 90%) and partly by the state governments. A pilot phase (INR 3,000 million from the central government) was followed by a regular phase (Total project cost projected INR 60,000 million from the central and local government) for the period 2007–2012. The scheme pertains to the restoration of 23,000 water bodies in almost all states with a target to create 17,000 km^2 of additional irrigation potential (Ministry of Water Resources, 2009).

The RRR scheme is part of the Bharat Nirman program. Bharat Nirman is covering improvement of rural infrastructure and two out of six parts are related to MAR, namely additional irrigation for 100,000 km^2 and drinking water supply for 55,000 habitations. Out of the 100,000 km^2 additional irrigated land, at least 28,000 km^2 should be irrigated with groundwater and 10,000 km^2 with water from the RRR scheme (Ministry of Water Resources, 2012). The additional water demand will be drawn partly from existing groundwater potential, but likely additional potential will also be created (MAR). Bharat Nirman was launched by the Ministry of Rural Development in 2005/2006. Under Bharat Nirman Phase I (2005 to 2009), funds utilized were INR 223,992 million (EUR 2,755 million) (Ministry of Rural Development, 2010). The National Rural Drinking Water Programme was performed with the objective to move away from over-dependence on a single drinking water source to multiple sources through conjunctive use of surface water, groundwater and rainwater harvesting; ensure sustainability in drinking water schemes.

The Mahatma Gandhi National Rural Employment Guarantee Act (MGNREGA) is supported under the Ministry of Rural Development, GoI. The activities that are supported under this act include water conservation, water harvesting and renovation of traditional water bodies among other things. Under the MGNREGA, a scheme on 'Artificial Recharge of Groundwater through dugwells' was launched in the year 2008 with a total outlay of INR 17,987 million (EUR 221.2 million), including subsidy component of INR 14,993 million (EUR 184.4 million). This project was implemented in seven states to recharge the existing dugwells, improve groundwater storage, increase the sustainability of groundwater during lean periods and improve the overall agricultural productivity. The total number of irrigation dug wells proposed for recharge is 4.45 million (Ministry of Water Resources, 2010).

The cost of the revised Master plan of the CGWB can be compared to the existing programs. CGWB proposes an implementation of the master plan (INR 792,000 million; EUR 9,741.6 million) over 10 years with joint financing in similar shares by the Ministry of Water resources, MGNREGA, stakeholder industries and state governments. The annual expenditure for such a plan would far exceed the sum of the water-related parts of the programs in Table 5.5. The implementation would thus require a complete change in Indian water policy.

Rainwater harvesting has been made mandatory in several cities and in some states of India with the aim to meet the increasing groundwater needs. A National Bank for Agriculture and Rural Development (NABARD) project is aiming at water resource conservation and management by rooftop rainwater harvesting (NABARD, 2012).

5.4 STATE-OF-THE-ART OF MAR IMPLEMENTATION IN INDIA

Although MAR has been implemented in millions of places in India, published results on the performance in terms of quantity (infiltration rates) and quality are scarce. In total, 27 publications as of March 2012 were found dealing with MAR and documenting field studies with quantitative data on different scales:

- 13 publications reported on field studies with less than 5 recharge structures,
- 8 publications gave examples of groups of structures with more than 5 and less than 100 recharge structures,

[2]Average currency exchange rate of year 2014: INR EUR = 0.0123 (Online Currency Converter, 2015). All amounts indicated in EUR were calculated with the same currency exchange rate.

- 2 publications gave overviews of recharge structures on a regional level with more than 100 structures,
- One publication took a theoretical approach only (groundwater modelling), and
- In 3 publications, the number of structures was not given.

The structures investigated can be categorized as given in Figure 5.6. In the small and medium scale investigations, recharge or injection wells represent the majority of investigated structures, whereas for large scale investigations most reported structures are check dams (incl. nala bunds and contour trenches) that were also studied in the small scale investigations to a considerable extent.

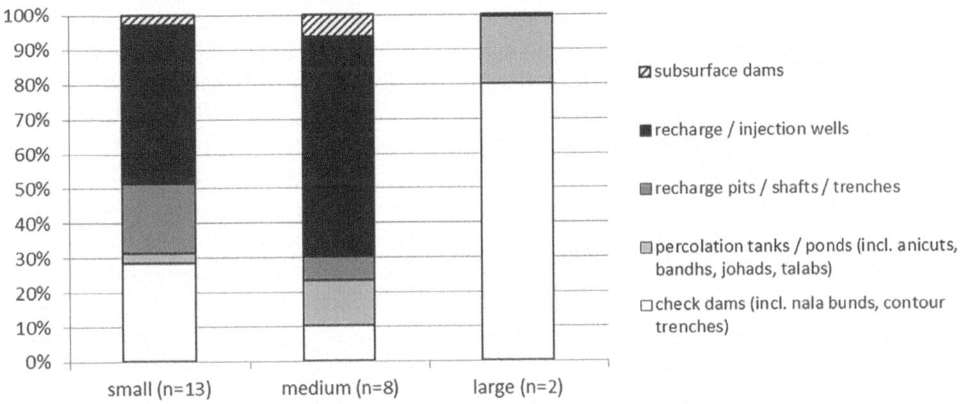

Figure 5.6 Aquifer recharge structures in publications as on March, 2012 on field studies with quantitative results. The field studies are divided in small (<5 structures), medium (5 to <100 structures) and large scale (>100 structures).

The abstraction is mainly carried out by bore-wells or dug wells, either with hand pumps or electrically equipped. The recharged water is usually used for irrigation, but in 10 of the 27 field studies with quantitative results domestic or drinking water use is also mentioned. Three urban case studies deal with water recharged for drinking water purposes only (Hyderabad, Bangalore and Chennai all mentioned in (UNESCO, 2006)). These are, however all direct rainwater harvesting structures from rooftops.

In the following sections the findings from field studies and related literature are reported in relation to the different activities related to MAR. Broadly MAR activities can be divided in:

- Planning and construction
- Operation and maintenance

After the need for modified or additional MAR structures has been quantified, the planning and construction of a structure can be addressed. The knowledge necessary for planning of a structure can be summarized as follows based on listings from Kumar *et al.* (2008) and the CGWB (2000: p. 52):

- Source water availability (see below)
- Topography
- Properties of soil
- Hydrogeological data (see below)
- Surface and groundwater quality over time (see below)

These factors are generally measured more precisely in the planning phase. Once these factors are known, the suitable structures for different topography, hydrogeology and rainfall and their percolation efficiency are quantified as given by CGWB (2000: p. 100).

5.4.1 Source water availability

The CGWB recommends using rainwater, run-off or treated waste water for recharge (CGWB, 2007: p. 15). For determining the availability of rainwater Kumar *et al.* (2008) show the importance of determining the rainfall distribution over the years, especially in arid regions. High rainfall in some rarely occurring years in these regions can only be captured with over-dimensioned structures. These will therefore only be partly utilized in most years and consequently have low percolation efficiency (volume of infiltrated water in relation to volume of the structure).

The run-off is calculated based on average rainfall, soil infiltration properties and topography (Zade *et al.* 2005). However, MAR activities may capture water that is planned to be utilized downstream, resulting in reallocation of water between users with little or no additional benefit (Kumar *et al.* 2008). Kumar *et al.* (2008) reported reduced inflow into the Ghelo-Somnath reservoir (Gujarat) because of intensive water harvesting in the upstream catchment of the reservoir. The authors calculated rainfall – run-off regression lines for the pre-/ and post- MAR intervention period. According to this calculation the rainfall amount which is needed to fill the reservoir increased from 320 mm/year to 800 mm/year. Rama *et al.* (2003) studied the redistribution of surface run-off in small catchments in Andhra Pradesh and Karnataka before, during and after groundwater recharge initiatives. The authors found strong evidence that extensive MAR interventions resulted in decreased run-off generation and, thus, reduced flow captured in the traditional water tanks situated downstream. This effect could be attributed to MAR interventions, and other factors such as deforestation or reduced rainfall could be ruled out (Rama *et al.* 2003). Once the run-off is captured in MAR structures, this water either evaporates or recharges the aquifer and is then pumped for irrigation. It is not clear if this reallocation of water, from traditional tank supply to decentralized groundwater recharge gives an additional value to the local communities. Run-off in urban areas, often referred to as storm water, is increasingly captured by rooftop rainwater harvesting schemes as mentioned in section 5.3.2 (CGWB, 2011a).

According to DK Chadha, former Chairman of the CGWB (Chadha, 2012), treated wastewater was not used up to now, partly because no quality guidelines for source water exist. However, in order to increase the available amounts of water, treated wastewater could also be considered for recharge. Treatment could take place in conventional wastewater treatment plants, constructed wetlands or soil aquifer treatment (SAT) type systems and need to be coupled with quality control to avoid contamination of aquifers. SAT is evaluated in India (Nema *et al.* 2001) and practiced in other countries, i.e. Israel, Australia and USA, with promising results (O'Connor *et al.* 2008). Negative social and religious views on applying treated wastewater for irrigation or drinking water purposes is often stronger than rational arguments based on water quality and risks. Using it for MAR transforms the water to more neutrally perceived groundwater and thus might be a way to overcome these reservations.

Finally, water from other catchments can be transported by canals over long distances. Some major projects have been implemented and others are planned (Central Water Commission, 2009). This is a costly option and might be considered in basins where no other sources are available.

5.4.2 Hydrogeological data

The 27 field studies with quantitative results cover a wide variety of natural settings: the average annual precipitation varied between 612 mm (Moga, Punjab: Bassian Drain, Block Nihalsisngh Wala (CGWB, 2011b)) and 1,788 mm (Balasore district + Field Site, Orissa (Hollaender *et al.* 2009)), with high inter-annual variations (long-term average minima: 331 mm; maxima: 1,424 mm reported for Delhi (UNESCO, 2006)). For those case studies for which hydrogeological information were available (20 studies), 10 were situated in a hard-rock environment (granite, gneiss, basalt) where the aquifer would probably be situated in the weathered/fractured zone or in alluvial deposits covering the hard-rock. The other hydrogeological settings can be summarized as sedimentary, mainly unconsolidated rocks, usually gravel or sand with sections of clay. The information on aquifer thickness, depth of the groundwater level or transmissivities is scarce (three, six and four case studies respectively report information on these parameters). Well yield is, however, a parameter that is frequently given in the publications. In sedimentary formations well yields vary between 1 and 115 m^3/h with the highest values in alluvial aquifers (Bhadrak, Orissa (CGWB, 2011b)) and the Tapi alluvial Belt Maharashtra (Jain, 2006). Maximum well yield in hard-rock environments, on the other hand, is always below 14 m^3/h (Deccan traps, Maharashtra (Jain, 2006)) and usually lie between 0.8 and 4 m^3/h. These figures give an idea of the hydraulic permeability encountered, but as data on draw-down and well design is lacking, quantitative information on specific capacities or transmissivities cannot be derived.

Aquifer properties as part of the hydrogeological data define the amount of water which can be infiltrated and stored in the aquifer. India's aquifers are broadly comprised of three groups of rock formations of different hydraulic properties (CGWB, 2006): unconsolidated porous, semi-consolidated porous and consolidated fissured formations. Unconsolidated formations have high transmissivity, hydraulic conductivity (Table 5.8) and also high storativity. They can thus rapidly absorb and store large amounts of water per unit volume, which make them well suited for MAR. The high transmissivity leads to high groundwater flow, redistributing water within the aquifer away from the infiltration point and along topographical gradients, which is not always desirable.

Table 5.8 Properties of aquifers in different groups of rock (Groundwater Estimation Committee, 2009).

Formation	Area Fraction [%]	Transmissivity [m²/d]	Hydraulic Conductivity [m/d]
Unconsolidated	30	250–4,000	10 to 800
Semi-consolidated	7	100–2,300	0.5–70
Consolidated	63	10–500	0.05–15

5.4.3 Surface and groundwater quality over time

The CGWB recommends use of rainwater and run-off or treated wastewater for recharge. As mentioned above recharge of wastewater is not practiced and it is commonly assumed that the source water is impure. This assumption is not always valid since the run-off may flush out accumulated contaminants on the way to the recharging point. Once recharged, the water will undergo changes in quality during the underground passage. Quality parameters of source water and known positive and negative effects from India and elsewhere are reviewed below.

Pathogens

Generally, subsurface passage is an effective medium for microbiological removal (Sharma *et al.* 2011). In case of sufficient flow path length and residence time during the subsurface passage, microbial contamination will be attenuated by physical straining and inactivation (or die-off) to levels below drinking water standards. Pathogens are critical for bank filtration systems because of the often short residence times. Bank filtration can achieve, under optimal conditions, several log removal over distances of few tens of meter travel distance for viruses (Tufenkji *et al.* 2002). Higher removal can be expected for larger particles i.e. protozoa and bacteria. For example the natural capacity for attenuation in bank filtration and lake filtration in Delhi and Naini Tal in India was shown to be effective and no breakthrough of bacteria was measured (Sprenger *et al.* 2008; Dash *et al.* 2008). Some countries have established a minimum subsurface travel time for recharge water (i.e. Germany 50 days) to ensure a certain removal. Thus it is important that sufficient travel times of contaminated surface water to wells and consequently a sufficient distance (e.g. >20 m, depending on geology, temperatures and water) to a recharge site are assured to avoid a pathogen breakthrough. In the Indian context a study of soil aquifer treatment of Nema *et al.* (2001) evaluates attenuation potential in relation to aquifer recharge.

Organic chemicals

Different schemes of MAR were found to remove organic trace contaminants, including pesticides, personal-care products, endocrine-disrupting compounds, and pharmaceutical active compounds to varying extents (Sharma *et al.* 2011; Maeng *et al.* 2010). Many of these substances are toxic, carcinogenic or suspected to be endocrine disruptors and therefore considered not only hazardous to the ecosystem but also to human health. The removal of these micro pollutants during subsurface passage depends on several factors such as: concentration level of the contaminant, redox conditions (Massmann *et al.* 2006; Patterson *et al.* 2002; Pavelic *et al.* 2005), residence time and the occurrence of organic matter in the aquifer (availability of electron donors) rather than the travel time (Schmidt & Lange, 2006). Removal capacity is very site-specific and general predictions are difficult to give. Anyhow, minimum travel time for a 30% removal of pharmaceutically active compounds is estimated to be at least 75 days (Maeng *et al.* 2010). Many MAR sites are characterized by the occurrence of a more or less developed redox sequence, providing oxic and anoxic conditions which in turn leads to the removal of many redox sensitive micro pollutants.

In India the data on organic micropollutants in the environment are limited. Warren *et al.* (2002) described the fate of organic contaminants (Lindane, benzo(a)pyrene) at the Rihand reservoir (Uttar Pradesh) and developed a mass balance model. Shukla *et al.* (2005) analysed the organochlorine pesticide contamination in groundwater in Hyderabad and detected several pesticides exceeding drinking water standards set by European countries.

Sampling in six bore wells in Chennai show fairly high quality compared to the WHO guidelines in terms of physical/chemical parameters (WHO, 2008). Out of eight chlorinated pesticides analyzed only Atrazine was detected in low concentrations (ng/L). On the other hand, various types of water borne pathogens were detected in all samples except those from a sealed well (Saph Pani Deliverable 4.3, 2014).

Samples from the Yamuna river in Delhi show presence of 12 respectively 18 of 39 selected micropollutants including pharmaceuticals and artificial sweeteners (Table 5.9). Further samples from bank filtration wells nearby show attenuation during the soil passage (Saph Pani Deliverable 4.4, 2014).

Table 5.9 Attenuation of 39 selected micropollutants in the Yamuna river and in nearby wells. Travel speed has been estimated to 0.9 m/d (Sprenger, 2011).

	Surface Water	Groundwater 200 m Distance	Groundwater 500 m Distance
Sum of concentrations July 2013 [ng/L]	11,133	2,393	1,123
Number of detected compounds July 2013	18	7	11
Sum of concentrations December 2013 [ng/L]	728	262	97
Number of detected compounds December 2013	12	4	3

Ionic contamination

As reported by CGWB (2012) the groundwater in numerous areas is unsuitable for drinking because of mineral contaminants such as fluoride, nitrate, arsenic or mineral salts. Indian researchers have identified mechanisms and sources for fluorine contamination (Rao, 2009; Reddy *et al.* 2010) and a recent review identifies filtration with magnesia as a suitable post-treatment in rural areas (Ibrahim, 2011). Reddy *et al.* (2011) showed how animal and human excrements can lead to rapid nitrate contamination under undiluted circumstances. Arsenic was shown to accumulate in shallow aquifers after desorption from sediments in Bengal rice cultivation (Farooq *et al.* 2010) and lakes (Acharyya & Shah, 2007). Pawar *et al.* (1998) showed the importance of protecting the aquifer from industrial effluents by analyzing the polluting effects of saline effluents of a sugar mill, whereas Garduno *et al.* (2011) listed contamination coming from industrial point sources, as well as geogenic contamination and agriculture all over India.

Mineral contaminants are in many cases present in the aquifer. Possible goals of MAR can be not to mobilize them or even to stabilize them by acting on the ion content and the redox potential of the infiltrated water. MAR can also help dilute mineral contaminants or provide pockets of water suitable for drinking in otherwise contaminated aquifers. And obviously a primordial goal would be to infiltrate water that does not add to the contamination.

In 11 of the 27 field studies with quantitative results, water quality information is given and mineral contamination is always a concern. In many cases it is not clear which issues are attributed to the influence of MAR and which are due to the background hydrochemistry of the groundwater. Stiefel *et al.* (2009), for example, investigated the qualitative impact of a check dam in Rajasthan and found only positive effects of the infiltrated water on ambient groundwater quality.

Salinity has been reported to be a problem in the state of Haryana (Malik *et al.* 2006) and in Chennai City, Tamil Nadu (UNESCO, 2006). In the first example a clear improvement was observed after the construction of 5 ASR wells (decrease in electrical conductivity from 9,000 to 1,500 μS/cm).

In other cases it is clearly stated that the implementation of MAR has led to an improvement of groundwater quality through dilution (Sivakumar *et al.* 2006; Sayana *et al.* 2010; Kaledhonkar *et al.* 2003). This was indicated by reduced levels of nitrate (112 ppm to 65 ppm (UNESCO, 2006)), fluoride (according to the CGWB (2011) values of >1.8 mg/L were reduced to <1 mg/L), hardness and sulphate).

On the other hand, Dwarakanath (UNESCO, 2006) reports an increase in potassium, chloride and fluoride due to MAR, though still within acceptable limits. Generally, elevated nitrate concentrations seem to be a problem: values above the permissible limit of 45 mg/L were reported in the Satlasana (Gujarat) and Coimbatore (Tamil Nadu) case studies (Gale *et al.* 2006) as well as in the vicinity of the Raj Bahwan premises (Bhubaneswar, Orissa) according to the CGWB (2011b). A connection to MAR is not clear and Gale *et al.* (2006) postulated agricultural influence.

To our knowledge, investigations on arsenic concentrations in artificially recharged groundwater are lacking, though implementation of MAR has been suggested to be a possible countermeasure in case of elevated concentrations in the groundwater (CGWB, 2011b).

Experience from case studies on SAT in India

Under Indian conditions only few studies of wastewater treatment using SAT technology exist. Primary treated municipal wastewater was used at the Sabarmati Riverbed in Ahmedabad (Nema *et al.* 2001). The authors found that SAT showed good removal of organic pollutants, nutrients and bacteria and was more efficient and economic than conventional wastewater treatment systems. Based on this pilot study a conceptual design of a 55,000 m^3/d SAT system using primary settled domestic wastewater was proposed for the city (CGWB, 2011b).

5.4.4 Infiltration rate and prevention of clogging

The CGWB (2011b) classifies a large number of case studies as success stories with respect to their impact on local groundwater level and/ or increased well yield. Annual volumes recharged per recharge structure ranges from 2 m^3 per m trench (Bhubaneswar, Raj Bhawan premises) to 24,000 m^3 per well (Bhadrak, Orissa) but are difficult to compare due to diverse hydrogeology, varying precipitation rates and a multitude of studied structures. Reported increase in groundwater level range from 0.2 to 1 m, but in some cases the number of abstraction wells has also increased considerably (18 additional wells resulting from the installation of 2 trenches and 3 recharge wells in Moga, Punjab Bassian Drain). The CGWB evaluated the performance of different MAR structures in different hydrogeological and meteorological contexts based on data from numerous pilot studies (section 5.3). The results were thoroughly documented (Chadha, 2012). Benchmark performances (e.g. 75% percolation efficiency (CGWB, 2013) and the suitability of structures for different contexts (CGWB, 2000: p. 100) were published. However, most results unfortunately remain inaccessible to the research community. Although a large amount of information on MAR systems in the different Indian states was found, it is difficult to derive general trends and transferable recommendations due to the above mentioned variability and lack of detailed scientific data.

The quantification of the recharged water is in the focus of the 27 field studies with quantitative results. This is either done by small scale observations (measuring water table fluctuations) or on catchment/sub-catchment scale.

Perrin *et al.* (2010), for example, balanced the volume of different percolation tanks and the evapotranspiration and concluded that between 5% and 8% of the monsoon rainfall (20 to 40 mm per annum) was infiltrated from these tanks on a small catchment scale 73% of the rainfall was lost to evaporation, leading to the conclusion that enhancing infiltration at existing structures (e.g. by desilting or pre-treatment) should be preferred to constructing new ponds. Both Perrin *et al.* (2010) and Palanisami *et al.* (2006) reported 90% and more of the rainfall was captured by the recharge structures – with potential negative effects for downstream users but beneficial to the water balance inside the (sub-) catchments. The amount of water evaporated in the study by Palanisami *et al.* (2006) was reported to be around 15% and thus significantly less than the 73% found by Perrin *et al.* (2010), most probably due to higher infiltration rates (percolation efficiency around 85%). For this reason, also the residence time of the water in the structures may be considerable: Gale *et al.* (2006) reported a surface water residence time of 5 months at a check dam in Coimbatore (Tamil Nadu).

Percolation efficiency, as the volume of infiltrated water in relation to the volume of a recharge structure can vary quite considerably. For some case studies, like one on check dams in Gujarat (Gale *et al.* 2006) efficiencies of >90% were reported whereas others give efficiencies below 20% (different structures on catchment scale in Rajastan reported by Glendenning and Verwoort (2010)).

This is attributed to two different factors:

- *The permeability of the subsurface*: infiltrated volumes of up to 1000 m^3/d were observed at gravity injection wells in a canal in Haryana, located in a coarse gravel aquifer (Kaledhonkar *et al.* 2003) – corresponding to infiltration rates of >10 m/d, whereas infiltration rates of a few centimetres per day are common for percolation tanks, check dams or trenches in weathered hard-rock areas (CGWB, 2011b; Perrin *et al.* 2010; Gale *et al.* 2006).
- Clogging of the recharge structure through high amounts of suspended solids (according to Palanisami *et al.* (2006) desilting improved the percolation efficiency from 83% to 87% in check-dams in Coimbatore and Hollaender *et al.* (2009) give clogging of ASR wells as a major issue, with TSS values of 800 mg/L even after pre-treatment for a field site situated in Balasore.

MAR measures often result in the development of a clogging layer at the area of recharge. The clogging layer has a lower hydraulic conductivity than the surrounding aquifer material and decreases infiltration rates. Clogging can be of physical (air entrapment in the aquifer, deposition of suspended solids), chemical (mineral precipitation, e.g. iron oxides) or biological nature (accumulation of organic matter). Physical clogging, e.g. silting, may be managed by treatment of the recharge water by simple sedimentation and filtration to remove suspended solids as described below. Chemical clogging of wells may be managed by frequent mechanical or chemical cleaning such as brushing or application of mild acids, respectively (McLaughlan, 1996). Periodic cleaning and re-development only delay the ageing process of the well. Biological clogging in ponds is often a result of algae die-off and can be managed by frequent removal and washing of the uppermost infiltration layer (Greskowiak *et al.* 2006). Algae growth and other biological clogging are reduced by minimizing nutrients (nitrogen and phosphorous) and organic carbon in the source water. This is in particular true where sewage is part of source water. Chlorine disinfection or other disinfectants with residual effects reduces biological activity at the infiltration interface. Finally, the clogging rates also depend on the infiltration rate, because with high infiltration rates higher amounts of nutrients and suspended solids arrive at the infiltration surface.

Generally silting is seen as a problem for MAR, especially for check dams or similar structures (Gale *et al.* 2006; Palanisami *et al.* 2006) and percolation tanks (Perrin *et al.* 2010). Chakrapani and Saini (2009) found that >75% of the annual sediment load was transported during the monsoon season. Pre-treatment is widely used, either through sedimentation tanks (UNESCO, 2006), sand filters (Kaledhonkar *et al.* 2003; Sivakumar *et al.* 2006; Tuinhof and Heederik, 2003) or metal screens (Kanhe & Bhole, 2006). Hollaender *et al.* (2009), for example, used different setups of gravel and rice straw to filter monsoon storm water at an ASR site in eastern India. The authors achieved a total removal rate of 70–90%, but TSS was still around 800 mg/L (See also chapter 12: Pre- and Post-treatment of BF and MAR in India: Present and Future). Panda (2002) tested gravel filters and embedded coconut fiber mats and achieved concentrations around 180 mg/L. Only one case study was found, in which silting did not seem to pose a problem: In ASR cavity wells in Haryana (northern India) the high TSS load (900 mg/L) did not result in reduced injection rates. This is attributed to a postulated process of flocculation of silt and particles that may then settle on the surface of the cavity and are then pumped back to the surface once the recovery cycle commences (Malik *et al.* 2006).

5.4.5 Maintenance of the structure and the surrounding area

Land-use activities in the vicinity of MAR structures need to be part of routine maintenance. For example a check dam constructed in 1975 collapsed in 1994 due to uncontrolled sand mining in the riverbed and the adjacent areas (Charalambous & Garratt, 2009).

MAR interventions reduce erosion, which is in general considered positive. However, prevention for movement of sediments with run-off and with river water may lead to reduced sediment influx into the sea, which may alter the erosional and depositional dynamics of the coast. To the best of our knowledge, studies on the impacts of MAR interventions such as check dams on changes in river sediment load and coastal stability in the Indian context are lacking.

5.5 CONCLUSION

In most parts of India, the monsoon lasts for about four months, followed by a dry period of about eight months. This rainfall pattern imposes huge seasonal variation in water availability. Aquifer recharge has been practiced to a large extent and for a long time to recharge the groundwater and ensure access to water all year. Experience with structures and groundwater management has developed in and been adapted to the various climatic and hydrogeological situations of India which is reflected by their variety and their presence at many historic sites.

India's use of groundwater has increased rapidly over the last fifty years. Today, groundwater covers about 43% of India's water demand. The demand exceeds the supply in many areas which leads to sinking groundwater tables. MAR is only one factor influencing the water balance. For example, water use for irrigation is of greater importance and consequently the type and the number of crops and the irrigation methods have a greater impact than MAR. Additional recharge through MAR could only make a minor contribution to the overall water balance (2%) according to rough estimations in the recent Master plan by CGWB (2013). However, it might contribute substantially (27%) to the groundwater consumption and relieve the situation in regions where water deficits exist. The Master plan of the CGWB (2013) proposes an additional 2.3 Million structures in rural areas to be constructed in the coming 10 years. There is no systematic inventory of existing structures today and numbers mentioned in literature range from 0.5 Million (Sakthivadivel, 2007) to 1.5 Million (Pandey *et al.* 2003).

The impact of aquifer recharge in the area on a watershed level and in India as a whole depends on the number of structures and also on their performance. But, from the review of field studies in literature, the quantitative scientific evidence for both positive and negative performance of MAR interventions is found scarce, an observation confirmed by Glendenning *et al.* (2012). As suggested by Glendenning *et al.* (2012), collection of quantifiable field data in combination with the development of new modelling tools is necessary in order to examine the wide range of potential positive and negative impacts of MAR measures on a watershed scale. In particular, as indicated by CGWB (2013: p. vii), data on the number, the performance and the effect of the existing structures would be necessary for future watershed management. This is especially important as the large amount of water used and the seasonal variations make it difficult to unambiguously identify non-committed water in a watershed. In several case studies new MAR structures recharge additional water upstream which is lacking downstream and lead to longer dry periods there.

Evaluation of the quantitative effect of recharge structures can show changes in performance over time. Monitoring of performance changes forms the decision basis for the operation and maintenance plans. MAR structures need regular maintenance to ensure stable long-term performance, but this is often lacking (UN Department of Technical Cooperation for Development, 1987; Palanisami *et al.* 2006; Gale *et al.* 2006; Glendenning *et al.* 2012; CGWB, 2013). This is in general more cost effective than construction of new structures and should thus be prioritized (CGWB, 2013: p. 195).

Up to now little attention was paid to the quality of recharge water; most of the Indian field studies do not measure quality of source water at all (section 5.4.3) and none of them measures pathogens. Up to now it was generally assumed that the used sources, rainfall and run-off, were safe to use (section 5.4.1). The potential positive and negative effects of MAR interventions on water quality are attracting increased interest. This is important, since almost all Indian districts have areas with nitrate contamination in the groundwater, and excessive concentration of other pollutants like arsenic, salt and fluoride is also widespread.

Generally, information on mixing ratios between naturally and artificially recharged water as well as travel times or redox conditions were found missing. In the case of critical parameters such as pathogens, fluoride or arsenic, this information could support the development of transferable guidelines for the safe implementation of MAR e.g. for drinking water supply.

This review covers knowledge and experience in India. It is mainly focused on how to plan, construct and operate MAR structures. There is also a larger context which is not treated, mainly social and economic factors. These need also be considered parallel to the technical aspects to find out whether an additional MAR structure is desirable, how to best organize the construction and maintenance and how to make best use of the recharged water.

5.6 REFERENCES

Acharyya S. and Shah B. (2007). Arsenic-contaminated groundwater from parts of Damodar fan-delta and west of Bhagirathi River, West Bengal, India: Influence of fluvial geomorphology and Quaternary morphostratigraphy. *Environmental Geology*, **52**, 489–501.

Allen R. G. (1998). Crop Evapotranspiration – Guidelines for Computing Crop Water Requirements. FAO Irrigation and Drainage Paper 56. http://www.fao.org/docrep/X0490E/X0490E00.htm (accessed 26 June 06 2012).

CGWB (1996). National Perspective Plan for Utilization of Surplus Runoff for Augmentation of Groundwater Resources in India. Report Central Groundwater Board, New Delhi, India.

CGWB (2000). Guide on Artificial Recharge to Ground Water. Report Central Ground Water Board, Ministry of Water Resources, Government of India, New Delhi, India.

CGWB (2002). Master Plan for Artificial Recharge to Ground Water. Report Central Ground Water Board, Ministry of Water Resources, Government of India, New Delhi, India.

CGWB (2006). Dynamic Ground Water Resources of India. Report Central Ground Water Board, Ministry of Water Resources, Government of India, New Delhi, India.

CGWB (2007). Manual on Artificial Recharge to Groundwater. Report Central Ground Water Board, Ministry of Water Resources, Government of India, New Delhi, India.

CGWB (2011a). Groundwater Scenario in Major Cities of India. Report Central Ground Water Board, Ministry of Water Resources, Government of India, New Delhi, India.

CGWB (2011b). Selected Case Studies, Rainwater Harvesting and Artificial Recharge. Report Central Ground Water Board, Ministry of Water Resources, Government of India, New Delhi, India.

CGWB (2012). Central Ground Water Board. http://cgwb.gov.in/aboutcgwb.html (accessed 25 June 2012).

CGWB (2013). Master Plan for Artificial Recharge to Groundwater. Report Central Ground Water Board, Ministry of Water Resources, Government of India, New Delhi, India.

Central Water Commission (2005). Water Resources Information System Data Book-2005. Report Central Water Commission, Ministry of Water Resources, Government of India, New Delhi, India.

Central Water Commission (2009). Water resources map of India.

Chadha D. K. (2012). Personal communication 15 August 2012.

Chakrapani G. and Saini R. (2009). Temporal and spatial variations in water discharge and sediment load in the Alaknanda and Bhagirathi Rivers in Himalaya, India. *Journal of Asian Earth Sciences*, **35**(6), 545–553.

Charalambous A. and Garratt P. (2009). Recharge abstraction relationships and sustainable yield in the Araniar Kortalaiyar groundwater basin, India. *Quarterly Journal of Engineering Geology and Hydrogeology*, **42**, 39–50.

Chaturvedi M. C. (1976). Water-Second India Studies. Macmillan Company of India Ltd, New Delhi.

Dash R. R., Mehrotra I., Kumar P. and Grischek T. (2008). Lake bank filtration at Nainital, India: Water-quality evaluation. *Hydrogeology Journal*, **16**, 1089–1099.

Deshpande N. and Singh N. (2010). Spatial and temporal variations in the occurrences of wet periods over major river basins in India. *J. Earth Syst. Sci.*, **119**, 561–578.

Dillon P., Gale I., Contreras S., Pavelic P., Evans R. and Ward J. (2009). Managing Aquifer Recharge and Discharge to Sustain Irrigation Livelihoods under Water Scarcity and Climate Change. IAHS Publications 330, Hyderabad, India.

European Environment Agency (2012). Average Annual Precipitation. http://www.eea.europa.eu/data-and-maps/figures/average-annual-precipitation (accessed 10 April 2012).

Farooq S., Chandrasekharam D., Berner Z., Norra S. and Stüben D. (2010). Influence of traditional agricultural practices on mobilization of arsenic from sediments to groundwater in Bengal delta. *Water Research*, **44**(19), 5575–5588.

Gale I. (2005). Strategies for Managed Aquifer Recharge (MAR) in Semi-Arid Areas. International Hydrological Programme, UNESCO, Paris.

Gale I. N., Macdonald D. M. J., Calow R. C., Neumann I., Moench M., Kulkarni H., Mudrakartha S. and Palanisami K. (2006). Managed aquifer recharge: An assessment of its role and effectiveness in watershed management. British Geological Survey Commissioned Report, CR/06/107N.

Garduño H., Romani S., Sengupta B., Tuinhof A. and Davis R. (2011). India, Groundwater Governance – A Case study. Report World Bank, New Delhi, India.

Glendenning C. J. and Vervoort R. W. (2010). Hydrological impacts of rainwater harvesting (RWH) in a case study catchment: The Arvari River, Rajasthan, India. Part 1: Field-scale impacts. *Agricultural Water Management*, **98**(2), 331–342.

Glendenning C. J., van Ogtrop F. F., Mishra A. K. and Vervoort R. W. (2012). Balancing watershed and local scale impacts of rain water harvesting in India – A review. *Agricultural Water Management*, **107**, 1–13. doi: 10.1016/j.agwat.2012.01.011.

Greskowiak J., Prommer H., Massmann G. and Nuetzmann G. (2006). Modeling seasonal redox dynamics and the corresponding fate of the pharmaceutical residue phenazone during artificial recharge of groundwater. *Environmental Science & Technology*, **40**(21), 6615–6621.

Groundwater Estimation Committee (2009). Report of the Groundwater Resources Estimation Committee, Ministry of Water Resources. Government of India, New Delhi, India.

Guhathakurta P. and Rajeevan M. (2006). Trends in the rainfall pattern over India. Report no. 2/200, National Climate Centre, India Meteorological Department, Pune, India.

Hollaender H., Mull R. and Panda S. (2009). A concept for managed aquifer recharge using ASR-wells for sustainable use of groundwater resources in an alluvial coastal aquifer in Eastern India. *Physics and Chemistry of the Earth, Parts A/B/C*, **34**(4–5), 270–278.

Ibrahim M. A. (2011). Effects of fluoride contents in ground water: A review. *International Journal of Pharmaceutical Applications*, **2**, 128–134.

India Infrastructure Research (2012). Water and Wastewater in India 2012. Report India Infrastructure Publishing Pvt. Ltd., New Delhi, India.

India Meteorological Department (1971). Scientific Report No. 136., India Meteorological Department, Government of India, New Delhi, India.

Indian Meterological Department (2004). Normal Annual Rainfall. http://www.imd.gov.in/section/climate/annual-rainfall.htm (accessed 26 June 2012).

Jain S. K. (2006). Artificial recharge to groundwater in an overdeveloped watershed – A case study of semi –arid region of Maharashtra, India. In: Final Report UNESCO G-WADI Meeting on Water Harvesting, Edmunds W. M. and Cardona C. (eds), UNESCO, United States, pp. 111–118.

Kaledhonkar M. J., Singh O. P., Ambast S. K., Tyagi N. K. and Tyagi K. C. (2003). Artificial groundwater recharge through recharge tube wells: A case study. *International Journal of Agricultural Research*, **84**, 28–32.

Kanhe N. M. and Bhole A. (2006). Study of Experiments on Roof Top Rainwater Harvesting and Surface Rainwater Harvesting for Recharging of Wells.

Kumar M., Patel A., Ravindranath R. and Singh O. (2008). Chasing a mirage: Water harvesting and artificial recharge in naturally water-scarce regions. *Economic and Political Weekly*, **43**(35), 61–71.

Kumar R., Singh R. and Sharma K. (2005). Water resources of India. *Current Science*, **89**(5), 794–811.

Maeng S., Ameda E., Sharma S., Gruetzmacher G. and Amy G. (2010). Organic micropollutant removal from wastewater effluent-impacted drinking water sources during bank filtration and artificial recharge. *Water Research*, **44**(14), 4003–4014.

Malik R., Yadav D., Jhorar B., Jhorar B. and Streck T. (2006). Physical and chemical interactions in an ASR cavity well in a brackish aquifer. In: Recharge Systems for Protecting and Enhancing Groundwater Resources, IHP/2006/GW-13, UNESCO, United States, pp. 121–126.

Massmann G., Greskowiak J., Dünnbier U., Zuehlke S., Knappe A. and Pekdeger A. (2006). The impact of variable temperatures on the redox conditions and the behaviour of pharmaceutical residues during artificial recharge. *Journal of Hydrology*, **328**, 141–156.

McLaughlan R. G. (1996). Water well deterioration, diagnosis and control. 2nd edn, National Centre for Groundwater Management, University of Technology, Sydney.

Ministry of Rural Development (2012). Annual Report 2011–2012. Ministry of Rural Development, Government of India, New Delhi, India.

Ministry of Water Resources (2009). Guidelines for Repair, Renovation and Restoration of Water Bodies with Domestic Support. Ministry of Water Resources, Government of India, New Delhi, India.

Ministry of Water Resources (2010). Press Information Bureau, GoI. http://pib.nic.in/newsite/erelease.aspx?relid=57646 (accessed 10 August 2012).

Ministry of Water Resources (2012). Ministry of Water Resources. http://mowr.gov.in/writereaddata/linkimages/arsenic14193577002.pdf (accessed 9 August 2012).

NABARD (2012). Model and Innovative Projects. http://www.nabard.org/ridf/model_rainwater.asp (accessed 15 September 2012).

Nema P., Ojha C., Kumar A. and Khanna P. (2001). Techno-economic evaluation of soil-aquifer treatment using primary effluent at Ahmedabad, India. *Water Research*, **35**(9), 2179–2190.

O'Connor G. A., Elliot H. and Bstian R. (2008). Degraded water reuse: An overview. *Journal of Environmental Quality*, **37**, 157–168.

Online Currency Converter (2015). http://www.freecurrencyrates.com/exchange-rate-history/EUR-INR/2014 (accessed 28 April 2015).

Palanisami K., Raviraj A., Thirumurthi S. and Sellamuthu K. (2006). Augmenting Ground Water Resources by Artificial Recharge AGRAR. Inception Report for the research site at Kodangipalayam village, Coimbatore District, Tamil Nadu, Water Technology Centre, Tamil Nadu, India.

Panda S. (2002). Rainwater Harvesting-cum-artificial Groundwater Recharge for Sustainable Water Resources Management. Report Submitted to Ministry of Human Resource Development, Department of Agricultural and Food Engineering, I.I.T Kharagpur, Kharagpur, West Bengal, India.

Pandey D. N., Gupta A. K. and Anderson D. M. (2003). Rainwater harvesting as an adaptation to climate change. *Current Science*, **85**(1), 46–59.

Parikh S., Shah T., Chadha D., Dubash N., Mohanty N., Jagawat H., Reddy P., Jha B., Moni M., Dhumal V., Mathew K., Palaniappan N., Takshi K. and Mishra A. (2007). Groundwater Management and Ownership – Report of the Expert Group. Planning Comission, Government of India, New Delhi, India.

Patterson B., Franzmann P., Davis G., Elbers J. and Zappia L. (2002). Using polymer mats to biodegrade atrazine in groundwater: Laboratory column experiments. *Journal of Contaminant Hydrology*, **54**, 195–213.

Pavelic P., Nicholson B. C., Dillon P. J. and Barry K. E. (2005). Fate of disinfection by-products in groundwater during aquifer storage and recovery with reclaimed water. *Journal of Contaminant Hydrology*, **77**(12), 119–141.

Pawar N. J., Pondhe G. M. and Patil S. F. (1998). Groundwater pollution due to sugar-mill effluent, at Sonai, Maharashtra, India. *Environmental Geology*, **34**, 151–158.

Perrin J., Massuel S. and Ahmed S. (2010). Contribution of percolation tanks to total aquifer recharge: The example of Gajwel watershed, southern India. Ismar Abu Dhabi. In: Achieving Groundwater Supply Sustainability and Reliability through Managed Aquifer Recharge, Proceedings of the Symposium ISMAR 7, California, United States, pp. 251–258.

Qader M. M. (2005). The ganges water diversion: Environmental effects and implications – An introduction. In: The Ganges Water Diversion: Environmental Effects and Implications, Qader M. M. and Singh V. (eds), Springer, The Netherlands, Vol. 49, pp. 1–12.

Rama M. S., Batchelor C. H., James A. J., Nagaraja R., Seeley J. and Butterworth J. A. (2003). Andhra Pradesh Rural Livelihoods Programme APRLP Water Audit Report, APRLP, Hyderabad, India.

Rao N. (2009). Fluoride in groundwater, Varaha River Basin, Visakhapatnam District, Andhra Pradesh, India. *Environmental Monitoring and Assessment*, **152**, 47–60.

Reddy D., Nagabhushanam P. and Peters E. (2011). Village environs as source of nitrate contamination in groundwater: A case study in basaltic geo-environment in central India. *Environmental Monitoring and Assessment*, **174**, 481–492.

Reddy D., Nagabhushanam P., Sukhija B., Reddy A. and Smedley P. (2010). Fluoride dynamics in the granitic aquifer of the Wailapally watershed, Nalgonda District, India. *Chemical Geology*, **269**(3–4), 278–289.

Rodell, M., Velicogna I. and Famiglietti J. (2009). Satellite-based estimates of groundwater depletion in India, *Nature* **460**, 999–1002.

Sakthivadivel R. (2007). The groundwater recharge movement in India. In: The Agricultural Groundwater Revolution: Opportunities and Threats to Development, Giordano M. and Villhot K. G. (eds), CAB International, Wallingford, United Kingdom, pp. 191–210.

Saph Pani D4.3 (2014). Post-treatment requirements of different natural treatment systems. *Saph Pani Project Deliverable*. http://www.saphpani.eu/downloads (accessed 19 August 2014).

Saph Pani D4.4 (2012). Recommended post-treatment options for water utilities employing natural treatment systems. *Saph Pani Project Deliverable*. http://www.saphpani.eu/downloads (accessed 19 August 2014).

Sayana V., Arunbabu E., Kumar L., Ravichandran S. and Karunakaran K. (2010). Groundwater responses to artificial recharge of rainwater in Chennai, India: A case study in an educational institution campus. *Indian Journal of Science and Technology*, **3**(2), 124–130.

Schmidt C. and Lange F. (2006). Ermittlung der Potenziellen Reinigungsleistung der Uferfiltration/Untergrundpassage Hinsichtlich der Eliminierung Organischer Schadstoffe unter Standortspezifischen Randbedingungen (Calculation of potential cleaning performance of bank filtration/subsoil with regard to the elimination of organic pollutants under site-specific conditions), Report BMBF-Project 02WT0280, BMBF-Projektverbund: Export-oriented research and development in the field of water supply and sanitation, Federal Ministry of Education and Research, Berlin, Germany.

Sharma S., Hussen M. and Amy G. (2011). Soil aquifer treatment using advanced primary effluent. *Water Science and Technology*, **64**(3), 640–646.

Shukla G., Kumar A., Bhanti M., Joseph P. and Taneja, A. (2005). Organochlorine pesticide contamination of ground water in the city of Hyderabad. *Environment International*, **32**, 244–247.

Singh D. and Singh A. (2002). Groundwater situation in India: Problems and perspectives. *Water Resources Development*, **18**, 563–580.

Sivakumar C., Elango L. and Gnanasundar D. (2006). Numerical modelling of groundwater flow in south Chennai coastal aquifer. In: MODFLOW and More 2006: Managing Ground-Water Systems, Poeter E., Hill M. and Zhen C. (eds), Colorado School of Mines, Colorado, USA, pp. 506–510.

Sprenger C., Lorenzen G. and Pekdeger A. (2008). Occurrence and Fate of Microbial Pathogens and Organic Trace Compounds at Riverbank Filtration Sites in Delhi, India. Techneau Integrated Project, D5.2.6, European Comission.

Sprenger C. (2011). Surface-/groundwater interactions associated with river bank filtration in Delhi (India) – Investigation and Modelling of Hydraulic and Hydrochemical Processes. PhD thesis, Department of Geosciences, Freie Universität Berlin, Berlin, Germany.

Tufenkji N., Ryan J. and Elimelech M. (2002). The promise of bank filtration. *Environmental Science & Technology*, **36**, 422A–428A.

Tuinhof A. and Heederik J. (2003). Management of aquifer recharge and subsurface storage. *Netherlands National Committee—International Association of Hydrogeology*, **4**(4), 1–94.

UN Department of Technical Cooperation for Development (1987). Hydrogeological and artificial recharge studies, Madras. Technical Report, UN Department of Technical Co-operation for Development, New York, United States.

UNESCO (2006). National Seminar on Rainwater Harvesting and Water Management. United Nations Educational, Scientific and Cultural Organization (UNESCO), Nagpur, India.

Warren C. S., Mackay D., Bahadur N. P. and Boocock D. G. (2002). A suite of multi-segment fugacity models describing the fate of organic contaminants in aquatic systems: Application to the Rihand Reservoir, India. *Water Research*, **36**(17), 4341–4355.

WHO (2008). Guidelines for Drinking-water Quality. Third Edition Incorporating the First and Second Addenda – Volume 1 Recommendations, World Health Organization, Geneva, Switzerland.

Zade M., Ray S., Dutta S. and Panigrahy S. (2005). Analysis of runoff pattern for all major basins of India derived using remote sensing data. *Current Science*, **88**(8), 1301–1305.

Chapter 6

Groundwater responses due to various MAR structures: Case studies from Chennai, Tamil Nadu, India

Raicy Mani Christy, Parimalarenganayaki Sundaram, Thirunavukkarasu Munuswamy, Thomas Lutz, Michael Schneider and Lakshmanan Elango

6.1 INTRODUCTION

Over-extraction of groundwater has resulted in seawater intrusion in coastal aquifers of many countries including India. Chennai, the fourth largest city in India, located on the coast of Bay of Bengal, is also affected by seawater intrusion. Furthermore, leaching of salts from marine deposits into the groundwater (Rao, 1959) and salt pan activities have caused high salinity in groundwater even at shallow depths. The current water requirement of Chennai city is met by desalination plants at Nemelli and Minjur, aquifers in Neyveli, Minjur and Panchetty, Cauvery water from Veeranam lake, the Krishna River from Andhra Pradesh, Poondi reservoir, and lakes at Red Hills, Chembarambakkam and Cholavaram (The Times of India, 2014). Rapid and heavy rains during short periods lead to a loss of large amounts of run-off to the sea and therefore natural rainfall recharge is very low. Recharge can be increased by different structures of Managed Aquifer Recharge (MAR).

As a part of Saph Pani, a detailed study was carried out to investigate the response of groundwater to the effect of recharge from three different structures, namely a percolation pond, two check dams and four temple tanks. The locations of the structures investigated are shown in Figure 6.1. The study area experiences a tropical monsoon climate. The average annual rainfall is around 1,200 mm/year, 35% falling during the southwest monsoon (June–September) and 60% during the northeast monsoon. The very dry period in this region is during March-May when the temperature rises above 40°C. Geologically, alluvial deposits are dominant in the northern part that comprises the Arani and Korattalaiyar rivers (Figure 6.1), whereas charnockites are exposed in the southern part along the Coovum and Adyar rivers (Suganthi *et al.* 2013). The percolation pond and check dam considered for this study are located in the Arani-Korattalaiyar river basin, and these locations mostly comprise of alluvium of about 60 m thickness overlying the impermeable formation. The coastal part where the percolation pond is located is characterized by the presence of marine sediments too. The groundwater level in the unconfined aquifer ranges from 2 to 6 m below ground level. In general, the regional groundwater flow leads towards the sea; however there may be variations in local hydraulic heads due to pumping. Groundwater recharge relies mainly on rainfall which feeds the non-perennial streams at the same time. At the temple tanks, groundwater occurs in shallow weathered charnockitic rocks. In this study, a comprehensive assessment was made of the role of MAR in coping with seawater intrusion and groundwater overexploitation. The salient aspects of the present study carried out in a percolation pond, two check dams and four temple tanks are discussed in this chapter.

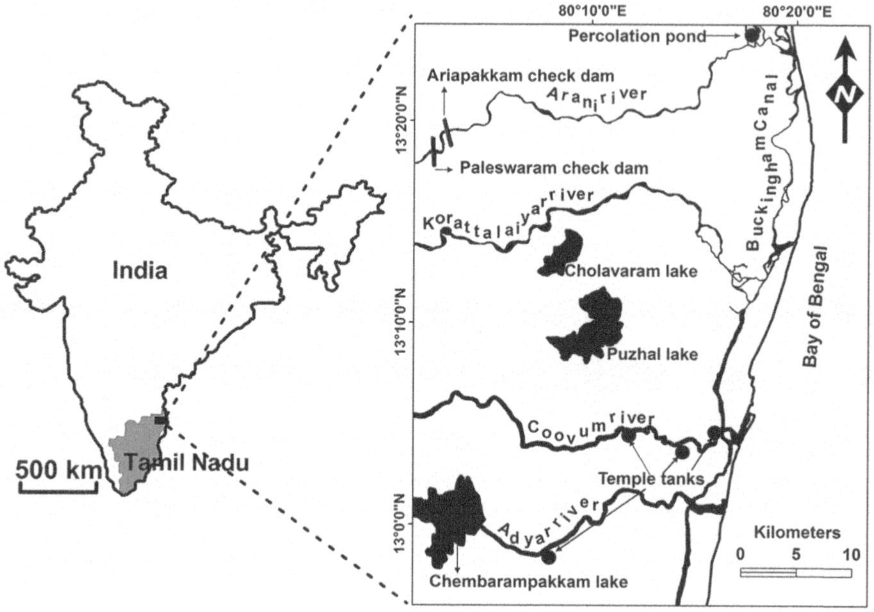

Figure 6.1 Location of check dam, percolation pond and temple tanks in the study area.

6.2 PERCOLATION POND

A percolation pond with a size of 8 m × 8 m × 1.5 m was constructed at Andarmadam in the Thiruvallur district of Tamil Nadu (Figure 6.1). It lies about 4 km west of the Bay of Bengal and 2 km south of Lake Pulikat. The Buckingham canal flows parallel to the Bay of Bengal on the eastern side of the pond.

6.2.1 Problem statement and objectives

The groundwater in the area is highly saline even at shallow depths due to the sediments of marine origin and due to seawater intrusion. Hence, the groundwater is not suitable for irrigation and domestic use. The objective of the study was to investigate the impact of the percolation pond on improving the groundwater quality and quantity in the area.

6.2.2 Results and interpretation

Water table rise

Three piezometers P_1, P_2 and P_3 of 2 m, 4 m and 6 m depth were installed at distances of 0.5 m, 1.0 m and 1.5 m from the pond (Figure 6.2a). The water levels in the pond and the nearby piezometers were monitored every three minutes by digital automatic water level indicators from September 2012 to May 2013 and from September 2013 to January 2014 (Raicy & Elango, 2015a). The water level in the pond was very high from September 2012 to February 2013 (Figure 6.2b), as the pond was filled with rain water during September-December (northeast monsoon). Afterwards, the water level gradually decreased and it dried up completely by May 2013. The groundwater level in the piezometer at 1.5 m from the pond gradually increased over the monitoring period and almost sustained at more than 30 cm above the water table measured before the construction of the pond (Raicy & Elango, 2015a).

Estimation of recharge

The amount of water recharged from the pond into the aquifer was estimated through a water balance approach. The approach considers the water level in the pond, surface area of the pond at different points in time, pan evaporation of the area and rainfall monitored by an automatic weather station in the area. It is assumed that the temporal decline in the water level of the pond is only due to evaporation and recharge of the nearby piezometers. As the water level in the pond was measured on daily basis and daily evaporation data was available, the groundwater recharge was estimated as equal to the change in storage in the pond minus the volume of water lost due to evaporation. The total volume of water recharged into the aquifer by the pond for a period from August 2012 to June 2013 was calculated to be between 250 and 300 m^3. The recharge rates were very high up to December 2012 and gradually decreased after that. Even though the water level in the pond was almost constant until February 2013, the volume of water recharged decreased after September 2012. This is attributed to the physical clogging of the bottom of the pond due to the presence of suspended particles in the pond water. The clogged layer was scrapped after the pond had dried up. Further, to overcome the problem of clogging, a recharge shaft of 0.15 m diameter and 9.5 m depth was constructedat the centre of the pond. The cross section of this arrangement is shown in Figure 6.3.

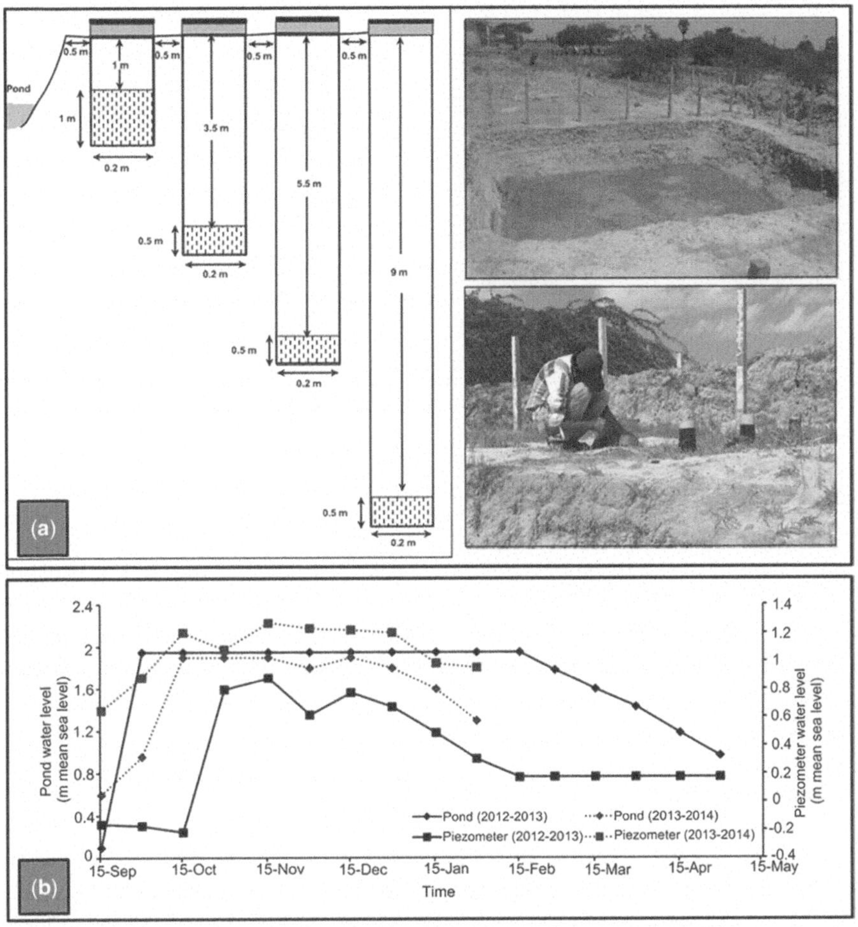

Figure 6.2 (a) Design of piezometer and field photographs (b) Water level in the pond and groundwater level in nearby piezometer (Raicy & Elango, 2015c).

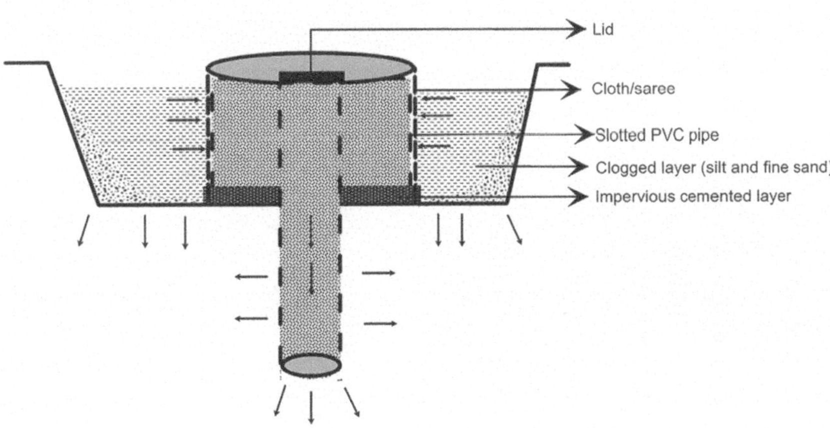

Figure 6.3 The conceptual design of the recharge shaft (not to scale).

The part of the recharge shaft above the bottom of the pond was slotted. Further, a slotted pipe of 0.254 m was installed around the recharge shaft. This outer slotted pipe was lined by an old saree to filter the fine suspended particles from entering into the recharge shaft. This has resulted in the enhancement of groundwater recharge in the following season (2013–2014) as shown in Figure 6.4.

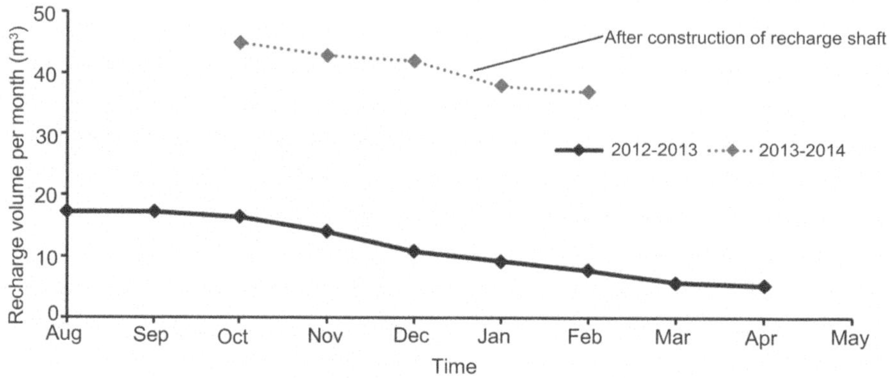

Figure 6.4 Volume of water recharged from the pond from August 2012 to June 2013 (Raicy & Elango, 2015c).

Estimation of physical clogging

The effect of clogging in the pond was studied in a laboratory experiment (Figure 6.5). This experiment was carried out to calculate both the clogging potential of the recharge structure and the rate of infiltration. For this purpose, pond water and sediments from the pond bottom were collected. A column filled with sediments collected from the pond and the pond water was allowed to pass through the column maintaining a constant head. After saturation of the sediments, the pond water was allowed to pass through the column by upward flow from the bottom and the discharge volume was measured at intervals of 10 minutes (Figure 6.5) (Raicy & Elango, 2015c). Similar studies on clogging potential in landfill cover systems were done by Reddy *et al.* (2008) and Reddy & Saichek (1998). They both used similar test setups. In a laboratory study of clogging processes and factors affecting clogging in a tailing dam at Shanxxi province China, Jun *et al.* (2007) concluded that ferrous iron oxidation and precipitation were major issues that led to completely clogged columns (Jun *et al.* 2007). However, such a problem was not faced in this study, which indicates the absence of iron in the pond water. Column experiments were performed to simulate infiltration of untreated river water by Bartak *et al.* (2014). In the study of Bartak *et al.* (2014) the hydraulic conductivity of the filter sand decreased exponentially due to external clogging by two orders of magnitude from 126 to 4 m/d. However, in the column experiment carried out in this study the hydraulic conductivity reduced only by four times due to clogging (Figure 6.5). A laboratory study on the influence of fine particle size and concentration on clogging of labyrinth emitters was carried out by Niu *et al.* (2012). The results obtained from this physical clogging study are comparable with the study carried out by Niu *et al.* (2012), since the particles that were smaller than 0.1 mm in diameter deposited and resulted in clogging as observed during the experiment until 0.25 days (Figure 6.5). To overcome this problem, a piece of saree/cloth was wrapped around the recharge well at the pilot site as shown in Figure 6.3.

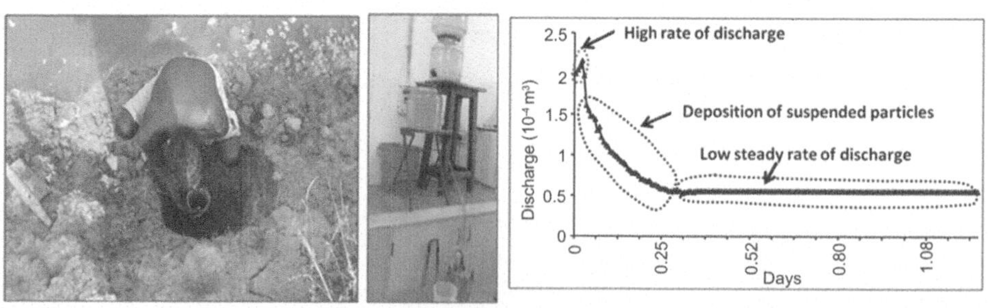

Figure 6.5 Photographs showing sediment collection from the field, the experimental setup for clogging assessment in the laboratory and the measured discharge rates in the laboratory experiment.

Groundwater quality

Pond water and groundwater from the piezometers was collected every two weeks and the major and minor ions were analysed by ion chromatography. The major hydro-chemical facies identified were Na-Cl, Ca-Mg-Cl and Ca-Cl. The type of water changed with the depth of the sampling point. Samples taken below 4 m from the surface fell in the Na-Cl type of water, whereas pond water and groundwater at shallow depths fell under the mixed water type category. This indicates that water from shallow depths was comparably fresh.

The concentrations of different ions in water samples from the pond and the piezometers are plotted in a Schoeller diagram (Figure 6.6). The concentration of major ions in the pond and piezometer P_1 were similar and both decreased during rainfall. In contrast, the concentration of ions in P_2 and P_3 remain the same even after a major rainfall event. The water in the pond and in P_1 had almost equal amounts of Ca^{2+} and Na^+ balanced by Cl^-, whereas the amount of Na^+ and Cl^- in the water of P_3 and P_4 varied (Raicy & Elango, 2015a). Figure 6.6 indicates that the concentration of major ions increase with respect to depth in samples collected before, during and after rainfall. Further, the concentration of ions in groundwater decreased after the commencement of recharge from the pond. Thus, the recharge of water resulted in improvement of groundwater quality up to a depth of about 5 m below the ground surface.

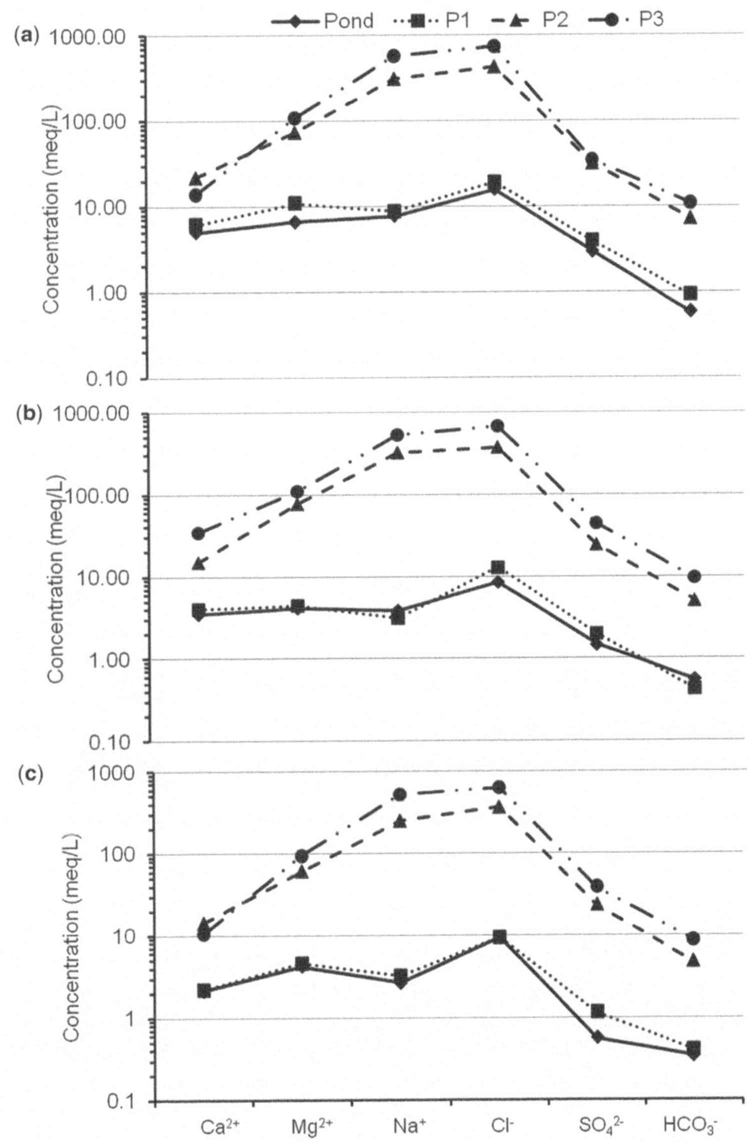

Figure 6.6 Schoeller diagram of samples collected (a) before rainfall, (b) during rainfall and (c) after rainfall to compare the ionic concentrations of different sets of samples (Raicy & Elango, 2015a).

6.2.3 Discussion

The percolation pond was effective in augmenting groundwater resources in this site. However, physical clogging was a major problem faced during this study. Several researchers (e.g. Rebhun & Schwarz, 1968; Behnke, 1969; Ripley & Saleem, 1973; Wood & Bassett, 1975; Vigneswaran & Suazo, 1987; Warner *et al.* 1994) reported the accumulation of suspended particulate matter that causes the progressive clogging of the soil in the ponds or trenches used for aquifer recharge. When the recharging water contains suspended solids of a size commensurate with that of the particles of the porous medium, suspended solids penetration does not occur to any significant extent, and accumulation at the surface leads to the formation of a filter cake, which reduces the overall hydraulic conductivity of the medium as reported by Baveye *et al.* (1998). To understand the process of clogging, a column experiment was carried out as a part of this project. However, Bouwer (1996) indicated that the parameters such as suspended solids content and biodegradable organic carbon and column flow tests, used to compare clogging potentials of recharge waters, are not useful predictors of plugging in injection wells. Even though, Bouwer (1996) reported that the column experiments are not very useful, in the present study the experiment helped to develop the innovative design of a recharge well with cloth screen as shown in Figure 6.3. This design considerably improved the recharge as the recharging water was made to bypass the clogged layer at the bottom of the pond. The groundwater quality also improved through recharge of water from the percolation pond up to a depth of about 6 m. The implementation of several such recharge structures in this area is expected to improve the groundwater potential of the area on a regional scale. The effect of physical clogging of the recharge shaft, however, needs to be assessed. The provision of few layers of saree at the outer slotted pipe of the recharge shaft, as carried out in this study, will overcome or at least reduce the problem of clogging. Such recharge shafts may not be needed in hard rock regions as the percolation efficiency (percolated fraction of stored water) of the pond ranged from 57% to 63% as estimated by another Saph Pani study carried out near Hyderabad, India (Massuel *et al.* 2014).

6.3 CHECK DAM

Check-dams are small barrier structures constructed across rivers and streams for the purpose of water harvesting. The small dams overflow when the river's water flow exceeds the river capacity. The increase in head due to storing of water helps to improve recharge, thereby replenishing the groundwater reserves. In this section, the effect of recharge from two check dams is described.

6.3.1 Problem statement and objectives

The area suffers from seawater intrusion and the Government of Tamil Nadu has constructed several MAR structures, including check dams, to enhance the groundwater availability and to improve the groundwater quality and quantity of the area. The objective of the study was to identify the impact of check dams in improving groundwater potential and quality.

6.3.2 Check dam at Paleswaram

The Paleswaram check dam is located north of Chennai in the Thiruvallur district (Figure 6.1). This check dam was constructed in the year 2010 across the Araniriver, near Paleshwaram village at a distance of about 35 km from the sea. The Arani River enters Tamil Nadu at Utthukottai and joins the Bay of Bengal near Pulicatlake. This check dam is 260 m long with a crest height of 3.5 m from the river bed and has a storage capacity of 0.8 million m^3 of water.

Results and interpretation

Estimation of recharge from the check dam

The water level fluctuation in the check dam is shown in Figure 6.7. The maximum water level in the check dam was observed during monsoon season (October to December), during which the check dam overflows for a few weeks per year. The lowest water level was reached during May to June, after which the river becomes dry. The water surface area and volume of water stored at various water levels in the check dam was estimated using the three dimensional capabilities of Arc GIS9.2 software. The topography of the river bed was measured using the Differential Global Positioning system (Leica GS09).

Based on the height of the check dam (3.5 m) and the river bed topography it was estimated that this check dam is capable of storing 0.8 million m^3 of water. Figure 6.8 shows the storage volume corresponding to various water levels in the check dam. The storage volume of the check dam is nearly 0.9 million m³ when it overflows or when the height of water in the check dam is above the crest (+23.8 m). These values are used for the estimation of groundwater recharge.

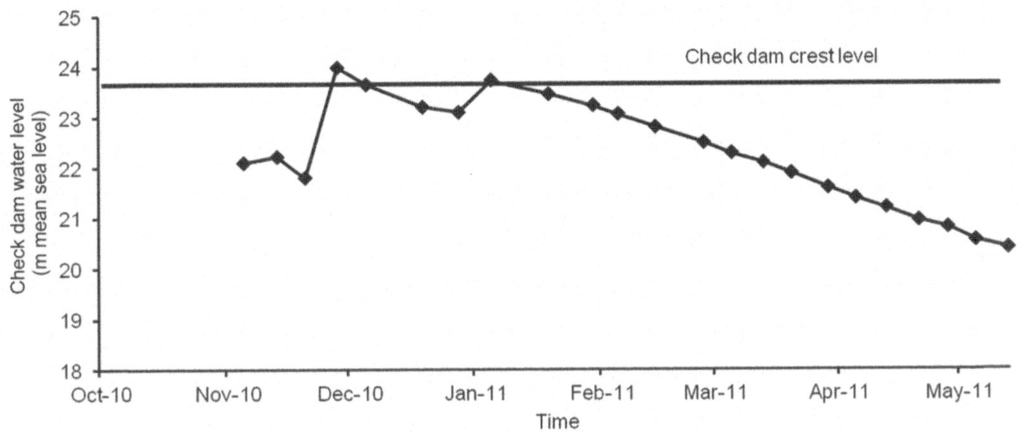

Figure 6.7 Temporal variation in water level fluctuations in the Paleswaram check dam.

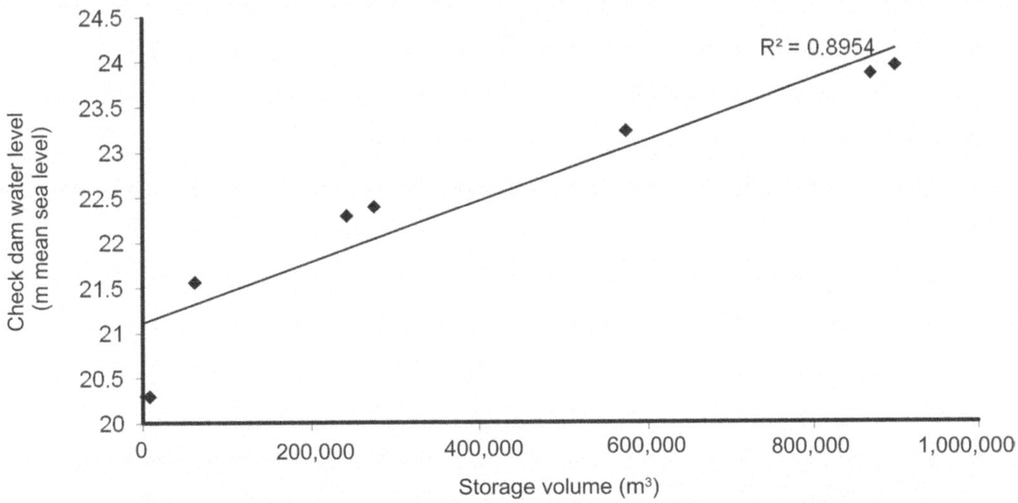

Figure 6.8 Storage volumes corresponding to water levels in the check dam.

Recharge was calculated during the period when there was no flow into the check dam either from the upstream side or as run-off from the nearby areas. As this is a newly constructed check dam, loss due to leakage through sluice gates was very minimal and hence considered negligible. Water stored in the check dam was not directly used for domestic and irrigation purposes, hence, the water lost from the check dam is due to groundwater recharge and evaporation (Eq. 6.1).

$$\text{Recharge}\,(\text{m}^3) = \text{Change in storage volume}\,(\text{m}^3) - \text{Water loss due to evaporation}\,(\text{m}^3) \quad (6.1)$$

Potential evaporation data was obtained from the Indian Meteorological Department, Chennai. Figure 6.9 gives the calculated monthly recharge from the check dam during the period when there was no inflow. The estimated volume of water stored by the check dam was about 1.7 million m³ during the year 2011–2012. This estimation indicates that during 2011–2012, out of 1.7 million m³ of water stored by the check dam, 1.14 million m³ (66%) was recharged into the aquifer (Parimalarenganayaki & Elango, 2013) and about 0.56 million m³ (34%) of water was lost due to evaporation from the check dam.

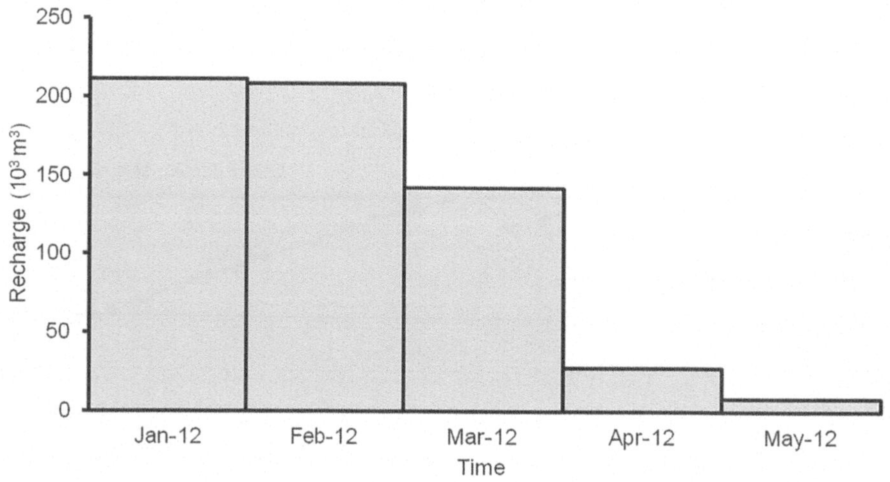

Figure 6.9 Volume of water recharged from the check dam.

Impact on groundwater level

The efficiency of the check dam in augmenting the groundwater recharge was assessed by comparing the temporal variation of the water level in the check dam and the groundwater level in the nearby wells used for drinking water purposes. The wells could be classified into two groups based on their similarity with the water level fluctuation in the check dam. Well number 1 and 12 are located within 1.5 km from the check dam towards the groundwater flow direction (west to east) and well number 11 and 13 are located within 1.5 km to 2.5 km in the flow direction (Figure 6.10). In the first group of wells (11, 13), the groundwater level rises twice a year and this coincides with the northeast and southwest monsoon. This indicates that rainfall is the major source of recharge at these locations. In the other group of wells (1, 12), the fluctuation in groundwater levels is very similar to the water level fluctuation in the check dam. This similarity indicates that surface water stored by the check dam is the major source of recharge at these locations. Taking into account the additional wells up to a distance of 1.5 km from the check dam towards east, an increase of groundwater levels between 1 m and 2.5 m could be observed after the construction of the check dam.

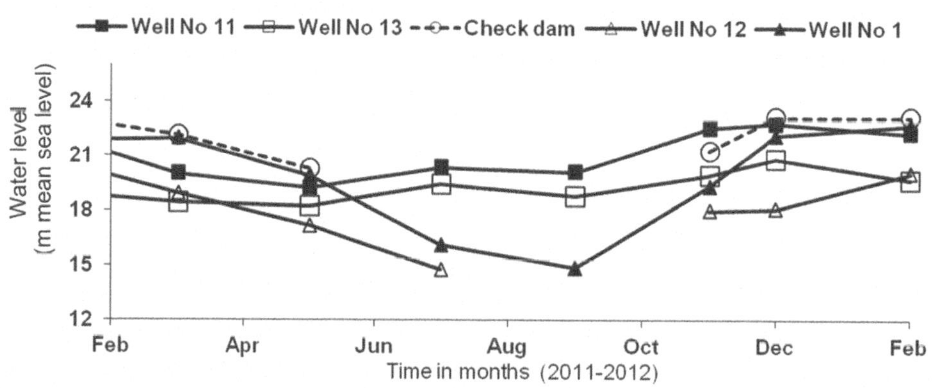

Figure 6.10 Water levels in the check dam and monitoring wells.

Improvement in groundwater quality due to recharge

In order to understand the effect of recharge on groundwater quality, correlations between electrical conductivity and chloride concentration of water in the check dam and groundwater of wells located at different locations were made (Figure 6.11). The wells that plot close to the data points of water from check dam have similar water compositions like the check dam, which indicates that the groundwater of these wells is mainly derived due to the recharge from the check dam. This group of wells are located within 1.5 km from the check dam. Wells located away from the check dam (>1.5 km) have higher electrical conductivity and chloride values, which indicates that these wells are mostly recharged by rainfall.

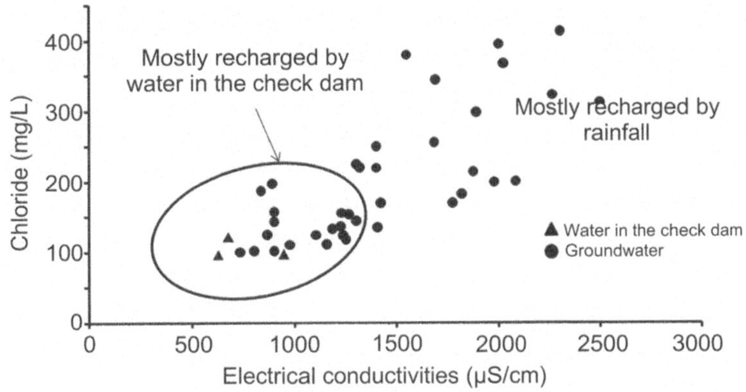

Figure 6.11 Plot on electrical conductivity and chloride concentration of groundwater and Paleswaram check dam water.

6.3.3 Check dam at Ariapakkam

The Ariapakkam check dam is located in the Thiruvallur district, north of Chennai, Tamil Nadu, about 2.5 km upstream of the Paleswaram check dam. Hydrogeochemical modelling using the thermodynamic computer program PHREEQC, Version 3 (Parkhurst & Appelo, 2013) indicates mineral dissolution within the aquifer. Although the exact mineral composition of the alluvial aquifer is unknown, the basic minerals can be estimated from studies of similar environments (Achyuthan & Thirunavukarasu, 2009). The phases used for the modelling of water-rock interaction within the aquifer are calcite, albite, anorthite, plagioclase, kaolinite/halloysite, gibbsite and Fe $(OH)_3$.

It was shown, using the method of inverse modelling, that dissolution and/or precipitation of these minerals can explain the hydrochemical characteristics of the groundwater within the study area.

Near the coast, other processes like ion exchange according to the following equation (Eq. 6.2) play a major role:

$$Ca^{2+} - X + 2Na^{+} \leftrightarrow 2Na^{+} - X + Ca^{2+} \tag{6.2}$$

in which '*X*' indicates the exchanger.

The general evolution of the groundwater, i.e. the resulting groundwater facies (in parenthesis), from inland to the coast and the corresponding hydrogeochemical processes (italic) are illustrated in Figure 6.12.

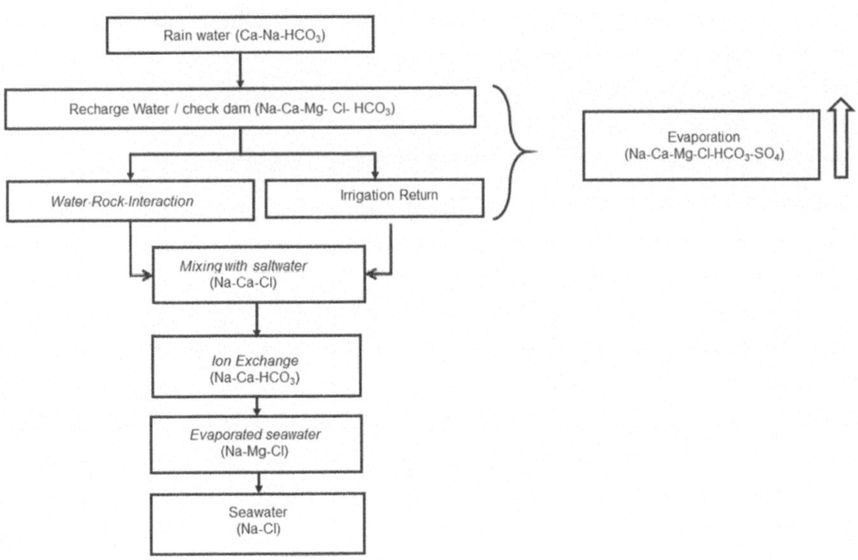

Figure 6.12 Hydrogeochemical evolution of the groundwater from inland to coast.

The wells around the Ariapakkam check dam all show the same ratio of equivalent concentrations of the major ions. This, as well as measured electrical conductivities from the wells, suggests that check dam water recharges groundwater leading to a similar hydrochemical composition. The same processes can be observed at the Paleswaram check dam (Figure 6.11). Near the coast, saltwater intrusion can be detected using the bromide/chloride ratio of groundwater. Bromide is a very good tracer to determine the ratio of saltwater in groundwater because it is a non-reactive ion and residual solutions are enriched in bromide in the course of evaporation.

Hydrogeochemical modelling shows that different mixing ratios of freshwater and seawater can explain the hydrochemical composition of groundwater, sampled in different wells in the coastal region (Table 6.1).

Table 6.1 Mixing of freshwater and seawater in the coastal region and comparison with groundwater sampled in wells using bromide as tracer ion, as well as analyzed chloride concentrations.

Sample	Mixing Ratio (Freshwater: Seawater)	Seawater [%]	Chloride [mg/L]	Water Type (Analysed Samples)
Freshwater (well)	1.0 : 0.0	0.0	52	Ca-Na-HCO_3-Cl
Well A	0.98 : 0.02	2.0	466	Na-Cl
Well B	0.97 : 0.03	3.0	620	Ca-Na-Mg-Cl-HCO_3
Well C	0.97 : 0.03	3.0	972	Na-Cl-SO_4
Well D	0.965 : 0.035	3.5	1,075	Na-Ca-Cl
Well E	0.74 : 0.26	26.0	4,800	Na-Cl
Seawater (Elliots Beach)	0.0 : 1.0	100.0	14,600	Na-Cl

Nevertheless, the analysed concentrations of calcium, magnesium and sodium form different water types, indicating that further processes like ion exchange and mineral dissolution or precipitation play an important role in the mixing zone of the coastal region. Wells in the southwest of the study area around the Ariapakkam check dam have low overall ion concentrations, whereas wells in the north have higher ion concentrations (Figure 6.13).

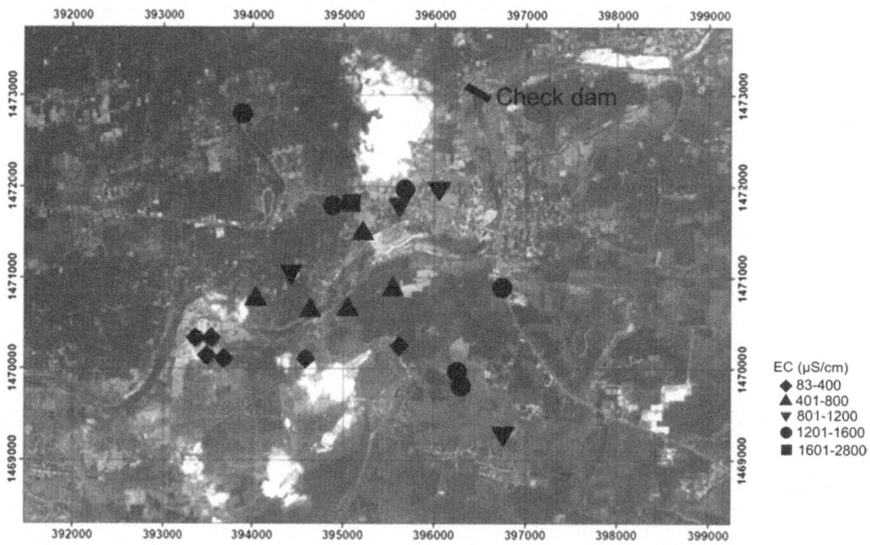

Figure 6.13 Measured electrical conductivities [µS/cm] in the groundwater near the Ariapakkam check dam (Base Map: Google Earth, 2012).

6.3.4 Discussion

The influence of the check dams at Paleswaram andAriapakkam in the Arani-Korattalaiyar aquifer on groundwater quality and usability at nearby areas was assessed using electrical conductivity and groundwater level measurements. Rain water is the most important recharge source. The infiltration of surface water from the check dam is indicated by similar groundwater chemistry and can be observed within a distance of about 1 km to the north and to the south of the check dam. Based on these measurements, the region that is positively influenced by this recharge structure could be delineated. The hydrochemical and hydraulic processes are identical at both locations.

At Paleswaram the groundwater within an area of 3 km^2 around the check dam has electrical conductivity values similar to the check dam water (indicating substantial recharge). The measurement of groundwater level before and after filling up of the check dam indicate that the groundwater level in the wells located within 500 m from the dam has increased by 2.5 m. About 66% of the total water harvested by the check dam has infiltrated into the aquifer during the year 2011–12. The approximate annual rainfall recharge is only about 15% and the increase in recharge is about 51%, which is much higher than the values reported by Alderwish (2010). Alderwish (2010) reported an increase of recharge by about 36% due to check dams in Yemen. Similarly, Gale (2006) carried out a study on check dams in three different hydro-meteorological and geological environments in the states of Gujarat (fractured granite rocks), Tamil Nadu (fractured charnockite rocks) and Maharashtra (Deccan basalt) in India and estimated an additional recharge of 4% to 16%, 23% and 13% due to the construction of check dams in Gujarat, Tamil Nadu and Maharashtra respectively. In comparison to these studies the estimated recharge in the Saph Pani case study was higher. This is due to the conducive geological, hydrogeological conditions at these locations, larger size of the dams and filling up of the dam two times a year.

6.4 TEMPLE TANKS IN CHENNAI CITY

Tanks are usually constructed to obtain water and to perform religious activities in several temples in India. They are usually constructed very close to temples or within a temple. There are about 2,359 temple tanks in Tamil Nadu. Out of these, 64 are in Chennai and its suburbs (Times of India, 2011). The temple tanks are usually not paved at the bottom and thus allow percolation of water. These tanks ensure the sustenance of groundwater supply in the surrounding areas. There is always an arrangement to collect rainwater into the tank from the sprawling temple complex as well as from surrounding streets and an outlet to drain any surplus (Raicy *et al.* 2014b). The stored water usually lasts until the next rainy season. There will not be any large scale direct extraction of water from the tank. The temple tanks are usually not paved at the bottom and thus allow percolation of water. So the loss of water from the tank is only due to evaporation and percolation.

6.4.1 Site description

Four of the temple tanks located within the urban region of Chennai city were selected for this study. The selection of these tanks was based on logistical convenience and their spatial distribution so as to understand their effect on a regional basis (Table 6.2).

Table 6.2 Name of the temple, tank size and location.

Temple Name and Location	Temple Tank Size Length × width × depth [m]	Lithology
Adipuriswarar, Chinthadiripet	35 × 33 × 3.45	Unconsolidated sediments
Agatheeswarar, Numgampakam	45.72 × 30.48 × 3.01	Alluvium
Kurungaleeswarar, Koyambedu	54.9 × 53 × 3	Sandyclay
Suriyamman, Pammal	180 × 100 × 3	Weathered/ fractured charnockite

The lithology of the first three tanks was ascertained by drilling boreholes using hand boring methods. During drilling, soil samples were collected from different depths. The fourth tank is known to fall in hard rock terrain.

6.4.2 Problem statement and objectives

As the temple tanks receive surface run-off and store it during most of the year, they interact with the groundwater. The objective of this study is to investigate the effect of recharge from temple tanks on groundwater quality.

6.4.3 Results and interpretation

Water samples from the temple tanks and the nearby wells were collected at periodical intervals and were analysed for pH and major ions. The pH of water from the temple tank and nearby groundwater varies from 6.8 to 9.5 indicating that the water is of alkaline nature. There was no good correlation between the quality of water in the tank and the nearby ground water in the case of Adipuriswarar, whereas there was reasonable correlation in the case of Agatheeswarar, Kurungaleeswarar and Suriyamman. The trend of temporal variations in the concentration of these ions in surface and groundwater confirms the interaction between them (Figure 6.14).

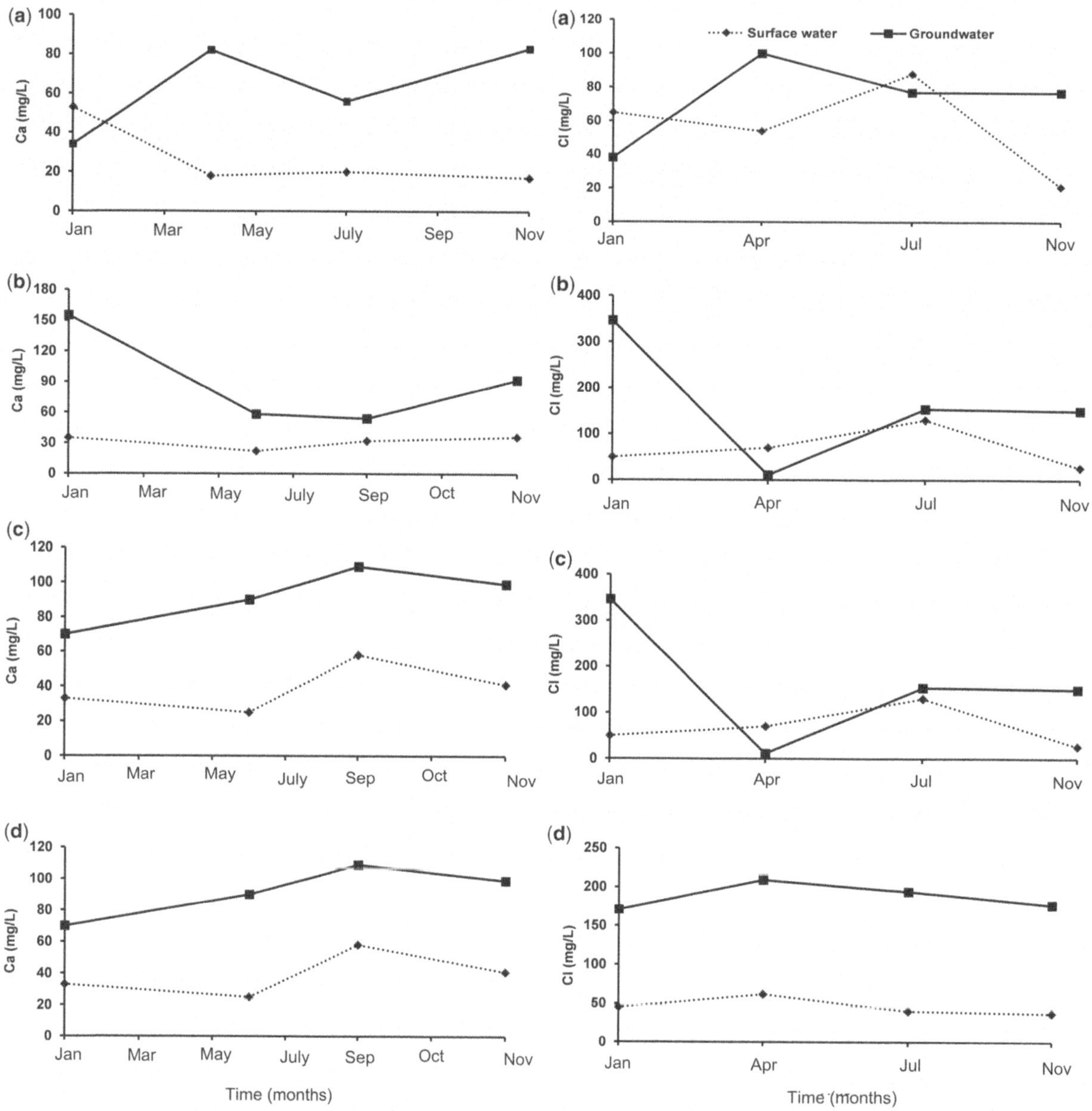

Figure 6.14 Temporal variations in concentration of calcium and chloride in water from temple tank and nearby groundwater at (a) Adipuriswarar, (b) Agatheeswarar, (c) Kurungaleeswarar and (d) Suriyamman.

6.4.4 Discussion

The major ions like Ca^{2+}, Mg^{2+}, Cl^- and SO_4^{2-} in surface water and groundwater are within the permissible limits of drinking water quality as per BIS standards (2012). The Ca^{2+} and Cl^- are the dominant ions in the water from the tank and groundwater. The similarity in the seasonal variation in the concentration of ions confirms the recharge of water from the tank. Thus, temple tanks acts as recharge structures and could improve the groundwater quality. However, great care has to be given to maintain the surrounding so that the rain water draining into the tank is reasonable clean. The recharge efficiency of the temple tanks will also diminish with time due to clogging as observed in the pilot study carried out by the construction of percolation pond as a part of Saph Pani. The temple tanks thus need to be cleaned by removing the finer materials settled at the bottom at least once in three years or when it becomes dry. Coarse sand may also be dumped after removing the clogged layer during the cleaning process.

6.5 CONCLUSION

In this study three different type of groundwater recharge structures were investigated to understand their benefits in terms of improving groundwater resources. In the case of the constructed pilot percolation pond, the groundwater head nearby increased about 30 cm and the quality of groundwater also improved in the top 4 m of the saturated zone (Raicy & Elango, 2015a). Percolation ponds are a simple method which can be implemented to recharge the aquifer. This method can easily be adopted by local farmers (Raicy & Elango, 2015a). Cost-effectiveness per recharged volume needs to be confirmed after additional studies on the influence of physical clogging on the percolation efficiency. If the costs, including the effect of clogging, are reasonable, construction of many such ponds could help improve the groundwater quality in the entire region. The comparison of groundwater levels, of water levels in the check dam and of rainfall assisted in demarcating the area that is highly benefited by the storage of water in the check dam. During the study period it was estimated that about 1.3 million m^3 of water was recharged out of the 2 million m^3 of water harvested by the check dam and the storage of water resulted in an increase in groundwater levels from 1 to 3.5 m within a radius of 2 km (Parimalarenganayaki & Elango, 2014). The recharge from the dam led to improvements in groundwater quality, which is evident from lower concentrations of major ions in the groundwater in the vicinity of the dam. The costs of 1 m^3 of water recharged by the check dam are about INR 1.20 (Parimalarenganayaki & Elango, 2014). Temple tanks act as recharge structures. The main advantage of such structures is that they are free from contamination by discharge of wastewater due to religious concerns and social aspects. This study also generated data for carrying out numerical modelling to study the impact of check dams and percolation ponds in improving groundwater level which is presented in Chapter 14.

6.6 REFERENCES

Achyutnan H. and Thirunavakarasu N. (2009). Quaternary stratigraphy of the koratallaiyar-cooum basins, Chennai. *Journal of Geological Society of India*, **73**, 683–696.

Alderwish A. M. (2010). Induced recharge at new dam sites—Sana'a Basin, Yemen. *Arabian Journal of Geosciences*, **3**, 283–293.

Bartak R., Grischek T. and Hoche D. (2014). MAR with untreated river water: Clogging of basins and coliform removal rates. *Journal of Hydrologic Engineering*, **20**, DOI:10.1061/(ASCE)HE.1943–5584.0000980.

Baveye P., Philippe V., Blythe L. H., Paul C. D. and Diego S. L. (1998). Environmental impact and mechanisms of the biological clogging of saturated soils and aquifer materials. *Critical Reviews in Environmental Science and Technology*, **28**(2), 123–191. DOI: 10.1080/10643389891254197.

Behnk J. J. (1969). Clogging in surface spreading operations for artificial ground-water recharge. *Water Resource Research*, **5**(4), 870–877.

BIS (2012). Indian Standard: Drinking Water Specifications IS-10500, Bureau of Indian Standards, New Delhi, India.

Gale I. N., Macdonald D. M. J., Calow R. C., Moench M., Kulkarni H., Mudrakartha S. and Palanisami K. (2006). Managed aquifer recharge: An assessment of its role and effectiveness in watershed Management. Rep. No.CR/06/107N (BGS). Available from: http://nora.nerc.ac.uk/7453/1/CR06107N.pdf.

Jun W., Yanqing W. and Jian L. (2008). Laboratory study of the clogging process and factors affecting clogging in a tailings dam. *Journal of Environmental Geology*, **54**, 1067–1074.

Massuel S., Perrin J., Mascre C., Mohamed W. and Boisson A. (2014). Managed aquifer recharge in South India: What to expect from small percolation tanks in hard rock? *Journal of Hydrology*, **512**, 157–167.

Niu W., Lu L. and Xue C. (2013). Influence of fine particle size and concentration on the clogging of labyrinth emitters. *Irrigation Science*, **31**, 545–555.

Parimala Renganayaki S. and Elango L. (2014). Assessment of effect of recharge from a check dam as a method of managed aquifer recharge by hydrogeological investigations. *Environmental Earth Sciences*, **73**(9), 5349–5361.

Parkhurst D. L. and Appelo C. A. J. (2013). Description of input and examples for PHREEQC version 3- A computer program for speciation, batch-reaction, one-dimensional transport, and inverse geochemical calculations. U.S. Geological Survey Techniques and Methods Book 6, chap. A43.

Pie Y., Yunkai L., Aoze L., Bo Z., Ze S. and Yinguang J. (2014). Eight emitters clogging characteristics and its suitability under on-site reclaimed water drip irrigation. *Irrigation Science*, **32**, 141–157.

Raicy M. C. and Elango L. (2015a). Percolation pond as an aquifer recharge method in a saline aquifer, north of Chennai: A perspective on the site selection criteria, implementation and its effects. *Environmental Earth Science* (submitted).

Raicy M. C., Parimalarenganayaki S. and Elango L. (2015b). Understanding the surface water-ground water interactions by geochemical measurements in three different groundwater recharge structures. *Journal of Environmental Science and Management* (submitted).

Raicy M. C. and Elango L. (2015c). Experimental determination of clogging potential and infiltration rates in percolation ponds: A case study from Andarmadam, Thiruvallur district of Tamil Nadu. *Environmental Monitoring Assessment* (submitted).

Rao P. (1959). Some observations in the marine geology, geology of Bay of Bengal (along the east coast of India). In: Proceedings of the Third Pan Indian Ocean Science Congress, Tananarive, Madagascar, section C, pp. 143–148.

Rebhun M. and Schwarz J. (1968). Clogging and contamination processess in recharge wells. *Water resource research*, **4**(6), 1207–1217.

Reddy K. R., Timothy D. S. and Aravind M. (2008). Clogging potential of tire-shred drainage layer in landfill cover systems. *International Journal of Geotechnical Engineering*, **2**, 407–418.

Reddy K. R. and Saichek R. E. (1998). Characterization and Performance Assessment of Shredded Scrap Tires as Leachate Drainage Material in Landfills. Proceedings of the Fourteenth International Conference on Solid Waste Technology and Management, Philadelphia, PA.

Ripley D. and Saleem Z. A. (1973). Clogging in simulated glacial aquifers due to artificial recharge. *Water Resource Research*, **9**(4), 1047–1057.

Suganthi S., Elango L. and Subramanian S. K. (2013). Groundwater potential zonation by remote sensing and GIS techniques and its relation to the groundwater level in the coastal part of the Arani and Koratalai river basin, Southern India. *Earth Sciences Research Journal*, **17**(2), 87–95.

Times of India (2011). Parched Temple Tanks to be Revived. http://timesofindia.indiatimes.com/city/chennai/Parched-temple-tanks-to-be-revived/articleshow/8894188.cms (accessed 26 November 2014).

The Times of India (2014). As Population Surges, Chennai's Water Needs to Double in 3 years. http://timesofindia.indiatimes.com/city/chennai/As-population-surges-Chennais-water-needs-to-double-in-3-years/articleshow/39454917.cms. (accessed 9 August 2014).

Vigneswaran S. and Suazo R. B. (1987). A detailed investigation of physical and biological cloggingduring artificial recharge. *Water Air and Soil Pollution*, **35**(1–2), 119–140.

Warner J. W., Gates T. K., Namvargolian R., Miller P. and Comes G. (1994). Sediment and microbial fouling of experimental groundwater recharge trenches. *Journal of Contaminant Hydrology*, **15**, 321–344.

Wood W. W. and Bassett R. L. (1975). Water quality Changes related to the development of anaerobic conditions during artificial recharge. *Water Resources Research*, **11**(4), 553–558.

Chapter 7

Percolation tanks as managed aquifer recharge structures in crystalline aquifers – An example from the Maheshwaram watershed

Alexandre Boisson, Marina Alazard, Géraldine Picot-Colbeaux, Marie Pettenati, Jérôme Perrin, Sarah Sarah, Benoît Dewandel, Shakeel Ahmed, Jean-Christophe Maréchal and Wolfram Kloppmann

7.1 INTRODUCTION

Managed aquifer recharge (MAR) through percolation tanks is a promising technique to increase local water availability. However, authors such as Dillon *et al.* (2009) point out the lack of data available for an accurate assessment and note that little evidence exists on the positive impact at local scale. Some authors even point out the possible negative impact at the watershed scale due to the enhancement of local water extraction (Calder *et al.* 2008; Glendenning *et al.* 2012; Sakthivadivel, 2007).

In the following study, a percolation tank located in the Maheshwaram watershed in Andhra Pradesh (India) is monitored to quantify its impact on water availability and quality. The objectives are to assess the potential of percolation tanks as managed aquifer recharge structures by:

1) quantifying the volume of water recharged to the aquifer by a percolation tank
2) developing a simple methodology for prediction of yearly benefits to the aquifer from percolation tanks
3) defining the hydrodynamic and hydrochemical specificities of percolation tanks in crystalline aquifers
4) assessing the impact of percolation tanks on the ground water quality

An overview of major results is presented in this chapter. More detailed information on methodologies and detailed results can be found in related publications (Boisson *et al.* 2014a; Boisson *et al.* 2014b; Pettenati *et al.* 2014; Alazard *et al.* 2015).

7.2 SITE DESCRIPTION

7.2.1 Maheshwaram watershed

The study was carried out in the Maheshwaram watershed, located 35 km south of Hyderabad (Andhra Pradesh State, India). This typical watershed is a good example of a south Indian context with semi-arid climate and crystalline aquifers. Moreover, previous hydrogeological studies in the watershed provided good baseline data (Dewandel *et al.* 2006; Maréchal *et al.* 2004, 2006; Wyns *et al.* 2004; Perrin *et al.* 2011). The watershed covers an area of 53 km^2 (Figure 7.1) and has a relatively flat topography ranging from 590 to 670 m above mean sea level. The climate is semi-arid with annual monsoon rains (rainy or 'Kharif' season from June to October). Mean annual precipitation is about 750 mm, of which more than 90% falls during the monsoon season (Maréchal *et al.* 2006). The mean annual temperature is about 26°C although during the summer

('Rabi' season from October to May), maximum temperature can reach 45°C (Maréchal *et al.* 2006). The resulting potential evapotranspiration is 1,800 mm/year. Due to the rapid growth of Hyderabad city, this watershed is now in transition from a rural to a suburban area. The aquifer is overexploited with more than 700 boreholes used for agriculture dominated by rice paddy fields (Dewandel *et al.* 2010). Currently, the water table is 15–25 m below ground and there is no surface water except for a few days subsequent to very heavy rain falls. Thus, no regular infiltration is observable in the watershed during most of the year.

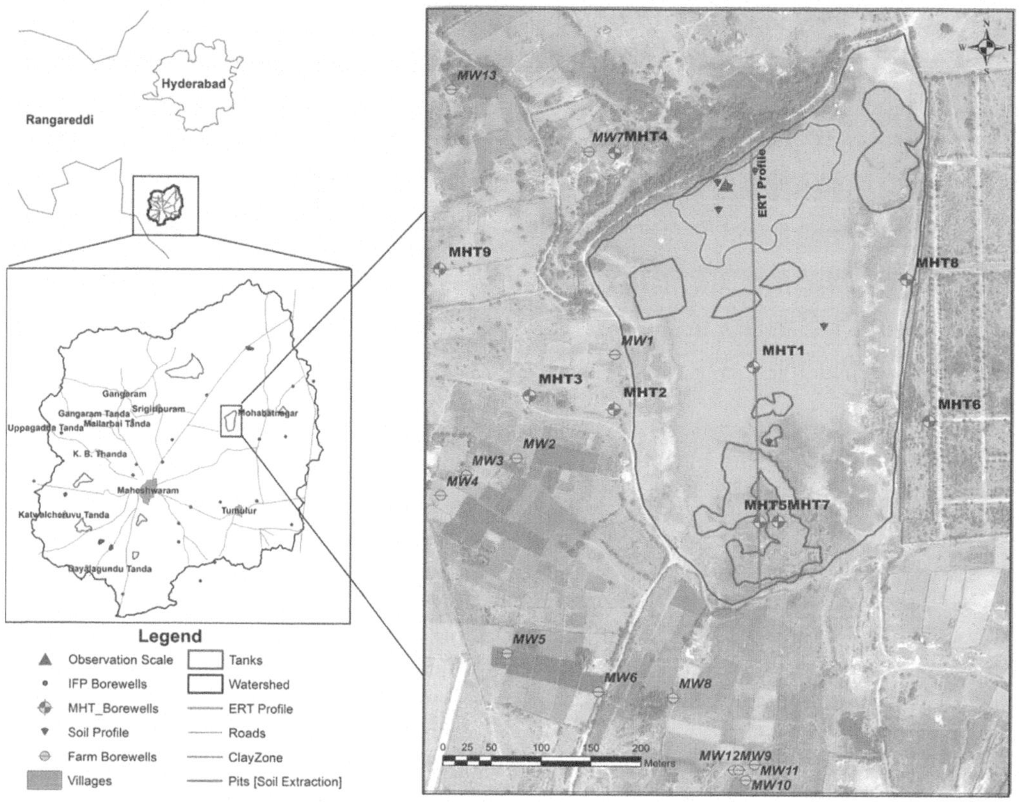

Figure 7.1 Study site map (from Boisson *et al.* 2014).

7.2.2 Main characteristics of the crystalline rock aquifer

Crystalline rock aquifers, which represent most of South India and about 66% of Andhra Pradesh state (G.S.I, 2005), present specificities which may limit the potential of MAR. The typical geological profile is described as follows (Acworth, 1987; Chilton & Foster, 1995; Dewandel *et al.* 2006; Figure 7.2): the first zone is the saprolite (or alterite or regolith) on top consisting of clay-rich material, derived from bedrock decomposition with a thickness of approximately ten meters. Because of its clayey–sandy composition, the saprolite zone has a high porosity, and a low permeability. This zone is the main storage capacity of the aquifer. The second zone is a fissured zone, generally characterized by dense horizontal fissuring in the first few meters and a depth-decreasing density of sub-horizontal and sub-vertical fissures (Maréchal *et al.* 2004). This zone mainly provides the transmissive function of the aquifer and is tapped by most of the wells drilled in hard-rock areas. The third zone is the fresh basement which is only permeable where tectonic fractures are present. Boisson *et al.* (2015) showed that the storage potential of the deepest fractures is very limited.

7.2.3 Tummulur tank monitoring program

Currently, three main percolation tanks are located within the watershed. The monitored tank is close to Tummulur village, in the downstream part of the watershed (Figure 7.1), and has been used for more than 10 years for water storage. An earth bund in its northern part dams the natural stream outlet, and consequently run-off water can be stored over an estimated maximum area of 158 000 m^2 and a maximum water depth of 3.5 m. Before the first significant rainfall events in 2012, the tank was dry for over 7 months with temperature ranging from 30 to 45°C which created shrinkage cracks at the entire tank area. This area is covered by silt loam soil on the surface underlain by sandy loam at a depth of 40–80 cm. Because of thin sedimentary deposits, there is an important clayey zone, at the foot of the bund, in the northern part of the tank. Clay pits are located on the southern part of the tank to feed the nearby brick industry. Nine monitoring boreholes (labelled MHT's) were implemented in 2012 and one staff gauge records the surface water level within the tank. Temperature and electrical conductivity logging are regularly performed in the boreholes, in addition to the long term piezometry, temperature and electrical conductivity records. Slugs tests were also performed (Boisson *et al.* 2015). The topography of the tank area was measured by DGPS (Differential Global Positioning System) and regular GPS tracking of the water contour give the tank area evolution. Within a radius of 500 m, at least 15 irrigation boreholes are in use (rice paddy and maize). Irrigation duration and times are controlled by the availability of electricity (7 h a day). The percolation tank constitutes a drinking water supply source for livestock (few goats and buffaloes) and no significant direct tank water extraction for irrigation purposes occurs. This tank system is representative of MAR practice in semi-arid southern India.

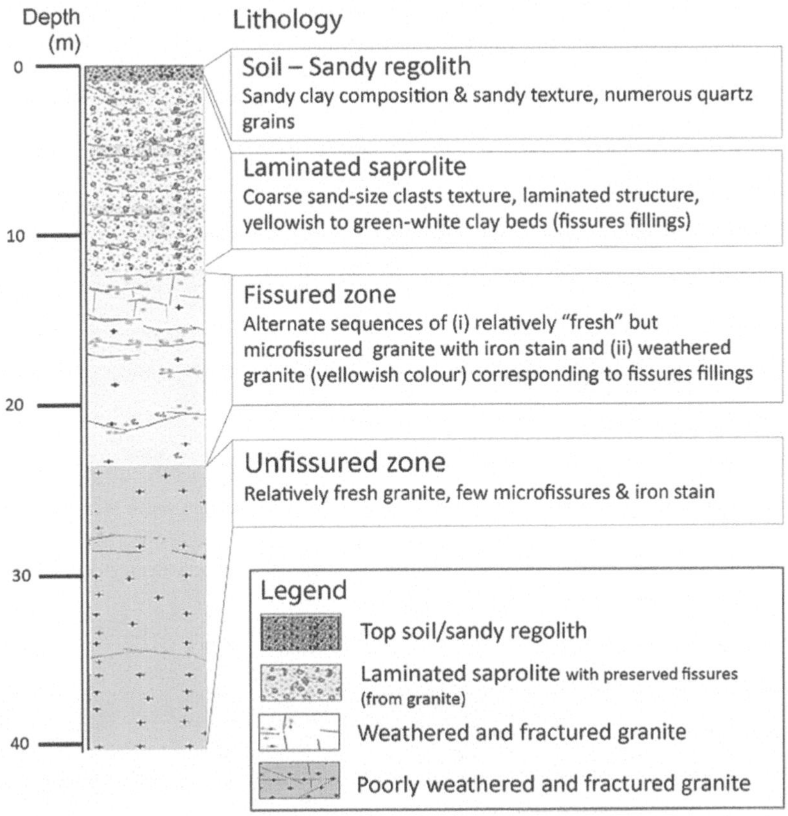

Figure 7.2 Geological log characteristic for the study area.

Eight sampling campaigns for major ions analyses and two campaigns for stable isotopes ($\delta^{18}O$ and δ^2H) analyses were performed in 2013 and 2014 during different hydrological conditions (Figure 7.3). Groundwater was sampled in 7 MHT boreholes and 3 farmer boreholes as well as the Tummulur tank water (when filled) (Figure 7.1). In situ electrical conductivity, temperature and pH were measured for each sample. Water samples were collected in clean, triple-rinsed polyethylene bottles. Raw water was sampled for isotopes and hydrogen carbonate analyses. Water for anion analyses was filtered (0.45 μm) in situ. Water for cation analyses was filtered (0.45 μm) and acidified (nitric acid, 1M) in situ. Bottles were completely filled, limiting air contact. Chloride, nitrate, sulphate and fluoride were analysed by ionic chromatography (Dionex) according to the NF ISO 10304 method. Hydrogen carbonate contents were determined using the NF ISO 9963–1 method based on potentiometric analysis. Calcium, potassium, sodium and magnesium were analysed by ICP-emission spectrometry according to the NF ISO 11885 method. Stable isotopes ratios ($\delta^{18}O$, δ^2H) were measured by mass spectrometry.

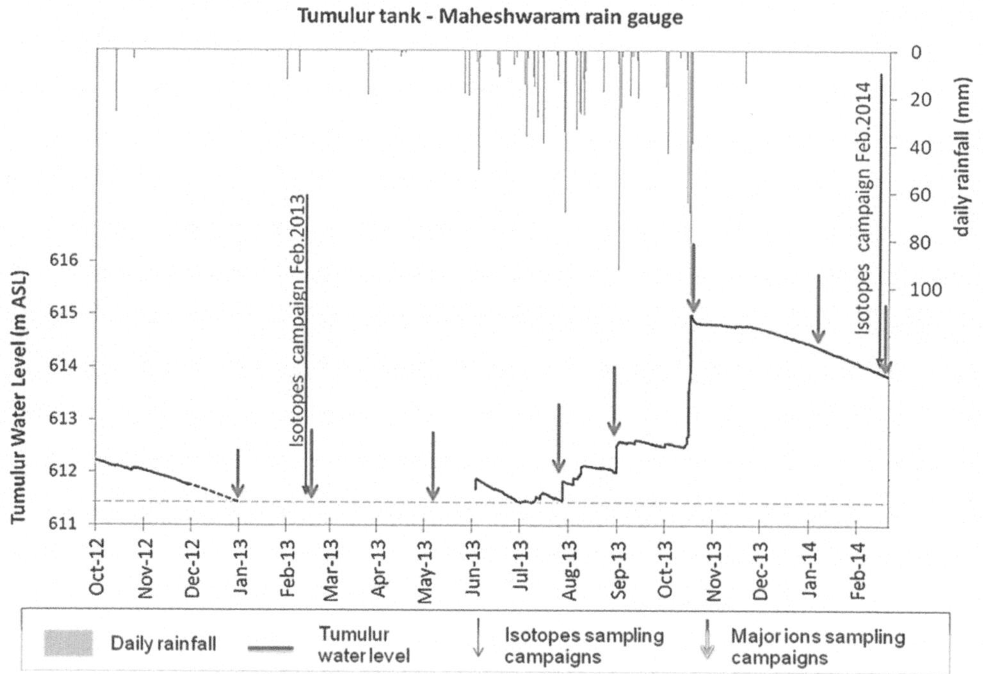

Figure 7.3 Tank water level (in m above sea level- ASL) evolution and daily rainfall with corresponding hydrological conditions on the Tummulur site for the sampling campaigns in 2013 and 2014.

7.3 RESULTS AND INTERPRETATION

7.3.1 Field results and observation

The Tummulur tank was monitored from 01/01/2012 to 11/04/2014. The variability of the monsoon (700 mm in 2012; 1 110 mm in 2013) had a large impact on water level and tank area evolution (Figure 7.4). From the water levels coupled with the DGPS topographic survey the maximum tank volume after the 2012 monsoon is estimated to be 8,100 m³ and above 90,000 m³ after the 2013 monsoon. During monsoon, a few extreme events appear to have a major impact on the tank's replenishment.

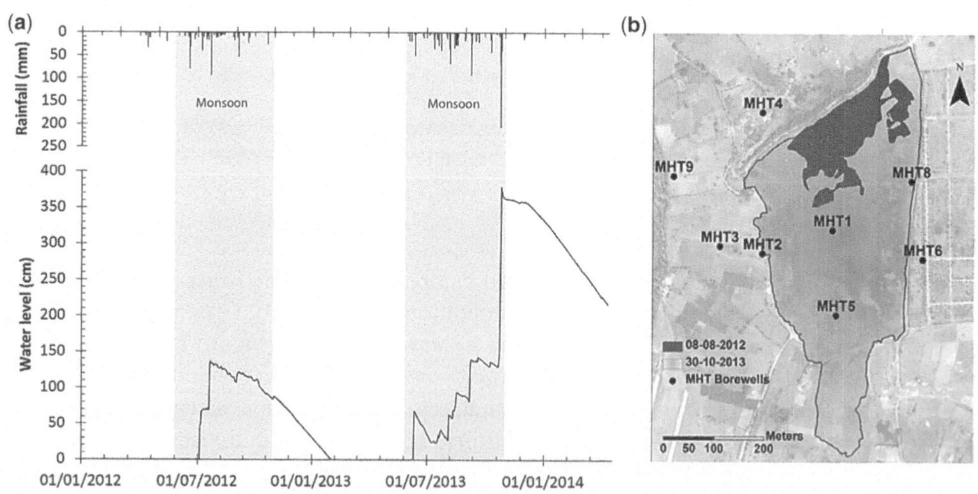

Figure 7.4 (a) Evolution of water levels in the tank versus rainfall; (b) Maximum water extent in tank are after 2012 monsoon (08/08/2012, dark grey) and 2013 monsoon (30/10/2013, light grey).

7.3.2 Tummulur tank water balance

A water balance for the Tummulur tank was established following the methodology developed in Massuel *et al.* (2014) computed on a daily basis. To sum up, the balance is based on the following equation:

$$\Delta V = F_{net} + R_{rain} - E - L_v - S_{eep} - P \qquad (7.1)$$

where

ΔV = change in tank water storage
F_{net} = net inflow to the tank i.e. run-off
R_{rain} = direct rainfall collected on the surface of the water tank
E = evaporation from the tank surface (estimated from Class-A pan data with a pan coefficient of 0.8)
L_v = water consumption by livestock
S_{eep} = seepage across the dam
P = percolation

From the local condition, i.e. limited cattle, the term L_v can be neglected. This assumption is justified since Massuel *et al.* (2014) show that in a similar context with more goats breeding this element represents a minor part of the water budget (<5%). F_{net} represent the run-off flowing to the tank from the surrounding area during important rainfall events.

A relation between tank water level and percolation was developed for the days without rainfall, run-off or seepage through the dam when the percolation can be defined as:

$$P = \Delta V - E \qquad (7.2)$$

This relation is then used to define percolation on a daily basis depending on the water level (Figure 7.5).

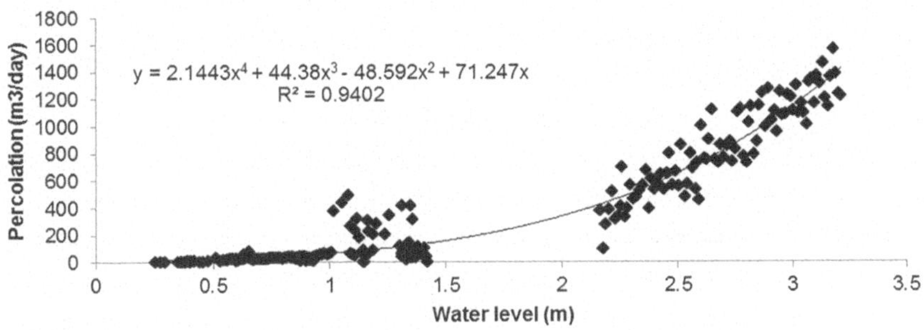

Figure 7.5 Relationship between the water level and the percolation rate in the tank.

Using this relation in eq. 7.1 a complete water balance at the tank scale can be established. The results are presented in Figure 7.6 and Table 7.1. Note that the balance for the 2013 monsoon ends on 11/04/2014 while percolation and evaporation would arise in the following months.

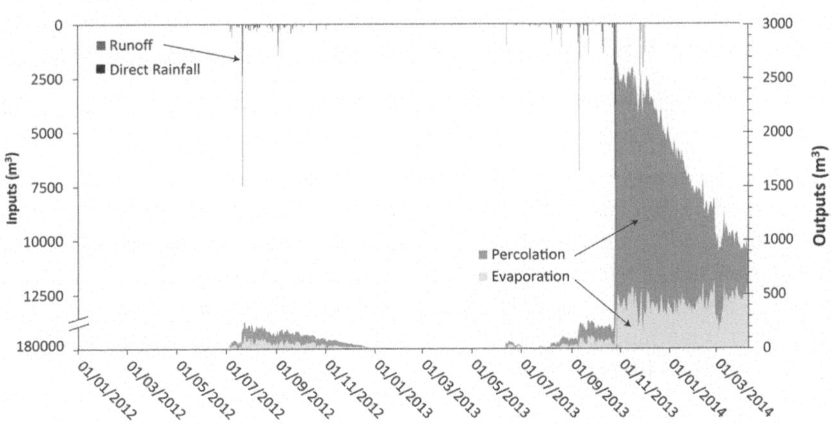

Figure 7.6 Temporal evolution of main components of a water balance at Tummulur percolation tank (Inputs on top x-axis and outputs in bottom x-axis).

Table 7.1 Main components of water balance of Tummulur percolation tank.

	2012 Monsoon	%	2013 Monsoon	%
Runoff [m³]	10,612	58	278,442	92
Direct rainfall [m³]	7,795	42	21,666	8
Input [m³]	18,407	100	300,108	100
Evaporation [m³]	7,281	38	75,970	25
Percolation [m³]	11,730	62	229,155	75
Output [m³]	19,010	100	305,124*	100
Balance in-out [m³]	−603		−5016	
Error (%)	3.22		1.66	

*Budget for the 2013 monsoon ends on 11/04/2014 without considering further evaporation and percolation.

The results of the water balance show a high variability in the tank infiltrated water from 11,730 m^3 during the 2012 monsoon to 229,155 m^3 during the 2013 monsoon. The results are in agreement with the predictive water balance performed in Boisson *et al.* (2014a) through a different approach for the 2012 monsoon. As well, the estimated infiltration rates and ratio percolation over inputs are in the same range as that observed in Massuel *et al.* (2014) in a similar context. It is noteworthy that tank refilling occurs mostly during a few intense events through run-off occurring on the watershed (up to 92.8% in 2013). On the contrary, outputs are evolving slowly during the entire flooding period.

Maréchal *et al.* (2006) estimate the mean annual groundwater abstraction at the watershed scale on Maheshwaram (53 km^3) to be 8.8 million m^3/year. Therefore, the percolated volume from the Tummulur tank represents 0.13% and 2.6% in 2012 and 2013 monsoons, respectively of the groundwater abstraction. Considering the 40% of return flow in average at the watershed scale (Maréchal *et al.* 2006), the percolated water from the tank represents 0.2% and 4.2% of the net groundwater abstraction for the two monitored periods.

Those calculations show that under low rainfall conditions, the impact of a percolation tank is negligible at the watershed scale, and very limited at the local scale. On the contrary, when it is filled, the tank has a significant impact on the total water balance and can be very helpful for the surrounding farmers. However, it should be noted that following an important rainy season, farmers tend to increase the irrigated surface in the tanks surroundings and therefore increase the water abstraction at the watershed scale.

7.3.3 Flow characteristics in crystalline aquifer

In crystalline rock aquifers, flow is constrained by the fracture network (distribution of fracture length, orientation and density and connectivity) (Bour & Davy, 1998; de Dreuzy *et al.* 2001). Studies on the Maheshwaram watershed have shown that the bulk permeability is anisotropic (Kh >> Kv; Maréchal *et al.* 2004) and that predominant water flow occurs in shallow fractures (Mayo *et al.* 2003) mostly at the contact between the saprolite and the top of the granite (Dewandel *et al.* 2006). Detailed investigations also show that the storage decreases drastically with depth and that the deepest fractures may have a semi-confined behaviour (Boisson *et al.* 2015). Example of a detailed log with pictures of fractures and fissures are shown in Figure 7.7. In this example, before monsoon (June) the water electrical conductivity profile is constant at a value of ~1,000 µS/cm. When the recharge starts (September–October), the borehole appears to cross two separate water bodies: the upper part of the profile appears to be diluted by rainfall and surface water (decrease of electrical conductivity); while in the deepest part, the conductivity increases. This behavior is observable in all monitored boreholes on the area (data not shown).

The crystalline rock aquifer structure implies a strong variability of the porosity with depth, ranging from 0.5 to 10% (White *et al.* 2001; Wyns *et al.* 1999) in the saprolite and from 1 to 2% in the fractured zone (Dewandel *et al.* 2012; Maréchal *et al.* 2004).

The higher storage capacity of the saprolite compared to the fractured zone can create strong discrepancies in the piezometric evolution. In MHT4 borehole, for example (Figure 7.8), two rain events of 18 and 19 mm (03 and 06/06/2013, when the water is in the fractured zone) induced water level rise of 8 m, while a rainfall event of 50 mm (12/06/2013, when the water table has reached the saprolite) induced a rise of 24 cm.

This specificity described in Boisson *et al.* (2015), should be taken into account while doing the monitoring of the boreholes, since a given rise of the water level may correspond to several water volumes depending of the local storage capacity.

Weathering profile maps made from Electrical Resistivity Tomography presented in Boisson *et al.* (2015) show that the storage potential of the tank is higher in the southern part of the tank due to a thicker weathering profile.

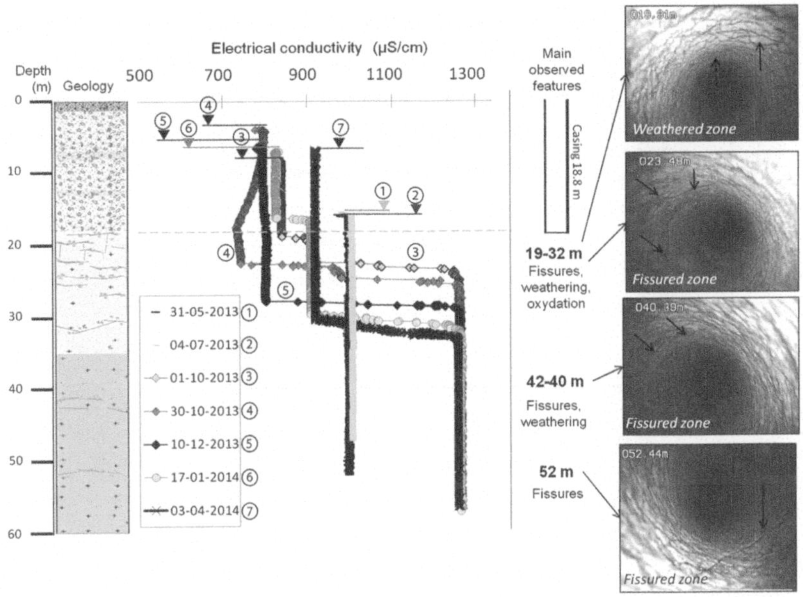

Figure 7.7 Geological log of MHT9 borehole; Electrical conductivity loggings evolution and fractures observed in borehole. (1) Pre-monsoon campaign, (2) early stage of the monson (i.e. July 2013), and (3 to 7) late stages of the monsoon and post-monsoon campaigns. Water level for each campaign are shown (triangles).

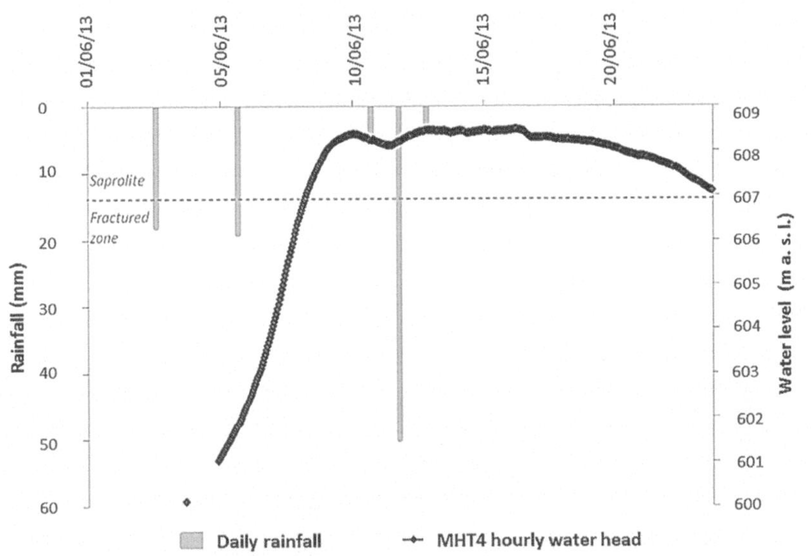

Figure 7.8 Evolution of ground water levels following rainfall event of different intensity in MHT4 borehole. The limit between the saprolite and the fractured zone (about 607 m above seal level-ASL) is shown as a dotted line.

7.3.4 Impact of Tummulur tank recharge on groundwater quality

Groundwater on the Tummulur site is characterised by high salinity (electrical conductivities between 800 and 1,700 µS/cm). Chemical groundwater facies is dominated by Na-HCO_3^- (sodium-hydrogen carbonate) water type (Figure 7.9) mainly caused by cation exchange, silicate weathering and leaching of fluid inclusions (Siva Soumya *et al.* 2013). The abundance of major anions is as hydrogen carbonate > sulphate$^-$ > nitrate or chloride > fluoride and that of major cations is sodium > calcium > magnesium > potassium. The sampled ground water shows concentrations of nitrate and fluoride above permissible limits for drinking water (50 mg/L and 1.5 mg/L, respectively according to the WHO guidelines or 50 mg/L and 1.2 mg/L respectively according to the Bureau of Indian Standards (BIS, 2012)).

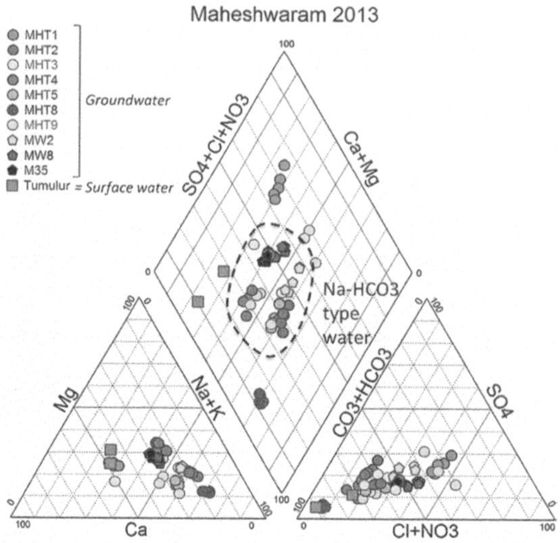

Figure 7.9 Piper diagram for groundwater (10 wells) and Tummulur tank surface water for the 2013–2014 campaign.

In the Maheshwaram context (i.e. strong evaporation rates, mineral dissolution of primary fluoride containing minerals such as fluorapatite, biotite and epidote, important irrigation return flow) fluoride is accumulated in groundwater (Pettenati *et al.* 2013). This accumulation is enhanced by the reduction of calcium activity due to calcite precipitation and by calcium/sodium exchange mechanism on clay minerals, thus impeding fluorite (CaF_2) precipitation, which is the only efficient mechanism controlling fluoride concentrations (Jacks *et al.* 2005). As a result of the strong cationic exchanges, most of the groundwater samples show a sodium excess, as seen on a scatter plot comparing the sodium + potassium (Na + K) excess vs. calcium + magnesium (Ca + Mg) excess ((Na + K)-Cl vs. (Ca + Mg)-(SO_4 + HCO_3) plot, Figure 7.10). In this geological context, hydrogen carbonate ions originate from the silicate weathering process (Rajesh *et al.* 2012), and organic matter mineralization leading to high carbon dioxide (CO_2) content in the soils can significantly increase the hydrogen carbonate content. High sulphate, nitrate and chloride contents do not come from the water-rock interaction; they mainly have anthropogenic and/or meteorological origin, enhanced by evaporation processes in soils (Rajesh *et al.* 2012; Siva Soumya *et al.* 2013).

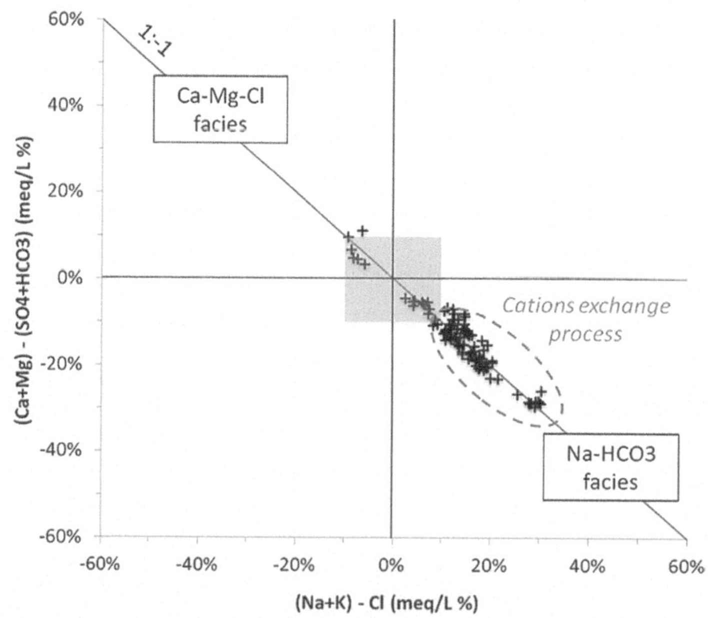

Figure 7.10 (Na + K)-Cl vs. (Ca + Mg)-(SO_4 + HCO_3) plot for Tummulur site groundwater (7 scientific boreholes and 3 farm boreholes) for the 2013–2014 sampling campaigns. NB: the ±10% area (grey area) does not allow accounting for ion exchange processes.

This incoming surface water has a wide range of effects on groundwater chemistry, depending on both time and space. Electrical conductivities between pre- and post-monsoon samples are quite constant, but have strongly varied during monsoon (Figure 7.11a).

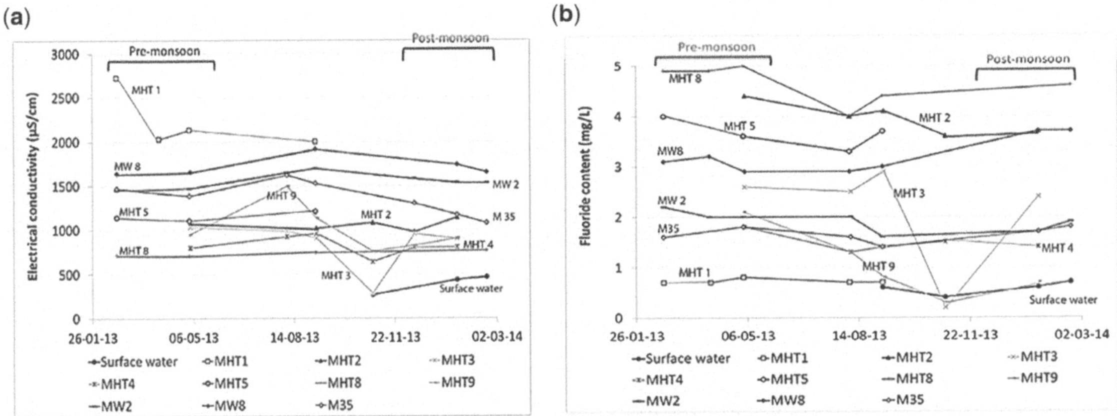

Figure 7.11 (a) Electrical conductivities in 10 boreholes and surface water on Tummulur site between February 2013 (pre-monsoon) and February 2014 (post-monsoon). (b) Fluoride content evolution in 10 boreholes and surface water on Tummulur site between February 2013 (pre-monsoon) and February 2014 (post-monsoon).

Nitrate, sulphate and chloride concentrations tend to decrease after the monsoon due to dilution, but can significantly increase due to leaching of soil processes (i.e. the chloride content increase in MHT9 during the early stage of monsoon season and the important input of nitrates at the beginning of monsoon for MHT4 and MHT9 and in the latest stage of monsoon for MHT2 and MHT3 – Figure 7.12).

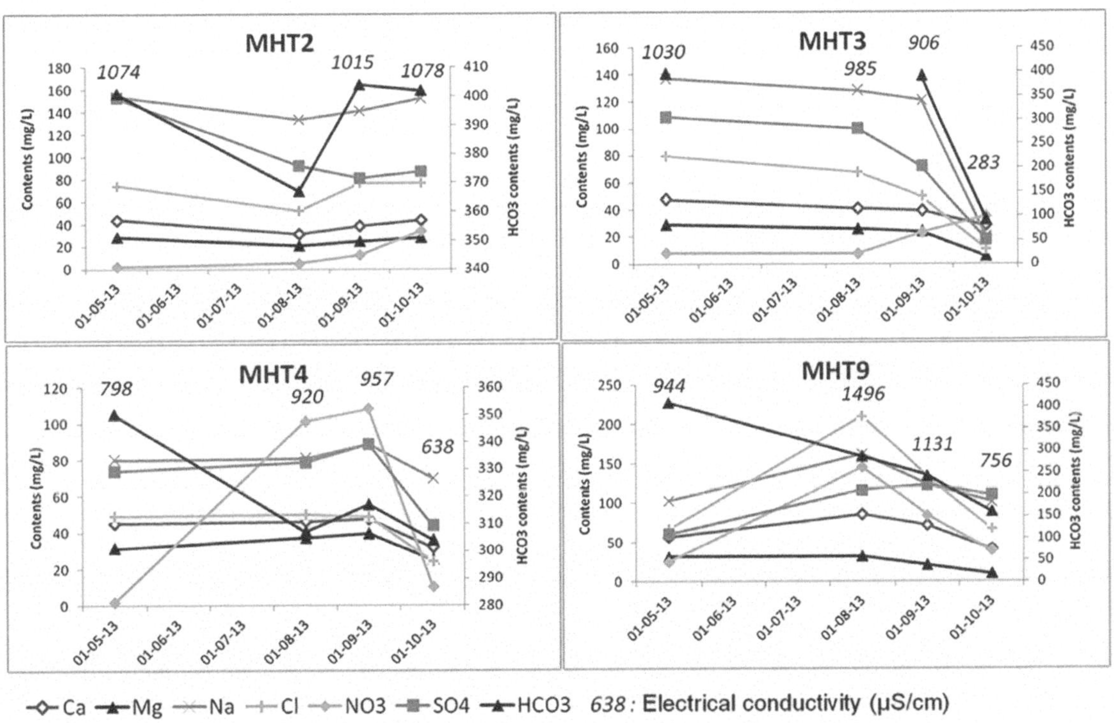

Figure 7.12 Evolution of calcium (Ca^{2+}), magnesium (Mg^{2+}), sodium (Na^{+}), chloride (Cl^{-}), nitrate (NO_3^-), sulphate (SO_4^{2-}) and hydrogen carbonate (HCO_3^-) contents (mg/L) for the campaigns of May (i.e. before monsoon), August and September (i.e. early stages of the monsoon), and October 2013 (i.e. late stage of the monsoon) with the corresponding electrical conductivity.

Except for MHT9, fluoride concentration do not decrease after the monsoon showing the fluoride content evolution complexity, i.e. an input of fresh water does not lead to its decrease (Figure 7.10b). Quite rapid chemical weathering of fluoride bearing minerals and cationic exchanges within clay minerals can occur during the percolation time, in these highly eroded environments (Alazard *et al.* 2015).

7.3.5 Stable isotopes

The recharge from the Tummulur tank is highlighted by the change in isotopic content ($\delta^{18}O$ and δ^2H) of surface and groundwater between pre- and post-monsoon samples (Figure 7.13). Pre-monsoon groundwater samples (i.e. 'dry conditions') form a line which deviates from the local meteoric water line (LMWL) according to an evaporation line with the same slope value (=4.2) found by Négrel *et al.* (2011). The LMWL was defined by Kumar and Pande (2010) for South India as $\delta^2H = 7.82 (\pm 0.17)^* \delta^{18}O + 10{:}23 (\pm 0.85)$ ‰ vs. Vienna Standard Mean Ocean Water (VSMOW).

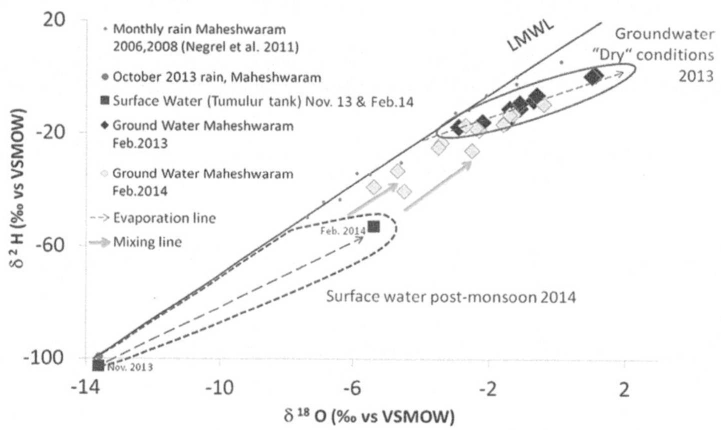

Figure 7.13 $\delta^{18}O$ vs. δ^2H (‰ vs. VSMOW) plot of monthly rain in Maheshwaram (Negrel *et al.* 2011), groundwater (February, 2013 and 2014) and Tummulur tank (November, 2013 and February 2014). Evaporation lines and mixing lines are shown as dashed lines and grey arrows.

The rain event of October 2013 was strongly depleted ($\delta^{18}O = -13.6$ and $\delta^2H = -99.3$‰ vs. VSMOW) and plot on the global meteoric water line (GMWL). As a result of this strong event, the surface water in November 2013 and February 2014 tuned out very low. Since October 2013, no significant input has occurred in the tank. As a consequence, the isotopic signature of the surface water in February 2014 results from the evaporation of the water body since October 2013 and strays from the GMWL.

Some post-monsoon groundwater samples deviate from the pre-monsoon values. Mixing lines between the surface water and the groundwater during 'dry conditions' appear. The post-monsoon groundwater samples, under the influence of surface water, move between the two poles, depending on the rate of surface water mixing (Figure 7.13).

7.4 DISCUSSION

This study is in agreement with the existing studies performed on percolation tanks in crystalline rocks aquifers performed in south India. It shows that under normal conditions the ratio percolation/stored water ranges in general between 56–63% (Massuel *et al.* 2014: 57% to 63%; Mehta *et al.* 1997: 57%; Perrin *et al.* 2012: 56%; Singh *et al.* 2004: 63%) and can be slightly higher in case of extreme monsoon and high water levels (in this study 75% for the 2013 monsoon). At the watershed scale (53 km^2) the tank impact can be negligible as for the 2012 monsoon but can reach 2.6% of the watershed abstraction in case of extreme monsoon, as in 2013. In the latter case presence of group of tanks may have a significant impact on the water balance of the watershed. Results of this study are in agreement with estimates from Perrin *et al.* (2012) on another watershed in a similar context stating that the multiple percolation tanks may contribute up to 33% of the recharge at the watershed scale (32 tanks on an 84 km^2 area near Gajwel AP). It is important to note that at a small scale the tank clearly enhances groundwater availability on the rainy years (i.e. rainfall are above average and when natural recharge is the most important) but cannot be considered as a solution to counteract directly dry years. Water stored during the rainy years should be partially kept for further possible dry years. This requires multiple-years management plans.

These management plans should include monitoring of water levels, limitations of water abstraction, maintenance of the tank to increase infiltration efficiency. From the two monitored years, the second monsoon event contributes to 95% of the percolated water of these two years. This highlights the strong variability of the monsoon and its obvious impact on the stored water volume. Often analyzed during one to three years-duration programs, tank efficiency studies do not allow assessing the variability related to the erratic monsoon behavior on a long term basis. Long term monitoring should be developed as well as methodologies to build reliable long term budgets. Those percolation tanks may also increase water abstraction at the watershed scale since the farmers adapt their crop to the water availability in crystalline rock aquifers (Fishman *et al.* 2011; Aulong *et al.* 2012).

An analysis of the flow dynamic through borehole logging highlights complex flow patterns during the recharge processes. The steep transition in electrical conductivity logs between pre and post-monsoon shows the complexity of the recharge mechanisms. It appears that the aquifer encompasses several water bodies, with distinct chemical features, which exchange evolving during the recharge processes. This latter point is further discussed in Alazard *et al.* (2015). This also highlights the existence of different flow paths and that recharge occurs not only vertically but that a strong horizontal flow has an impact on the change in water level.

Geological information coupled with hydraulic tests point out the contrast between storage capacities in the different layers of the aquifer. The deepest fractures may rapidly respond but do not represent a large water volume even with an important rise in water level. This specificity can lead to confusion since, most of the time, as observed by Batchelor *et al.* (2003), the monitoring of percolation tank efficiency is based on the rise of water level in nearby boreholes, without taking into account the properties of the local media. In hard rock aquifers it should also be noted that the geological structure may create inequity between farmers in the recovery of water since accessibility is constrained by the connectivity of the fractures which decreases with depth, leading in case of low-water levels to compartmentalisation of the aquifer (Guihéneuf *et al.* 2014). For example, Boisson *et al.* (2014b) show that in the Tummulur case in 2012, only 2 farmers appear to pump 53% to 88% of the stored water during years of low rainfall while the impact for the farmers at the south side of the tank is negligible. For large scale planning it is also important to take into account the externalities for the downstream users since the water collected in a given watershed will not flow downstream (Calder *et al.* 2008). The water balance presented here is based on observation and cannot be used for prediction. Simple predictive modelling based on water budget as the one developed from the same site presented in Boisson *et al.* (2014a) are efficient, easy to perform and require limited monitoring. However, as observed by Boisson *et al.* (2014b), the contribution of surface water to the real effective recharge of the aquifer tends to be overestimated due to possible storage in the unsaturated zone, especially in cases of a weak monsoon. More investigation should be carried out at this point to develop guidelines and management policies.

A tank's impact on water quality is usually considered as positive since it appears to dilute contaminants. However, in the case of geogenic contaminants such as fluoride, the infiltration of water with a different chemistry may change equilibriums and may tend to release fluoride in the aquifer (Pettenati *et al.* 2014). Also, in case of polluted water or leaching soils, infiltration of surface water may decrease water quality (bacteria or solute).

This percolation tank assessment shows that those structures may, under certain conditions (i.e. heavy monsoon) be a possible solution to enhance ground water availability but is of limited impact during average monsoon conditions. However, these percolation tanks need to be carefully considered and guidelines should be defined for assessment and management.

7.5 CONCLUSION

This book chapter synthesizes the investigations performed on the Tummulur tank through Saph Pani. Details may be found in related publications. The main conclusions of these investigations are:

1) Percolation tanks can be efficient to recharge water in case of high rainfall but have negligible impact on groundwater replenishment in case of low to average monsoon years (i.e. below ~750 mm/year).
2) Tank replenishment occurs during a few limited rainfall events (2–3) per year through run-off and is therefore sensitive to surrounding changes in land use.
3) Water recovery in a hard rock aquifer may create inequity between farmers depending on the borehole's location and connectivity to conductive fractures due to the geological media heterogeneity (details in Boisson *et al.* 2014b; Alazard *et al.* 2015).
4) Percolation tanks may locally and temporarily enhance water extraction due to an increase of paddy field cultivated areas and, hence, can have a limited impact on the watershed water balance.
5) Hydrogeological studies and dense piezometric networks should be performed to ensure accurate monitoring, taking geological heterogeneity (both horizontal and vertical) into account.

6) Simple guidelines for MAR implementation and monitoring should be developed taking in account the conceptual models presented here.
7) Infiltration does not systematically improve water quality by dilution (e.g. fluoride) (See details in Pettenati *et al.* 2014). Therefore careful monitoring of water quality with regard to fluoride should be performed regularly to ensure adequate water quality.

7.6 REFERENCES

Acworth R. I. (1987). The development of crystalline basement aquifers in a tropical environment. *Q J. Eng. Geol.*, **20**, 265–272.

Alazard M., Boisson A., Pettenati M., Dewandel B., Perrin J., Picot-Colbeaux G., Maréchal J. C., Kloppmann W. and Ahmed S. (2015). Recharge dynamics and flow structure investigation through borehole logging in south Indian fractured crystalline aquifer under semi-arid climate. Implication on MAR and water quality. *Submitted.*

Aulong S., Chaudhuri B., Farnier L., Galab S., Guerrin J., Himanshu H. and Reddy P. P. (2012). Are South Indian farmers adaptable to global change? A case in an Andhra Pradesh catchment basin. *Regional Environmental Change*, **12**(3), 423–436.

Batchelor C. H., Rama Mohan Rao M. S. and Manohar Rao S. (2003). Watershed development: A solution to water shortages in semi-arid India or part of the problem? *Land Use Water Resour. Res.*, **3**, 1–10.

Boisson A., Villesseche D., Baisset M., Perrin J., Viossanges M., Kloppmann W., Chandra S., Dewandel B., Picot-Colbeaux G., Rangarajan R., Maréchal J. C. and Ahmed S. (2014a). Questioning the impact and sustainability of percolation tanks as aquifer recharge structures in semi-arid crystalline context. *Environmental Earth Sciences*, 1–11. DOI:10.1007/s12665–014–3229–2.

Boisson A., Baisset M., Alazard M., Perrin J., Villesseche D., Kloppmann W., Chandra S., Dewandel B., Picot-Colbeaux G., Ahmed S. and Maréchal J. C. (2014b). Comparison of surface and groundwater balance approaches in the evaluation of managed aquifer recharge structure: Case of a percolation tank in hard rocks aquifer in India. *Journal of Hydrology*, **519**, 1620–1633. DOI:10.1016/j.jhydrol.2014.09.022.

Boisson A., Guihéneuf N., Perrin J., Bour O., Dewandel B., Dausse A., Viossanges M., Ahmed S. and Maréchal J. C. (2015). Determining the vertical evolution of hydrodynamic parameters in weathered and fractured south Indian crystalline-rock aquifers: Insights from a study on an instrumented site. *Hydrogeol J.*, **23**: 757–773. DOI:10.1007/s10040–014–1226-x.

Bour O. and Davy P. (1998). On the connectivity of three-dimensional fault networks. *Water Resources Research*, **34**, 2611–2622.

Calder I., Gosain A., Rama Mohan Rao M. S., Batchelor C., Snehalatha M. and Bishop E. (2008). Watershed development in India. 1. Biophysical and societal impacts. *Environ Dev Sustain*, **10**(4), 537–57. DOI:10.1007/s10668–006–9079–7.

Chilton P. J. and Foster S. S. D. (1995). Hydrogeological characteristics and water-supply potential of basement aquifers in Tropical Africa. *Hydrogeol. J.*, **3**(1), 3–49.

De Dreuzy J., Davy P. and Bour O. (2001). Hydraulic properties of two-dimensional random fracture networks following a power law length distribution 1. Effective connectivity. *Water Resources Research*, **37**, 2065–2078.

Dewandel B., Lachassagne P., Wyns R., Maréchal J. C. and Krishnamurthy N. S. (2006). A generalized 3-D geological and hydrogeological conceptual model of granite aquifers controlled by single or multiphase weathering. *Journal of Hydrology*, **330**, 260–284.

Dewandel B., Perrin J., Ahmed S., Aulong S., Hrkal Z., Lachassagne P., Samal S. and Massuel S. (2010). Development of a tool for managing groundwater resources in semi-arid hard rock regions: Application to a rural watershed in South India. *Hydrol. Process*, **24**, 2784–2797. DOI:10.1002/hyp.7696.

Dillon P., Gale I., Contreras S., Pavelic P., Evans R. and Ward J. (2009). Managing aquifer recharge and discharge to sustain irrigation livelihoods under water scarcity and climate change. In: Improving Integrated Surface and Groundwater Resources Management in a Vulnerable Changing World, IAHS – The International Association of Hydrological Sciences (ed.), IAHS Publishing, Wallingford, United Kingdom, pp. 1–12.

Fishman R. M., Siegfried T., Raj P., Modi V. and Lall U. (2011). Over-extraction from shallow bedrock versus deep alluvial aquifers: Reliability versus sustainability considerations for India's groundwater irrigation. *Water Resour Res*, **47**(6), 1–15. DOI:10.1029/2011WR010617.

G.S.I. (2005). Geological and mineral map of Andhra Pradesh 1:500 000. Geological Survey of India.

Glendenning C. J., van Ogtrop F. F., Mishra A. K. and Vervoort R. W. (2012). Balancing watershed and local scale impacts of rain water harvesting in India – A review. *Agric. Water Manage.*, **107**, 1–13.

Guihéneuf N., Boisson A., Bour O., Dewandel B., Perrin J., Dausse A., Viossanges M., Chandra S., Ahmed S. and Maréchal J. C. (2014). Groundwater flows in weathered crystalline rocks: Impact of piezometric variations and depth-dependent fracture connectivity. *Journal of Hydrology*, **511**, 320–334.

Jacks G., Bhattacharya P., Chaudhary V. and Singh K. P. (2005). Controls on the genesis of some high-fluoride groundwaters in India. *Applied Geochemistry*, **20**, 221–228.

Kumar S. R., and Pande N. G. (2010). Isotopic characteristics of Indian precipitation. *Water Resour. Res.*, **46**, W12548, DOI:10.1029/2009WR008532.

Maréchal J. C., Dewandel B. and Subrahmanyam K. (2004). Use of hydraulic tests at different scales to characterize fracture network properties in the weathered-fractured layer of a hard rock aquifer. *Water Resour. Res.*, **40**(11), 1–17. DOI:10.1029/2004WR003137.

Maréchal J. C., Dewandel B., Ahmed S., Galeazzi L. and Zaidi F. K. (2006). Combined estimation of specific yield and natural recharge in a semi-arid groundwater basin with irrigated agriculture. *Journal of Hydrology*, **329**, 281–293.

Massuel S., Perrin J., Mascre C., Mohamed W., Boisson A. and Ahmed S. (2014). Managed aquifer recharge in South India: What to expect from small reservoirs in hard rock. *J. Hydrol.*, **512**, 157–167.

Mayo A. L., Morris T. H., Peltier S., Petersen E. C., Payne K., Holman L. S., Tingey D., Fogel T., Black B. J. and Gibbs T. D. (2003). Active and inactive groundwater flow systems: Evidence from a stratified, mountainous terrain. *GSA Bull*, **115**(12), 1456–1472.

Metha M. and Jain S. K. (1997). Efficiency of artificial recharge from percolation tanks. In: Recharge of Phreatic Aquifers in (semi-) Arid Areas, I. Simmers (ed.), IAHS Publishers, Hyderabad, pp. 271–277.

Négrel P., Pauwels H., Dewandel B., Gandolfi J. M., Mascré C. and Ahmed S. (2011). Understanding groundwater systems and their functioning through the study of stable water isotopes in a hard-rock aquifer (Maheshwaram watershed, India). *Journal of Hydrology*, **397**, 55–70.

Perrin J., Mascre C., Pauwels H. and Ahmed S. (2011). Solute recycling: An emerging threat to groundwater quality in southern India? *Journal of Hydrology*, **398**(1–2), 144–154.

Perrin J., Ferrant S., Massuel S., Dewandel B., Maréchal J. C., Aulong S. and Ahmed S. (2012). Assessing water availability in a semi-arid watershed of southern India using a semi-distributed model. *Journal of Hydrology*, **460**, 143–155.

Pettenati M., Perrin J., Pauwels H. and Ahmed S. (2013). Simulating fluoride evolution in groundwater using a reactive multicomponent transient transport model: Application to a crystalline aquifer of Southern India. *Applied Geochemistry*, **29**, 102–116.

Pettenati M., Picot-Colbeaux G., Thiéry D., Boisson A., Alazard M., Perrin J., Dewandel B., Maréchal J. C., Ahmed S. and Kloppmann W. (2014). Water quality evolution during managed aquifer recharge (MAR) in India crystalline basement aquifers: Reactive transport modeling in critical zone. *Procedia Earth and Planetary Science*, **10**, 82–87.

Rajesh R., Brindha K., Murugan R. and Elango L. (2012). Influence of hydrogeochemical processes on temporal changes in groundwater quality in a part of Nalgonda district, Andhra Pradesh, India. *Environ. Earth. Sci.*, **65**, 1203–1213.

Sakthivadivel R. (2007). The groundwater recharge movement in India. In: The Agricultural Groundwater Revolution: Opportunities and Threats to Development, Giordano M. and Villholth K. H. (eds.), CAB International, Wallingford, United Kingdom, pp. 195–210.

Singh J., Awashi M. K. and Sharma R. K. (2004). Quantification of percolation from percolation tank. *J.Soil Wat. Conserv. India*, **3**(3–4), 128–132.

Siva Soumya B., Sekhar M., Riotte J., Banerjee A. and Braun J. J. (2013). Characterization of groundwater chemistry under the influence of lithologic and anthropogenic factors along a climatic gradient in Upper Cauvery basin, South India. *Environ. Earth. Sci.*, **69**, 2311–2335.

White A. F., Bullen T. D., Schulz M. S., Blum A. E., Huntington T. G., and Peters N. E. (2001) Differential rates of feldspar weathering in granitic regoliths. *Geochem Cosmochem Acta*, **65**(6), 847–869.

Wyns R., Baltassat J. M., Lachassagne P., Legtchenko A. and Vairon J. (2004). Application of proton magnetic resonnace soundings to groundwater reserves mapping in weathered basement rocks (Brittany, France). *Bull Soc Géol France*, **175**, 21–34.

Chapter 8

Constructed wetlands and other engineered natural treatment systems: *India status report*

Dinesh Kumar, Saroj Kumar Sharma and Shyam R. Asolekar

8.1 INTRODUCTION

India has one of the largest numbers of small and medium enterprises (SMEs) engaged in a variety of sectors including petrochemical, fertilizer, fine chemicals, pharmaceuticals and intermediates, dyes, paints, pigments as well as automobile and mechanical jobbing industry. Subsequent to the liberalization of economy and industry in the later part of last century, India has seen a tremendous growth in industry as well as urbanized population. As a result, the Government of India (GoI) has been investing in the expansion and strengthening of urban infrastructure sectors over the past three decades. In that context, there have been systematic efforts of up-scaling of facilities for treatment and distribution of drinking water as well as collection, treatment and disposal of wastewater (Arceivala & Asolekar, 2006; Asolekar *et al.* 2013).

The task of treatment and disposal of effluents generated by large-scale industries appear to be relatively under control when compared to the challenge posed by SMEs. Barring the few exceptions, the situations pertaining to collection, treatment and disposal of wastewater in cities and towns continue to be inadequate in most of the municipalities in India (Arceivala & Asolekar, 2012; Kalbar *et al.* 2013). Furthermore, the challenge of treating wastewater generated by rural communities has not even been addressed.

8.1.1 Significance of natural treatment systems in the context of India

As reported by Asolekar (2002), disposal of untreated or partially treated effluents into rivers and lakes as well as run-off from urban and agricultural areas are the two main reasons responsible for deterioration of drinking water resources in India. It is clear that less than 10% of the generated wastewater are treated effectively, while the rest of the wastewater find their ways into the natural ecosystems in the vicinity. In addition, excessive withdrawal of water for agricultural and municipal utilities as well as use of rivers and lakes for religious and social practices and perpetual droughts limit the capacity of natural water sources to provide adequate dilution (Asolekar, 2002; Asolekar *et al.* 2013; Starkl *et al.* 2013; Chaturvedi *et al.* 2014).

According to the statistics of year 2005, presented by Chaturvedi and Asolekar (2009) on wastewater management in India, about 26,000 MLD of wastewater was reportedly collected cumulatively in two mega cities (population above 5 million), 11 large metro cities (population from 2 to 5 million) and 26 small metro cities (population from 1 to 2 million), 384 class I cities (population from 100,000 to 1 million) and 498 class II cities (population between 50,000 and 100,000). These urban centres are inhabited by more than 70% of India's 500 million urban population. Overall, merely 27% of urban wastewater received some kind of treatment.

The statistics of year 2009 revealed a similar trend. In all, 38,254 MLD of wastewater were generated from class I cities and class II towns but only a treatment capacity of 12,000 MLD existed (Central Pollution Control Board (CPCB), 2009). The class I cities of India are contributing about 93% of total wastewater generated by class-I cities and class-II towns. The wastewater generated in class-I cities was estimated to be 35,558 MLD and treatment capacity exists for only 11,553 MLD in these cities, *i.e.*, only 32% of wastewater is being treated, whereas the rest is disposed untreated. In India, there are 35 metropolitan cities (with a population of more than 1 million) which are generating wastewater of 15,644 MLD but the existing treatment capacity is 8,040 MLD, which is only 51% of the total wastewater generated in these cities. The generated wastewater in class-II towns was estimated as 2,696 MLD and only 233 MLD treatment capacities exist in these cities, which show that only 8% of wastewater is being treated. Thus, there is a large gap between the amount of wastewater generation and treatment in India. Due to disposal of the untreated wastewater into water bodies, both surface and groundwater are being contaminated. The Central Pollution Control Board (CPCB) (2009) also reported unsatisfactory operation and maintenance (O&M) of existing wastewater treatment plants (WWTPs) and pumping stations, as nearly 39% WWTPs are not conforming to the minimum standards prescribed under the prevailing regulatory standards meant for disposal of treated wastewater into rivers and lakes (receiving water bodies).

The use of natural treatment systems (NTSs) for treatment of domestic wastewater was practiced in ancient India. The community tanks in villages, water bodies maintained by temples for performance of religious functions and crimination rites, irrigation systems installed and maintained in community-joint forests invariably received controlled flows of wastewater. These were some of the noteworthy examples of sustainable wastewater management in India's village ecosystems (Jana, 1998; Chaturvedi & Asolekar, 2009).

At the level of the Central Government, the Ministry of Rural Development, the Ministry of Urban Development, Ministry of Environment and Forests (MoEF) as well as Ministry of Water Resources (MoWR) and Ganga Rejuvenation have been incorporating the strategy of providing low-cost eco-centric treatment to wastewater for correcting the pollution of natural water courses in India. The Jawaharlal Nehru National Urban Renewal Mission and several programs have been implemented by the GoI over the past three decades. Similarly, the State Governments in the Union of India have also been complimenting efforts in the respective states and favouring the decentralized treatment technologies to address issues associated with disposal of wastewater.

8.1.2 Scope and objectives

Clearly, there exists a looming challenge of inadequate and insufficient infrastructure for treatment of wastewater throughout India, both in urban as well as rural communities. The Union of India has exhibited a serious commitment to fulfilling this basic necessity of rural and urban communities – responding to the political pressure exerted by them. For example, as reported by Asolekar (2013), in the context of rejuvenation and ecological up-gradation of the Ganga River, the entire North of India (almost 400 million people) has forged an alliance on political and social platforms.

Currently, the Honourable Supreme Court of India has ordered the responsible State Governments in the Union of India, including the MoEF, to ascertain that the untreated and partially treated wastewater shall not be disposed into the tributaries and main stream of Ganga River. Already, over the past two decades, there have been concerted efforts in the direction of up-scaling of infrastructure for wastewater treatment all over India. On one hand, there are several communities waiting eagerly to be included in the programme for improving sanitation, while on the other hand, the budgets allocated to wastewater treatment facilities are not adequate.

At such a crossroad, identification and adoption of so-called "appropriate technological solutions" will become more critical than ever – especially in a developing economy like India (Kalbar *et al.* 2012). As argued by Kumar and Asolekar (2015) the broad class of engineered NTSs- including horizontal sub-surface flow constructed wetlands (HSSF-CWs) – will continue to dominate the platform of favoured technologies for treatment and recycling of wastewater in warmer climates and increasing of the unmet demand for waters for irrigation and industry.

In this context, an attempt has been made to assess the status of engineered NTSs, including HSSF-CWs installed all over India, in order to manage wastewater and in some cases mixed with biodegradable industrial effluents. In this study, however, the other NTSs such as riverbank filtration (RBF), soil aquifer treatment (SAT), managed aquifer recharge (MAR) or some other riparian zone technologies to address agricultural and urban run-off have not been included and are addressed in other chapters.

Currently, in India, there are substantial efforts to incorporate NTSs into wastewater treatment facilities in smaller communities. However, not all the facilities are working satisfactorily nor meet the design and regulatory expectations. Typologies of the reasons behind their failure and success have also been articulated in this chapter. It is hoped that the assessment presented in this chapter may also be helpful for the planners and implementing agencies in developing countries like India in meeting the challenge of wastewater treatment in the years to come with the help of low-cost and eco-centric technological interventions.

8.2 METHODOLOGY

The survey focused on HSSF-CWs and other NTSs currently employed in treatment of domestic wastewater in India. The prior experience of Asolekar and co-workers influenced this survey (Chaturvedi & Asolekar, 2009; Asolekar *et al.* 2013).

In all, five technologies practiced at various locations in India were selected for site assessment, namely: HSSF-CWs, duckweed ponds (DPs), waste stabilization ponds (WSPs), polishing ponds (PPs) and Karnal-type constructed wetlands (KT-CWs) for on-land disposal of wastewater.

8.2.1 Questionnaire for the survey and identification of the sites

Before starting the activities related to assessment of the potential of existing HSSF-CWs and other NTSs for wastewater treatment and reuse across India, a questionnaire was developed for collecting data from the field. The questionnaire was developed after a broad discussion with experts working in the area of natural treatment technologies as well as with the partners from the Saph Pani Project to incorporate various data requirements. Appendix 8.1 exhibits the questionnaire utilized during the field survey. As can be noticed the questionnaire includes three kinds of data about a given field site; *viz*. first, the technical data, second, the data on economics and finally, the social and consumer related data.

The primary aim of identification of prospective sites for field investigation was to identify and seek permission from the respective municipal authorities to investigate if the potential of the NTSs installed for treatment of the wastewater generated by the respective communities is met in reality. In addition, it was hoped that the choice of sites would be representative of the actuality of the technology practiced today in the context of municipal wastewater management. After all, the real proof of the utility of the survey proposed in the research lies in the fact that the concerned development and implementing agencies would find the learnings from the survey useful for development and monitoring of wastewater treatment facilities. Numerous sites of HSSF-CWs and other NTSs were found all over India (Appendix 8.2).

A closer look at the 108 sites listed in the Appendix 8.2 and in the light of information collected in the reconnaissance survey; it was decided to study around one third of the sites through a questionnaire survey so that the conclusions drawn from the survey would be of practical relevance. Thus, 41 sites were finally shortlisted for administering the questionnaire survey for this study.

8.2.2 Data collection and assessment

A tentative list of engineered HSSF-CWs and other NTSs was prepared after discussion with various water and wastewater practitioners as well as governing and regulatory bodies, including state pollution control boards, public health engineering departments of different states, and water and sewerage boards. A literature review was also carried out in order to select the most appropriate and representative sites for assessment. After identifying the potential representative sites of HSSF-CWs and other NTSs across the country, the identified sites were visited in order to obtain the relevant information mentioned in the questionnaire. The specifications and the data related with the identified sites of HSSF-CWs and other NTSs were cross-checked with plant operators onsite during the visit and were documented in the database. It aimed at understanding technical and management related facts as well as obtaining qualitative description of the issues faced while providing the treatment at the respective wastewater treatment facility.

The 41 identified sites of WWTPs were visited during the survey and secondary data were collected by interviewing the operating staff of the respective WWTPs as well as by utilizing the literature, log books, and progress reports supplied by the respective personnel. The data were logged into the questionnaires in the sections covering technical, physical, geographical as well as social aspects of the respective engineered systems. The assessment of selected WWTPs were planned and performed by visiting the shortlisted sites at least once (in some cases even twice or thrice). A two-step approach was adopted during the field work: first, the rapid national survey of identified engineered HSSF-CWs and other NTSs and second, the detailed assessment of selected representative sites.

In summary, 41 WWTPs based on engineered natural treatment technologies (WSPs, PPs, DPs, HSSF-CWs and KT-CWs) were surveyed between December 2011 and June 2014. In addition to collecting the technical data, views of the personnel related to the difficulties faced in routine and episodic O&M were also recorded. This assessment intends to highlight some of the more intricate and counter-intuitive lessons which can potentially be used for up-grading of technologies and effectiveness of NTSs in the Indian context as well as for articulation of policy and regulatory reforms in India.

8.3 RESULTS AND DISCUSSION

The data collected during the field survey was analysed from mainly three perspectives:

1) Performance of WWTPs based on engineered natural treatment technologies in India,
2) Problems associated with O&M of NTSs across India, and
3) Issues associated with management of wastewater in India

8.3.1 Performance of WWTPs based on engineered natural treatment technologies in India

Secondary data on performance of the WWTPs, reported by the respective operators of 41 shortlisted facilities across India, were collected through field visits over the period of 2.5 years. Table 8.1 summarizes the indicative statistics on efficacy of the technologies (values for Biological Oxygen Demand (BOD_5), Chemical Oxygen Demand (COD) and Faecal Coliforms (FC)) expressed as the ratios of the typical outlet to inlet concentrations.

Table 8.1 Summary of the performance reported by the respective operators of 41 shortlisted NTS-based WWTPs across India. The values for BOD_5, COD and FCs, indicative of efficacy of the technologies, were expressed as the ratios of the typical outlet to inlet concentrations.

Types of NTSs	Number of Sites	Average Annual Performance (% Removal)		
		BOD_5	COD	Cs
WSPs	23	50–96	62–82	90–99.3
PPs	7	33–69	17–46	90–98.75
DPs	3	89–95	NA	94.5–99
HSSF-CWs	6	61–93	64–90	99–99.99
KT-CWs	2	NA	NA	NA
Total	41			

The national survey of HSSF-CWs and other NTSs indicates that nearly 76% of the WWTPs investigated were generally achieving the *Minimum National Standards* stipulated by the MoEF (GoI) for disposal of treated wastewater into legally permitted surface water bodies; or for the purposes of land irrigation as prescribed in the *Water (Prevention and Control of Pollution) Act*, 1974 and companion regulations.

It is to be noted that the *Minimum National Standards* stipulated by the MoEF for disposal of treated wastewater into ambient aquatic environment were meant to be the guideline for ensuring the "minimum" performance expected from a given municipality. There are several communities, however, who believe in achieving much higher performance with respect to the quality of their treated wastewater so as to minimize the impacts on surrounding aquatic bodies. The local self-governments as well as the regulatory authorities are fully empowered under the prevailing environmental regime to make such determinations and implement these stringent standards at local levels on a case-to-case basis in consultation with the community and the stakeholders. Also, several communities (especially the ones that are land-locked) have no receiving water bodies for disposal of their treated effluents. There are several other locations where farms and city-spaces have been facing acute shortages of water for irrigation. In such instances, the MoEF and MoWR have been permitting land irrigation with treated wastewater meeting certain norms acceptable to the regulatory framework and have judiciously monitored the crops and vegetation in agriculture and commercial agro-forests. Thus, the GoI has developed and implemented the *Minimum National Standards* for disposal of treated wastewater into ambient aquatic environment as well as for on-land application for irrigation (as shown in Table 8.2) in conjunction with several other standards and safe-guards built into the prevailing regulatory framework.

Table 8.2 The Minimum National Standards stipulated by the MoEF (GoI) for disposal of treated wastewater into legally permitted surface water bodies or for the purposes of land irrigation through the Water (Prevention and Control of Pollution) Act, 1974. (Information based on CPCB, 2005).

Parameters	pH	BOD_5 [mg/L]	COD [mg/L]	TSS* [mg/L]	TDS** [mg/L]
Standards for discharge in streams	5.5–9	30	100	100	2100
Standards for land irrigation	5.5–9	100	–	200	–

* Total Suspended Solids (TSS)
** Total Dissolved Solids (TDS)

Local standards and guidelines, as discussed above, play a crucial role in making decisions with respect to the extent of treatment to be adopted as well as in determining the type of technology to be implemented for treatment of wastewater in a given community. It was clear after the national survey that nearly all the administrators and decision makers in the respective communities had thoughtfully gravitated to "engineered natural treatment systems" for treatment of wastewater generated by their communities. Responding to the local requirements, a variety of NTSs have apparently been chosen by the 41 communities investigated in the present survey. The relative distribution of numbers of facilities based on technologies employed by them is depicted in Figure 8.1. Out of the 41 selected WWTPs based on engineered natural treatment technologies, 23 plants had WSPs, three plants had DPs, seven plants had PPs and eight plants employed HSSF-CWs or KT-CWs.

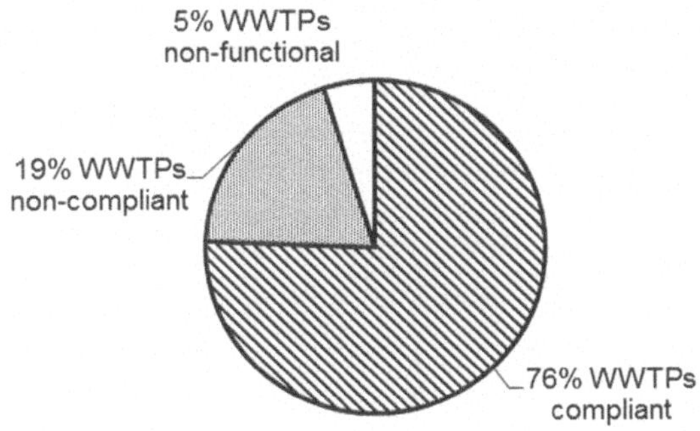

Figure 8.1 Distribution of cumulative number of WWTPs in various compliance categories among the 41 WWTPs surveyed during December 2011 and June 2014.

The relative distribution of numbers of WWTPs in various compliance categories among the 41 facilities surveyed during December 2011 and June 2014 is displayed in Figure 8.1. Similarly, Figure 8.2 presents the relative distribution of WWTPs in various compliance categories among the 41 facilities surveyed all over India. Typically, technologies like HSSF-CWs, KT-CWs as well as DPs seem to cater to the communities, generating relatively smaller flow rates of wastewater compared to technologies including WSPs and PPs. These data clearly suggest that size of a given community has a lot to do with selecting centralized versus decentralized technology for management of their wastewater.

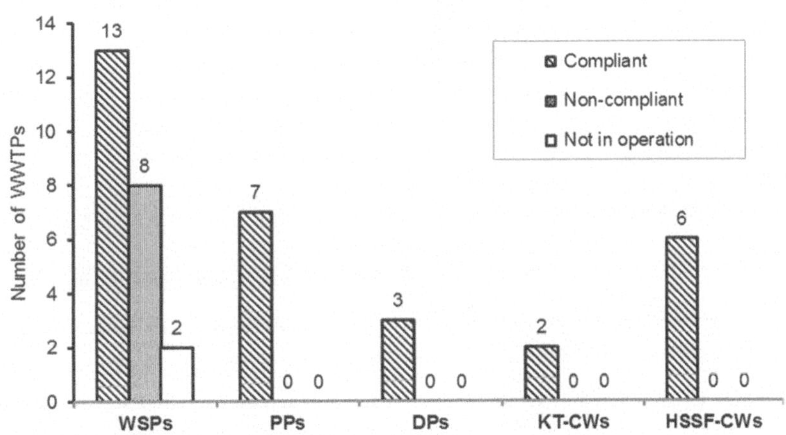

Figure 8.2 Compliance status of individual technologies among the 41 WWTPs.

8.3.2 Natural treatment technologies practiced in India

A detailed review of a variety of NTSs practiced in Asia in general, and India in particular, is presented in various chapters of this Handbook and also by Arceivala and Asolekar (2006). Most of the NTSs consist of a train of individual unit processes, set-up in series, with the output of one process becoming the input of the next process. The first stage usually comprises physical processes that take out pollutants in a physico-chemical manner. After this, biological processes generally further treat the remaining pollutants. These may 1) convert dissolved or colloidal impurities into a solid or gaseous form, so that they can be removed physically, or 2) convert them into dissolved materials, which remain in solution and typically are not as undesirable as the original organic pollutants. The solids (residuals or sludge) which result from these processes form a side-stream and are typically treated for further stabilization and desirably converted into manure or soil conditioners and disposed of into the farms and commercial agro-forests and green city-spaces in the vicinity. These practices, however, are customarily regulated by the empowered agencies so that the stabilized sludge does not introduce trace toxic metals and other pollutants typically emitted by industrial activities into farms and soils and thereby contaminate food.

Typically, WWTPs based on WSPs consist of a cascade of ponds. These ponds can be classified into three classes: 1) anaerobic ponds, 2) facultative ponds and 3) aerobic ponds. Alternately, on the basis of water depths, ponds may also be divided into two classes: a) shallow ponds and b) deep ponds. Shallow ponds (typical water depths <2.5 m) include conventional aerobic ponds as well as polishing or maturation ponds with marginal facultative conditions near the sediment-zone. The deep ponds (typical water depths >2.5 m) include facultative ponds having aerobic, facultative and anaerobic layers. The ponds are also at times anaerobic owing to their greater depths of 5 to 10 m. The generalized treatment processes adopted at most of the NTSs surveyed in this study (based on WSPs) are shown in Figure 8.3(a).

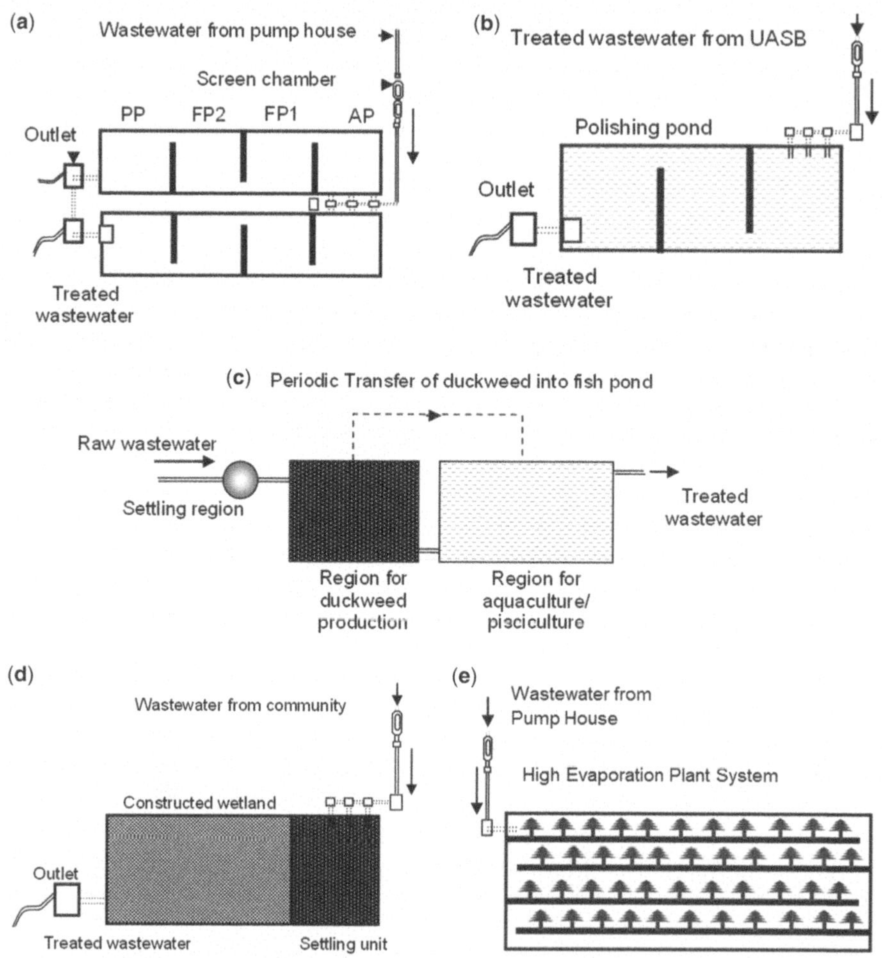

Figure 8.3 Schematic representation of the five natural treatment technologies typically installed in the 41 WWTPs surveyed across India: (a) Waste stabilization pond, (b) Polishing pond, (c) Duckweed pond, (d) Horizontal sub-surface constructed wetland and (e) Karnal-type constructed wetland.

The WWTPs based on the eco-centric technology of the DP has typically three treatment units, namely: 1) settling tank, 2) DP and 3) fishponds. After settling, the primary treated wastewater is subjected to a DP, where major reduction in carbon, nitrogen and phosphorus from the wastewater take place. The DPs are known for their combined action of phytoplankton, zooplankton and bacteria. Thus, the secondary treated wastewater from the DP finally is let into a fishpond to provide further polishing. The fishponds typically perform two functions. First, they provide some kind of polishing to the secondary treated wastewater. Second and more importantly, they consume the duckweed and algae and in response produce more fish – which could be harvested and sold in the market to make a profit and earn a livelihood. It is interesting to note that the duckweed generated in response to treatment in the DPs need to be routinely harvested and transferred into the associated fishponds to feed the fishes. The duckweed typically doubles its mass in two to three days under supportive conditions of nutrients, sunlight and temperature. The algae, however, are developed in fishponds in response to algal-bacterial "polishing" of secondary treated wastewater. Thus, treated wastewater emerging from DPs can be safely used for irrigation. Sizes of different treatment units in such systems are customarily estimated on the basis of biological kinetics of degradation of duckweed pond and fishpond, life cycle of fishes, climatic conditions and feasibility of land available.

The typical flow sheet for duckweed-fed aquaculture for wastewater treatment adopted at most of the places is depicted in Figure 8.3(c). At many places in India, DP systems have been found to be quite effective for treatment and reuse of rural wastewater. Also, they seem to perform well in various climatic conditions across India as well as meeting the prescribed regulatory standards.

One of the most commonly encountered systems for treatment of wastewater in rural areas and small urban communities across India is the so-called "constructed wetland" (CW). There are, by and large, two variants of CWs encountered among the present installations in India. As a part of the shortlisted WWTPs investigated in this survey, six HSSF-CWs and two KT-CWs were studied.

CWs, first developed in 1960s by Dr K. Seidel in Germany, are now accepted to be the low-cost, eco-centric technology especially beneficial for small towns that typically cannot afford expensive conventional treatment systems (Billore *et al.* 1999; Billore *et al.* 2001; Vymazal, 2010). A CW is a simple and effective wastewater treatment approach, which consists of a shallow depression in the ground with a levelled bottom. With incorporation of sophisticated flow controls and monitoring devices, it is possible to build the WWTPs with CW-technology to exercise a higher degree of control over the process and performance (Brix, 1997; Vymazal, 2013a). CWs seem to cater for nearly any combination of wastewater and biodegradable industrial effluent.

The CWs appear to perform all of the biochemical transformations related to the degradation of a variety of pollutants present in domestic and industrial wastewater including carbonaceous, nitrogenous and pathogenic constituents (Vymazal, 2013a; Vymazal, 2013b). The CWs can be employed in place of the commonly practiced conventional wastewater treatment strategy – which is not favoured on account of it being energy intensive and ineffective in removing pathogens. In a typical rural setting, CWs appear to treat wastewater to a higher degree when compared with the more conventional rural alternatives including septic tanks, drain fields and other forms of land treatment.

The HSSF-CW requires a primary treatment to raw wastewater before treating it into the wetland-bed. A primary treatment unit is normally installed in most of the treatment systems incorporating CWs to minimise the complications normally arising due to larger debris, garbage, floating polymeric wastes and fragments of packaging materials carried with the raw wastewater. It has become clear to the operators of the CW-systems that the life of wetland-beds would prolong if the superior primary treatment units are installed to remove even fine suspended solids in the influents to the WWTPs. Three plant species were most commonly found in CWs across India, namely: *Canna indica*, *Phragmites karka* and *Typha latifolia* .The typical flow sheet for HSSF-CW for wastewater treatment adopted at most of the places is depicted in Figure 8.3(d).

It is interesting to note here that HSSF-CW-systems have been found to be quite effective for treatment and reuse of wastewater generated by rural and town communities. The engineered wetland systems seem to be quite robust and versatile in a variety of climatic conditions across India as well as meet the prescribed regulatory standards. Communities seem to prefer them even more in the recent time owing to the innate advantage offered by HSSF-CWs in the context of minimizing mosquito breeding and thereby minimizing the threat of cerebral malaria, dengue and several vector-based diseases.

The other variant of CWs, KT-CWs have been installed in some places in India – especially in land-lock regions where there was typically no option for disposal of treated effluents. These systems were found to be quite effective for achieving complete evapotranspiration of wastewater subjected to them. In addition, the KT-CWs generate fuel-wood as well as feedstock for pulp and paper industry and thus provide an opportunity to engage in commercial agro-forestry to the community. The nutrients as well as the buffer capacity typically present in wastewater can potentially create a novel opportunity of application onto acidified and infertile wastelands. Thus, KT-CWs could probably become the most appropriate and economically viable proposition for the rural areas interested in restoring wastelands as well as generate biomass.

Those tree species which are fast growing and can transpire high amounts of moisture through evapotranspiration processes and are typically able to withstand high moisture contents in their root-zones are the most suitable for KT-CWs. For example, *Eucalyptus* is one such species. It has the capacity to transpire large amounts of water, and grows rather fast – thereby giving high yield of timber and green biomass. Raw wastewater is normally applied through furrow sand trees are planted on the ridges. The typical flow sheet for KT-CWs for wastewater treatment adopted in the systems investigated in this study is shown in Figure 8.3(e).

8.3.3 Problems associated with operation and maintenance of NTSs across India

Results of the national survey of HSSF-CWs and other NTSs across India indicated that at some places NTSs have failed in achieving the prescribed standards of treated wastewater. During the field visits, many reasons responsible for failures of NTSs were identified – which are summarized in this section.

It appears that there are several problems associated with mixing of industrial toxic effluents with domestic wastewater before subjecting into the treatment facility (Figures 8.4 and 8.5). For example, as shown in Figure 8.4, the KT-CW facility, in the City of Ujjain in central India, failed due to mixing of the industrial effluent generated by dyeing industry of cotton fabrics with urban wastewater. Reportedly, the colour of the mixture of domestic wastewater and textile effluents flowing into the KT-CW was visible frequently. Clearly, as seen from Figure 8.5, the trees (especially the foliage) were found to be wilting due to the toxic effects of industrial effluents. Reportedly, among the two KT-CW systems catering to the City of Ujjain, one KT-CW received only domestic wastewater (about half of the flow of wastewater from the City). The other KT-CW received wastewater flow that was mixed with textile industry wastewater. The operator of the facilities showed the difference in vitality of vegetation in the two KT-CWs. Similar observations were also made during field visits to WSPs wherein very poor performance was found (indicated by lower average% BOD_5 removal) due the mixing of industrial effluents with domestic wastewater entering the WWTPs.

Figure 8.4 Effluent from the textile industry was mixed with domestic wastewater at the inlet of KT-CW in the City of Ujjain, central India.

Figure 8.5 Wilting of vegetation resulting from toxicity of industrial pollutants was observed.

By and large, poor O&M of primary treatment units (or absence of it) was found to be one of the major causes of failure of NTSs based on CW-technology. Figures 8.6 and 8.7 exhibit two examples of such lapses found while visiting the *Ekant Park* HSSF-CW and the HSSF-CW installed at *City WWTP*, respectively (both located in the City of Bhopal, State of Madhya Pradesh in Central India). It was evident during the site visit that there was no periodic cleaning of sludge accumulated in primary treatment unit.

Figure 8.6 Poor maintenance of the primary treatment unit prior to HSSF-CW in Ekant Park, City of Bhopal, central India.

Figure 8.7 Chocking of wetland bed was observed in the CW at Ekant Park.

Even the CW-beds faced similar negligence on part of the respective civic authorities and the failure of NTSs based on HSSF-CW was feared by the operators of the facility of CW-bed (located in the City of Bhopal) and in the outskirts of the City of Ropar, State of Punjab (in northern India), respectively, as depicted in Figures 8.8 and 8.9. At both sites, the CW-beds were chocked with weeds and unwanted growth of planted vegetation was evident. Though both WWTPs were giving satisfactory quality of treated effluents at the time of the survey, the need for systematic and disciplined harvesting of biomass as well as implementing de-weeding programme thoroughly from time-to-time cannot be overemphasized.

Similarly, in Figures 8.10 and 8.11, the poor maintenance of primary treatment units carries forward the unsettled and floating debris to the first pond (anaerobic pond) in the WSP-system installed in the *Municipal WWTP* of the City of Vrindavan, State of Uttar Pradesh in northern India. Figures 8.12 and 8.13 exhibit extremely poor maintenance of the primary treatment units in the *Municipal WWTP* of the City of Agra, State of Uttar Pradesh and in the *Municipal WWTP* of the City of Miraj, State of Maharashtra in western India. Clearly, in the case of WSPs there too is a need for sedimentation and removals of particulate matter in primary treatment unit before wastewater are subjected to WSPs. In the absence of adequate primary treatment the ponds in WSPs could not perform properly. Such WSPs were found to develop short-circuiting and bypassed untreated or partially treated wastewater through channels in sludge beds – thereby reaching the final outlets and thus resulting in non-compliance.

Figure 8.8 No primary treatment unit was provided prior to the HSSF-CW in the City of Bhopal, central India.

Figure 8.9 Poor maintenance of primary treatment unit lead to carry-forward of garbage and solids in the wetland bed of HSSF-CW in the City of Ropar, northern India.

Figure 8.10 Poor maintenance of the primary treatment unit prior to WSP in the *Municipal WWTP* of the City of Vrindavan, northern India.

8.3.4 Issues associated with management of NTSs in India

In the Indian context, water, wastewater and the associated utility services is the "state subject" *i.e.* the funding for development of sanitation projects, O&M of the facilities, monitoring of performance, general administration and revenue collection related to the utility. The important agencies involved in these functions can typically be divided into four groups, namely: 1. Urban

Local Bodies (ULBs; comprising of Municipal Corporation, Nagar Palika and Parishad and Village Council), 2. State and Central Governments (comprising of respective state governments, the Government of India, National River Conservation Directorate in the MoEF, Yamuna Action Plan and Public Health Engineering Departments in various states and in GoI), 3. Water Boards (comprising of State Jal Boards and Water Authorities, Water and Sewerage Boards and Environmental Planning & Coordination Organization) and 4. United Nations Development Programme (UNDP). Table 8.3 summarizes the number of WWTPs which received capital costs as well as number of WWTPs which are being operated by the respective agencies corresponding to the above-mentioned four groups of agencies. A detailed account of the reasons for failure (or success) has been presented in the last column of Table 8.3. Clearly, it appears that the agencies that built, commissioned and transferred the WWTPs to the ULBs for O&M were the glaring success stories. If the operating agencies plan and allocate adequate funds for O&M, the chances of success were even greater. In summary, providing capital investments to the community is as important as helping them in planning to provide adequate O&M costs.

Figure 8.11 Poor maintenance of primary treatment unit lead to carry-forward of garbage and solids in WSP in the *Municipal WWTP* of the City of Vrindavan.

Figure 8.12 Poor maintenance of the primary treatment unit prior to WSP in the *Municipal WWTP* in City of Agra.

8.3.5 Post-treatment and reuse of effluents from NTSs in India

Out of 41 NTSs investigated across India, very few (only two WWTPs) have the post-treatment facility (disinfection using chlorine gas). The summary of available post-treatments and reuse in the context of different eco-centric technologies surveyed in the present study are presented in Table 8.4. The two WWTPs employing UASB followed by PPs were found practicing chlorination, namely: in the *Municipal WWTP* in City of Kapoorthala, State of Punjab and *Municipal WWTP* in City of Agra, State of Utter Pradesh. These WWTPs reused their treated effluents for irrigation and the leftover excess treated effluents were disposed into the Yamuna River. Typically, 1–2 mg/L doses of dissolved molecular chlorine were applied at the outlet of PPs before the effluents were reused (or disposed of). The downstream reuse options practiced by various communities among the 41 WWTPs surveyed in this study were also summarized in Table 8.4. It appeared that the most commonly practiced reuse option was irrigation and leftover treated effluent is disposed of into nearby rivers or lakes.

Figure 8.13 Poor maintenance of the primary treatment unit prior to WSP in the *Municipal WWTP* in City of Miraj, western India.

8.4 CONCLUSIONS AND LESSONS LEARNT

Over the past three decades, the GoI has made several efforts to supply drinking water to communities in urban as well as rural India. Though there was a large investment concurrently made in creating infrastructure for wastewater across India, the shortfall between water supply and wastewater treatment continues to grow at steep rates. Thus, there exists a large gap between the amount of wastewater generated and treated in urban and peri-urban communities. It is alarming that the water bodies of both surface and groundwater are currently found to be severely contaminated by untreated or partially treated wastewater.

Clearly, there exists a looming challenge of inadequate and insufficient infrastructure for treatment of wastewater throughout India, both in urban as well as rural communities. The Ministry of Urban Development, the MoEF as well as the MoWR and Ganga Rejuvenation have been incorporating the strategy of providing low-cost eco-centric treatment to wastewater for correcting the pollution of natural water courses in India.

In India, engineered NTSs are currently installed at 108 sites for treatment of mixtures of wastewater (biodegradable industrial effluents were also mixed in some situations). Through questionnaire surveys, one third of those sites (41 WWTPs) were shortlisted and visited during December 2011 to June 2014. The salient conclusions and learnings from the national survey are summarized below:

1) Out of the 41 selected WWTPs based on engineered natural treatment technologies, 23 plants had WSPs, three plants had DPs, seven plants had PPs and eight plants employed HSSF-CWs or KT-CWs.
2) Nearly 75% of the WWTPs investigated in this study were generally achieving the *Minimum National Standards* stipulated by the MoEF (GoI), for disposal of treated wastewater into legally permitted surface water bodies or for the purposes of land irrigation as prescribed in the *Water (Prevention and Control of Pollution) Act*, 1974 and companion regulations.
3) Technologies like HSSF-CWs, KT-CWs as well as DPs seemed to cater to the communities which generated relative smaller flows of wastewater when compared with other technologies including WSPs and PPs.
4) Local standards and guidelines play a crucial role in making the decisions with respect to the extent of treatment to be adopted as well as in determining the type of technology to be implemented for treatment of wastewater in a given community.
5) The FC removal is normally the slowest process and for that reason it becomes the main design criterion for a PP. In India depth of UASB PP has been kept at 1–1.5 meter with an average HRT of 24 hrs. In most of the places, such short HRT is insufficient to remove FCs to a desirable extent.

6) The DP systems were found to be quite effective for treatment and reuse of rural wastewater. Sizes of different treatment units of DP systems are customarily estimated on the basis of biological kinetics of degradation of pollutants and extinction of pathogens in DP and fishpond, life cycle of fishes, climatic conditions and land availability.
7) The most commonly encountered problems during successful operation of NTSs across India include mixing of industrial effluents and poor O&M of the treatment facilities, which cause malfunctioning. The agencies that financed, built and commissioned the WWTPs and subsequently transferred the WWTPs to the respective ULBs for O&M are the successful examples. If the operating agencies planned and allocated adequate funds for O&M, the chances of success would be even higher.
8) The engineered wetland systems were found to be quite robust and versatile in a variety of climatic conditions across India and met the prescribed regulatory standards.
9) Communities seem to prefer HSSF-CWs even more in recent times owing to the innate advantages offered by them in the context of minimizing mosquito breeding and thereby minimizing the threat of cerebral malaria, dengue and several vector-based diseases.

Table 8.3 Agencies responsible for providing capital investments for the communities to establish the WWTPs based on the eco-centric technologies as well as O&M of the facilities.

Sr. No.	Empowered Agencies	Number of WWTPs Funded (Capital and O&M Costs)	Number WWTPs Operated and Maintained	Observations and Comments
1	ULB[a]	5	22	At most of the places, ULBs are operating the facilities rather well. In some cases, however, the village councils did not have funds to perform adequate O&M and as result, the systems were not functioning well.
2	State or Central Government[b]	19	4	The four WWTPs, which were funded and operated by the same agencies, seemed to be operating satisfactorily. However, the WWTPs which were transferred to ULBs for O&M were facing difficulties on account of perpetual delays in releasing funds for O&M year-after-year.
3	Water Boards[c]	16	15	In context of the Indian administrative setup, the Water Boards were supposed to fund the capital costs, build those WWTPs and then transfer to the respective ULBs for O&M. However, the Water Boards do not have rapport with the respective urban and rural communities. As a result, they would fail in transferring the WWTPs after building and commissioning to the respective ULBs. Thus, Water Boards in different states in India are in possession of such WWTPs that were not transferred to ULBs and end up running them with no or minimal allocation of O&M funds. Obviously, all such WWTPs have been chronically facing problems with respect to O&M.
4	UNDP	1	0	The WWTP based CW-technology was found working well after the UNDP built and commissioned it for the Agra Municipal Corporation, City of Agra, State of Uttar Pradesh. In due course, the WWTP was transferred to the Agra Municipal Corporation; who were found to be operating it satisfactorily. The UNDP had provided for the O&M costs to cover the initial years. Subsequently, the Agra Municipal Corporation was supposed to allocate own funds through their revenue collection.

[a] comprising of Municipal Corporation, Nagar Palika and Parishad and Village Council.
[b] comprising of respective state governments, the Government of India, National River Conservation Directorate in the MoEF, Yamuna Action Plan and Public Health Engineering Departments in various states and in GoI.
[c] comprising of State Jal Boards and Water Authorities, Water and Sewerage Boards and Environmental Planning & Coordination Organization.

Table 8.4 Summary of the post-treatment and downstream reuse of treated effluent practiced in the 41 shortlisted NTS-based WWTPs across India.

S. No.	Technology	Number of WWTPs	Capacity Range [MLD]	Post-treatment (after NTS before Disposal or Reuse)	Downstream Reuse of Treated Effluent
1	WSPs	23	0.5–52.7	None	Irrigation of agricultural fields, river and lake discharge
2	DPs	3	0.5–1.0	None	Irrigation of agricultural fields
3	PPs	7	14–78	Two WWTPs perform chlorination (1–2 mg/L doses of dissolved molecular chlorine) at the outlet of PP before the effluent is reused	Irrigation of agricultural fields, river and lake discharge
4	HSSF-CWs + KT-CWs	8	0.05–7.8	None	Lake discharge

In summary, the broad class of engineered NTSs, including CWs, will continue to dominate the platform of "favoured technologies" for treating and recycling wastewater, taking advantage of the warm climate in India and thereby satisfying (at least partially) the unmet demand for waters for irrigation and industry. The CWs can potentially be the alternative to the commonly practiced conventional wastewater treatment strategy – which is not favoured on account of it being energy intensive and ineffective in removing pathogens. In a typical rural setting, CWs appear to treat wastewater to a greater degree when compared with more conventional rural alternatives like septic tanks, drain fields and other forms of land treatment.

ACKNOWLEDGEMENTS

The authors acknowledge the co-funding obtained from Saph Pani Project by the European Commission (7th Framework Programme) under Grant Number 282911, India as well as from the Indian Institute of Technology Bombay for this work. Authors are grateful to the various ministries and agencies of the Government of India and municipalities for facilitating this survey and providing data.

8.5 REFERENCES

Arceivala S. J. and Asolekar S. R. (2006). Wastewater Treatment for Pollution Control and Reuse, 3rd edn, Tata McGraw Hill Education India Pvt. Ltd., New Delhi, India.

Arceivala S. J. and Asolekar S. R. (2012). Environmental Studies: A Practitioner's Approach. Tata McGraw Hill Education India Pvt. Ltd., New Delhi, India.

Asolekar S. R. (2002). Greening of industries and communities: Rhetoric vs. action, In: Rio to Johannesburg: India's Experience in Sustainable Development, LEAD India (ed.), Orient Longman, Hyderabad, India, pp. 125–166.

Asolekar S. R. (2013). Report on experiences with constructed wetlands and techno-economic evaluation. Report Saph Pani: Enhancement of Natural Water Systems and Treatment Methods for Safe and Sustainable Water Supply in India, D3.1, Project supported by the European Commission within the Seventh Framework Programme Grant agreement No. 282911. www.saphpani.eu

Asolekar S. R., Kalbar P. P., Chaturvedi M. K. M. and Maillacheruvu K. Y. (2013). Rejuvenation of rivers and lakes in India: Balancing societal priorities with technological possibilities. In: Comprehensive Water Quality and Purification, S. Ahuja (ed.), Elsevier, Waltham, United States, pp. 181–229.

Billore S. K., Singh N., Ram H. K., Shrama J. K., Singh V. P., Nelson R. M. and Dass P. (2001). Treatment of molasses based distillery effluent in a constructed wetland in central India. *Water Science and Technology,* **44**(11), 441–448.

Billore S. K., Singh N., Sharma J. K., Dass P. and Nelson R. M. (1999). Horizontal subsurface flow gravel bed constructed wetland with *Phragmiteskarka* in Central India. *Water Science and Technology*, **40**(3), 163–171.

Brix H. (1997). Do macrophytes play a role in constructed treatment wetlands? *Water Science and Technology,* **35**(5), 11–17.

Chaturvedi M. K. M. and Asolekar S. R. (2009). Wastewater treatment using natural systems: The Indian experience. In: Technologies and Management for Sustainable Biosystems, J. Nair and C. Furedy (eds), Nova Science Publishers, Inc., United States, pp. 115–130. ISBN: 978-1-6087-104-3.

Chaturvedi M. K. M., Langote S. D., Kumar D. and Asolekar S. R. (2014). Significance and estimation of oxygen mass transfer coefficient in simulated waste stabilization pond. *Ecological Engineering*, **73**, 331–334.

CPCB (2005). General Standards for Discharge of Environmental Pollutant: Effluents. http://cpcb.nic.in/GeneralStandards.pdf (accessed 30 November 2014).

CPCB (2009). Status of water supply, wastewater generation and treatment in class-I cities and class-II towns of India. Control of Urban Pollution Series: CUPS/70/2009–10.

Jana B. B. (1998). Sewage-fed aquaculture: The Calcutta model. *Ecological Engineering*, **11**, 73–85.

Kalbar P. P., Karmakar S. and Asolekar S. R. (2012). Selection of an appropriate wastewater treatment technology: A scenario-based multiple-attribute decision-making approach. *Journal of Environmental Management*, **113**, 158–169.

Kalbar P. P., Karmakar S. and Asolekar S. R. (2013). The influence of expert opinions on the selection of wastewater treatment alternatives: A group decision-making approach. *Journal of Environmental Management*, **128**, 844–851.

Kumar D. and Asolekar S. R. (2015). Significance of engineered natural treatment systems for treating sewages: India case study. *Manuscript in Preparation*.

Starkl M., Amerasinghe P., Essl L., Jampani M., Kumar D. and Asolekar S. R. (2013). Potential of natural treatment technologies for wastewater management in India. *Journal of Water, Sanitation and Hygiene for Development*, **3**(4), 500–511.

Vymazal J. (2010). Constructed Wetlands for Wastewater Treatment. *Water*, **2**, 530–549. DOI:10.3390/w2030530.

Vymazal J. (2013a). Emergent plants used in free water surface constructed wetlands: A review. *Ecological Engineering*, **61**, 582–592.

Vymazal J. (2013b). Plants in constructed, restored and created wetlands. *Ecological Engineering*, **61**, 501–504.

8.6 APPENDIX

Appendix 8.1 Questionnaire utilized during the field survey of 41 shortlisted NTS-based WWTPs across India during December 2011 and June 2014.

1 General information				
Contact Details	Contact person: Name and Address of WWTP: Phone no: Fax: E-mail:			
Legal Status				
Type of wastewater treated				
Mode of conveyance				
Year of WWTP's commissioning.				
Treatment technology				
Treatment chain/mode of operation				
Type of plant/Fish species				
2 Financial details				
Capital cost of the WWTP (Rs. 100,000 (hundred thousand))				
Cost of treatment (O&M Cost /month)				
Funding agency for wastewater treatment cost				
Revenue generated per month				
Agency bearing wastewater collection costs				
3 Design details				
Primary treatment units				
Screen chamber	Type of screen	Number of screen	Unit size	Other details
Grit chamber	Unit size	Number of units	Hydraulic Retention Time (HRT)	Other details
Secondary treatment units	Unit 1 (LxBxD)	Unit 2 (LxBxD)	Unit 3 (LxBxD)	Unit 4 (LxBxD)
Design basis				
BOD_5 (mg/L)	BOD_{inlet} BOD_{outlet}			
COD (mg/L)	COD_{inlet} COD_{outlet}			
pH	pH_{inlet} pH_{outlet}			
TKN (mg/L)	TKN_{inlet} TKN_{outlet}			
TP (mg/L)	TP_{inlet} TP_{outlet}			
TSS (mg/L)	TSS_{inlet} TSS_{outlet}			

Appendix 8.1 Questionnaire utilized during the field survey of 41 shortlisted NTS-based WWTPs across India during December 2011 and June 2014 (*Continued*).

Total Coliform (Count/100 mL)	Inlet				
	Outlet				
Faecal Coliform (Count/100 mL)	Inlet				
	Outlet				
HRT					
4 Actual performance					
Unit name		Unit 1	Unit 2	Unit 3	Unit 4
BOD_5 (mg/L)	BOD_{inlet}				
	BOD_{outlet}				
COD (mg/L)	COD_{inlet}				
	COD_{outlet}				
pH	pH_{inlet}				
	pH_{outlet}				
TKN (mg/L)	TKN_{inlet}				
	TKN_{outlet}				
TP (mg/L)	TP_{inlet}				
	TP_{outlet}				
TSS (mg/L)	TSS_{inlet}				
	TSS_{outlet}				
Total coliform (Count/100 mL)	Inlet				
	Outlet				
Faecal coliform (Count/100 mL)	Inlet				
	Outlet				
HRT					
5 Downstream destination of treated wastewater					
Downstream reuse		Agriculture	Landscape irrigation	Groundwater recharge	Others
Downstream disposal		River	Open drain	Estuary	Others
6 Post treatment before reuse					
Type of treatment given					
		pH			
		DO			
		Electric Conductivity			
		TOC			
		Ammonium			
		Nitrate			
		Phosphate			
		Turbidity			
		Total coliform			
Water quality before post treatment		Faecal coliform			
		DO			
		Electric Conductivity			
		TOC			
		Ammonium			
		Nitrate			
		Phosphate			

(*Continued*)

Appendix 8.1 Questionnaire utilized during the field survey of 41 shortlisted NTS-based WWTPs across India during December 2011 and June 2014 (*Continued*).

	Turbidity
	Total coliform
	Faecal coliform
	Others
Water quality after post treatment	pH
	DO
	Electric Conductivity
	TOC
	Ammonium
	Nitrate
	Phosphate
	Turbidity
	Total coliform
	Faecal coliform
	Others
Cost of post treatment in Rs./m^3	
If effluent not being reused now, is there any potential for reuse? If yes, for which purpose	
7 Health and environmental risks	
Are there any incidences of source pollution, which occurred in the past?	
Is there any risk for the person operating the system?	
Is there any risk for people involved in the disposal handling?	
Is there any risk for people living in the surrounding area of the system?	
For which purposes is the water used?	
If water is used for irrigation, what plants are irrigated?	
If vegetables are planted, are they eaten raw?	
How many people are exposed to the wastewater before treatment and after treatment?	
Are there any wells near the area where the treated water is reused?	
Are there any other possible risks to the environment	
Additional remarks	
8 Flow sheet of WWTP	

Appendix 8.2 List of known existing sites of engineered constructed wetlands and other NTS-based WWTPs across India (information based on the present field survey as well as from CPCB, 2009).

SN	Types of NTSs	Capacity (MLD)	Year of Commissioning	Type of Post-treatment	Down Streams Use of Treated Effluent	Location
1	WSP	14	2003	No Post-treatment	Godavari River	Ramagundam I, Andhra Pradesh
2	WSP	4	2003	No Post-treatment	Godavari River	Ramagundam II, Andhra Pradesh
3	WSP	4	2003	No Post-treatment	NA	Bhadrachalam, Andhra Pradesh
4	WSP	14	2004	No Post-treatment	Godavari River	Ramagundam IV, Andhra Pradesh
5	WSP	4	1988	No Post-treatment	Punpun, Ganga	Kermallichak, Bihar
6	WSP	2	1988	No Post-treatment	Ganga River	Chapra, Bihar
7	WSP	46	1965	No Post-treatment	Seonath River	Kutelabhatavill, Bhilai Nagar, Chhatisgarh
8	WSP	14	1965	No Post-treatment	NA	Risalivillage, Bhilai Nagar, Chhatisgarh
9	WSP	9	1965	No Post-treatment	NA	Bhilai House, Bhilai Nagar, Chhatisgarh
10	WSP	27.27	2003	No Post-treatment	Yamuna River	Timarpur, Delhi
11	PP	20	2000	No Post-treatment	Yamuna River	Faridabad I, Haryana
12	PP	45	2000	No Post-treatment	Yamuna River	Faridabad II, Haryana
13	PP	50	2000	No Post-treatment	Yamuna River	Faridabad III, Haryana
14	WSP	8	2000	No Post-treatment	Yamuna River	Karnal II, Haryana
15	WSP	1	2001	No Post-treatment	NA	Chhchhrauli, Haryana
16	WSP	1.5	2001	No Post-treatment	NA	Indri, Haryana
17	WSP	1	2001	No Post-treatment	NA	Radaur, Haryana
18	WSP	9	2003	No Post-treatment	Agricultural Field	Palwal, Haryana
19	WSP	3	2004	No Post-treatment	NA	Gharaunda, Haryana
20	WSP	3.5	2004	No Post-treatment	NA	Gohana, Haryana
21	WSP	19.45	2001	No Post-treatment	Tungabhadra	Davanagere, Karnataka
22	WSP	5.83	2001	No Post-treatment	Bhadra River	Bhadravati, Karnataka
23	WSP	1.47	2001	No Post-treatment	NA	Nanjagud, Karnataka
24	WSP	1.36	2001	No Post-treatment	NA	Sri Rangapatna , Karnataka
25	WSP	18.16	2003	No Post-treatment	Tunga River	Shimoga, Karnataka
26	WSP	1.45	2004	No Post-treatment	NA	K R Nagar, Karnataka
27	WSP	4.5	2007	No Post-treatment	NA	Pamba, Kerla
28	WSP	8	NA	No Post-treatment	NA	Bherkheda, Bhopal, Madhya Pradesh
29	WSP	52	2001	No Post-treatment	Shipra River	Ujjain, Madhya Pradesh
30	KT	1.67	2001	No Post-treatment	Shipra River	Barogarh, Ujjain, Madhya Pradesh
31	KT	1.67	2001	No Post-treatment	Shipra River	Barogarh, Ujjain, Madhya Pradesh
32	KT	1.2	2001	No Post-treatment	NA	Chapara, Madhya Pradesh
33	KT	0.75	2001	No Post-treatment	NA	Keolari, Madhya Pradesh

(*Continued*)

Appendix 8.2 List of known existing sites of engineered constructed wetlands and other NTS-based WWTPs across India (information based on the present field survey as well as from CPCB, 2009) (*Continued*).

SN	Types of NTSs	Capacity (MLD)	Year of Commissioning	Type of Post-treatment	Down Streams Use of Treated Effluent	Location
34	KT	9	2004	No Post-treatment	Betwa River	Vidisha, Madhya Pradesh
35	WSP	6	2005	No Post-treatment	Tapi River	Burhanpur, Madhya Pradesh
36	WSP	2.5	1995	No Post-treatment	Sina, Bhima River	Aurangabad, Maharashtra
37	WSP	5	NA	No Post-treatment	Salim Ali Lake	JNEC, Aurangabad, Maharashtra
38	OP	18.9	1995	No Post-treatment	Gima River	Jalgaon, Maharashtra
39	OP	12.87	Pre 95	No Post-treatment	Manjeera River	Latur , Maharashtra
40	WSP	26/8.9	2000	No Post-treatment	Godavari River	Nanded-Waghala, Maharashtra
41	WSP	1	2003	No Post-treatment	NA	Trimbakeshwar, Maharashtra
42	WSP	23.82	2004	No Post-treatment	Krishna River	Sangli-Miraj and Kupwad, Maharashtra
43	WSP	33	2003	No Post-treatment	Mahanadi River	Cuttak, Orissa
44	WSP	2	2005	No Post-treatment	NA	Talcher, Orissa
45	WSP	2.6	2003	No Post-treatment	NA	SultanpurLodhi, Punjab
46	WSP	2.56	2004	No Post-treatment	Satluz river	Phillaur, Punjab
47	PP	25	NA	Chlorination	Agricultural Field	Kapoorthala, Punjab
48	PP	22.73	2005	Information Not Available	NA	Raipur Kalan,Chandigarh,
49	DP	0.5	NA	No Post-treatment	Agricultural Field	Bais Village, Ludhiana, Punjab
50	DP	0.5	NA	No Post-treatment	Agricultural Field	Village Saidpur, Ludhiana, Punjab
51	DP	0.5	NA	No Post-treatment	Agricultural Field	Village Sandhuan, Roop Nagar, Punjab
52	WSP	0.5	NA	No Post-treatment	Agricultural Field	Village Dedwal, Ludhiana, Punjab
53	WSP	0.5	NA	No Post-treatment	Agricultural Field	Village Sandhuan, Roop Nagar, Punjab
54	DP	1	NA	No Post-treatment	Agricultural Field	Village Uncha, Roop Nagar, Punjab
55	WSP	20	2007	No Post-treatment	Agricultural Field	Village Nanded, Jodhpur, Rajasthan
56	WSP	20	2007	No Post-treatment	Agricultural Field	Vallabh Garden Bikaner, Rajasthan
57	PP	111	2004	No Post-treatment	Agricultural Field	Ludhiana, Zone B, Punjab
58	PP	152	2004	Information Not Available	Agricultural Field	Ballok, Ludhiana
59	PP	48	2005	Information Not Available	Agricultural Field	Jmalpur, Ludhiana
60	WSP	28	2003	No Post-treatment	Kaveri	Tiruchirappalli II, Tamil Nadu
61	WSP	3.94	2003	No Post-treatment	NA	Bhawani, Tamil Nadu
62	WSP	58	2004	No Post-treatment	Kaveri	Tiruchirappalli, Tamil Nadu
63	WSP	20	2004	No Post-treatment	Kaveri	Erode I, Tamil Nadu
64	WSP	3.96	1988	No Post-treatment	NA	Farrukhabad, Uttar Pradesh
65	WSP	9	1999	No Post-treatment	Yamuna River	Noida III, Uttar Pradesh
66	WSP	10	2001	No Post-treatment	Yamuna River	PeelaKhar, Agra, Uttar Pradesh
67	PP	14	NA	Chlorination	Yamuna River	Dayal Bag, Agra, Utter Pradesh

68	PP	78	NA	No Post-treatment	Agricultural Field	Dhandpur, Agra, Utter Pradesh
69	WSP	2.5	2001	No Post-treatment	Yamuna River	BurhikaNagla, Agra, Uttar Pradesh
70	WSP	32	2001	No Post-treatment	Kali River	Muzaffarnagar, Uttar Pradesh
71	PP	70	2001	Information Not Available	NA	Hindone I, Ghaziabad, Uttar Pradesh
72	PP	56	2001	Information Not Available	NA	Hindone II, Ghaziabad, Uttar Pradesh
73	PP	34	NA	Information Not Available	NA	Noida I, Uttar Pradesh
74	PP	27	NA	Information Not Available	NA	Noida II, Uttar Pradesh
75	PP	27.5	NA	No Post-treatment	NA	Mirzapur, Uttar Pradesh
76	WSP	14.5	2001	No Post-treatment	Agricultural Field / Yamuna River	Bangalighat dairy farm, Mathura, Uttar Pradesh
77	WSP	4	NA	No Post-treatment	Agricultural Field	Baba Temple, Vrindavan, Uttar Pradesh
78	WSP	12.5	2001	No Post-treatment	Agricultural Field / Yamuna River	Masani, Mathura, Uttar Pradesh
79	WSP	0.5	NA	No Post-treatment	Agricultural Field	Kali Deh, Vrindavan, Uttar Pradesh
80	WSP	10.45	2001	No Post-treatment	Yamuna River	Etawah Uttar Pradesh
81	WSP	10	1987	No Post-treatment	Ganga River	E (Madrail),Bhatpara, West Bengal
82	WSP	30	1987	No Post-treatment	Ganga River	S.Sub-E, Kolkata, West Bengal
83	WSP	4.54	1987	No Post-treatment	Ganga River	Chandannagar II, West Bengal
84	WSP	8	1987	No Post-treatment	Beel	Baharampur, West Bengal
85	WSP	16.5	1988	No Post-treatment	Irrigation, Pissiculture	Panihati, West Bengal
86	WSP	45	1988	No Post-treatment	Irrigation, Pissiculture	Bally, West Bengal
87	WSP	14.1	1988	No Post-treatment	Irrigation, Pissiculture	Bandipur, West Bengal
88	WSP	4.54	1988	No Post-treatment	Irrigation, Pissiculture	Titagarh, West Bengal
89	WSP	10	1988	No Post-treatment	Ganga River	Nabadwip, West Bengal
90	WSP	3	2003	No Post-treatment	Ganga River	Khardaha, West Bengal
91	WSP	3.93	2003	No Post-treatment	Ganga River	Maheshtala, West Bengal
92	WSP	5.9	2003	No Post-treatment	Ganga River	Barrackpur, West Bengal
93	WSP	1	2003	No Post-treatment	Ganga River	Barrackpur, West Bengal
94	WSP	10.9	2003	No Post-treatment	Ganga River	Barrackpur, West Bengal
95	WSP	4.35	2003	No Post-treatment	Ganga River	Barrackpur, West Bengal
96	WSP	1.9	2005	No Post-treatment	NA	Murshidabad, West Bengal
97	WSP	0.52	2005	No Post-treatment	NA	Diamond Harbour, West Bengal
98	WSP	1.39	2006	No Post-treatment	NA	JiaganiAjimganj, West Bengal
99	CWs	21.25m × 5.5m	NA	No Post-treatment	NA	Kakatiya Musical Garden of Warangal City, Andhra Pradesh
100	CWs	NA	NA	No Post-treatment	NA	Mahindra Mahindra, Igatpuri, Nashik.
101	CWs	NA	NA	No Post-treatment	NA	Presidency Kid Leather Ltd. Kannivakkam Tamil Nadu

(*Continued*)

Appendix 8.2 List of known existing sites of engineered constructed wetlands and other NTS-based WWTPs across India (information based on the present field survey as well as from CPCB, 2009) (*Continued*).

SN	Types of NTSs	Capacity (MLD)	Year of Commissioning	Type of Post-treatment	Down Streams Use of Treated Effluent	Location
102	CWs	NA	NA	No Post-treatment	NA	Guru govindsingh Park (Ekant Park) Southern area Bhopal
103	CWs	1	NA	No Post-treatment	NA	Kankhal, Haridwar, UttaraKhand
104	CWs	NA	NA	No Post-treatment	NA	Sainik School Bhuneshwar, Orissa
105	CWs	0.5	NA	No Post-treatment	NA	village PipalMajra, District Ropar, Pumjab
106	CWs	2.5 acres	NA	No Post-treatment	NA	village Shekhupur in District Patiala, Punjab
107	CW	7.8	2008	No Post-treatment	Mansagar Lake (Recreational)	Mansagar Lake, Jaipur, Rajasthan
108	CW	NA	NA	No Post-treatment	NA	Ujjain

Chapter 9

Experiences with laboratory and pilot scale constructed wetlands for treatment of sewages and effluents

Dinesh Kumar and Shyam R. Asolekar

9.1 INTRODUCTION

Constructed wetlands (CWs) have been implemented as wastewater treatment facilities in many parts of the world, but to date, the technology has been largely ignored in developing countries in general and Indian sub-continent in particular, where effective, low-cost wastewater treatment strategies are urgently needed (Arceivala & Asolekar, 2006; Asolekar *et al.*, 2013; Chaturvedi *et al.*, 2014). CWs are used extensively to treat domestic (Billore *et al.*, 1999; Kadlec & Knight, 1996) and industrial wastewaters (Hammer, 1989; Billore *et al.*, 2001). They have also been applied to passive treatment of diffuse pollution including mine wastewater drainage (Hammer, 1989; Kadlec & Knight, 1996; Jing *et al.*, 2001), and highway runoff following storm events (McNeill & Olley, 1998). Besides, CWs, being a model ecosystem, can serve as wildlife habitats and can be perceived as natural recreational areas for the local community (Hawke & José, 1996).

Most recently, it has been envisioned that CWs can be applied in place of or in combination with some appropriate post-treatment technologies to provide techno-economically feasible and socially acceptable way for wastewater management (Arceivala & Asolekar, 2012). The reuse, or reclamation, of wastewater using CW technology also provides an opportunity to create or restore valuable wetland habitat for wildlife and environmental enhancement. Kadlec and Knight (1996) have discussed in detail the various advantages of using wetland technology for wastewater treatment. An additional benefit gained by using wetlands for wastewater treatment is the multi-purpose sustainable utilization of the facility for uses such as swamp fisheries, biomass production, seasonal agriculture, water supply, public recreation and wild life conservation. In the appropriate climatic condition of India, CWs could be successfully established with plant species acclimated to the tropical environment and harvested for use in secondary functions like fuel production.

CWs include free water surface, and sub-surface flow systems. Based on the type of flow, sub-surface flow systems are further classified into vertical sub-surface flow constructed wetlands (VSSF-CWs) and horizontal sub-surface flow constructed wetlands (HSSF-CWs). The sub-surface flow systems involve sub-surface flow through a permeable medium. The "root-zone method" and "rock-reed-filter" are other names for these systems that have been used in the literature. Because emergent aquatic vegetation is used in these systems, they depend on the same basic microbiological reactions for treatment. The pollutants in such systems are removed through a combination of physical, chemical and biological processes including sedimentation, precipitation, and adsorption to soil particles, assimilation by the plant tissue, and microbial transformations.

The performance of CW depends on many factors including its type and design, organic loading rate and hydraulic retention time (Karpiscak *et al.*, 1999). In spite of having significant nutrient removal capability, due to the effect of changing temperatures, the treatment efficiency of CWs tends to change throughout the year (Bachand & Horne, 2000; Healy & Cawley, 2002). The macrophytes growing in CWs have several properties in relation to the treatment process that make them an essential component of the design (Brix, 1997). Selection of plant species for treatment of wastewater by CWs always remains a difficulty to scientists working in this area because metabolism of the macrophytes affects the treatment processes to different extents depending on the type of the CW. The plants species used in CWs designed for wastewater treatment should therefore: (1) be tolerant of high organic and nutrient loadings, (2) have rich below-ground organs (*i.e.* roots and rhizomes) in order to provide substrate for attached bacteria and oxygenation (even very limited) of areas adjacent to roots and rhizomes and (3) have high above-ground biomass for winter insulation in cold and temperate regions and for nutrient removal via harvesting (Koncalova *et al.*, 1996). Vymazal (2011) has reviewed the plants used in HSSF-CWs and concluded that local species which are easily available and grow well under local climatic conditions are most suitable. Among those plants, many ornamental species have been used, especially for on-site treatment where aesthetics is often a part of the design (Vymazal, 2011).

9.1.1 Scope and objectives

The possible ways to improve the efficiency of natural treatment systems (NTSs) in general and engineered CWs in particular comprise of incorporating the most common and the best practices into wastewater treatment plants (WWTPs) at the design-stage itself. Also, a knowledge-based approach will have to be systematically implemented during construction, commissioning as well as operating and maintaining the facility. The research and technology development activity undertaken in Work-Package-3 of the Saph Pani Project were planned and executed with this overall idea. It was also recognised at the outset that the enhancement of the performance of a given WWTP based on the engineered CW-technology can be achieved when the eco-centric technology implemented in the project performs according to the intended functions in the treatment train (Asolekar, 2013). Furthermore, the natural treatment technology based WWTPs will be suitably adopted, operated and maintained by the given community if the treatment train incorporates suitable tertiary treatment to produce recyclable quality of treated sewage. Some of the important factors that should be considered while deciding upon a strategy to improve the treatment efficiency of NTSs include rate, extent and variability of wastewater reaching the system, climate changes, population changes, pattern of urban and industrial development, changes in agricultural practices, soil erosion and sedimentation, scope of construction activities in nearby areas, and nutrient loading.

A multi-pronged experimental and modelling approach was planned and implemented under these areas. Accordingly, the following outcomes, addressing the specific objectives pertaining to enhancement of the performance of engineered CWs are presented in this chapter:

I. Interventions leading to improvement of treatment efficiency
II. Enabling strategies for successful operation

9.2 METHODOLOGY

The knowledgebase required for planning, designing, constructing, operating and maintaining of HSSF-CWs has been one of the focuses in the Saph Pani Project. Accordingly, the methodological details pertaining to the studies on media and vegetation as well as the kinetic experiments are presented in the following sub-sections:

1. Studies on media and vegetation
2. Kinetic studies using laboratory CW-reactors
3. Studies in pilot-scale HSSF-CW

In addition, the significance of the above studied parameters has also been elaborated in their respective sections.

9.2.1 Studies on media and vegetation

Since the packing (medium) in CW-bed plays an important role, several media were investigated from a number of materials available locally for study and characterization. The media were subjected to examination of sieve analyses, porosity, bulk density, scanning electron microscopy (SEM analysis) with energy dispersive X-ray spectroscopy (SEM-EDX) and Fourier transform infrared spectroscopy (FTIR) for characterizing the micro-structure and chemical characterization of solid surface. The standard methods and protocols, as outlined by Gee and Bauder (1986), were employed for sample preparation and analyses of the media. The instruments used for characterization of media are listed in Table 9.1.

In order to identify the appropriate species of vegetation for CW-bed, six plant species were selected for characterization in this study, namely: *Canna indica, Typha latifolia, Colocasia esculenta, Sagittaria latifolia, Justicia americana* and *Hymenachne amplexicaulis*. The plant samples were collected from their natural habitats (typically natural wetlands or wetlands created due to land disposal of untreated or partially treated wastewater) in the communities surrounding Mumbai – especially the peri-urban suburbs and the adjoining villages. The plant samples were gently washed with tap water to remove any sticking soils attached to the roots and stems. Further, the plants were air dried at room temperature to evaporate any excess water droplets attached with the biomass.

Table 9.1 The instruments and references for methods used for characterization of media employed in laboratory CW-reactors.

Sr. No.	Description of Sample & Parameter	Sample Preparation	Analytical Method	Equipment
1	Porosity of packed media	Gee and Bauder (1986)	Gee and Bauder (1986)	Measuring cylinder
2	Bulk density of packed media	Gee and Bauder (1986)	Gee and Bauder (1986)	Measuring cylinder
3	SEM-EDX analysis of packed media	Teršič (2011)	SEM-EDX spectrometer	LEO-1530VP
4	FTIR analysis of packed media	KBr technique	FTIR spectrometer	Bruker IFS 66 vs-1 spectrometer

At least five plants were sampled from each pile of plants and subjected to further tests. Each species was separated into roots, stems, leaves and flowers by chopping the plants samples (Figure 9.1). The fractions were weighed to record the wet weight-fractions of the given plant species. Further, the samples were subjected to hot air drying at 101°C and the corresponding dry weight-fractions were also recorded.

Figure 9.1 The whole plants sampled from their natural habitats and further processed and chopped for estimation of dry and wet weight-fractions. (a) Whole *Typha latifolia* plants, (b) Chopped *Typhala tifolia* plants, (c) Whole *Canna indica* plants, (d) Chopped *Canna indica* plants.

9.2.2 Kinetic studies using laboratory CW-reactors

Box-type open crate (of PVC) was used for holding the randomly packed media in the laboratory CW-reactors. The reactors were devised with an outlet flow control valve fitted at the bottom of the crate so that the reactor could be drained without disturbing media. The dimensions of the crate are given in Table 9.2.

Table 9.2 Dimensions of the CW-reactors used in the study.

Dimension	Size
Length (inner edges)	0.605 m
Breadth (inner edges)	0.405 m
Depth	0.235 m (packed bed) and 0.075 m (free board above media)
Plan area (top surface)	0.245 m^2
Bulk volume of the packed bed (solids and pore volume)	0.0576 m^3

The laboratory CW-Reactors, used in this study, was setup and commissioned at the outsets on IIT Bombay Campus. In case of the unplanted "control" reactor, raw sewage settled in equalization-well at the pumping station was deposited in each reactor (24 L volume) and retained for 24 hours. At the end of this batch process, the sewage was drained completely and fresh sewage was deposited in the laboratory CW-Reactor. This cycle was repeated for 14-days and the media were conditioned. A similar 14-day condoning routine was implemented in case of the reactors to be planted. At that point, the conditioned laboratory CW-Reactors were planted with *Canna indica* seedlings having the stem lengths of typically 300 to 400 mm. Hereafter, both, the planted as well as un-planted control laboratory CW-Reactors were subjected to similar routine of sewage deposition.

In order to expose the young seedlings in a progressive manner through the sewage, all the laboratory CW-Reactors were first subjected to diluted sewage (50% raw sewage and 50% bore-well water) for a period of 14-days. Subsequently, the laboratory CW-Reactors were subjected to 100% raw sewage settled in equalization-well for a period of 60 days. It was observed that the seedlings planted in the reactors had grown nearly to 600–700 mm height and appeared lush green, luxurious and firmly rooted in the media with healthy growth of rhizomes. The media in, both, the planted as well as un-planted control laboratory CW-Reactors at the end of 90–100 days of conditioning and acclimatization process were found to have rather uniform coating of bio-film and the bed were odor-free and exhibited the typical smell of aerobic packed-bed sewage treatment system (like trickling filter). At this point the laboratory CW-Reactors (control and planted) were subjected to kinetic studies.

The laboratory CW-reactors were placed in a garden open to atmosphere in an area not covered by tree shade. Each batch run was conducted in triplicates in planted CW-reactors as well as one reactor was run analogously without plants (control). Experiments were conducted in batch mode in the laboratory CW-reactors by charging 24 L of raw sewage. The laboratory reactors were setup to receive sewage from the equalization well where the entire IITB Campus wastewater is collected and settled. The laboratory CW-reactors had manufactured sand as the packing material and *Canna indica* (yellow flower variety) as well the *Canna indica* (red flower variety) were the two plant species experimented with. All the planted reactors including the reactor without plants were filled with 24 L of wastewater (HSSF-CW conditions) and sampled at the beginning of batch run (time $t = 0$) and subsequently at the end of 24, 48 and 72 hour reaction periods. At the time of sampling, the entire water (24 L) contained in the given laboratory CW-reactor was drained using the outlet valve and the sample was collected from the collection bucket after mixing the contents. Afterwards, the contents of bucket were poured back into the reactor to continue degradation.

Typically, the raw sewage had the concentrations of 5-day biochemical oxygen demand (BOD_5) of 80 ± 20 mg/L, chemical oxygen demand (COD) of 180 ± 25 mg/L, total kjeldahl nitrogen (TKN) of 15 ± 5 mg/L, and total phosphorous (TP) of 3 ± 1 mg/L. The initial, intermittent and final samples from the CW-reactors were collected after specified time intervals and analysed for pH, temperature, turbidity, conductivity, BOD_5, COD, TKN, total phosphorous and suspended solids. Sewage samples and treated effluents were tested using the standard methods (APHA *et al.*, 2005). In a given batch experiment, the samples were typically collected after 24, 48 and 72 hour reaction periods and analysed for the above stated parameters. Finally, the removal efficiencies for targeted pollutants in different laboratory CW-reactors were estimated and interpreted in the context of vegetation in the reactors and the associated media. The rate constants were also estimated by interpreting the experimental data to gain insights into the kinetics of reactions that represented removal of pollutants in the HSSF-CW.

9.2.3 Studies in pilot-scale HSSF-CW facility

IIT Bombay has designed, constructed and commissioned a pilot-scale HSSF-CW having dimensions 13m × 3m × 0.6 m for investigating some of the significant issues associated with design, operation and maintenance of engineered CWs. The plant species *Canna indica*, which was selected as one of the suitable plant species based on the laboratory CW-reactor studies, was planted in the pilot CW. The raw sewage from IITB Campus (settled in equalization-well) was fed to the pilot-scale

CW facility. The schematic of the pilot-scale CW-facility is shown in Figure 1.7 of Chapter 1 and the pictures of laboratory CW-reactors as well as pilot scale HSSF-CW are shown in Figure 9.2.

Figure 9.2 Laboratory CW-reactors and pilot-scale CW facility used in the research on IIT Bombay Campus, Mumbai. (a) Laboratory CW-rectors, (b) Pilot-scale HSSF-CW at IIT Bombay Campus.

9.3 RESULTS AND DISCUSSION

Studies were conducted using laboratory CW-reactors operated in batch mode. Also, a pilot-scale HSSF-CW facility was designed, built and commissioned for demonstration and research purposes. Locally available sewage from the campus of IIT Bombay was used as the feed to the wetland. The objectives of this research were to determine suitable plant species to be applied in the wetland, the assessment of performance of the system and to explore innovative ways to utilize the harvested biomass. Some of the salient results from this research have been categorized and reported in the following sub-sections:

- characterization of media and vegetation,
- biodegradation kinetics using laboratory CW-reactors,
- performance assessment using pilot-scale HSSF-CW and
- strategies for performance enhancement.

9.3.1 Characterization of media and vegetation

The physical characteristics of the various wetland bed media were estimated (porosity 45–70%, bulk density 700–1,400 kg/m^3). The microstructure and morphology of manufactured sand was characterized with the help of micrographs obtained from SEM, as depicted in Figure 9.3a. The micrograph indicated that the micro-cracks were approximately of 5–10 micrometer width and the attached fine particles were of 1–10 micrometer diameter. The FTIR spectroscopic analysis indicated that silica, calcium, aluminium and iron were the significant elements dominating surface composition as seen in Figure 9.3b. Similarly, the surface morphology and composition of several media including natural sand, quartz sand, manufactured sand etc. were investigated. The manufactured sand, when used as packing medium in laboratory CW-reactor, was found to be effective in removal of phosphorus owing to presence of calcium, iron and aluminium in the surface composition of the mineral present in the rock.

As stated earlier, six plant species from the locally available natural wetlands were identified and sampled for further characterisation. Those plant species (*Canna indica*, *Typha latifolia*, *Colocasia esculenta*, *Sagittaria latifolia*, *Justicia americana* and *Hymenachne amplexicaulis)* were analysed for wet and dry weights. The whole plants were sampled from their natural habitats and further processed, chopped and dried for estimation of the respective weight-fractions (Table 9.3).

It should be noticed that the weight fractions of total biomass above ground and total biomass below ground have the variability among the plant species with respect to their dry weight-fractions (as depicted from the range given). This information plays an important role in the selection of plant species in the context of the pollutants to be addressed as well as the extent of removal to be achieved in the treatment facility. It is well known that on one hand the foliage assists in photosynthesis and thereby production of oxygen (Uzman, 2001). On the other hand, the rhizosphere (the sub-surface volume around roots) supports a healthy and diverse consortium of aerobic microorganisms – which thrives in case of those plant species, which have larger network of roots. It is well known that the overall performance of a given natural wetland or engineered CW depends on synergistic interaction of biotic and abiotic components of the system – especially the media, vegetation and pollutants (Gyssels *et al.*, 2005). The plant roots provide necessary surfaces for attachment of beneficial microorganisms as well as provide oxygen for their metabolism (Brix, 1994; Reed *et al.*, 1995). The carbonaceous and nitrogenous pollutants in wastewater subjected to the root zone in constructed wetland are thus typically processed (degraded) by microbes and also through plant uptake as well as through interaction with soils.

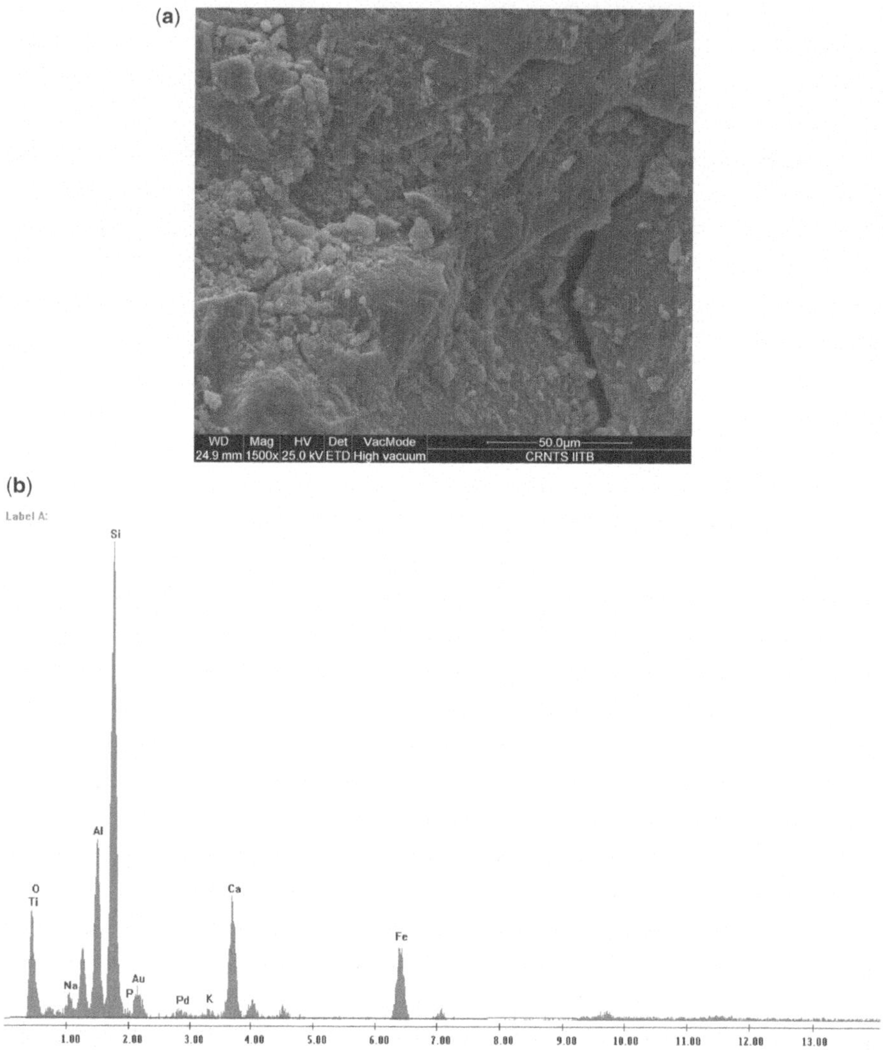

Figure 9.3 Characterization of manufactured sand with the help of (a) image obtained from the scanning electron microscope (SEM) and (b) the Fourier transform infrared (FTIR) spectra indicating chemical composition of mineral surface.

Table 9.3 Comparison of the average range of wet and dry weight-fractions of the six plant species included in this research.

Sr. No.	Plant Part	% Range of Dry Biomass	% Range of Moisture Content
1	Root	9.3–16.5	83.5–90.7
2	Stem	6.2–7.2	92.8–93.8
3	Leaf	7.2–21.6	78.4–92.8
4	Flower	14.4	85.6
5	Total biomass above ground	10.5–16.6	83.4–89.5
6	Total biomass below ground	11.6–16.6	83.4–88.4

9.3.2 Biodegradation kinetics using laboratory CW-reactors

The biodegradation kinetics in CW is conventionally assumed to be of first order and it is used for the design of CWs. Although, the first order model provides a very simple method of designing; it has many limitations in explaining the biodegradation phenomenon especially when the rate constant varies with time. The first order decay model is unable to describe the flow and removal process occurring in the wetland, mainly due to the strong interdependence between the hydraulics and kinetics. To tackle this limitation, Monod kinetics or time-based retardation kinetics for biodegradation is suggested in literature (Rousseau *et al.*, 2004). In Monod kinetics, biodegradation follows a Monod-type equation and in time-based retardation kinetics, the biodegradation rate constant is claimed to vary with time along with temperature.

The three plant species namely: *Canna indica*, *Typha latifolia* and *Phragmites australis* were found to be the most suitable in the Indian context because they are typically found abundantly in "natural wetlands". Our nation-wide survey (see chapter 8) had reported that those species have also been successfully employed in most of the working CW-based sewage treatment plants across India. Batch experiments were conducted using the laboratory CW-reactors planted with *Canna indica*. Experimental runs were conducted and the data were analysed for assessment of treatment efficiencies and estimation of degradation rate constants. Efficiencies of the reactors were also interpreted in the context of different media used in packed beds. Figure 9.4 presents kinetic data of a set of experiments conducted to study degradation of pollutants using laboratory CW-reactors. Figures 9.4a, 9.4b and 9.4c depict pollutant removal in control (no plants) batch runs, pollutant removal in *Canna indica* (yellow flower) in the batch runs and pollutant removal in *Canna indica* (red flower) in the batch runs.

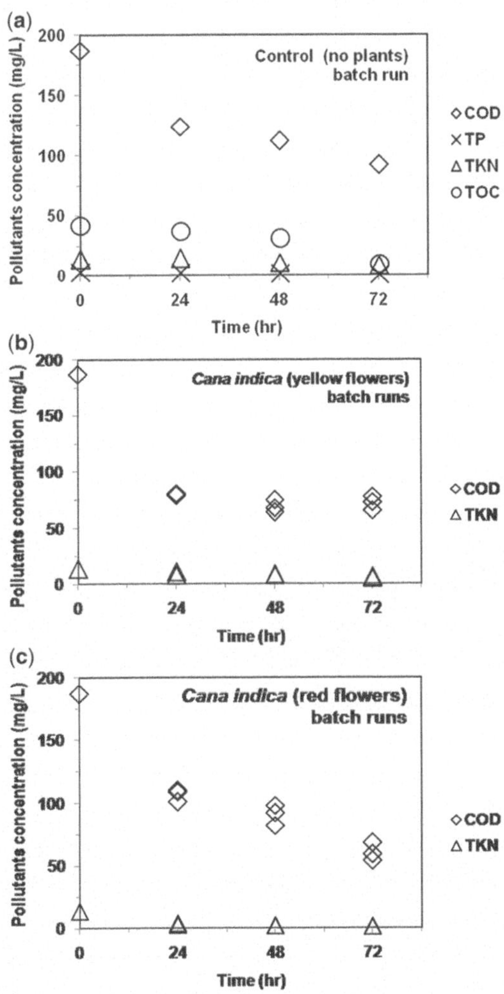

Figure 9.4 Assessment of pollutant degradation kinetics using laboratory CW-reactors. (a) pollutant removal in control (no plants) batch runs, (b) pollutant removal in *Canna indica* (yellow flower) in the batch runs and (b) pollutant removal in *Canna indica* (red flower) in the batch runs.

Percentage removal of COD, TP, TKN and total organic carbon (TOC) in 72 hours of reaction time in batch runs conducted outdoor using laboratory CW-reactors packed with manufactured sand treating sewage from IIT Bombay Campus is depicted in Table 9.4. The control reactor showed lesser %-removal of COD when compared with the %-removal in case of reactors with plants (51% when compared with 61 or 68%). Kumar *et al.* (2015c) hypothesized that the difference between the CW-reactors with and without plants in a way demonstrate the enhancement of COD removal achieved on account of facilitation of photosynthesis-related production of oxygen introduced by the plants in the respective reactors and thereby promoting elevated aerobic degradation in the rhyzosphere. In other words, clearly, the shoot of the plants (especially through photosynthesis in the foliage) as well as the root system of the plant (rhyzosphere) contribute in enhancement of the performance of CW-reactors with plants. This can roughly be estimated as nearly 10%. Similarly, the results displayed in Table 9.4 showed the plant-mediated enhancement in case of removal of COD, TP, TKN and TOC.

Table 9.4 Percentage removals of COD, TP, TKN and TOC in 72 hours of reaction time in batch runs conducted outdoor using laboratory CW-reactors packed with manufactured sand treating sewage from IIT Bombay Campus.

Description of Reactors	COD Removal	TP Removal	TKN Removal	TOC Removal
Control reactor (without plants)	51%	60%	41%	40%
Reactors with *Canna indica* (yellow flower)	61% [10% enhancement due to plants]	80% [20% enhancement due to plants]	56% [15% enhancement due to plants]	62% [22% enhancement due to plants]
Reactors with *Canna indica* (red flower)	68 [17% enhancement due to plants]	89 [29% enhancement due to plants]	90 [49% enhancement due to plants]	80 [80% enhancement due to plants]

It is clear from Table 9.4 that compared to control reactors (without plants) the presence of both plant species enhanced the removal of all the pollutants studied. Furthermore, Table 9.4 also highlights the effects of two different plant species belonging to the similar genus (Kumar *et al.*, 2015c). The *Canna indica* (red flower) species appears to remove more COD, TP, TKN and TOC by about 10%, 11%, 60% and 29%, respectively; when compared with the *Canna indica* (yellow flower) species. Although the magnitude of the numbers need confirmation with the help more experimental data, it can nevertheless be concluded that the performance of laboratory CW-reactors consisting of manufactured sand packed beds planted with *Canna indica* (red flower) species showed noticeable better performance when compared with *Canna indica* (yellow flower) species in nearly all the pollutants.

The kinetic data thus, obtained were also interpreted by Kumar *et al.* (2015c) using the so-called "first-order" kinetics owing to the simplicity of the kinetic model. It was argued by them that in several analogous situations, biological degradation does confirm to the first-order kinetics (even referred to as pseudo-first order kinetics). The first-order rate expression can be represented as:

$$\frac{dC(t)}{dt} = -k \cdot C(t) \tag{9.1}$$

Initial condition: at $t = 0$, $C_{(0)} = C_0$
Solution of this rate expression is:

$$C(t) = C_0 \cdot e^{-k \cdot t} \tag{9.2}$$

By rearranging the above closed-form solution on the initial value problem, one can linearise the above solution as follows to interpret the kinetic data:

$$\ln\left(\frac{C(t)}{C_0}\right) = -k \cdot t \tag{9.3}$$

Thus, if $\ln(C/C_0)$ is plotted against batch reaction time "t", the slop of the linear regression (fitted line) would be k *i.e.* pseudo-first order rate constant [1/d] for the given pollutant.

$C(t)$ = outlet pollutant concentration, [mg/L]
C_o = inlet pollutant concentration, [mg/L]
k = pseudo first-order reaction rate constant, [1/d]

Based on this first-order kinetic model, Kumar *et al.* (2015c) the experimental data from the laboratory CW-reactors having packed beds of manufactured sand were interpreted and the pseudo first-order reaction rate constant was estimated to be in the range of 0.35/d to 0.55/d (R^2 values for linear regression were in the range 0.60–0.95) corresponding to different species of *Canna indica*.

9.3.3 Performance assessment using pilot-scale HSSF-CW

A pilot-scale constructed wetland at IIT Bombay campus having dimensions 13 m × 3 m × 0.6 m is made for the study purpose. The media used is the construction debris having porosity of 0.45. The plant species named *Canna indica* was grown fully for six months. The influent wastewater is being fed from sump that receives the campus wastewater. In order to assess the influence of operational parameters, a group of experiments were conducted using the pilot-plant of HSSF-CW. The parameters taken into consideration included: effective reaction time in wetland bed (24–72 hrs), depth of water column (200 cm–600 cm), recirculation of wastewater, dry periods (12 hrs–24 hrs) in-between two consecutive pilot-plant runs. During performance assessment, around 12 cubic meters of wastewater was filled in the system for 600 cm waster column in packed bed (maximum water holding capacity of system through packed medium).

As depicted in Figure 9.5a and 9.5b, the values for COD and faecal coliforms, which are indicative of the efficacy of HSSF-CW, were expressed as the ratios of the typical outlet to inlet concentrations in the respective locations. The engineered CWs are apparently relatively more effective in removing the biodegradable organic pollutants in sewages (indicated by COD). However, the systems are not as effective in removal of feacal coliforms – 3 to 4 log-reduction as it was observed for sewage treatment plants (Asolekar, 2013). The NTS (particularly CWs) are also capable of removing pathogenic entities relatively more effectively when compared with the technologies typically employed in the conventional sewage treatment plants (e.g. activated sludge process, trickling filters, extended aeration, sequential bio-reactor etc.).

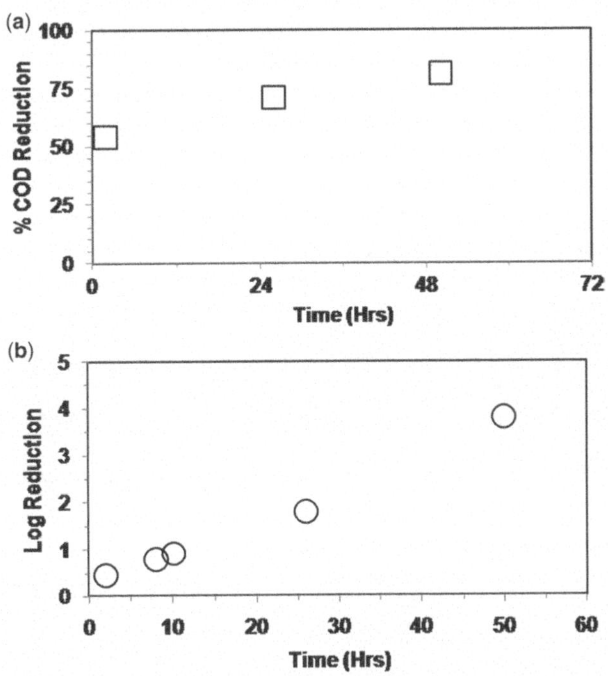

Figure 9.5 Assessment of pollutant degradation using pilot-scale constructed wetland using *Canna indica*. (a) COD removal in batch runs and (b) total coliform log removal batch runs.

Removal of organics and coliform bacteria in pilot-scale HSSF-CW exhibited the so-called pseudo-first order decay kinetics. The pilot-scale constructed wetland was filled with partially settled sewage from sewage collecting well of IIT

Bombay Campus. The raw wastewater represents the actual sewage composition of IIT Bombay Campus and the daily composition found to be quite similar except the seasonal variation. The effective reaction time in wetland bed, depth of saturated zone in the bed, recirculation of wastewater from downstream to upstream position have shown desirable effects and the overall performance of HSSF-CW did improve (Kumar & Asolekar, 2015c). Dry periods (12–24 hrs) in-between the consecutive pilot-plant runs did not seem to influence the removal of coliform bacteria. More experimental work is in progress to investigate the kinetics of degradation and operational issues in this context. The results helped in suggesting measures for improving operational stability, minimising the clogging propensity as well as for determining best practices for operation and maintenance of constructed wetlands.

9.3.4 Strategies for performance enhancement

Inadequate treatment and disposal of sewages as well as the loads brought in by the so-called non-point source pollution emerging from farm runoff and unsewered urban and rural drainages pose the severe challenge of contamination of surface and sub-surface waters in India. The soil aquifer treatment, especially the engineered CWs as well as managed aquifer recharge and riverbank filtration have been concluded to be the useful and relevant candidate technologies having the eco-centric character and competencies for addressing some of the critical problems of aquatic contamination.

Based on the learnings from the research conducted in the Saph Pani Project, the following six-pronged strategy has been proposed:

1) Based on the national survey and keeping the significance of NTSs in general and constructed wetlands in particular in mind; it is recommended that the reuse-oriented technological options should be favoured for public investment in the coming future (refer to Chapters 1, 8 and 10 for further details). Such deliberate choices are likely to achieve cost-effective treatment of sewages and thereby will achieve up-gradation of contaminated ambient waters. It has become important that relatively higher quality of waters be made available for the purposes of agriculture, process industry as well as uses in recreation and groundwater replenishment (Kumar & Asolekar, 2015a,b; Kumar *et al.*, 2015b).
2) Merely compliance-driven investments are being seen as ecosystem damaging and wasteful. It is concluded in this research that the most appropriate sewage treatment system in India could incorporate excellent primary treatment unit followed by secondary treatment unit based on NTS (Kumar & Asolekar, 2015b; chapter 10).
3) Depending on the reuse option prescribed by the community, a high–class tertiary treatment unit followed by disinfection should also be combined with the NTS so that treated wastewater can be gainfully reused (Asolekar *et al.*, 2013; Kumar *et al.*, 2015a; Kumar & Asolekar, 2015b; Chapter 10). The possible ways to improve the efficiency of engineered CWs comprise incorporating the most common and the best practices into WWTPs at the design-stage itself.
4) The engineered CW in conjunction with adequate primary treatment and suitable tertiary treatment presents the possibilities of producing treated effluents of rather high quality. Such treated effluents can be used for irrigation, gardening and even for recharging into contaminated urban lakes and ponds (Asolekar *et al.*, 2013; Kumar & Asolekar, 2015b; Chapter 10; present chapter).
5) CWs can be applied in place of or in combination with conventionally used wastewater treatment technologies to provide techno-economically feasible and socially acceptable way for wastewater management. The CWs are simple to operate and can be easily combined with cultivation of fodder, production of recyclable water, production of fuel, timber for pulp and paper industry as well as up-gradation of lake or river ecosystem and develop habitats for fishes and birds (Kumar & Asolekar, 2014a; Chapter 8 & 10).
6) Strengthening institutional arrangements and financial provisions, which are conducive for incorporating engineered CWs in WWTPs as well as motivating community to own and operate such decentralized systems, is going to be a task to be addressed by the municipalities in the years to come (Starkl *et al.*, 2012; Starkl *et al.*, 2013; Asolekar *et al.*, 2013; Starkl *et al.*, 2015; Kumar & Asolekar, 2015b; Chapter 10).

9.4 CONCLUSIONS AND LESSONS LEARNT

CWs are used to treat domestic and industrial wastewaters world-wide. The engineered CWs are not the isolated example of traditional systems proving to be misfit in the modern times. However, some of the drivers that proved to be favourable to the traditional systems and methods during the yesteryears need to be identified and analysed and efforts should be made to implant those elements into the systems and solutions of the modern times. A knowledge-based approach should be systematically implemented during construction, commissioning as well as operating and maintaining the CW-facilities. Some of the salient conclusions of this chapter can be summarized as follows:

- The weight fractions have the variability among the plant species with respect to their dry weight-fractions of roots and stems (as depicted from the range given in Table 9.3). The rhizosphere (the sub-surface volume around roots) supports a healthy and diverse consortium of aerobic microorganisms – which thrives in case of those plant species which have larger network of roots (Decamp *et al.*, 1999).
- *The three plant species namely*: *Canna indica*, *Typha latifolia* and *Phragmites australis* were found to be the most suitable in the Indian context because they are found abundantly in "natural wetlands". These species have also been successfully employed in most of the working CW-based sewage treatment plants across India.
- The shoot of the plants (especially through photosynthesis in the foliage) as well as the root system of the plant (rhizosphere) contribute in enhancement of the performance of CWs. This can roughly be estimated as nearly 10%.
- The *Canna indica* (red flower) species appears to remove more COD, TP, TKN and TOC by about 10%, 11%, 60% and 29%, respectively; when compared with the *Canna indica* (yellow flower) species.
- Removal of organics and coliform bacteria in pilot-scale HSSF-CW CW exhibited the so-called pseudo-first order decay kinetics.
- The effective reaction time in wetland bed, depth of saturated zone in the bed, recirculation of wastewater from downstream to upstream position have shown desirable effects and the overall performance of HSSF-CW did improve. More details has been given in Kumar *et al.* (2014).
- Dry periods in-between the consecutive pilot-plant runs did not seem to be influencing the removal of coliform bacterial. More experimental work is in progress to investigate the kinetics of degradation and operational issues in this context.

ACKNOWLEDGEMENTS

Authors acknowledge the co-funding obtained from Saph Pani Project by the European Commission (7th Framework Programme) under Grant Number 282911, India as well as from Indian Institute of Technology Bombay for this work.

9.5 REFERENCES

APHA, AWWA and WEF (2005). Standard Methods for the Examination of Water and Wastewater. 21st edn, American Public Health Association, Washington, D.C.

Arceivala S. J. and Asolekar S. R. (2006). Wastewater Treatment for Pollution Control and Reuse. 3rd edn, Tata McGraw Hill Education India Pvt. Ltd., New Delhi.

Arceivala S. J. and Asolekar S. R. (2012). Environmental Studies: A Practitioner's Approach. Tata McGraw Hill Education India Pvt. Ltd., New Delhi.

Asolekar S. R. (2013). Report on Experiences with Constructed Wetlands and Techno-Economic Evaluation, Deliverable D3.1 *SaphPani* Project: Enhancement of Natural Water Systems and Treatment Methods for Safe and Sustainable Water Supply in India, Project supported by the European Commission within the Seventh Framework Programme Grant Agreement No. 282911. www.saphpani.eu.

Asolekar S. R., Kalbar P. P., Chaturvedi M. K. M. and Maillacheruvu K. Y. (2013). Rejuvenation of rivers and lakes in India: Balancing societal priorities with technological possibilities, In: *Comprehensive Water Quality and Purification*, S. Ahuja (ed.), Elsevier, Waltham, United States, pp. 181–229.

Bachand P. A. M. and Horne A. J. (2000). Denitrification in constructed wetland free-water surface wetlands: II. Effects of vegetation and temperature, *Ecological Engineering*, **14**, 17–32.

Billore S. K., Singh N., Ram H. K., Shrama J. K., Singh V. P., Nelson R. M. and Das P. (2001). Treatment of molasses based distillery effluent in a constructed wetland in central India. *Water Science and Technology*, **44**(11), 441–448.

Billore S. K., Singh N., Sharma J. K., Dass P. and Nelson R. M. (1999). Horizontal subsurface flow gravel bed constructed wetland with *Phragmites karka* in Central India. *Water Science and Technology*, **40**(3), 163–171.

Brix H. (1994). Use of constructed wetlands in water pollution control: Historical development, present status, and future perspectives. *Water Science and Technology*, **30**, 209–223.

Brix H. (1997). Do macrophytes play a role in constructed treatment wetlands? *Water Science and Technology*, **35**(5), 11–17.

Chaturvedi M. K. M., Langote S. D., Kumar D. and Asolekar S. R. (2014). Significance and estimation of oxygen mass transfer coefficient in simulated waste stabilization pond. *Ecological Engineering*, **73**, 331–334.

Decamp O., Warren A. and Sandoz R. (1999). The role of ciliated protozoa in subsurface flow wetlands and their potential as bioindicators. *Water Science and Technology*, **40**, 91–98.

Gee G. W. and Bauder J. W. (1986). Particle size analysis. In: *Methods of Soil Analysis*. Part 1. Physical and Mineralogical Methods, A. Klute (ed.), 2nd edn, Amercian Society of Agronomy, Madison, United States, pp. 383–411. http://dx.doi.org/10.1016/S0167-2991(04)80157-5.

Gyssels G., Poesen J., Bochet E. and Li, Y. (2005). Impact of plant roots on the resistance of soils to erosion by water: A review. *Progress in Physical Geography*, **29**(2), 189–217.

Hammer D. A. (1989). Constructed Wetlands for Wastewater Treatment – Municipal, Industrial and Agricultural. Lewis Publ., Chelsea, Michigan, United States.

Hawke C. J. and Jose P. V. (1996). Reed Bed Management for Commercial and Wildlife Interests (RSPB Management Guides), The Royal Society for the Protection of Birds, Sandy, United Kingdom.

Healy M. G. and Cawley A. M. (2002). The nutrient processing capacity of a constructed wetland in western Ireland. *Journal of Environment Quality*, **31**, 1739–1747.

Jing S. R., Lin Y. F., Lee D. Y. and Wang T. W. (2001). Using constructed wetland systems to remove solids from highly polluted river water. *Water Science and Technology: Water Supply*, **1**(1), 89–96.

Kadlec H. R. and Knight R. L. (1996). Treatment Wetlands. Lewis, Boca Raton, New York, London, Tokyo.

Karpiscak M. M., Freitas R. J., Gerba C. P., Sanchez L. R. and Shamir E. (1999). Management of dairy waste in the Sonoran Desert using constructed wetland technology. *Water Science and Technology*, **40**(3), 57–65.

Koncalova H., Kvet J. and Lukavska J. (1996). Response of *Phragmites australis, Glyceria maxima* and *Typha latifolia* to addition of piggery sewage in a flooded sand culture. *Wetlands Ecology and Management*, **4**, 43–50.

Kumar D. and Asolekar S. R. (2015a). Significance of Engineered Natural Treatment Systems for Treating Sewages: India Case Study. (Manuscript in preparation).

Kumar D. and Asolekar S. R. (2015b). Integrated Assessment of Natural Treatment Systems. (Manuscript in preparation).

Kumar D. and Asolekar S. R. (2015c). Pilot-Scale Horizontal Sub-Surface Flow Constructed Wetland to Treat Sewage. (Manuscript in preparation).

Kumar D., Asolekar S. R. and Sharma S. K. (2015a). Post-Treatment and Reuse of Secondary Treated Effluents using Natural Treatment Systems: The Indian Practices. (Manuscript under review).

Kumar D., Chaturvedi M. K. M., Sharma S. K. and Asolekar S. R. (2015b). Typologies of Classification of Engineered Natural Treatment Systems in India. (Manuscript in preparation).

Kumar D., Sutar R. B., Kumar D., Kamble K. A., Reddy B. L. and Asolekar S. R. (2015c). Kinetics of Degradation in Laboratory and Pilot-Scale Engineered Constructed Wetlands Treating Raw Sewage. (Manuscript under preparation).

McNeill A. and Olley S. (1998). The Effects of Motorway Runoff on Watercourses in South-West Scotland. *J. CIWEM*, **12**, 433–439.

Reed S. C., Crites R. W. and Middlebrooks E. J. (1995). Natural Systems for Waste Management and Treatment. 2nd edn, McGraw Hill Inc., New York.

Rousseau D. P., Vanrolleghem P. A. and De Pauw, N. (2004). Model-based design of horizontal subsurface flow constructed treatment wetlands: A review. *Water Research*, **38**(6), 1484–1493.

Starkl M, Brunner N., Amerasinghe P., Mahesh J., Kumar D., Asolekar S. R., Sonkamble S., Ahmed S., Md. Wajihuddin A., Pratyusha S. and Sarah (2015). Stakeholder views, financing and policy implications for reuse of wastewater for irrigation: A case from Hyderabad, India. *Water*, **7**(1), 300–328; doi: 10.3390/w7010300.

Starkl M., Amerasinghe P., Essl L., Jampani M., Kumar D. and Asolekar S. R. (2012). Potential of Natural Technologies for Decentralised Wastewater Management in India, Decentralised Wastewater Management, International Water Association, November 20–22, 2012, Nagpur, India.

Starkl M., Amerasinghe P., Essl L., Jampani M., Kumar D. and Asolekar S. R. (2013). Potential of natural treatment technologies for wastewater management in India. *Journal of Water, Sanitation and Hygiene for Development*, **3**(4), 500 511.

Teršič T. (2011). SEM/EDS analysis of soil and roasting vessels fragments from ancient mercury ore roasting sites at Idrija area. *Geologija*, **54**(1), 31–40.

Uzman A. (2001). Book review: Molecular cell biology (4th edition): Harvey Lodish, Arnold Berk, S. Lawrence Zipursky, Paul Matsudaira, David Baltimore and James Darnell; Freeman & Co., New York, NY, 2000, 1084 pp., ISBN 0–7167–3136–3. In: *Biochemistry and Molecular Biology Education*, **29**(3), 2001, 126–128, ISSN 1470–8175.

Vymazal J. (2011). Constructed wetlands for wastewater treatment: Five decades of experience. *Environment Science and Technology*, **45**(1), 61–69.

Chapter 10

Significance of incorporating constructed wetlands to enhance reuse of treated wastewater in India

Dinesh Kumar, Saroj Kumar Sharma and Shyam R. Asolekar

10.1 INTRODUCTION

Rural, semi-urban and urban communities in India have been looking for augmentation of their water supplies. Almost none of the communities have adequate water supply and infrastructure for collection and disposal of wastewater across the country. Reportedly, nearly every developing country worldwide has been facing water scarcity during the past four decades. The challenge of water shortage and its consequences remains a major point of discussion in all the key international water, environment and development related meetings and conferences. For example, the issues associated with drinking water, wastewater and contamination of rivers and oceans were debated to formulate a collective action plan in three recently held international conferences, namely: United Nations Conference on Environment and Development, 3rd to 14th June 1992, in Rio de Janeiro, Brazil; World Summit on Sustainable Development, 26th August to 4th September 2002, in Johannesburg, South Africa and United Nations Conference on Sustainable Development, 13th to 22th June 2012, in Rio de Janeiro, Brazil.

Several international platforms have been urging that the world will have to come together to address the challenge of pollution of our surface waters and marine coastal ecosystems. It has been recognised that urban and rural communities worldwide are contaminating water resources by disposing their domestic wastewater into nearby water bodies. The threat has reached such proportions, especially in developing countries, that the communities are now forcing local self-governments and national governments to solve the crises through social and political actions.

During the first four decades of urban development in free India (1947 to the early nineties) the emphasis was laid on fetching potable water from 50 to 100 km distances from pristine rural settings (water reservoirs of dams and lakes). Such water supply schemes cannot be planned and implemented anymore because they are not considered to be politically defensible. The Environmental Impact Assessment Regulation was passed in 1986 by the Ministry of Environment and Forests (MoEF) of the Government of India (GoI) and large development projects need to be categorically approved by the team of experts at the MoEF as well as scrutinised and cleared by the stakeholders in a public hearing. During the past two decades, it has become more complicated because government policies not only favour inclusive growth of rural and tribal communities, but also factor in environmental and ecological costs in the impact analyses and cost benefit analyses performed before approving any development projects.

Presently, the GoI does not support exploitation of tribal and remote rural locations and forests for the benefit of urban and peri-urban communities. Besides, as stated earlier, the tribal and rural communities have begun to exert their political pressure onto growth-related policies and programmes formulated and implemented by the Central and State Governments in the Union of India. Clearly, a time has come when alternate suitable technological solutions that are concurrent with the capabilities of local agricultural and natural ecosystems must be favoured.

India's commitment to global warming related actions and the Kyoto Protocol obviously has challenged the conventional approach of water supply and wastewater management in rural and urban communities. It is now expected that all municipal authorities will have to prepare their respective "resource consumption and environment management plan" and after deliberating on the short-term and long-term "sustainability" of their proposals funding would be released by the respective ministries.

For example, the recent guidelines of the Ministry of Rural Development, the Ministry of Urban Development, MoEF as well as the Ministry of Water Resources (MoWR) and the Ganga Rejuvenation for development of infrastructure for wastewater management in rural and urban communities lay emphasis on decentralized and low-energy consuming solutions. Clearly, greener eco-centric solutions will typically be favoured in the coming years. One more factor that is likely to influence the solutions implemented in the near future is the shortage of funds. These socio-economic and political realities are influencing the technological choices of municipalities. In this context, natural treatment systems (NTSs) are indeed emerging as the preferred solution – especially in rural and small communities in India.

10.1.1 The potential of constructed wetlands for treatment of wastewater

The engineered NTSs have been incorporated into wastewater treatment plants (WWTPs) to treat wastewater since the early seventies – especially in the developing countries of Asia and Africa. Several WWTPs employing a variety of NTSs have been studied and reported in literature (Arceivala & Asolekar, 2006, 2012; Chaturvedi & Asolekar, 2009; Starkl *et al.*, 2013; Asolekar *et al.*, 2013; Chaturvedi *et al.*, 2014). Chapter 8 of this Handbook presents the lessons learned from the national survey of engineered NTSs currently functioning in India – which was one of the outputs of the Saph Pani Project. Reportedly, there are108 sites across India where NTSs are used for treating mixtures of wastewater and in some cases biodegradable industrial effluents. Among those, the 41 WWTPs were studied in-depth by the authors during December 2011 to June 2014. The details of these 41 sites have been presented in the Report No. D3.1 of the Saph Pani Project (Asolekar, 2013).

As described in Chapter 8, 23 plants had waste stabilization ponds (WSPs), three plants had duckweed ponds (DPs), seven plants had polishing ponds (PPs) and 8 plants employed horizontal sub-surface flow constructed wetlands (HSSF-CWs) or Karnal-type constructed wetlands (KT-CWs). The constructed wetlands (CWs) were preferred by small rural and semi-urban communities, especially to treat wastewater and industrial biodegradable effluents to achieve removal of carbonaceous and nitrogenous organic pollutants. In some cases, the treated effluents from CWs were put to reuse for the irrigation and rejuvenation of lakes (Asolekar *et al.*, 2013; refer also to Chapter 8 of this Handbook). Vymazal and Brezinová (2014) also reported similar observations from Czech Republic. The CWs were also found to be favoured in situations wherein evaporation of treated effluents needed to be achieved. Several researchers, too, have reported the preference for CWs in several communities in the world (Burken & Schnoor, 1998; Mara & Pearson, 1998; Metcalf & Eddy Inc., 2003; Kamath *et al.*, 2004; Mara, 2004; Arceivala & Asolekar, 2006; Asolekar *et al.*, 2013; Chaturvedi *et al.*, 2014; Vymazal, & Brezinová, 2014; Starkl *et al.*, 2015).

The WWTPs based on engineered CWs are found to be comparably or better performing than the conventional treatment technologies, which include activated sludge plants, sequential batch rectors, trickling filters, oxidation ditches or extended aeration basins–especially when compared to consumption of electrical energy and chemicals. Kumar and Asolekar (2014a and 2014b) have also highlighted the preferences of tribal communities settled in remote locations as well as rural and peri-urban communities surrounded by agricultural and commercial agro-forestry for NTSs in general and CWs in particular. In developed countries, CWs have been used for treating a variety of wastewater including domestic wastewater (Cooper *et al.*, 1997), acid mine drainages (Wenerick *et al.*, 1989), agricultural run-off, landfill leachates (Staubitz *et al.*, 1989), urban storm water run-off and for polishing treated effluents to be returned to freshwater resources.

The HSSF-CWs are typically employed for treatment and reuse of treated wastewater and sometimes even for treatment of industrial effluents. Such wetlands include 'Reed beds' and 'Root-zone' treatment methods devised to obtain environmental duty from the macrophytes cultivated in trenches or on beds having been saturated with wastewater. Wetlands have also been suggested as an alternate for treating nitrate bearing contaminated aquifers, denitrification of nitrified domestic effluents and irrigation return flows (Baker, 1998). Furthermore, the denitrification efficiency in the presence of low organic carbon was shown to depend on the C:N ratio (carbon:nitrogen ratio), with peak efficiencies occurring at a C:N ratio of 5:1.

CWs have also been used for treating the eutrophic water from Lake Taihuin China (Li *et al.*, 2008), and for providing make-up water for Lake Mansagar in Jaipur, the State of Rajasthan, India as well as for the conservation of ecosystems (Asolekar *et al.*, 2013). The habitat of endangered bird species was created on vegetated silt mounds in Lake Mansagar, which received treated effluents from the City of Jaipur (Asolekar *et al.*, 2013).

Mandi *et al.* (1998) reported a study on the treatment of domestic wastewater under semi-arid conditions in Morocco. At a hydraulic loading rate of 0.86–1.44 m^3/d to a reed bed planted with *Phragmites australis*, organic removal of 48–62%, total suspended solids (TSS) removal of 58–67% and 71–95% removal of parasites were reported. In Egypt, Stott *et al.* (1999)

achieved a 100% removal of parasitic ova from domestic wastewater intended for agriculture use. In Iran, a subsurface flow reed bed (*Phragmites australis*) of 150 m^2 was reportedly employed for treating municipal wastewaters. At an organic loading rate of 200 kg/(ha $*$ d), which is higher than the previously recommended rate of 133 kg/(ha $*$ d), (Metcalf and Eddy Inc., 2003), removal efficiencies for Chemical oxygen Demand (COD) 86%, biological oxygen demand (BOD_5) 90%, TSS 89%, total nitrogen (TN) 34%, total phosphorous (TP) 56%, and faecal coliform (FC) bacteria 99%, were obtained.

Okurut *et al.* (1999) demonstrated the viability of CWs with indigenous *C. papyrus* and *Phragmites mauritianus* in Uganda for treating municipal wastewater. In the *C. papyrus* systems, average mass removal rates for COD, TSS, ammonium (NH_4-N), TN and o-phosphorus were 15.32, 6.62, 6.5, 1.06, 0.06 g/($m^2 *$ d), respectively. In *Phragmites mauritianus* systems, the rate for the same parameters was 2.25, 0.9, 0.66, 0.65, and 0.058 g/($m^2 *$ d), respectively. The level of BOD_5 and TSS in the effluents was below 20 and 25 mg/L. A higher degree of FC removal was reported for the CW planted with *C. papyrus*.

The potential of CWs for application by small communities for wastewater treatment was examined in Nepal (Laber *et al.*, 1999). A hybrid CW system comprised of horizontal flow and vertical flow beds (140 m^2 bed area for horizontal flow and 120 m^2 bed area for vertical flow) with *Phragmites karka* was tested for one year on full-scale for treatment of hospital wastewater. At a hydraulic loading rate of 107 mm/d, removal efficiencies for COD 93%, BOD_5 97%, ammonium 99.7%, TP 74%, total coliforms 99.99%, *Escherichia coli* 99.99%, *Streptococcus* 99.97% and TSS 98% were obtained.

10.1.2 Scope and objectives

The WWTPs based on engineered NTSs in general, and CWs in particular, have been adopted worldwide for treatment of wastewater and biodegradable industrial effluents – especially in developing countries. Among several variants of engineered CWs, the four varieties are typically practiced all over the world, which include HSSF-CWs, vertical flow constructed wetlands (VF-CWs), free floating constructed wetlands (FF-CWs) and hybrid systems (Hybrid-CWs). The HSSF-CWs are gaining increasing acceptance among rural, peri-urban and remotely located small communities to treat domestic wastewater and reuse it to augment irrigation water as well as conservation and sustenance of lakes and rivers in India – which is also the scope of the present chapter. The specific objectives pursued in the present analysis are:

- In-depth assessment of three significant CWs in the context of India
- Lessons from selected case studies for in-depth assessment
- Typologies of failure of CW and remedial measures

10.2 IN-DEPTH ASSESSMENT THROUGH CASE STUDIES

Based on the survey of engineered NTSs in India (described in Chapter 8), the following three HSSF-CWs that have been providing various kinds of services were selected for in-depth assessment:

1) HSSF-CW at Mansagar Lake, in the City of Jaipur, State of Rajasthan in northern India
2) HSSF-CW in Katchpua slum, City of Agra, State of Uttar Pradesh in northern India
3) HSSF-CW in Pipar Majra, a rural community in the District Ropar, State of Punjab in northern India

The focus for this in-depth assessment included some of the critical considerations regarding public health and sanitation, environmental sustainability, fulfilment of societal aspirations and implementation of certain out-of-box ideas on institutional arrangements.

10.2.1 HSSF-CW at Mansagar lake, in the city of Jaipur, state of Rajasthan in Northern India: Case study 1

The City of Jaipur, popularly known as the *Pink City*, with a population of ≈ 3.1 million, is situated in the State of Rajasthan on the north western frontier of India in the midst of the *Thar Desert*. The City is placed in the "golden quadrilateral highway network" because of its popularity as a tourist destination. Historically, it has been the hub for arts and crafts including bandhani, block printing, stone carving and sculpture; tarkashi, zari, gota, kinari and zardozi, silver jewellery, gems, *kundan*, *meenakari*, miniature paintings; blue pottery; ivory carving; shellac work and leatherwear (The World Public Library Association, 2015).

Among the many archaeological and cultural monuments in the City of Jaipur, Lake Mansagar, a 300-acre man-made water body surrounded by the Nahargarh Hills, attracts visitors. The artificial lake was created behind the dam on River Darbhawati in the 18th century. Today it is the only significant water body in the City of Jaipur. The *Jalmahal* (literally means "palace in water"), an architectural monument, is situated in the midst of Lake Mansagar – which served as the summer palace for the Royal Family since the 18th century (Jal Mahal Innovation Report, 2005).

Historically, the Lake served the Royal Family in the summers and the residents of Jaipur throughout the year for recreation. An even more important attraction of the Lake has been its outflow for the benefits of farmers downstream. The cultural and economic uses of the Lake, however, came into conflict with rather exploitative usage of the water by the residents of the City of Jaipur and surrounding communities. Much of the untreated wastewater (and some partially treated wastewater) was disposed of into the Lake year-round. The problem was aggravated further by letting the storm water run-off into the Lake – which carried municipal solid wastes, silt and other urban pollutants. As a result, the Lake witnessed a steady decline of its health during the past three decades and reached apathetic extent of deterioration. At any point during the year, if the Lake had any water, it was due to the deposited wastewater. Not surprisingly then, the surrounding community and the City of Jaipur had endemic malaria and water borne diseases owing to the menace of mosquitoes and contamination of water in dug-wells and river.

Asolekar *et al.* (2013) has reported a brief account of the flagship project aimed at rejuvenating Lake Mansagar, which was conceived, planned, designed and executed by Asolekar and co-workers under the supervision of Prof. Soli J. Arceivala. This project has significance because it has three relatively uncommon features. First, it employed eco-centric technology to create make-up water for the Lake from treated wastewater of the City of Jaipur. Second, it aspired to up-grade the lake ecosystem to create favourable conditions for thriving biodiversity, habitats for fishes and nesting sites and hideouts for various species of birds. Third, this project experimented with a novel approach of the so-called public-private partnership (PPP) to attract capital investments and involvement of commercial implementing agencies. A four-step plan devised and implemented before implementing the ecosystem-related intervention was as follows:

1) A settling zone was created at the mouth of the storm water drain entering the Lake. Mechanical devices and operators were devised to skim off the floating plastic and any municipal garbage carried through the storm water. Also, arrangements were made to dredge out the settled sludge in the settling zone.
2) The wastewater generated northward of the ridge, dividing the City of Jaipur, was treated in the late nineties with the help of activated sludge treatment facility. Over the years, the WWTP received larger flows of wastewater. Therefore, the WWTP was modernized and expanded to give adequate primary treatment followed by conventional activated sludge process (ASP)-based secondary treatment, as depicted in Figure 10.1 and Figure 10.2–10.5.
3) A detailed hydro-geological study of the lake bed and the peripheral land was done and the mathematical model was developed by Prof G. C. Mishra from the National Institute of Hydrology (NIH), Roorkee. Based on the results, a water balance of Lake Mansagar was developed. It was concluded that at least 7 MLD of water would be needed as a daily make-up flow. Accordingly, 7.8 MLD flow was tapped from the outlet of the "WWTP-North at Brahampuri", which was situated nearly 3 km south of Lake Mansagar at a slightly higher altitude. As seen in Figures 10.1 and 10.2–10.5, the tapped flow was further subjected to tertiary treatment comprised of phosphorus removal and HSSF-CWs and the resulting tertiary treated effluent was sent to Lake Mansagar. In order to get an adequate area for CWs, three separate patches of land at different locations were used (location 1, 2 & 3).
4) The balance-treated wastewater from the modernized WWTP facility was diverted (nearly two-thirds flow) through an under-ground sewer and disposed into the Darbhavati River. The treated wastewater meets the regulatory standards to be disposed safely into river.

Over a period of eight years of sustained efforts, the aquatic ecosystem and the terrestrial vegetation on the periphery have reached a stage where these now survive without any special protection or interventions. For example, the surface feeders and bottom grazing fishes have been thriving in a sustainable manner so that the water body seems to have a balance between fishes and plants. Needless to emphasize the additional advantages Lake Mansagar continuously derives from the presence of various life forms. It is well known that such aquatic life polishes organic micro-pollutants from a given water body. Another benefit from the efforts of rejuvenation of Lake Mansagar resulted in the creation of a conductive aquatic environment for breeding of endemic and migratory species of birds. Reportedly, more than 20 species of birds have been residing on the various islands designed to support the habitats of the respective species.

One such experiment on sustainable development and business generation has been completed in Jaipur. This unique project involved environment conservation, heritage restoration and business promotion all rolled into one. To achieve a good quality of Lake water, a water budget study was carried out and the quantity of water required as make-up water estimated was 7.8 MLD. This amount of water requirement was being fulfilled by the tertiary treated wastewater from HSSF-CW installed in the vicinity of Jalmahal. The secondary treated wastewater first undergoes phosphorus removal followed by tertiary treatment in the HSSF-CW (total 40,000 m^2). The tertiary treated wastewater is then led into Lake Mansagar, its BOD_5 has come down from 150 mg/L to 10–15 mg/L.

The Rejuvenation of Lake Mansagar through application of 7.8 MLD was designed to give a secondary treatment consisting of a conventional ASP followed by chemical phosphorus removal units, as depicted in Figure 10.2 and 10.3. The tertiary treated effluent was led to HSSF-CW as seen from Figure 10.4 and the treated wastewater from CW-bed was used for rejuvenation Lake Mansagar as evidenced in Figure 10.5.

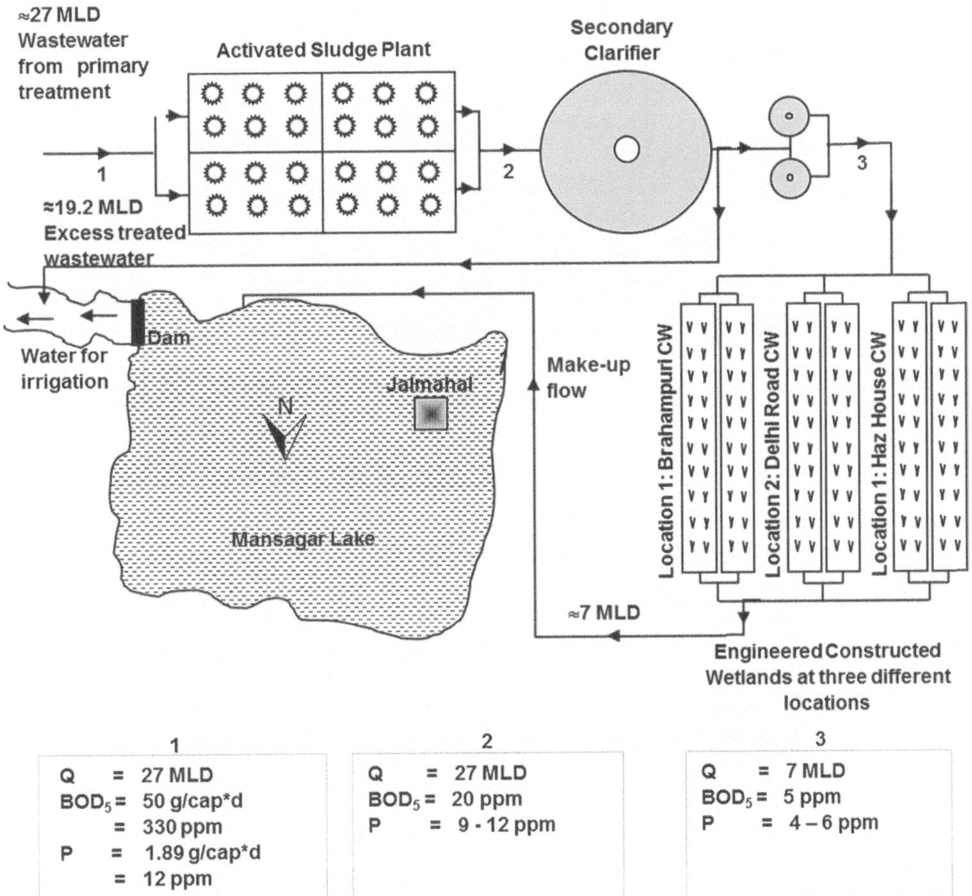

Figure 10.1 Flow diagram of the secondary WWTP of approximately 27 MLD capacity followed by two-stage tertiary treatment to generate 7 MLD make-up water for Lake Mansagar, City of Jaipur and the salient design objectives.

Figure 10.2 Conventional ASP to treat wastewater from the City of Jaipur (≈ 27 MLD).

Figure 10.3 Removal of phosphorus using lime precipitation technology (first step in tertiary treatment).

Figure 10.4 HSSF-CW (second step in tertiary treatment.

Figure 10.5 Rejuvenated Mansagar Lake that receives approximately 7 MLD of tertiary treated wastewater.

Prior to this rejuvenation work, the lake had received the untreated or partially treated wastewater from the City – which caused threats of several vector borne diseases in the surroundings of the Lake. Also, the accumulated untreated or partially treated wastewater posed a problem of ground water contamination. Rejuvenation of Mansagar Lake involved the private sector in sustainable development of the project area, and created a PPP of the State Government and Jalmahal Resorts Pvt. Ltd. (JMRPL). A grant of INR 24.72 crores (INR 247.2 million; ≈EUR 3.04 million[1]) was allocated for the restoration under the National River Conservation Program in 2002. Jaipur Development Authority was appointed as the nodal agency for the Lake Restoration part of the project. In 2007, JMRPL took the whole area of the Lake (310 acres) and its surrounding (totalling 432 acres) on a 99-year lease from the Government of Rajasthan for restoration.

10.2.2 HSSF-CW in katchpura slum, city of Agra, state of Uttar Pradesh in Northern India: Case study 2

The City of Agra in the State of Uttar Pradesh (population ≈ 1.8 million) is the most famous tourist attraction in India. Another interesting fact about the City of Agra and its peri-urban and sub-urban communities is that the region hosts one of the largest agglomerations of cottage and small-scale industries in India. Traditionally, Agra and the surrounding villages have a workforce engaged in artistic handicraft, *zarizardozi,* stone carving and in lay work as well as glass-blowing, bangle-work, foundry and brick kilns.

The City of Agra has several wastewater treatment facilities. One of the interesting WWTPs is the engineered CW in Kachpura slum. It was established in 2010 by the *Crosscutting Agra Program* intended for low-income communities with financial assistance from the *Water Trust UK* and *London Metropolitan University* and technical support from the *Vijay Vigyan Foundation.* This decentralized WWTP is operated by the Agra Municipal Corporation. The WWTP was designed to receive domestic wastewater from the five clusters of Kachpura slum *via* gutters. A part of the flow of wastewater (approximately 0.05MLD) was subjected to the CW created in the slum – right behind the *Taj Mahal.* The treated wastewater was typically utilized by the surrounding communities for irrigation. To balance untreated wastewater was led through gutters to the Yamuna River. The flow diagram adopted for treatment of wastewater is shown in Figure 10.6.

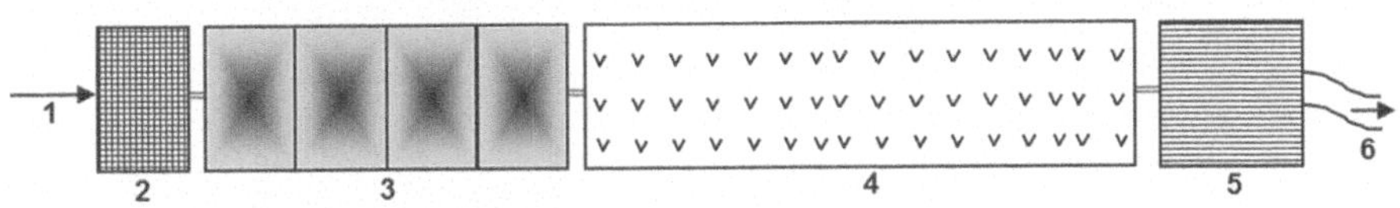

1 = Wastewater from the Katchpura (peri-urban suburb of Agra)
2 = Pre-treatment unit (bar screen and grit chamber)
3 = Baffled septic tank
4 = Engineered constructed wetland (HSSF-CW)
5 = Tank for collection of treated wastewater
6 = Channel carrying overflow from tank for gardening and irrigation

Figure 10.6 Flow diagram of the pre-treatment followed by septic tank and HSSF-CW of approximately 0.05 MLD capacity in Katchpura slum, City of Agra, State of Uttar Pradesh in northern India.

The CW in Katchpura slum was designed to give a first-rate primary treatment consisting of the settling chamber followed by a baffled septic tank, as depicted in Figure 10.7 and 10.8. The primary treated effluent was led to HSSF-CW as seen from Figure 10.9 and the treated wastewater from the sump was used to irrigate the garden as evidenced in Figure 10.10.

The primary treatment is comprised of a grit chamber, bar-screen and a baffled septic tank. The bar-screen prevents the floating solids from entering the septic tank. After pre-treatment, the wastewater is led to the baffled septic tank (nine baffles). The CW bed is filled with three layers of media, namely white river pebbles at the bottom, red stones in the middle and gravel of size 30 to 50 mm nominal diameter on the top. The bed is planted with locally abundant *Canna indica* plants. The overall operation and maintenance (O&M) of the system was found to be satisfactory. Reportedly, the capital cost incurred for establishing this WWTP was ≈EUR 15,000 and the O&M costs are currently of the order of ≈EUR 3,500 per year. The local residents reported that prior to the establishment of CW, the canal was wider carrying untreated wastewater, which created a number of problems including mosquito breeding and foul odour. The CW system has noticeably improved the sanitation in the community.

[1]Average currency exchange rate year 2014: INR EUR = 0.0123 (Online Currency Converter).

Figure 10.7 Pre-treatment unit comprising of bar-screen and grit removal.

Figure 10.8 Primary treatment unit (baffled septic tank).

Figure 10.9 HSSF-CW having *Canna indica*.

Figure 10.10 Tank for collection of treated wastewater to be sent for gardening and irrigation.

10.2.3 HSSF-CW in Pipar Majra, a rural community in the district Ropar, state of Punjab in northern India: Case study 3

Ropar (also called as Rupar), is a small city and a municipal council in Rupnagar district, the State of Punjab, in northern India. The City of Ropar has several wastewater treatment facilities. One of the interesting WWTPs is the engineered HSSF-CWs in the outskirts of the City (≈50 km away from the city) with a capacity of 0.5 MLD.

It was established in 2006 by the *Ropar Municipal Corporation* to improve the sanitation in the village community (Pipar Majra), with financial assistance from the Water and Sewerage Board as well as from the National Rural Employment Guarantee Act (MGNREGA) 2005. The respective village council has been operating this decentralized WWTP. The WWTP was designed to receive domestic wastewater from the village community *via* gutters.

The primary treatment is comprised of a grit chamber, bar-screen and aseptic tank. The bar-screen prevents the floating solids from entering the septic tank. After the pre-treatment, the wastewater is led to the septic tank. The process used for treatment of wastewater at the treatment site is depicted in Figure 10.11. The CW bed is filled with locally available river sand. The bed is planted with locally abundant *Typha latifolia* plants (i.e. common cattail) (Figure 10.12). The primary treated wastewater from the septic tank further undergoes secondary treatment through HSSF-CW. The treated effluent from the CW bed is discharged into the adjacent fishpond for pisciculture (Figure 10.13).

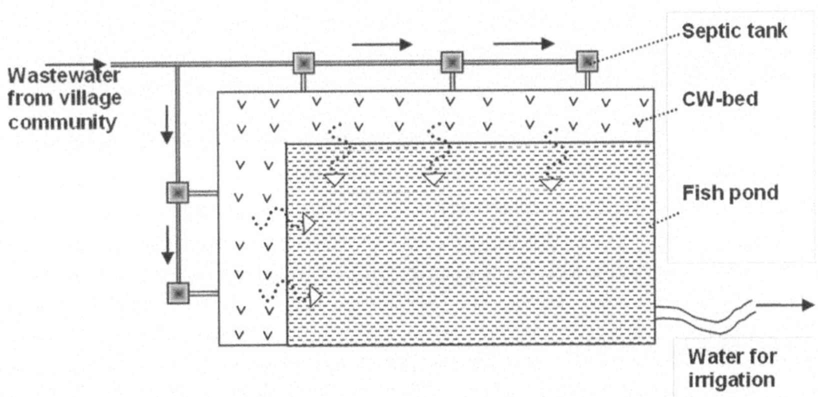

Figure 10.11 Flow diagram of the septic tanks overflowing into the HSSF-CW of approximately 0.5 MLD capacity percolating treated wastewater into a fish pond for further polishing in Pipar Majra, a rural community in the District Ropar, State of Punjab in northern India.

Figure 10.12 Outflow of septic tank connected to the HSSF-CW having *Typha latifolia*.

Figure 10.13 Treated wastewater percolated from wetland bed into the fish pond.

The overall O&M of the system was found to be satisfactory. Reportedly, the capital cost incurred for establishing this WWTP was ≈EUR 25,000. Presently, the WWTP is under stress clearly reflected by the obvious sign of clogging in the bed. The problem of bed clogging arises because of improper functioning of the primary treatment unit. The improper functioning of the septic tank arises due to negligence of operating agencies, because de-slugging of the septic tank is not performed regularly. Due to the improper functioning of the settling unit, the floating sludge from septic tank continues to enter the CW bed which results in clogging. The system may degrade beyond possible recovery if proper action, particularly immediate cleaning of septic tank is not given.

The local residents reported that prior to the establishment of CW, the untreated wastewater from the village accumulated in many ponds around the village – which created a number of problems including mosquito breeding and foul odour. The CW system has noticeably improved the sanitation in the community.

10.3 RESULTS AND DISCUSSION

There is high potential to (re)use the treated effluent from CW systems. For that, infrastructure should be in place for transferring the treated effluents from the treatment plants to the site. The effluent from CWs could also be used in some industrial processes after suitable post-treatment. Finally, artificial recharge of the treated effluent from NTSs is another attractive option to polish the effluent quality and to replenish the depleting groundwater levels in different places of India if groundwater quality is not endangered by the infiltration. However, there exists a potential danger of infiltrating persistent toxic organic pollutants, trace metals, pathogenic bacteria and viruses into the receiving aquifer. Therefore, it could be desirable to take a balance view in the context of adequate scientific information before such steps are implemented on large scale.

In most of the cases, properly operated systems of wastewater treatment based on CWs are able to achieve up to 3–4 log reduction in pathogenic bacteria count. In some cases, complete removal of pathogenic bacteria has also been reported. More importantly, natural die-off of pathogenic bacteria may be the best way because it does not require adding any potentially harmful substances like chlorine into the wastewater (the conventional practices used for disinfection). Most chemical and physical methods used to disinfect the wastewater (except chlorine) are either costly or ineffective for long-term practice, therefore CWs provide one of the most appropriate ways of reducing the pathogenic count without adding any harmful by-products to the wastewater.

There are different lessons that could be articulated based on the three case studies presented in Section 10.2. It must be noted that the three locations were deliberately chosen for developing a deeper understanding of strength and weaknesses of establishing engineered CW as the "technology of choice" for addressing the issues associated with management of wastewater in the context of a variety of situations. For example, it was recognized at the outset that the challenges faced by an urban local body (ULB), while restoring and rejuvenating a lake or a river subjected to the input of wastewater in the respective city or town, are completely different when compared with the challenges faced while treating domestic wastewaters generated by any peri-urban or sub-urban community. Likewise, both the techno-economical and the socio-cultural considerations play a crucial role when wastewater generated in a rural community need to be managed successfully.

There are two aspects to consider when assessing the effectiveness of CW as a technology during planning and designing a WWTP for a given community. First, the techno-economic evaluation of CW – especially if the technology is capable of delivering suitable quality of treated wastewater acceptable to the regulatory agency and the municipal administration. Second, the equally pertinent consideration should be the affordability and manageability of the technology. Arceivala and Asolekar (2006) have emphasized these two aspects as the most relevant and critical for determining if a specified technology is "appropriate" in a given technological and socio-economic context.

10.3.1 Highlights of the performance of selected case studies

Kumar and Asolekar (2014a, 2014b) have described the detailed results of nation-wide survey of NTSs in general and CWs in particular. The summary of the performance of the three CWs selected for in-depth study across India, as described in the prior section, is depicted in Table 10.1.

Table 10.1 Summary of the performance of the three shortlisted HSSF-CW across India based on analyses of the data reported by the respective operators. The values for BOD_5, COD and Faecal coliforms (FC), indicative of efficacy of treatment of the respective WWTPs (expressed as the ratios of the typical outlet to inlet concentrations).

Location of WWTP	Capacity	% Removal (Average Annual Performance)		
		BOD_5	COD	FC
HSSF-CW at Mansagar Lake, City of Jaipur,State of Rajasthan, India	7.8 MLD	50–96%	75–87%	99.99%
HSSF-CW in Katchpua slum, City of Agra, State of Uttar Pradesh, India	0.05 MLD	61%	64%	90 to 99%
HSSF-CW in Pipar Majra, a rural community in the District Ropar, State of Punjab, India	0.5 MLD	85–87%	Not available	99.90%

The values for BOD_5, COD and FC, which are indicative of the efficacy of HSSF-CWs, were expressed as the ratios of the typical outlet to inlet concentrations in the respective locations. It can be concluded from Table 10.1 that the engineered CWs are apparently relatively more effective in removing the biodegradable organic pollutants in wastewater (indicated by BOD_5 and COD). However, the systems are not as effective in removing of FC (3–4 log-reduction) as was observed for WWTPs. It was argued by Kumar and Asolekar (2014a, 2014b) that the NTSs (particularly CWs) are known for removing pathogenic entities relatively more effective when compared with the technologies typically employed in conventional WWTPs (e.g. ASP, trickling filters, extended aeration, sequential bio-reactor etc.).

The performance of CWs with respect to the removal of FC (Table 10.1) is typically better by one to two orders of magnitude than conventional wastewater treatment technologies. However, there is, as pointed out by Kumar and Asolekar (2014c) and Kumar *et al.* (2014), still room for further development of techniques to enhance the performance of CWs through targeted research and development. Some of the details of those strategies and the lessons learned from pilot-scale experimentation were also highlighted in Chapter 9. Clearly, the need for effective post-treatment is vital if one intends to reuse treated wastewater for irrigation applications involving physical contact and for other higher end-uses.

10.3.2 Lessons learnt from rejuvenation of Lake in the city of Jaipur

In case of the Mansagar Lake, the City of Jaipur was disposing nearly half of the wastewater generated by the metropolis into the lake. It continued for over two decades and the lake deteriorated to a condition of a big cesspool. The surrounding peri-urban community and the visitors of Jalmahal monument were troubled by stench, ugly sight and mosquito menace. As described earlier, droughts and excessive abstraction from the lake for irrigation had converted the lake in a sludge bed. It was in this context, the engineered CW (of HSSF-CW type) was devised to provide a first-rate tertiary treatment providing daily make-up flow of 7.8 MLD to the lake.

Interesting to note that the original ecosystem of the lake and the surroundings had been completely destroyed as a result of the long-term onslaught of untreated wastewater and urban run-off into the lake. Therefore, the real challenge was to revive the aquatic ecosystem in the desert environment. This was achieved through implementing the two kinds of interventions. The first challenge was to bring make-up water of suitable quality and quantity to the lake, desirably every day, so as to maintain the adequate depth of water in the lake throughout the year. The second challenge was the propagation of the aquatic and terrestrial ecosystem within the water-body and on its periphery. Both these challenges were, in fact, interdependent.

It was not practical to hope that the community would support the ecosystem in the lake without ensuring the sustained outflow for benefit of their farms downstream. Thus, the survival of lake as well as rejuvenation of the ecosystem had to be engineered with the help of "appropriate technology" that was capable of providing adequate quantity and quality of water (acceptable levels of BOD_5, TSS, nitrogen and phosphorus). Incorporation of the engineered CW for providing the tertiary treatment to upgrade the output from the adjoining WWTP, worked-out to be the practical solution.

Clearly, the efforts directed to up-gradation of the Lake ecology has been one of the most spectacular benefits – both, for the ecosystem as well as tourism. The downstream community of the farmers was probably the other spectacular beneficiary. Before the rejuvenation of Lake, the farmers were using untreated or partially treated wastewater intermittently available somehow to satisfy their irrigation needs. The lake now is capable of discharging adequate quantity and quality of water for irrigation throughout the year.

10.3.3 Lessons learnt from decentralized treatment of wastewater from a peri-urban community in Agra

Resulting from rapid industrialization over the past three decades in India, the workforce has been migrating to urban centres in an unprecedented manner. Concurrently, the number of peri-urban and slum communities situated on the periphery of such industrialized urban centres is escalating exponentially across India (Asolekar *et al.*, 2013). The inadequate infrastructure for providing sanitation in such highly populated peri-urban and slum communities have typically worked synergistically to aggravate the public health-related challenges. The Katchpura Slum, a peri-urban community within the limits of Agra Municipal Corporation, is a typical community facing challenges arising from its location, population density and status of sanitation.

Nearly 7,500 cities in India are struggling to manage resource consumption and sanitation-related challenges. Clearly, ULBs that provide sustainable solutions and that are successful in motivating local communities to get involved while addressing these challenges are thriving. In that sense, the intervention in Katchpura Slum may be seen as a winning example of achieving integrated development through adoption of inclusive approaches. Kumar and Asolekar (2014c) have emphasized the significance of the decentralized approach adopted in the Katchpura Slum as one of the valuable examples of bringing about the so-called "integrated development" of a slum community with the help of an inclusive management approach.

10.3.4 Lessons learnt from decentralized treatment of wastewater from a rural community

Yet another interesting application of engineered CW was found in the village of Pipar Majra in the district of Ropar, State of Punjab. The HSSF-CW was employed to treat the wastewater generated by the rural community in Pipar Majra. The WWTP has been treating domestic wastewater to such an extent that nearly every drop of the treated wastewater is put to some gainful economic as well as ecological use. In that sense, the WWTPs can be termed as so-called "zero liquid discharge" facilities.

The treated wastewater at the outlet of HSSF-CW is collected in a pond having a natural clay liner installed at the bottom to minimize percolation losses. Interestingly, the pond also serves as a flood protection installation by virtue of its location within the rural community and special care was taken in designing the facility to ensure that the surface water run-off could easily be drained into the pond in the event of flash floods. A lift-irrigation facility has also been devised for the benefit of the agricultural community. There are three noteworthy features of this pond, namely:

1) The storm water run-off collection into the pond brings a lot of water to the pond providing dilution to the pond water as well as flood protection for the surrounding community.
2) The daily addition of treated wastewater ensures an adequate make-up of the water, both in terms of quality and quantity.
3) The abstraction of water through the lift-irrigation system provides blow-down of the residual water of the pond – thereby minimizing salinity, suspended solids, total phosphorus and any toxic micro-pollutants.

In addition to the special features described above, the pond has been serving the community by harbouring a sustained population of fishes, which has in fact become an income source for fishermen. The Ministry of Rural Development, Government of India has been encouraging the creation of ponds in rural communalities through the "MGNREGA Schemes"–especially in land-locked settings (like village Pipar Majra) to ensure increased flood protection and to provide water for irrigation. Arguably, as highlighted by Kumar and Asolekar (2014, 2014b), the WWTP facility in Pipar Majra combines some of the most desirable features, including use of eco-centric technology, production of a better quality of treated wastewater and creation of a pond within the village which provides irrigation water and flood protection for the surrounding community.

10.3.5 Typologies of failures of constructed wetlands and remedial measures

Based on the survey of NTSs in general and CWs in particular (especially those from sites visited for in-depth study) several insights into the typologies of failure of engineered CWs were articulated. As depicted in Table 10.2, the three sites investigated for the in-depth study formed the basis of the analyses presented in the table. However, the lessons learned should not be viewed as restricted to the respective sites because the case studies have addressed a diverse variety of institutional situations and technology related issues. It is hoped that the lessons learned in this analyses will prove to be significant and helpful during future efforts of implementation and replication.

As summarized in Table 10.2, the poor O&M of the eco-centric technology typically results from inadequate primary treatment in almost all cases. Clogging of the porous media can have a domino effect on the efficacy of all the unit operations included in the treatment train. Another common challenge has been insufficient funds for O&M. One of the cardinal principles used in India's environmental jurisprudence is: "the polluter pays!" Unfortunately, the institutional arrangements are either too weak or non-existent when it comes to collecting and utilizing the fees from the "users or polluters" who are sending their wastewater to the WWTP (Arceivala & Asolekar, 2012).

In that sense, the CWs at Lake Mansagar, as described in sections 10.2.1 and 10.3.2, did not face any financial crises because the quality of secondary treated wastewater is ensured by the Jaipur Municipal Corporation and the O&M of the phosphorus precipitation plant and CWs has been taken care of through the PPP arrangements.

As regards to the social challenges, it is now clear that the modalities of access to the harvested biomass and entitlement of the community owning the WWTP based on CWs will need to be worked out in more detail and mutually agreed upon. In absence of such systematic efforts, it has been observed that communities feel alienated from the wetland beds. Such negative impression discourages the operators of the WWTP even further to the extent that no user fees are charged at all and O&M deteriorates further.

Finally, the success of the WWTP largely depends on the balance between the realistic inlet quality and quantity of wastewater and the expectation of the community to treat it for certain kinds of reuse applications. The three selected case studies presented have categorically underscored these facts. Over the years, the demography, land-use pattern, economic activity and the extent of industrialisation has been transformed to a new state of equilibrium. Thus, the proportion of domestic and industrial effluents will vary and the WWTP may become relatively obsolete or redundant. This, however, is not only true for WWTPs based on CWs but also for other wastewater treatment technologies.

Table 10.2 The typologies of failure based on the analyses of data and experiences collected from site visits to the three shortlisted HSSF-CW across India.

Sr. No.	Typology of Failure	Jaipur	Agra	Ropar
1	Poor O&M	No issue was observed	No issue was observed	Poor maintenance of primary treatment unit leads to carry-forward of garbage and solids in the wetland bed and system experiences clogging. Harvesting of vegetation is not done. This has led to the dead plants falling on the wetland bed.
2	Financial crises	No issue was observed	It is well known that the system performance can be improved by ensuring release of sufficient funds for the O&M periodically.	It is well known that the system performance can be improved by ensuring that adequate work-force is employed and sufficient funds are provided for the O&M.
3	Social issues	Due to lack of fencing around the CW-bed, domestic animals usually eat the biomass which affects the performance of the systems.	No issue was observed	No issue was observed
4	Mixing of industrial toxic effluents with wastewater	The facility is suffering with the mixed type of industrial toxic wastewater from dyestuff industry, which has toxic effects.	No issue was observed	No issue was observed

10.4 CONCLUSIONS AND LEASSONS LEARNT

Over the past two decades municipalities have been making efforts to address the challenge posed by inadequate and insufficient infrastructure for treatment of wastewater throughout India, both, of urban as well as rural communities. The Ministry of Urban Development, the MoEF as well as the MoWR and Ganga Rejuvenation have been incorporating the strategy of providing low-cost eco-centric treatment to wastewater to counteract the pollution of natural watercourses in India. In the recent past, many states in the Union of India have been taking steps to develop and implement plans for strengthening systems to provide sanitation in sub-urban, peri-urban and small communities. As reported in Chapter 8, it was observed that the shortfall between the quantities of wastewater generated *versus* quantity treated has been increasing. It is in this context, municipalities serving small communities are looking for alternatives that are eco-friendly and inexpensive when it comes to capital and O&M costs. Several CWs were surveyed during the national survey from December 2011 to June 2014. Three WWTPs based on CWs were studied in more detail. The salient conclusions and learnings from these case studies are summarized below:

1) HSSF-CWs have been adopted worldwide for treatment of wastewater and biodegradable industrial effluents, especially in developing countries.
2) The WWTPs based on CW technology have been found to be quite effective for the treatment and reuse of wastewater generated by rural and town communities across India.
3) The engineered CW systems seem to be quite robust and versatile in a variety of climatic conditions across India and meet the prescribed regulatory standards.
4) Communities seem to prefer them even more in the recent times due to the CWs' innate advantages of minimizing mosquito breeding and thereby minimizing the threat of cerebral malaria, dengue and several vector-borne diseases.

5) Several CW-based WWTPs have been designed and implemented for wastewater treatment and reuse across India including the remarkably successful system in the City of Jaipur, which has transformed Lake Mansagar and rejuvenated its ecosystem, the CW in Katchpura Slum of City of Agra and the CW-pond in the rural setting of Pipar Majra of the Ropar District.
6) Primary treatment plays an important role in trouble-free O&M of HSSF-CWs. Therefore, adequate primary treatment should be designed, installed as well as operated in any WWTP based on CW.

In summary, the engineered CWs, in conjunction with adequate primary treatment and suitable tertiary treatment, present the possibility of producing treated effluents of rather high quality. Such treated effluents can be used for irrigation, gardening and even for recharging into contaminated urban lakes and ponds. In addition, the CWs are simple to operate and can be easily combined with cultivation of fodder, production of recyclable water, production of fuel, production of timber for the pulp and paper industry as well as up-gradation of lake or river ecosystem and to create habitats for fishes and birds. Strengthening institutional arrangements and financial provisions, which is conducive for incorporating engineered CWs in WWTPs as well as for motivating communities to own and operate decentralized systems, is going to be a vital task to be addressed by the municipalities in the years to come.

ACKNOWLEDGEMENTS

The authors acknowledge the co-funding obtained from Saph Pani Project by the European Commission (7th Framework Programme) under Grant Number 282911, India as well as from the Indian Institute of Technology Bombay for this work. Authors are grateful to the various ministries and agencies in the Government of India and municipalities for facilitating this survey and providing data.

10.5 REFERENCES

Arceivala S. J. and Asolekar S. R. (2006). Wastewater Treatment for Pollution Control and Reuse. 3rd edn, Tata McGraw Hill Education India Pvt. Ltd., New Delhi, India.

Arceivala S. J. and Asolekar S. R. (2012). Environmental Studies: A Practitioner's Approach. Tata McGraw Hill Education India Pvt. Ltd., New Delhi, India.

Asolekar S. R. (2013). Report on Experiences with Constructed Wetlands and Techno-Economic Evaluation, Saph Pani: Enhancement of Natural Water Systems and Treatment Methods for Safe and Sustainable Water Supply in India, Report No. D3.1, Project supported by the European Commission within the Seventh Framework Programme Grant agreement No. 282911. www.saphpani.eu.

Asolekar S. R., Kalbar P. P., Chaturvedi M. K. M. and Maillacheruvu K. Y. (2013). Rejuvenation of rivers and lakes in India: balancing societal priorities with technological possibilities. In: *Comprehensive Water Quality and Purification*, S. Ahuja (ed.), Elsevier, Waltham, pp. 181–229.

Baker L. A. (1998). Design considerations and applications for wetland treatment of high-nitrate waters. *Water Science Technology*, **38**, 389–395.

Burken J. and Schnoor J. (1998). Predictive relationships for uptake of organic contaminants by hybrid poplar trees. *Environ. Sci. Technol.*, **32**, 3379–3385.

Chaturvedi M. K. M. and Asolekar S. R. (2009). Wastewater treatment using natural systems: the Indian experience. In: Technologies and Management for Sustainable Biosystems, J. Nair and C. Furedy (eds.), Nova Science Publishers, Inc., United States, pp. 115–130. ISBN: 978–1–60876–104-3.

Chaturvedi M. K. M., Langote S. D., Kumar D. and Asolekar S. R. (2014). Significance and estimation of oxygen mass transfer coefficient in simulated waste stabilization pond. *Ecological Engineering*, **73**, 331–334.

Cooper P., Smith M. and Maynard H. (1997). The design and performance of a nitrifying vertical-low reed bed treatment system. *Wat. Sci. Tech*, **35**, 215–221.

Jal Mahal Innovation Report (2005). Jal Mahal Tourism Project, Infrastructure Leasing & Financial Services (IL&FS). http://www.ilfsindia.com/downloads/bus_rep/jalmahal_tourism_rep.pdf (accessed 12 December 2014).

Kamath R., Rentz J. A., Schnoor J. L., Alvarez P. J. J. (2004). Phytoremediation of hydrocarbon-contaminated soils: principles and applications, In: Studies in Surface Science and Catalysis, R. Vazquez-Duhalt and R. Quintero-Ramirez (eds.), Elsevier, pp. 447–478. ISSN 0167–2991.

Kumar D. and Asolekar S. R. (2014a). Significance of engineered natural treatment systems for treating sewages: India Case Study. (*Manuscript in Preparation*).

Kumar D. and Asolekar S. R. (2014b). Integrated assessment of natural treatment systems. (*Manuscript in Preparation*).

Kumar D. and Asolekar S. R. (2014c). Pilot-scale horizontal sub-surface flow constructed wetland to treat sewage. (*Manuscript in Preparation*).

Kumar D., Chaturvedi M. K. M., Sharma S. K. and Asolekar S. R. (2014). Typologies of classification of Engineered Natural Treatment Systems in India. (*Manuscript in Preparation*).

Laber J., Haberl R. and Shrestha R. (1999). Two-stage constructed wetland for treating hospital wastewater in Nepal. *Water Science and Technology*, **40**(3), 317–324.

Li L., Li Y., Biswas D. K. Nian Y. and Jiang G. (2008). Potential of constructed wetlands in treating the eutrophic water: evidence from Taihu Lake of China. *Bioresource Technology*, **99**, 1656–1663.

Mandi L., Bouhour K. and Ouazzan N. (1998). Application of constructed wetlands for domestic wastewater treatment in an arid climate. *Water Sci. Technol.*, **38**, 379–387.

Mara D. (2004). Domestic Wastewater Treatment in Developing Countries. Earthscan, United States.

Mara D. and Pearson H. (1998). Design manual for waste stabilization ponds in Mediterranean countries. Leeds Lagoon Technology International Ltd., Leeds, United Kingdom.

Metcalf and Eddy Inc. (2003). Waste water engineering treatment and reuse. 4th edn, Tata Mc Graw Hill publication, New Delhi, India.

Online Currency Converter (2015). Exchange rate history EUR INR 2014. http://www.freecurrencyrates.com/exchange-rate-history/EUR-INR/2014 (accessed 28 April 2015).

Okurut T. O., Rijs G. B. J. and Van B. J. J. A. (1999). Design and performance of experimental constructed wetlands in Uganda, planted with *Cyperus papyrus* and *Phragmites mauritianus*. *Water Science and Technology*, **40**(3), 265–271.

Starkl M., Amerasinghe P., Essl L., Jampani M., Kumar D. and Asolekar S. R. (2013). Potential of natural treatment technologies for wastewater management in India. *Journal of Water, Sanitation and Hygiene for Development*, **3**(4), 500–511.

Starkl M., Brunner N., Amerasinghe P., Mahesh J., Kumar D., Asolekar S. R., Sonkamble S., Ahmed S., Wajihuddin M., Pratyusha A. and Sarah S. (2014). Stakeholder views, financing and policy implications for reuse of wastewater for irrigation: a case from Hyderabad, India. *Water*, **7**(1), 300–328. doi:10.3390/w7010300.

Staubitz W. W., Surface J. M., Steenhuis T. S., Perverly J. H., Lavine M. J., Weeks N. C., Sanford W. E. and Kopka R. J. (1989). Potential use of constructed wetlands to treat landfill leachate. In: Constructed wetlands for Waste Water Treatment: Municipal, Industrial and Agricultural, D. A. Hammer (ed.), Lewis Publishers, Chelsea, Michigan, pp. 735–742.

Stott R., Jenkins T., Bahgat M. and Shalaby I. (1999). Capacity of constructed wetlands to remove parasite eggs from wastewater in Egypt. *Water Sci. Technol.*, **40**(3), 117–123.

The World Public Library Association. Jaipur State. http://netlibrary.net/article/whebn0007585804/jaipur%20state (accessed 10 July 2015).

Vymazal J. and Brezinová T. (2014). Long term treatment performance of constructed wetlands for wastewater treatment in mountain areas: four case studies from the Czech Republic. *Ecological Engineering,* **71**, 578–583.

Wenerick W. R., Stevens S. E., Webster H. J., Stark L. R. and deVeau E. (1989). Tolerance of three wetland plant species to acid mine drainage: a greenhouse study. In: Constructed Wetlands for Waste Water Treatment: Municipal, Industrial and Agricultural, D. A. Hammer (ed.). Lewis Publishers, Chelsea, Michigan, pp. 801–807.

Chapter 11

Characterization and performance assessment of natural treatment systems in a Wastewater Irrigated Micro-watershed: Musi River case study

Priyanie Amerasinghe, Mahesh Jampani, Sahebrao Sonkamble, Md. Wajihuddin, Alexandre Boisson, Md. Fahimuddin and Shakeel Ahmed

11.1 INTRODUCTION

In India, many agricultural landscapes close to the cities have been irrigated with wastewater for many years. Fresh water scarcity, easy access, ready markets for high value crops have been the main reasons for its common reuse practice (Scott *et al.*, 2004; Qadir *et al.*, 2010). The water demand can only become worse in the future, with demands exceeding the supply where more than 40% of the population live. It is estimated that in around 12 years, nearly 60% of the world's population might face scarcity of water. Being an agriculture-based economy, India's irrigation water demand is the highest compared to all the other sectors, and given the increasing demand for fresh water, there is an urgent need to concentrate on efficient water resource management through enhanced water use efficiency and safe use of wastewater in agriculture (WHO, 2006). Wastewater reuse can be seen in diverse settings, and therefore, context specific. For example, when used for agriculture, the water can be simply diverted from storm water drains, or lifted from polluted rivers. Thus, the quality of this source water can vary significantly and requires analysis and treatment prior to safe reuse in agriculture (WHO, 2006).

Wastewater can be treated by land application, under natural conditions, and there are also other Natural Treatment Systems (NTS), like sedimentation ponds and wetlands that are effective (Arceivala & Asolekar, 2006). Many types of NTS that are effective in the treatment of wastewater have been described in the literature (Reed *et al.*, 1995; Shipin *et al.*, 2005; Crites *et al.*, 2014). The low-cost nature of the systems has been particularly appealing to those that find the traditional sewer network systems expensive. Where the soils and the groundwater conditions are appropriate, treatment of wastewater has been achieved for artificial recharge of groundwater, and the infiltration through the soils have facilitated the upgrading of the quality of water (Pescod, 1992). Thus, the soils and aquifers can treat wastewater to improve the quality, and where continuous wastewater irrigation takes place, the action of land application of wastewater can have a treatment effect. In conventional terms this is referred to as soil aquifer treatment (SAT). Naturally occurring wetlands are also effective in treating wastewater, but sometimes poorly investigated.

A number of previous studies on the Musi River catchment have documented the landscape dynamics (Mahesh *et al.*, 2015), hydrogeology and water quality (Perrin *et al.*, 2011a; Schmitt, 2010; Amerasinghe *et al.*, 2009), socio-economics (Buechler *et al.*, 2002), health impacts of wastewater use (Srinivasan & Reddy, 2009; Wakode *et al.*, 2014), soil and salinity implications (Biggs & Jiang, 2009), and willingness to pay for cleaner water (Mekala *et al.*, 2009). However, no studies have been undertaken to examine the natural treatment potential of the site where long term wastewater irrigation has been practiced. In this study, we attempted to characterise and assess the treatment performance of NTS in the micro-watershed.

11.2 STUDY SITE

The Musi River is part of the Krishna River basin, and is associated with an ancient irrigation system comprising a large wetland system. The river starts from the Ananthagiri hills and feeds the Krishna River after passing through the city of Hyderabad (6.8 million population, Census, 2011), picking up over 1.2 million m^3/d of wastewater (both domestic and industrial) from the city, which is a mixture of partially treated or untreated water (NRCD, 2001; Van Rooijen *et al.*, 2005: Amerasinghe *et al.*, 2012; Starkl *et al.*, 2013). The wastewater is used downstream for irrigation, either directly via a system of irrigation canals or after storage in tanks. The wastewater is a significant resource in this semi-arid peri-urban environment where the cultivation of fodder grass, paddy and vegetables provide economic benefits to many poor inhabitants of the area (Jacobi, 2009). Year round cultivation, which generates large return flows from irrigated fields, also contribute to a large share of the aquifer recharge (Perrin *et al.*, 2011b). Within the micro-watershed shallow groundwater is also pumped for irrigation in areas where canal water is not accessible, or too polluted to irrigate certain crops according to farmers, especially rice.

The present study was carried out in a micro-watershed Kachiwani Singaram (KSMWS), comprising an area of 274 ha which is close to Hyderabad and is within the Musi River watershed (Figure 11.1). Along its 274 km length, 22 weirs provide temporary storage of water for a series of irrigation canals that run parallel to the river. The KSMWS receives the river water at the first weir, through an irrigation canal which runs approximately 15 km and ends in a village tank. Canal discharge rates were highly variable in time and space, and were generally low, i.e. less than 2 m^3/s. The micro-watershed's climate is semi-arid, with a mean annual rainfall of about 750 mm (most of the rainfall occurs between June to October) with both high spatial and temporal variability. Over the last 18 years, annual rainfall in Hyderabad varied from 535 to 1473 mm respectively for years 2000 and 2010. Most of the rain events occurred during the monsoon, which was from June to October. The mean annual temperature was about 26°C, although during the summer time the maximum temperature can reach up to 45°C. The Musi riverbed shows a flat topography (mean slope <1%) (Massuel *et al.*, 2007). The watershed geology consists of a basement made of orthogneissic granite also known as "pink granite" with granite, quartz and dolerite intrusions and showed a well-developed weathering profile. From a landscape view the area under study comprises agriculture land (irrigated with wastewater and groundwater), a small reed pond (wetland) in the middle, barren land and built-up areas in the northern region. The village Kachiwani Singaram after which the micro-watershed was named, is at the eastern border of the delineated area and is situated adjacent to the irrigation canal. The major crops grown in the area were rice, paragrass and vegetables. Both canal water and groundwater were used for cultivation, depending on the availability and access. Therefore, within the micro-watershed irrigation practices varied widely.

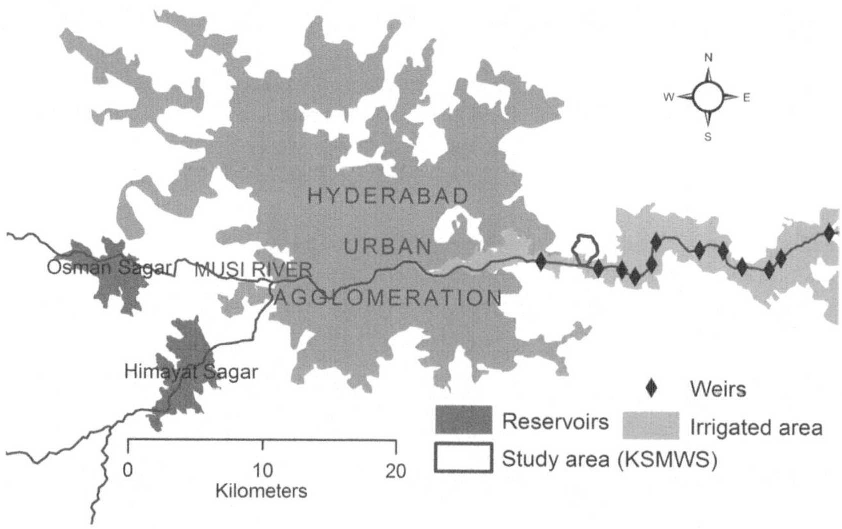

Figure 11.1 Location of the Kachiwani Singaram micro-watershed (KSMWS).

11.3 STUDY APPROACH

The micro-watershed was characterised using a number of activities, including land use surveys, discharge measurements, water level monitoring (continuous and piezometric campaigns), water budgets, pumping tests, geophysical surveys (electrical resistivity tomography), and water quality analysis (major ions, trace elements, biological oxygen demand, microbiology and pesticides). A performance assessment of the micro-watershed was carried out using past and current field studies spanning over 10 years. The flow and transport model for the local hard-rock aquifer system was developed to understand the treatment capacity within the micro-watershed.

11.4 MATERIALS AND METHODS

The methods followed for land use mapping, status of surface and groundwater, geophysical and hydrogeological characterization, and water quality assessments are given below.

Land use mapping was carried out using high resolution satellite imagery (1–12 m resolution, Google Earth images) and interpretations from a previous study (Amerasinghe *et al.*, 2009). Micro-level spatial land use variability was assessed using digital globe satellite data (dataset used: 21 January 2012) from Google Earth and validated with ground truth data, field observations and farmers interviews. Digital globe data covering the study area were used to investigate the spatial variability in cropping patterns and other land cover classes. The datasets were geo-referenced with UTM projection and WGS 84 datum. The datasets on cropping patterns, built-up areas, topographical maps, ground truth data and farmer interviews were used as inputs for classification and accuracy assessment.

River discharge measurements were carried out using a standard float method. A distance of 20 m, free of vegetation was selected for the measurements. If the flow was very slow, shorter distances were considered. Oranges were used as a floating object and a stopwatch was used to monitor the time taken to reach the specified distances. The experiment was repeated three times and the average was used for calculations.

Application of geophysical parameters such as electrical resistivity with geological perception was utilised to resolve the hydrogeological complexity of the near surface (Sonkamble *et al.*, 2013). Geophysical investigations such as electrical resistivity tomography (ERT) were carried out to delineate the subsurface lithological layers and saturated thickness. SYSCAL Junior Switch multi-node computer-controlled imaging system (IRIS make, France) was used with 48 electrodes connected to a multi-core cable. Wenner–Schlumberger and Dipole-Dipole configurations were selected to scan the subsurface profile of lengths ranging from 96 to 470 m (as per the availability of space) with a unit electrode spacing of 2.0 m to 10.0 m. A total of 17 ERT profiles were carried out at 11 different locations (Figure 11.2) to represent various stages of the weathering processes. A continuous ERT profile line A–A′ of 1.62 km distance was chosen to decipher the spatial variations of the regolith (saprolite) thickness and saturated zone, and also to determine the subsurface contamination from north to south orientation up to the Musi River. In addition to the A–A′ profile line, the ERT investigations were also performed at other locations in close proximity of observation wells. Surface modelling based on topography was also performed using ArcGIS software where the topography was estimated from the ASTER DEM with 30 m resolution datasets. Detailed information encompassing site geology, geomorphologic and hydrogeological conditions at each ERT profile was noted during the ERT investigations and was utilised in the geophysical interpretation of the images.

Monthly measurements of water levels were carried out in the four monitoring piezometers (W1, W5, W6 and W8) (Figure 11.3). The W1 was located south of the irrigation canal in the paddy field and captures a wastewater contaminated shallow unconfined aquifer. W5 was located 400 m north of the irrigation canal and located in a paragrass field irrigated with wastewater. W6 was at 700 m north from the irrigation canal, located in a paddy field irrigated with groundwater. Similarly, W8 was placed 1 km north of the irrigation canal, and positioned at an elevated topography, located in a vegetable farmland irrigated with groundwater. Automatic level-loggers (Solinst® Levelogger), recording water level and water temperature at a 30 min time interval were installed in W8 and W6. Four extended piezometric campaigns over the catchment were carried out in May, June (pre-monsoon), September (monsoon), and October (post-monsoon) of 2010. Through interpolation using inverse distance weighting, annual (pre-/post-monsoon 2010) and inter-annual comparison (post-monsoon 2010/post-monsoon 2013) maps were prepared and further analysed to highlight particular patterns of the groundwater movement.

Hydrodynamic properties were assessed by conducting hydraulic tests, i.e. pumping test, in the four piezometers (W1, W5, and W8). Drawdown was monitored during the entire pumping duration (recorded at three minute interval). Then the recovery was recorded every minute up to reaching the initial water level. These data were used to estimate transmissivity of the aquifer using the software WinIsape© developed by BRGM. The Theis method was used for interpretation (Theis, 1935), using a best-fitting procedure. Variations in water velocity inside the bore well during pumping were measured. For this, a flow meter was used with a propeller which turns when it comes into contact with the water flow (measure of flow velocity). Measurements were made at the bottom of the casing and extended to the base of the hole. Each productive fissure was indicated by a drop in flow velocity. Results were correlated with geological and observations of cuttings, to identify the productive zones.

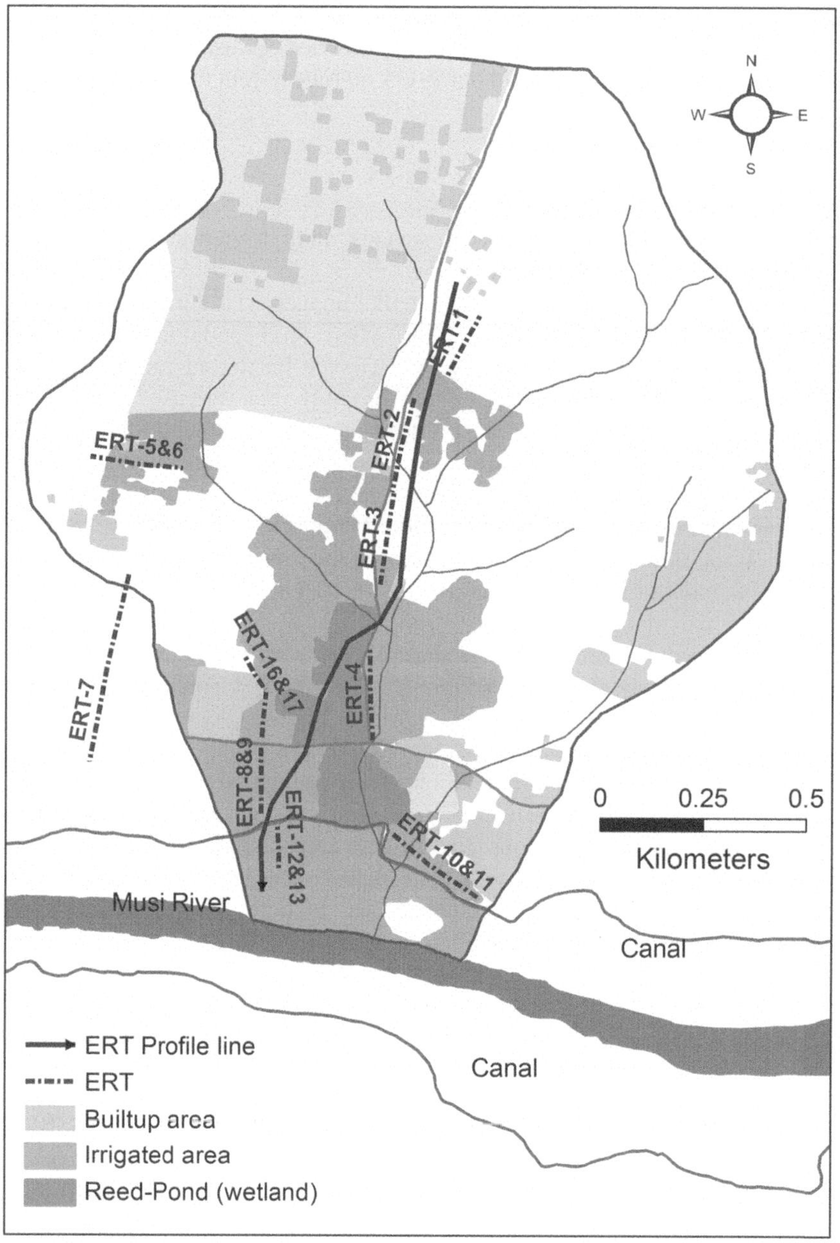

Figure 11.2 Geophysical measurement (ERT) locations in the KSMWS (N 17.390914. E 78.627036).

Two hydro-chemical water sampling campaigns were carried out in 2012 during the pre-monsoon (June) and post-monsoon seasons (November). Fourteen water samples were collected from canals and wells (Figure 11.3) and analysed for major anions and cations, heavy metals, selected microbiological parameters and pesticides. Eleven groundwater, 2 canal and a wetland water sample were analysed. The groundwater samples were from piezometers (4), bore wells (3 agriculture wells and 1 domestic well) and open dug wells (3). Groundwater and surface water samples were collected in 1 L polyethylene bottles for major cations and anions and before sampling the bottle was cleaned three times with sample water. Analysis was performed for major cations and anions as per the guidelines by APHA, 2005. The pH was measured using a pH meter; electrical conductivity using a conductivity meter; carbonates, bicarbonates, calcium and magnesium using a titration method; sulphate using a turbidity meter; fluoride using an ion meter with specific electrodes; sodium and potassium using a flame photometer and nitrate using an atomic absorption spectrophotometer. Pesticides samples were collected in 1 L amber coloured high density polyethylene bottles and kept below 4°C until analysis. Organochlorine, Organophosphorous and Carbamate pesticides were measured using a liquid-liquid extraction and method and analysed in a GC-MS (Gas chromatography – Mass

Spectroscopy) instrument. Results of previous sampling campaigns during the period 2006–2008 (Amerasinghe *et al.*, 2009; Perrin *et al.*, 2011a) were also used for comparison, and to understand the groundwater dynamics in the watershed along the Musi River as well as the impact of wastewater on groundwater quality.

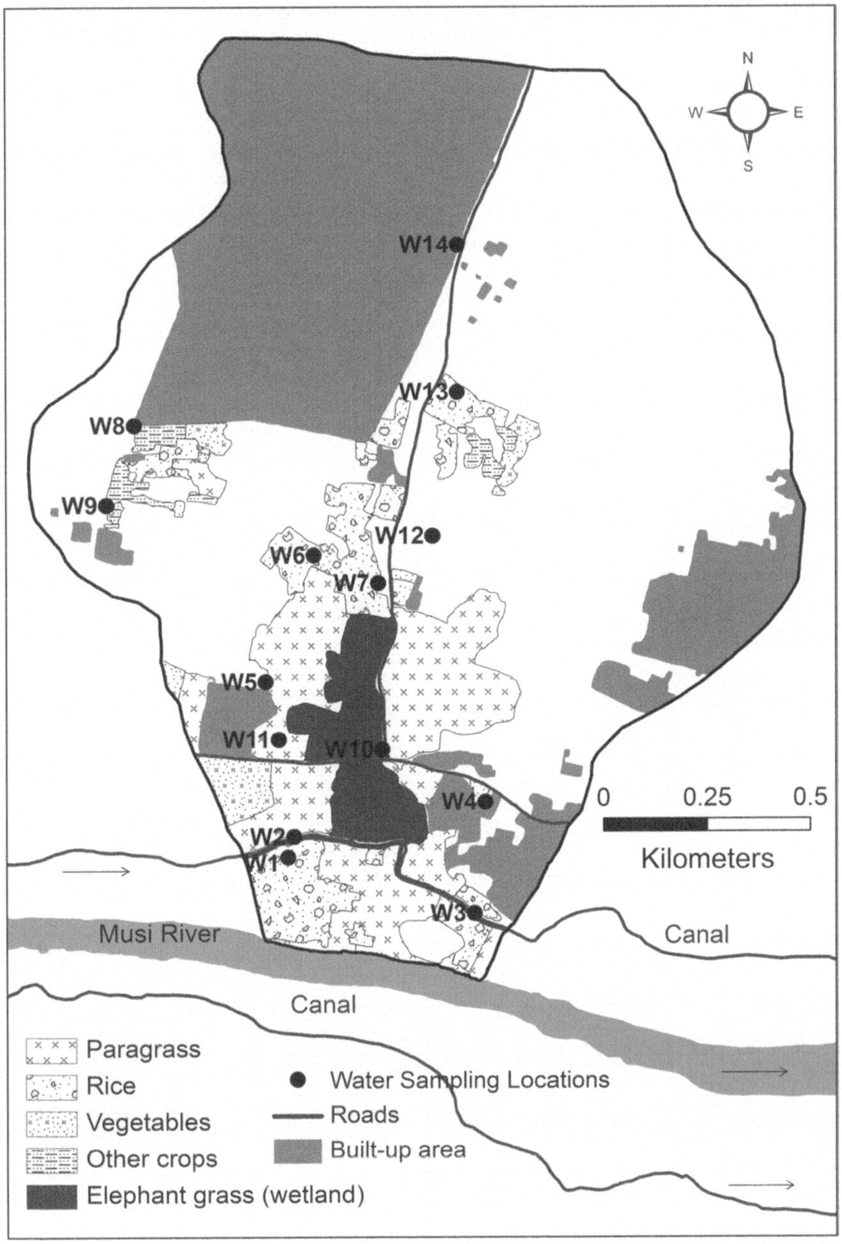

Figure 11.3 Land use classes and water sampling points in the KSMWS. W1-W14 are sampling wells.

11.5 RESULTS AND DISCUSSION

Characterization and performance assessment of the micro-watershed was based on a number of features, namely, land use, land cover, hydrogeology, geophysical attributes, water balance, aquifer characteristics and water quality. This agriculture-based micro-watershed depended on both wastewater and groundwater irrigation for its production systems. Closer to the canal the water levels in piezometers were high (up to 3 m below ground level), due to the continuous wastewater irrigation activities that occurred year round. In the north of the watershed, the water levels fluctuated over the seasons, due to high pumping rates of the irrigation and domestic wells during production times, and being the only source of water. Thus, across the micro-watershed, the performance appears to be strongly influenced by the wastewater irrigation practices, groundwater pumping and seasonal rainfall. Here, we discuss the SAT process and the performance of the natural wetland.

11.5.1 Land use, geomorphology, water balance and aquifer characteristics

The Musi River is a perennial river due to the urban wastewater discharges from the city of Hyderabad. Based on the current water supply, the wastewater generated from the city was estimated to be around 1.2 million m^3/d (Mahesh *et al.*, 2015), which was channelled into irrigation canals, for agriculture use within the micro-watershed.

Flow rate measurement campaigns (river and canal) showed an increase in irrigation rates during the wet season (Figures 11.4 & 11.5). However, there was no clear correlation between the rainfall and river discharge because the influence of urban wastewater was far greater (Figure 11.4). Thus, at a local level the performance of the aquifer appears to be strongly influenced by wastewater irrigation, groundwater pumping and seasonal rainfall.

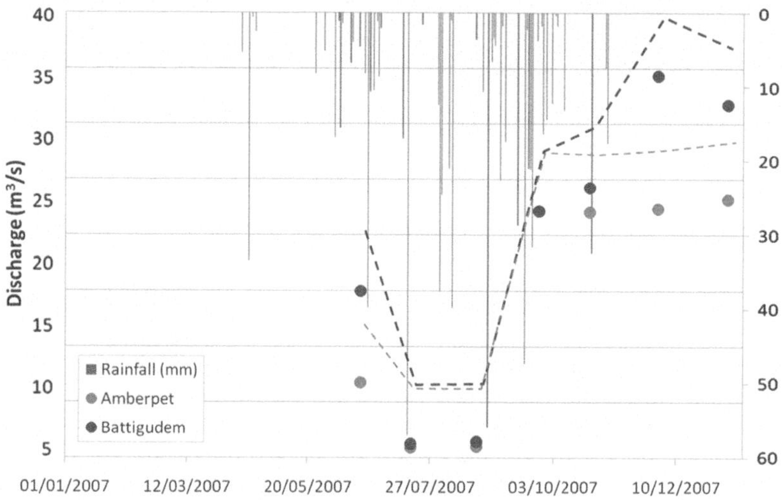

Figure 11.4 Discharge rate (m^3/s) measurements of the Musi River at Amberpet and Battigudem from May 2007 to January 2008 and daily rainfall (mm) in Hyderabad from January 2007 to January 2008.

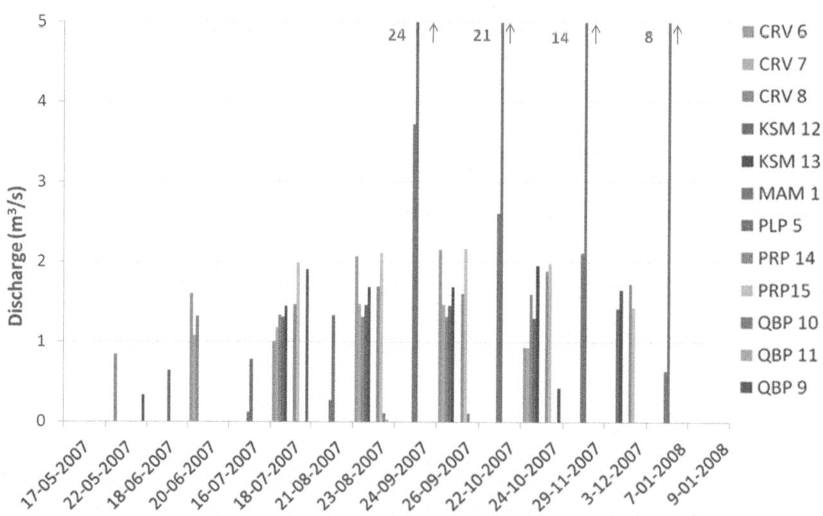

Figure 11.5 Discharge rate (m^3/s) measurements of the irrigation canals at 5 locations from May 2007 to January 2008. CRV = Chinnaravirala, KSM = Kachiwani Singaram, PLP = Pillaipalli, PRP = Parvathapuram, QBP = Quthbullapur are names of villages through which the canals pass.

A number of land use classes were identified and they were agriculture, areas under development, and barren lands. The irrigated area in the micro-watershed was approximately 48 ha and the major crops grown in the area were paragrass (56%), paddy rice (32%) and vegetables (8%). In terms of types of irrigation, 74% of the areas were under wastewater irrigation, and the remainder was groundwater. Paragrass was the dominant crop in the watershed, and wastewater was the main source of irrigation water. The micro-watershed showed visible signs of growth in infrastructure development and population, but the area under cultivation remained the same. The only difference was that as development processes engulfed the cultivated areas, new production sites were established (Mahesh *et al.*, 2015). We assumed that these changes would not affect the run-off or water balance and any increases in domestic use would be negligible compared to irrigation use.

As in many semi-arid environments, spatial and temporal interactions between surface water and groundwater were complex. The surface water percolation occurred mainly through preferential paths governed by topography and subsurface hydraulic properties. An analysis of the topography of the area using GIS tools indicated more than 10 small preferential flow areas within the micro-watershed. Localised recharge was also possible due to the uneven terrain. All these investigations revealed that the major surface run-off flow direction was towards the canal and then finally into the Musi River, which is in the north to south direction.

The hydro-geomorphology and the groundwater potential showed that the potential recharge conditions in the area varied from poor to good in the north to south direction of the micro-watershed. For example, the well yield in the northern part (pediment zone) was low, i.e. 30–40 m^3/d, and the recharge potential was also low, and restricted to the fissure zones. However, the recharge potential was high (up to 40%) in the shallow flood plain close to the Musi River (Figure 11.6).

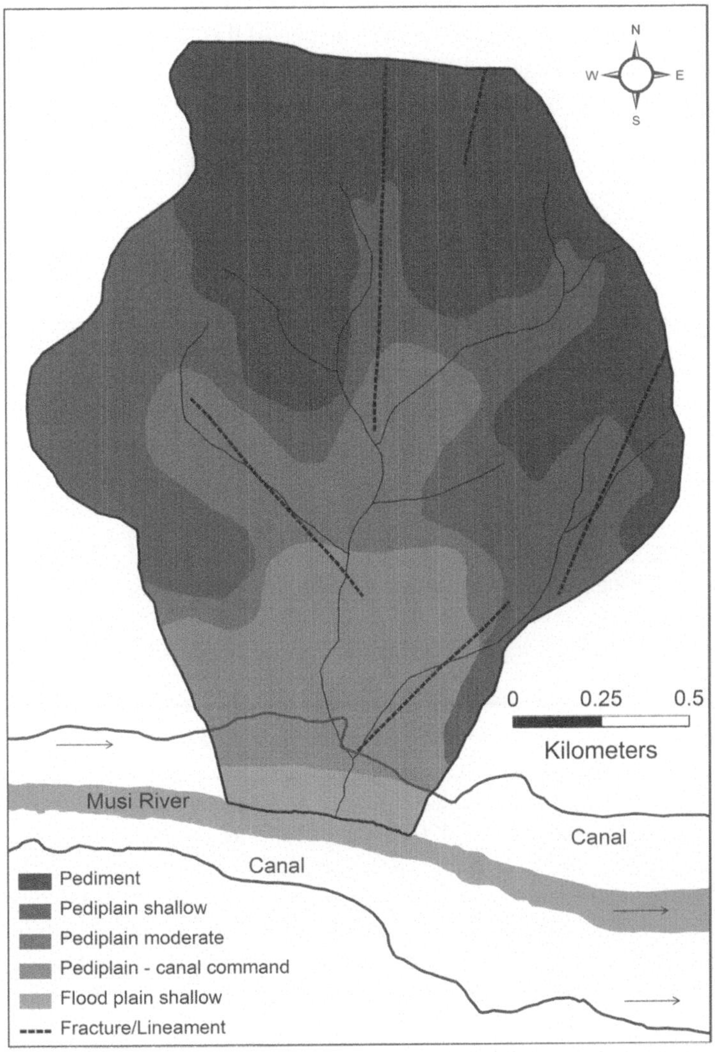

Figure 11.6 Geomorphology of the KSMWS.

The piezometric maps were significantly influenced by irrigation pumping during the pre-monsoon season (May and June). During the monsoon (October and November), the pumping influence was reduced, and consequently, a rise in the water table (5 meters) was observed in the central part of the study area. The high variability in the piezometric levels is understandable given the strong heterogeneities of the aquifer and the irrigation activities. Significant variations in the water level were captured daily, and even hourly, due to the periodic pumping action for irrigation. Piezometric levels (W1 and W5) in the southern part close to the canal water, did not vary significantly during the year and were within 2 m below ground surface. The water levels in this southern zone were controlled by the river which constitutes a boundary of the aquifer. The northern region piezometric levels (W6 and W8) were strongly influenced by the monsoon recharge that occurred from end of June as expected, in these unconfined crystalline aquifers with vertical recharge. The rapid rise of the water level after each rainfall event indicates the presence of rapid preferential flow paths in the saprolite zone due to the existence of preserved fractures (rise of groundwater levels less than one day after rainfall event). After the monsoon, the decrease of water level was fairly constant (dependant of natural flow and pumping) and between the years in those two bore wells.

When piezometric levels were compared between post-monsoon periods of 2010 (from a previous study) and 2013, two different zones of hydrologic activity could be delimited: A typical continuously decreasing water level due to over-exploitation of groundwater, especially during the dry season, and a gaining water level in the south influenced by abundant irrigation with wastewater.

The aquifer characteristics derived from the pumping tests of piezometers (W1, W5, W6 and W8) are given in Table 11.1. The pumping tests performed on four tube wells showed transmissivity values ranging from 9.9×10^{-4} to 3.4×10^{-3} m^2/s as expected in this geological context (Dewandel *et al.*, 2006). Flow meter tests in the fractured zone showed that transmissivity values were controlled by a few productive fractures zones in each tube well (1 to 3) mostly on the upper part of the weathering profile. These results are consistent with previous studies (Dewandel *et al.*, 2006; Maréchal *et al.*, 2004) in the region showing that transmissivity was mainly constrained by the fracture zone at the contact between the poorly transmissive saprolite which ensure the storage capacity of the aquifer and the deeper fresher granite where only a few fractures may provide water in limited amount. The decrease of fracture density with depth may induce compartmentalisation of the aquifer in low water level conditions due to a decrease of connectivity (Guihéneuf *et al.*, 2014).

Table 11.1 Aquifer characteristics of the KSMWS.

Piezometer	Discharge [L/s]	EC [mS/cm]	T [m^2/s]	Max Drawdown [m]	K [m/s]	S [–]
W1	0.63	1.4–1.55	9.9×10^{-4}	0.78	1.7×10^{-5}	3.02×10^{-4}
W5	0.61	1.55–1.63	1.1×10^{-3}	0.59	3.4×10^{-5}	3.03×10^{-4}
W8	0.55	0.95–1.15	3.4×10^{-3}	0.32	7.4×10^{-5}	8.90×10^{-8}
W6	0.18	1.65–1.7	–	2.35	–	

EC: Electrical Conductivity, T: Transmissivity, K: Hydraulic conductivity (K = T/e, with e, aquifer thickness), S: Storativity. Storativity values are indicative as tests were carried out in pumping wells (i.e., no observation wells) and during a short period.

11.5.2 Water quality

Surface water quality

The salinity and conductivity (EC) in the Musi River was high and showed an increasing trend downstream (McCartney *et al.*, 2008; Amerasinghe *et al.*, 2009; Biggs & Jiang, 2009). The EC in the canal water samples indicated an increase in the post-monsoon samples though not significant (1,217–1,490 µS/cm). This was contrary to the general expectation that monsoon would result in lowering the EC. In the wetland sample (W10) the EC was higher than in the canal reaching 1,750 µS/cm during post-monsoon period. The EC values in groundwater were higher than that for canal water. High temporal and vertical variability in EC logs were observed depending on the monsoon dilution or inversion of hydraulic gradient or placed close to the wastewater canal. Abrupt EC changes at around 30 m and 44 m below ground level in some wells suggested the presence of active flow fractures. Multiple factors may have contributed, amongst which weathering, silicate hydrolyses process and the hydraulic gradient may play an important role.

Major ions (sodium and chlorides), that contributed to salinity, did not vary significantly in the pre- and post-monsoon samples, however, values were indicative of anthropogenic influences. The wetland showed a greater variation in its constituents compared to the canal water. The canal water samples had variable nitrate concentrations ranging from 0 to greater than 150 mg/L. The high values could be due to the influence of agricultural activities as well as the source water, which carries urban run-off. Nitrate content in the Musi River and wetland (W10) were low (<20 mg/L) probably due to denitrification process in surface water linked to high organic matter loads (Reeds) (Lofton *et al.*, 2007).

The total dissolved solids varied with time and decreased during the monsoon period, probably due to a combined effect of dilution from rainfall and run-off water. High concentrations of hydrogen carbonate, chloride, sodium and sulphate ions were common and with significant enrichments of sodium in the Musi River and chloride and nitrate in groundwater and an excess of magnesium for some samples of groundwater. In the surface water samples from the canal and wetland, the pH ranged from 7 to 8.3 and did not show much variation between the pre and post-monsoon samples

In 2007, the biological oxygen demand (BOD) levels in the Musi River exceeded 200 mg/L in the samples close to urban areas, but declined further downstream (around 40 km), except when some local activities like livestock bathing led to elevated levels. In comparison, for the same year, the BOD in the irrigation canal at the KSMWS ranged from 105–150 mg/L, over a period of 4 months (August to December). Apart from the biological contaminants of source water, livestock wallowing, and open defecation may have also contributed to the elevated levels. In the present study (2012), the BOD values decreased considerably, and ranged from 15–65 mg/L, probably associated with the improvements in sanitation infrastructure (Wastewater treatment plant rehabilitation and construction of two new ones) at the city level, during the past few years. Faecal coliforms were detected in the post-monsoon samples (>1,600 MPN/100 mL) however, *E. coli* O157:H7 was not detected. At present, local contamination may not be contributing to these values significantly. However, the hamlet of Kachiwani Singaram is expanding and will have an impact on the irrigation channels and therefore, there is a possibility that these canals may become wastewater drains, unless the wastewater disposal systems are set in place.

Groundwater quality

Groundwater quality showed a strong spatial variability in groundwater chemistry (i.e. mineralisation, long term wastewater irrigation, agriculture practices etc.). Two main poles of EC were visible: one representative of fresh groundwater with EC < 1,000 μS/cm, and one pole with groundwater influenced by canal water return flows with EC > 1,000 μS/cm. It is also clear that additional sources of groundwater contamination (e.g. agriculture, sewerage) existed with localised points showing quite high EC (even higher than 2,000 μS/cm). In the sector where groundwater is impacted by canal water return flows, groundwater EC was higher than raw canal water, most likely as a result of re-concentration by evapotranspiration processes and dissolution of ions by water-rock interactions. EC in post-monsoon 2012 is in general higher than the pre-monsoon samples in the southern part of the watershed. This may be due to the increase in hydrogen carbonate and calcium content or the strong influence of the wastewater from the canal.

Rainfall appeared to impact on the chloride, nitrate and sulphate concentrations. This may be due to the complex interactions with the local hard rock aquifer system. Except a slight increase in W1 post-monsoon samples, there was not much variation in the fluoride concentrations in groundwater. The concentration of nitrate and fluoride were above the permissible limits for drinking water (50 mg/L and 1.5 mg/L respectively according to the WHO guidelines, WHO, 2011). Sulphate reached values up to 411 mg/L (W4). It is evident that the groundwater chemistry with respect to each of the elements was influenced by long term wastewater irrigation, application of synthetic fertilizers, soil salinity, and ionic contributions from rock-water interactions and rainfall.

Increases in hydrogen carbonate ions in the post-monsoon groundwater samples were attributed to enhanced organic matter mineralization as a result of carbon dioxide, associated with the run-off processes. The high chloride content does not have a lithological origin in this hard rock terrain. It could be of anthropogenic and/or meteoric origin enhanced by evaporation processes in soils. The groundwater samples in the southern part of the catchment (W1 and W5) showed a sodium excess, indicating a strong cationic exchange process within the clay minerals in the soil. High contents of fluoride were due to rock-water interactions enhanced by irrigation return flows (Pettenati *et al.*, 2013). Levels of nitrate and sulphate were attributed to agricultural practices and wastewater irrigation.

Hydro-chemical water facies showed that the temporal variations of major ions due to rainfall were not significant, however, anthropogenic activities like wastewater irrigation, application of fertiliser, soil salinity could contribute to high levels of nitrates and sulphates based on the level of activity. Excessive irrigation return flows, through solute recycling may have led to significant aquifer salinization further exacerbated by the presence of thick clay soils, which favour strong cationic exchanges in this area (Perrin *et al.*, 2011a).

Of the 14 samples tested, only 7 samples were positive for pesticides. The pesticides fell into three families, namely, organochlorine, organophophorus and carbamate. Organochlorine pesticides like butachlor, organophosphorous pesticides like malathion and carbamate pesticides like carbofloro nuclon granules were used by farmers. While it is expected that the monsoon rains would dilute pesticide concentrations, in some samples there were increased levels after the monsoon, due to leaching and possibly agriculture run-off. Some pesticides like propoxure was 'not in use', but tested positive and cannot be fully explained. The permissible limit for drinking water as per BIS 10500 standards (Bureau of Indian Standards, 2012) for each pesticide is 0.01 μg/L, and those samples that were positive showed values higher than 0.01 μg/L. Even though people are not drinking the water either from the canal or local groundwater wells that were tested, the elevated concentrations of pesticide elements raise a question of ecological impacts and food production in the area. Faecal coliform range was higher in the post-monsoon samples (2–170 MPN/100 mL) than in the pre-monsoon samples (0.3–21 MPN/100 mL). A degree of soil remediation may have contributed to the reduction in levels.

Water budget

The water budget was expressed as follows (Perrin *et al.*, 2011a)

$$R + R_f + Q_{\text{in}} \pm L_c - P - E = Q_{\text{out}} \tag{11.1}$$

where R is the natural recharge during the monsoon, R_f is the return flows (mainly from wastewater/groundwater irrigation but also from domestic water uses), L_c is canal losses or gains, Q_{in} is groundwater inflow across groundwater reservoir limits, P is groundwater pumping, E is the evaporative discharge from the water table, Q_{out} is the groundwater contribution to Musi river base flow.

Then, the groundwater budget of the watershed can be expressed as follows: The final estimated groundwater outflow from the watershed or groundwater contribution towards Musi river base flow was 136 mm/yr. This means that the groundwater contribution towards the Musi River was 0.4% of the river flow in the study area. Thus, the wastewater irrigation contributed to the base flows of the Musi River significantly 70% of the irrigation return flow came from wastewater irrigation (113 mm/yr) and 30% was from groundwater extraction (29 mm/yr) and a small part of the return flow at the watershed scale (0.9 mm/yr) came from domestic water recharges (Perrin *et al.*, 2011a).

Natural wetland

Within the study site a natural wetland, comprising marsh grass, was also studied for its treatment potential. The wetland received return flows from the east and west sides where paragrass cultivation was carried out using canal water. A comparison of water quality between canal water (W2 = inlet) and the outlet point of the wetland (W10), showed that nitrate-N (W2 = 10 mg/L and W10 = 1 mg/L), sulphates (W2 = 305 mg/L and W10 = 68 mg/L), chemical oxygen demand (W2 = 32–248 mg/L and W10 = 16–144 mg/L) and BOD (W2 = 10–65 mg/L and W10 = 5–32 mg/L) were reduced during both seasons, indicating the natural potential for contaminant attenuation. EC was however high compared to all other samples and was attributed to the weathering within the wetland. The study indicated that the natural wetlands could have great potential for contaminant attenuation, and more detailed studies are required to understand hydro-geochemistry of this natural wetland.

11.6 CONCLUSION

Land use data, hydrodynamic monitoring, hydraulic tests, hydro-geophysical surveys, water chemistry data were useful to develop a conceptual model for flows and transport for the KSMWS, which consists of a hard rock aquifer (Figure 11.7). The study showed that there is a north-south gradient in the micro-watershed, which together with the different hydrological flows impacts the groundwater. However, when excessive pumping occurred, the water levels in the northern part decreased during certain periods, reversing the water flow direction. This added to the complexity of the hydrodynamics of the site. The water budget indicated that canal irrigation was the main recharge flux at the basin scale and had a strong impact on groundwater quality. In such hard rock aquifers the saprolite plays a storage role due to its porosity, and the fissured zone provides the transmissive functions. When the water levels are shallow, the saprolite layer allows a regional groundwater flow.

Of all the wells monitored, W1 and W5 wells share similar hydrogeological conditions with the Musi River, where minimal variations in relation to water levels were observed. Thus, the canal plays only a passive role on the hydrogeology in the southern part of the watershed. The return flows induced mixing of the groundwater and canal water that was pumped for irrigation, especially in the middle potion of the watershed, which also contributed to a high groundwater recharge. Long term wastewater irrigation in the area has resulted in high salinity, compounded by agriculture run-off (e.g. nitrates, pesticides).

A number of pollution sources linked to agricultural activities (nitrates and pesticides in the northern part of the micro-watershed), water rock interactions (fluoride), and intensive agricultural practices with city wastewater were evident.

Given the number of contaminants found in groundwater, SAT has not contributed significantly to improving the groundwater quality, although the removal of some elements may have taken place. Further, while the geophysical studies performed showed zones with relatively deep weathering (potentially inducing high permeability), the limited size of the area, and continuous flooding due to agricultural activities may not allow sufficient retention time for effective treatment. Return flows may further enhance mineralization and facilitate fluoride release from the rocks.

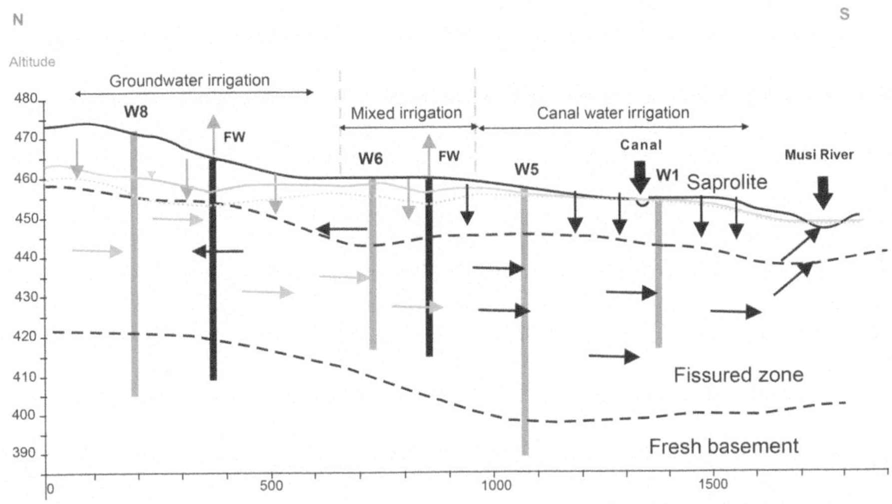

Figure 11.7 Conceptual model of groundwater flow and transport in the KSMWS. W8, W6, W5 and W1 are piezometric wells. The irrigation wells were used for agriculture. Dotted light grey line – pre-monsoon (June) water level; Light grey line – post-monsoon water level (November).

Overall, it is clear that in this natural setting, SAT may not be effective due to the flow patterns and short retention periods. Some preliminary tests for nitrates have shown that biogeochemical reactivity is dependent on hydrogeological and hydrogeochemical heterogeneity (Boisson *et al.*, 2013; McGuire *et al.*, 2002). Therefore, more studies under controlled conditions are required to understand the soil behaviour in remediation, and design small-scale engineered treatment systems that can be effective locally. Finally, although the potential for SAT may exist in the Musi River watershed, the complex hydrogeological conditions in this site were not conducive for the elimination of pollutants in a sustainable manner.

ACKNOWLEDGEMENTS

The authors would like to thank Aruna Jyothi for the GIS products and Judith Christiana for proof reading the chapters at various stages.

11.7 REFERENCES

Amerasinghe P., Weckenbrock P., Simmons R., Acharya S., Drescher A. and Blummel M. (2009). An Atlas of Water Quality, Health and Agronomic Risks and Benefits associated with "Wastewater" Irrigated Agriculture: A Study From the Banks of the Musi River, India. Albert-Ludwigs-Univ., Inst. für Physische Geographie. http://wwiap.iwmi.org/Data/Sites/9/DOCUMENTS/PDF/bmz_india_finalatlas_27oct09.pdf. (Accessed 25 October 2014)

Amerasinghe P., Jampani M. and Drechsel P. (2012). Cities as Sources of Irrigation Water: An Indian Scenario. IWMI-Tata Water Policy Research Highlight 53, International Water Management Institute, Colombo, Sri Lanka.

APHA (2005). American Water Works Association and Water Pollution Control Federation. Standard Method for the Examination of Water and Wastewater, 21st edn, American Public Health Association, New York, pp. 5–55.

Arceivala S. J. and Asolekar S. R. (2006). Wastewater Treatment for Pollution Control and Reuse, 3rd edn, Tata McGraw Hill, New Delhi, pp. 55–60.

Biggs T. W. and Jiang B. (2009). Soil salinity and exchangeable cations in a wastewater irrigated area, India. *J. Environ. Qual.*, **38**(3), 887–896.

Bureau of Indian Standards (BIS) 10500 (2009). Drinking Water – Specification. Indian Standard, 2nd Revision, Bureau of Indian Standards, New Delhi, India, pp. 10–11.

Boisson A., de Anna P., Bour O., Le Borgne T., Labasque T. and Aquilina L. (2013). Reaction chain modeling of denitrification reactions during a push–pull test. *Journal of Contaminant Hydrology*, **148**, 1–11. doi: 10.1016/j.jconhyd.2013.02.006

Buechler S., Devi G. and Raschid L. (2002). Livelihoods and Wastewater Irrigated Agriculture: Musi River in Hyderabad City, Andhra Pradesh, India. UA Magazine, **8**(h032354), 14–17.

Census of India (2011). Provisional Population Totals, Urban Agglomerations/Cities.censusindia.gov.in/2011-prov-results/paper2/data_files/India2/Table_3_PR_UA_Citiees_1Lakh_and_Above.pdf

Crites R. W., Middlebrooks E. J. and Bastian R. K. (2014). Natural Wastewater Treatment Systems. CRC Press, Taylor & Francis Group, Florida, pp. 1–549.

Dewandel B., Lachassagne P., Wyns R., Maréchal J. C. and Krishnamurthy N. S. (2006). A generalized 3-D geological and hydrogeological conceptual model of granite aquifers controlled by single or multiphase weathering. *Journal of Hydrology*, **330**, 260–284. doi: 10.1016/j.jhydrol.2006.03.026

Guihéneuf N., Boisson A., Bour O., Dewandel B., Perrin J., Dausse A., Viossanges M., Chandra S., Ahmed S. and Maréchal J. C. (2014). Groundwater flows in weathered crystalline rocks: Impact of piezometric variations and depth-dependent fracture connectivity. *Journal of Hydrology*, **511**, 320–334. doi: 10.1016/j.jhydrol.2014.01.061.

Jacobi J. (2009). Agricultural Biodiversity as a Livelihood Strategy? The Case of Wastewater-Irrigated Vegetable Cultivation along the Musi River in Peri-urban Hyderabad, India. MSc Thesis, Freiberg University, Freiberg, Germany, pp. 65–66.

Lofton D. D., Hershey A. E. and Whalen S. C. (2007). Evaluation of denitrification in an urban stream receiving wastewater effluent. *Biogeochemistry*, **86**, 77–90.

McCartney M., Scott C., Ensink J., Jiang B. and Biggs T. W. (2008). Salinity implications of wastewater irrigation in the Musi River catchment in India. *Cey. J. Sci. (Bio. Sci.),* **37**(1), 49–59.

Mahesh J., Amerasinghe P. and Pavelic P. (2015). Elucidating changes in land use patterns, crop choices and farmer perceptions in a wastewater-dominated peri-urban agricultural system in India. In: *Development Trends in Urban and Peri-Urban Agriculture*, R. S. Deshpande, P. Nath, P. G. Chengappa, E. Kannan, K. R. M Swamy and C. P. A. Iyer (eds), Westville Publishing House, New Delhi. ISBN 978–93–83491–22-3. *In press.*

Maréchal J. C., Dewandel B. and Subrahmanyam K. (2004). Use of hydraulic tests at different scales to characterize fracture network properties in the weathered-fractured layer of a hard rock aquifer: Fracture network properties in hard rock. *Water Resources Research*, **40**(11), W11508. doi: 10.1029/2004WR003137.

Massuel S., George B., Gaur A. and Nune R. 2007. Groundwater modelling for sustainable resource management in the Musi Catchment, India. In: *Proceedings, International Congress on Modelling and Simulation*, 10–13 December 2007, Christchurch, New Zealand, p. 1425–1439.

McGuire J. T., Long D. T., Klug M. J., Haack S. K. and Hyndman D. W. (2002). Evaluating behavior of oxygen, nitrate, and sulfate during recharge and quantifying reduction rates in a contaminated aquifer. *Environmental Science & Technology*, **36**(12), 2693–2700.

Mekala G. D., Samad M., Davidson, B. and Boland A. M. (2009). Valuing a clean river: A case study of Musi River, Hyderabad, India (No. 48164). Australian Agricultural and Resource Economics Society. http://econpapers.repec.org/paper/agsaare09/48164.htm (accessed 10 November 2014)

National River Conservation Directorate (NRCD) (2001). Musi river conservation project, holistic detailed project report. Project Report 1, HMWSSB & NRCD, Ministry of Environment, Forests and Climate Change, Government of India, New Delhi, India.

Perrin J., Ahmed S., Dinis L., Aellen V., Amerasinghe P., Pavelic P. and Schmitt R. (2011a). Groundwater Processes in a Micro-Watershed Influenced by Wastewater Irrigation, Peri-Urban Hyderabad. Scientific report National Geophysical Research Institute (NGRI), Bureau de recherches géologiques et minières (BRGM), International Water Management Institute (IWMI), Hyderabad, India.

Perrin J., Mascré C., Pauwels H. and Ahmed S. (2011b). Solute recycling: An emerging threat to groundwater quality in southern India? *Journal of Hydrology*, **398**, 144–154. doi: 10.1016/j.jhydrol.2010.12.024.

Pescod M. D. (1992). Wastewater Treatment and Use in Agriculture. Food and Agricultural Organization (FAO), Irrigation and Drainage Paper 47, FAO, Rome, Italy.

Pettenati M., Perrin J., Pauwels H., Ahmed S. (2013). Simulating fluoride evolution in groundwater using a reactive multicomponent transient transport model: Application to a crystalline aquifer of Southern India. *Applied Geochemistry*, **29**, 102–116.

Qadir M., Wichelns D., Raschid-Sally L., McCornick P. G., Drechsel P., Bahri A. and Minhas P. S. (2010). The challenges of wastewater irrigation in developing countries. *Agricultural Water Management*, **97**(4), 561–568.

Reed S. C., Crites R. W. and Middlebrooks E. J. (1995). Natural Systems for Waste Management and Treatment, 2nd edn, McGraw-Hill Inc., New York, pp. 1–433.

Scott C. A., Faruqui N. I. and Raschid-Sally L. (2004). Wastewater Use in Irrigated Agriculture: Management Challenges in Developing Countries. CAB International. http://www.crdi.ca/un_focus_villes/ev-68323-201-1-DO_TOPIC.html (Accessed 10 November 2014)

Shipin O., Koottatep T., Khanh N. T. T. and Polprasert C. (2005). Integrated natural treatment systems for developing communities: Low-tech N-removal through the fluctuating microbial pathways. *Water Science & Technology*, **51**(12), 299–306.

Schmitt R. (2010). Wastewater Reuse in Indian Peri-Urban Agriculture: Assessment of Irrigation Practices in a Small, Peri-Urban Catchment in Hyderabad, Andhra Pradesh, India. MSc report ETHZ, Zurich, Switzerland.

Sonkamble S., Chandra S., Nagaiah E., Dar F. A., Somvanshi V. K. and Ahmed S. (2013). Geophysical signatures resolving hydrogeological complexities over hard rock terrain—a study from Southern India. *Arabian J Geosci.*, **7**(6), 2249–2256. doi: 10.1007/s12517-013-0931-4

Srinivasan J. T. and Reddy V. R. (2009). Impact of irrigation water quality on human health: A case study in India. *Ecological Economics*, **68**(11), 2800–2807.

Starkl M., Amerasinghe P., Essl L., Jampani M., Kumar D. and Asolekar S. R. (2013). Potential of natural treatment technologies for wastewater management in India. *Journal of Water, Sanitation and Hygiene for Development*, **3**(4), 500–511.

Starkl M., Brunner N., Amerasinghe P., Mahesh J., Kumar D., Asolekar S.R., Sonkamble S., Ahmed S., Wajihuddin Md., Pratyusha A. and Sarah S. (2015) Stakeholder views, financing and policy implications for reuse of wastewater for irrigation: A case from Hyderabad, India. *Water*, 7, 300–328; doi: 10.3390/w7010300

Theis C. V. (1935). The relation between the lowering of the Piezometric surface and the rate and duration of discharge of a well using ground-water storage. Eos, *Transactions American Geophysical Union*, **16**(2), 519–524.

Van Rooijen D. J., Turral H. and Biggs T. W. (2005). Sponge city: Water balance of mega-city water use and wastewater use in Hyderabad, India. *Irrigation and Drainage*, **54**(S1), S81–S91.

Wakode H. B., Baire K., Jha R., Ahmed S. and Azzam R. (2014). Assessment of impact of urbanization on groundwater resources using GIS techniques – case study of Hyderabad, India. *Int. J. Environ. Res.*, **4**, 1145–1158.

World Health Organization (WHO) (2006). Guidelines for the Safe Use of Wastewater, Excreta and Greywater. Volume 2: Wastewater Use in Agriculture. WHO, Geneva, Switzerland, pp. 1–196.

World Health Organization (WHO) (2011). Guidelines for Drinking-Water Quality. 4th edn, WHO, Geneva, Switzerland, pp. 1–564.

Chapter 12

Pre- and post-treatment of bank filtration and managed aquifer recharge in India: Present and future

Saroj Kumar Sharma, Cornelius Sandhu, Thomas Grischek, Ankush Gupta, Pradeep Kumar, Indu Mehrotra, Gesche Grützmacher, P. J. Sajil Kumar, Lakshmanan Elango and Narayan C. Ghosh

12.1 INTRODUCTION

Soil aquifer-based natural treatment systems namely bank filtration (BF), artificial recharge and recovery (ARR) and soil aquifer treatment (SAT) are managed aquifer recharge (MAR) systems that are robust, reliable, capable of removing multiple contaminants and sustainable (Dillon, 2005; Amy & Drewes, 2007; Ray, 2008). In addition to replenishing groundwater aquifers and depending on the quality of the water source used for recharge (river or lake water, storm water, wastewater treatment plant effluents) and local hydrogeological conditions, these MAR systems can serve at least as a pre-treatment or sometimes even as a total treatment system (Sharma & Amy, 2010; Sharma *et al.*, 2012).

Very often the treated water from these natural treatment systems (NTSs) may not meet the required local water quality guidelines or standards for intended use and thus require additional post-treatment. Furthermore, some contaminants present in the source water may pollute the aquifer or influence the performance of NTSs and therefore pre-treatment of source water is often carried out before the application of natural systems. Pre-treatment and post-treatment thus form an integral part of the NTSs. Depending upon the raw water quality, local hydrogeological conditions, process conditions applied and intended use of the treated water a NTS can have pre-treatment or post-treatment or both.

Pre-treatment refers to removing or reducing the concentrations of some of the critical contaminants in the source water to enhance the performance of subsequent treatment systems. Pre-treatment may be required in NTSs to avoid clogging and contamination of the aquifers, to increase the run time, and to enhance the removal efficiencies of different contaminants. Post-treatment refers to further upgrading the quality of the "treated water" from different NTSs so that it meets the water quality requirements for different applications. Requirements for post-treatment of "product water" from natural systems vary significantly depending on the quality of the source water used, type, design, and operation of NTSs employed, process conditions applied and applicable water quality guidelines or standards for intended use. The type of pre-treatment and post-treatment that should be applied, however, depends mainly on the source water quality, the type of NTSs being used, the process conditions applied as well as the intended use of the treated water from the NTSs.

Table 12.1 provides an overview of the main water quality concerns for different natural systems used in India and pre- and post-treatment applied.

Table 12.1 Pre-treatment and post-treatment applied to different NTSs in India.

NTS	Pre-Treatment Applied	Main Water Quality Concerns in Abstracted Water	Post-Treatment Methods Applied
BF	Not applicable	Pathogens, hardness, ammonium, nitrate (organic micropollutants)	Disinfection, lime softening, aeration, coagulation, sedimentation, Rapid sand filtration
Artificial recharge	Sedimentation, sand filtration	Iron, manganese, fluoride and arsenic in local groundwater (of geogenic origin)	Disinfection, aeration + sand filtration, several adsorption and coagulation-based systems for treatment of specific contaminants like arsenic and fluoride

12.2 PRE- AND POST-TREATMENT OF BF AND MAR IN INDIA: PRESENT STATUS

12.2.1 Present status of post-treatment of BF in India

BF (river or lake) has been utilized as a technology for water abstraction and treatment in some water supply systems in India. Very often wells are constructed on the riverbank as an intake to facilitate water collection from the rivers with varying water depth and quality. Table 12.2 summarizes the main water quality concerns and post-treatment applied at some selected sites in India (based on literature review and field data collection).

At some other BF sites investigated in 2013 and 2014 in the states of Andhra Pradesh and Jharkhand as part of the Saph Pani project, the BF wells are of a radial collector design and are located within the riverbed (Chapter 2, Table 2.1 and Saph Pani D1.4, 2014). Due to the relatively shallow depth (3–6 m) of the radial collector pipes and consequently short travel time of the filtrate especially during monsoon, a breakthrough of pathogens and turbidity can occur. Thus, at all these sites the abstracted filtrate is post-treated by aeration, flocculation, rapid sand filtration and finally disinfection. Although an elaborate post-treatment is applied at these sites, the existing BF schemes are in some areas the only viable means of obtaining water compared to direct surface water or even groundwater. Thus, BF buffers the quantity of water required through bank-/bed-storage and can thus be considered as an element of MAR and integrated water resources management (Sandhu *et al.*, 2015).

Post-treatment at BF case study sites

Detailed descriptions of the Saph Pani Project BF case study sites as well as results of the water quality analysis from different sampling campaigns during the project are included in the previous chapters. This section summarizes the post-treatment aspects of these case study sites.

Haridwar: The BF system in Haridwar (Uttarakhand) consists of 22 bottom-entry caisson wells located at varying distances between 4 and >490 m, along the Ganga River and the Upper Ganga Canal, which accounts for at least two-thirds of the total drinking water production for the city of Haridwar. Additionally, groundwater is abstracted using 56 vertical wells from the deeper confined aquifer to meet the water demand of the city. Post-treatment at the case study site Haridwar is limited to disinfection of the water abstracted from the production wells by using sodium hypochlorite. Each production well (caisson well or tube well) has its own sodium hypochlorite dosing system/pump. Adequate stock of 12.5% sodium hypochlorite solution (for 2–3 weeks) is maintained at each production well or dosing point. The operators are provided with a dosing chart/table for estimating the pumping rate of the hypochlorite dosing pump depending on the capacity of the well or size of the reservoir. Sodium hypochlorite is injected (using a dosing pump) directly into the distribution pipeline immediately after the abstraction pump (see Figure 12.1). However, when the disinfectant dosing pumps are defunct or non-existent, sodium hypochlorite is poured manually directly into the caisson wells or into water storage tanks/reservoirs.

Srinagar: Srinagar (in Uttarakhand) is located on the south bank of the meandering Alaknanda River. The combined drinking water production for Srinagar and the town of Pauri (the water for which is abstracted and treated in Srinagar before being pumped 29 km to Pauri) was around 3,750 m^3/d in 2010 while the demand was estimated as 4,880 m^3/d (Kimothi *et al.*, 2012). More than 80% of the total raw water for the drinking water supply of Srinagar and Pauri is abstracted upstream of the town directly from the Alaknanda River. The abstracted surface water is first coagulated with alum and then flows to sedimentation tanks followed by rapid sand filters and is finally chlorinated before being supplied to the distribution network. In May 2010, one river bank filtrate well was constructed in the South-West part of the town, which abstracts 852 – 937 m^3/d of water, depending upon the operating hours of the well, to supplement the existing surface water supply (Kimothi *et al.*, 2012). The well is located 170 m from the riverbank and was drilled down to a depth of 18 m. After abstraction of water from the production well and on-site disinfection using sodium hypochlorite, the water is pumped into a storage reservoir where it is mixed with the conventionally treated surface water and then supplied into the distribution network by gravity. Water quality investigations of the production well up to September 2012 have shown that the nitrate concentration was in the range of 53–123 mg/L, with a mean concentration of 86 mg/L, in the abstracted water (see Chapter 2). Although the mean hardness concentration monitored was 439 mg/L as calcium carbonate ($CaCO_3$), in the absence of an alternative drinking water source it is still within the permissible limit of 600 mg/L (Bureau of Indian Standards (BIS) 10500, 2012). Thus the main parameters of concern for post-treatment are the occasional presence of coliforms in very low numbers (prior to disinfection), and high nitrate concentrations (>45 mg/L; BIS 10500, 2012) in the abstracted water.

Table 12.2 Summary of the post-treatment applied in selected BF systems in India.

BF Site	Source of Water for BF	Main Water Quality Concern after BF	Post-Treatment Applied	References
Haridwar (Uttarakhand)	Ganga River and Upper Ganga Canal	Occasional presence of pathogen indicators in some wells in very low concentrations	Chlorination only	Sandhu *et al.* (2011a), Dash *et al.* (2010), Saph Pani D4.3 (2014)
Nainital (Uttarakhand)	Nainital Lake	As above and occasional hardness	Water softening and chlorination	Saph Pani D4.3 (2014), Dash *et al.* (2008)
Srinagar (Uttarakhand)	Alaknanda River	Nitrate >45 mg/L in abstracted water due to geogenic origin	Chlorination of river bank filtrate water and subsequent mixing with conventionally treated surface water (coagulation-sedimentation-filtration-chlorination)	Saph Pani D4.3 (2014)
Satpuli (Uttarakhand)	East Nayar	Occasional presence of pathogen indicators in very low concentrations	Chlorination only	Ronghang *et al.* (2012), Chapter 2
Mathura (Uttar Pradesh)	Yamuna River	Organic matter (Dissolved organic carbon) Hardness; Pathogens; Arsenic; Organic micropollutants (OMPs)	Aeration- filtration-chlorination	Singh *et al.* (2010), Kumar *et al.* (2012), Saph Pani D1.4 (2014)
Patna (Bihar)	Ganga River	Occasional presence of pathogen indicators in some wells in very low concentrations	Chlorination only	Sandhu *et al.* (2011a & b)
Ahmedabad (Gujarat)	Sabarmati River	Pathogens; Organic matter	(i-a) SW abstraction in monsoon: Chlorination (2 times)-filtration-chlorination (i-b) Abstraction from river bank filtrate wells is discontinued when breakthrough of turbidity is high (ii) Non-monsoon: Chlorination (of bank filtrate) only	Sandhu *et al.* (2011a), Saph Pani D1.4 (2014)
Medinipur (West Bengal)	Kangsabati River		Chlorination only	Sandhu *et al.* (2011a)
Muzaffar Nagar (Uttar Pradesh)	Kali River	Pathogens	Chlorination only	Thakur *et al.* (2009)
Delhi – Palla (National Capital Territory)	Yamuna River	Iron, manganese, fluoride (present in deeper aquifer) Pathogens during monsoon	Chlorination only	Sprenger *et al.* (2008), Lorenzen *et al.* (2010)

Figure 12.1 Typical sodium hypochlorite dosing system for RBF wells in Haridwar.

Nainital: Uttarakhand Jal Sansthan (Government of Uttarakhand) utilizes water from the Naintal Lake employing BF technology to supply water to the city of Nainital. Disinfection using bleaching powder is the main post-treatment step in Nainital. The operators are provided with a chart/table for estimating the amount of bleaching powder to be added per day. There are standard size tanks for feeding bleaching powder and the prepared solution which is dosed by a hydraulic method with an overflow weir (without a pump). 2 kg of bleaching powder is added every hour to a 2,000 L tank to obtain a bleaching powder concentration of 2 mg/L. With 25% chlorine content in the bleaching powder, the estimated chlorine at the initial stage is 0.5 mg/L. Bleaching powder is dosed at 5 locations in the water supply system, namely in the (i) Main pump house, (ii) Children park pump house, (iii) Phasi gadera, (iv) Sukhatal tube well and (v) Sukhatal old water works. After bleaching powder solution dosing, the residual chlorine content in the reservoirs (before supplying water to the distribution system) is not measured. However, samples are taken from different points in the distribution system to monitor residual chlorine concentrations. There is an ion-exchange system for water softening at Mallital pumping station (near the Children Park) where about half of the total production at this site is treated and then the two streams are mixed again before disinfection and supply (see Figure 12.2). It was reported by the operators that the ion exchange system is not in operation regularly because of the high costs. A new ion-exchange treatment system for hardness removal is under construction (near the main pump house) under an Asian Development Bank project.

Delhi: Previous studies at the BF case study site located on the East bank of the Yamuna River in East Delhi opposite the Nizamuddin area on the West bank of the river (Chapter 1, Figure 1.4) as well as sampling campaigns during Saph Pani have shown that presence of elevated levels of ammonium, nitrate, arsenic, fluoride, iron and manganese are the main water quality concerns for wells in the area. Furthermore, source water samples (Yamuna river) also showed the presence of some organic micropollutants (OMPs) in significantly high concentrations compared to other BF sites, which are, however, substantially removed during the soil passage. Traces of some OMPs were also found in samples from observation wells and hand pumps. Very low concentrations of some OMPs were found in the production wells compared to significantly higher concentrations in the river water (Saph Pani D1.4, 2014).

The water abstracted from the production wells (radial collector wells) around this site is mixed with the surface water from other sources and treated extensively at full-scale water treatment plants (e.g. Okhla and Common Wealth Games water treatment plants). No major water quality problems are expected. People using hand pumps or tube wells in this area are, however, advised to use household level ammonium, nitrate, arsenic and fluoride removal systems to ensure that water from these local wells are meeting the drinking water quality requirements. Furthermore, water authorities using production wells around the case study sites for municipal water supply are advised to monitor the presence of OMPs in the well waters and in the treated water, to ensure that these pollutants are sufficiently removed during the treatment process.

Table 12.2 clearly shows that removal of pathogens, hardness, and organic matter are the key elements of post-treatment systems for bank filtrates in India. Furthermore, as the rivers/lakes in India are often polluted with untreated or poorly treated sewage and industrial wastes, the presence of bulk organic matter and micropollutants could be one of the main requirements of post treatment systems. Limited information is available on the concentrations of OMPs in the raw water and filtrates at the BF sites in India. Nevertheless, a snap-shot screening of dissolved organic carbon and 54 OMP compounds (of environmental relevance in Europe) in surface water and BF well water was conducted for the Saph Pani project BF case study sites and some other existing and potential BF sites in the states of Bihar, Jharkhand, Andhra Pradesh, Madhya Pradesh, Gujarat and the city of Jammu in the state of Jammu and Kashmir during the dry pre-monsoon season in May-June 2014 and during the monsoon in June-July 2013 (Saph Pani D1.4, 2014). In this context, of all the investigated sites, it was found that the stretch of the Yamuna River starting in Central Delhi (ITO Bridge) up to ~200 km downstream of Agra had the highest occurrence and also the near highest concentrations of OMPs comprising pharmaceutical, medical contrast media, personal care products, corrosion inhibitors, insecticide and herbicide compounds. The removal efficiency of OMPs by BF can be demonstrated by taking the Mathura BF site as an example where the Yamuna River has comparably higher concentrations of OMPs than Delhi and Agra. Consequently, the concentrations of some OMPs in the BF well's (Radial collector, fast travel time) water were 13–99% lower than in river water, whereas others were not present in well water (Saph Pani D1.4, 2014). On the other hand, in the surface water in Haridwar, Srinagar and Nainital in Uttarakhand and in Gumla, Ray Bazaar and Daltonganj in Jharkhand most of the 54 OMPs were either not present at all or only detectable in very low concentrations (for a few types OMPs) (Chapter 2, Table 2.1; Sandhu *et al.*, 2015). None were detectable in the BF wells at these sites. Another potential source of contamination is the dissolution of iron, manganese, arsenic or fluoride from the aquifer due to low, dissolved oxygen in river water and/or anoxic conditions created by the infiltrating river or lake water.

Figure 12.2 Ion-exchange water softening system treating bank filtrate at Nainital.

At a majority of the BF sites at perennial surface water bodies (snow-melt and spring-fed) in India's hilly or foothill regions chlorination is the only treatment applied (Chapter 2, Table 2.1; Sandhu *et al.*, 2015). In BF wells of radial collector design constructed within the riverbeds of very polluted rivers with fast travel times (e.g. Yamuna River in Mathura), relatively high concentrations of natural organic matter in (river) bank filtrates is usually found. Because of this formation of trihalogenmethanes or other disinfection by-products is another major water quality concern (Kumar *et al.*, 2012). In summary, depending on the raw water quality (of rivers or lakes), local hydrogeological conditions and the contaminant removal efficiency at a particular site, the post-treatment of filtrates of the BF sites in India would require improvement of one or more of the following group of parameters:

- Pathogens
- Hardness
- Iron, manganese, ammonium, nitrate, arsenic and fluoride
- Bulk organics and OMPs.

12.2.2 Present status of pre- and post-treatment of MAR systems in India

India has a long tradition in water harvesting and artificial recharge. A variety of ARR structures are used either in connection with rooftop rainwater harvesting (RWH) or with normal surface run-off. The structures used for MAR are recharge pits, open wells, Aquifer storage and recovery wells, injection wells, gravity wells, recharge shaft and tube wells (Shivkumar, 2006; Holländer *et al.*, 2009).

MAR systems in India are mainly designed for and constructed toward using rainwater or flood water for groundwater augmentation and therefore pre-treatment is mostly limited to sedimentation basins and sand filters to avoid clogging and to maintain the infiltration rate. Pre-treatment, if any, often forms a part of the recharge structure. Consequently, clogging is widely encountered in a majority of the MAR systems. Identified pre-treatment methods for surface run-off are sedimentation, sand filters, wrapped polyvinyl chloride (PVC) pipes and metallic filters:

- *Sedimentation*: A trapezoidal shaped grassed water way of 40 m length and 0.63 m depth was used to store excess water and to increase the sedimentation by reducing the flow velocity in the Balasore district (Holländer *et al.*, 2009).
- PVC pipes of 20 cm diameter with slots (3 mm × 75 mm) used in recharging water at Sirsa branch canal. These slots were wrapped with the coconut coir to prevent the entry of suspended solids. An annular space between the bore hole and the pipe was filled with gravel of 9 mm to 12 mm in diameter (Kaledhonkar *et al.*, 2003).
- Sand filtration is a common pre-treatment method in many of the rooftop RWH schemes. In RV College of Engineering Bangalore, at the bottom of a gravity recharge settling tank a sand bed for a depth of 150 mm and stone aggregate for a depth of 200 mm were filled as filter media (Shivakumar, 2006).
- Kanhe and Bhole (2006) used metallic filters made up of copper were used as filter media.

In cases where rainwater is used for recharge, differences in the rainwater chemistry can be observed between rural and urban areas. Sulphate, sea salt (i.e. sodium, chloride) and ammonium are present in higher concentrations in urban regions than in rural areas. Microbial concentration was also found to be much higher in water from roof top for urban areas. In RWH and recharging systems, elimination of the first flush from the roof tops is a common practice to minimize negative impacts on source water quality (Vasudevan & Tandon, 2006; Shivkumar, 2006). The aim of this method is to remove the first rain with lots of impurities from the interaction between atmosphere and also from the dirty roof tops. There is an inbuilt filter system in most of the rooftop harvesting systems practiced in India. This is fixed immediately after the first flush separator and acts as a primary treatment method.

Reuse of treated wastewater to recharge groundwater is not a common practice in India. Considering the high potential of this method as a future option, several pilot studies on this topic are in progress. In the case of wastewater effluent recharge systems, pre-treatment may be preliminary with settling and aeration or secondary treatment including physical, chemical and biological process. Jamwal and Mittal (2010) reported a primary and secondary treatment for the sewages from Delhi city. They used the activated sludge process as the major primary treatment step together with an oxidation pond for a selected sewage treatment plants. As secondary treatment a fluidized bed, BIOFORE (i.e. aerobic or anaerobic biological reactors) was used for bacteriological removal.

The results of the pre-treatment structures used in the 8 investigated case studies (literature-based) are presented in Table 12.3. Analysis of MAR systems studied revealed that a high percentage (38%) of the recovered water was used for drinking purposes. Among the other uses, 37% were used for irrigation and 25% were used for other domestic purposes. It should be noted that all the harvested rooftop rainwater is used for the either drinking or domestic purpose after MAR. Still, no post-treatment was mentioned.

Table 12.3 Pre-treatment in selected MAR systems in India (based on 8 case studies).

Location and Reference	Structures	Pre-Treatment	Use of Effluent
Ranga Reddy district, Hyderabad (Dwarakanath, 2006)	Rooftop rainwater harvesting (RWH), Recharge pit with boulders and sand	Boulder and sandy filter in recharge pit	Drinking water
Osmania University Campus, Hyderabad (Dwarakanath, 2006)	Rooftop RWH, Infiltration with 5 pits and 5 recharge wells	Sand and metal filters	Domestic
Padmavathi Nagar, Chennai (Jebamalar & Ravikumar, 2006)	Rooftop RWH connected to open wells	Filter, possibly sand filter	Drinking water
Balasore district (Holländer *et al.*, 2009)	Aquifer storage and recovery-wells, channels for catchment	Sedimentation and desilting Filter	Irrigation
RV College of Engineering Campus, Bangalore (Shivakumar, 2006)	Rooftop RWH to gravity injection well	Two settling tanks with sand bed and stone aggregate as a filter	Drinking water
Dhuri Drain, Punjab (Chadha, 2003)	Vertical Shafts and Injection Wells	Sand and gravel pack as filter	Not specified, presumably used for irrigation
Sirsa branch canal (Kaledhonkar *et al.*, 2003)	Two recharge tube wells	Filter pit and coconut coir wrapped on the slotted PVC pipe	Presumably for irrigation
CSV, Wardha, Maharashtra (Kanhe & Bole, 2006)	RWH tank bore well, dug well, soak pit	Ground filters and metallic filters	Mainly for domestic purposes

Groundwater is considered as safe drinking water source in many parts of India. Often people use untreated or inappropriately treated groundwater for drinking. The major water quality issues identified in groundwater in India are elevated concentrations of salinity, iron, fluoride, arsenic and nitrate. The treatment of fluoride deserves special attention due to its hazardous effects on human health. In general, rainwater recharge (normally acidic to neutral pH) will dilute the fluoride-rich water and reduce the concentration. However, this depends entirely on the chemical composition of the source water. In Maheswaram, both rain water and surface run-off contribute to the total recharge. In case the acidic pH of rainwater changes to alkaline during the interaction with the aquifer, the fluoride from the source rock will be mobilized. Moreover, the sodium:calcium ratio in the study area is higher than 1, which also an important factor in elevating the fluoride concentration. A possible treatment method would be to elevate the calcium concentration to a higher level than sodium in the source water, so that the fluoride will precipitate as calcium fluoride (CaF_2) and reduce the fluoride in groundwater.

Pre- and post-treatment needs at MAR case study sites

Details of the Saph Pani MAR case study sites, groundwater quality data and their suitability for drinking water supply and irrigation as well as treatment requirements are included in previous chapters and presented in detail in Saph Pani D4.1 (2013). Summaries of the pre- and post-treatment needs at selected sites are presented in the following paragraphs.

Chennai: Salinity and high magnesium concentrations are the major quality problems when using Chennai groundwater for drinking purposes. Concentrations of sulphate and nitrate exceeded the guideline value in 20% of the wells, raising problems for use as drinking water. This suggests that the groundwater needs pre-or post-treatment in terms of these parameters prior to distribution, although widespread implementation of MAR structures might improve this problem. As in Raipur, the microbial quality is not monitored. If needed, chlorination is a cheap and effective post-treatment option to eliminate microbial contamination. In the case of irrigation suitability, salinity and total dissolved solids are relatively high in a few samples. High magnesium levels/concentrations are observed in the majority of the samples (80%) in terms of MAR. Although data on turbidity in the source water is lacking, experience from Anna University shows that reduction of suspended solids improves infiltration rates. A pilot study conducted at Anna University, Chennai using SAT showed high removal efficiency for nitrate (up to 98%) after 15 cycles (Deepa & Krishnaveni, 2012) under reducing conditions. Under these conditions SAT can be a cost-efficient treatment method for the removal of nitrate, which is encountered in few locations in the study area.

Groundwater samples were taken from two dug-cum bore wells and four deep bore wells around the Periapalayam Check Dam area (Chennai). A relatively high number of pathogen indicators are the main water quality concern for water from wells. Surface water stored by the check dam had very high turbidity and a high a number of pathogens due to the collection of run-off from various land uses. It is likely that these contaminants were not sufficiently removed during soil passage. It is recommended that people using water from these wells should employ household-level, low-cost disinfection methods like boiling, chlorine tablets or chlorine solutions, solar disinfection, ceramic filters or bio-sand filters to ensure that the water is microbiologically safe to drink and meets the drinking water quality requirements.

Maheswaram: The available data shows that, in general, most of the water quality parameters of the groundwater are suitable for drinking purposes. However, a few parameters, like calcium and magnesium, were high at certain times. At the watershed scale the major health hazard is due to the year-round high fluoride concentration. This situation is not an isolated one as high levels of fluoride are reported from 16 of 28 states (Mariappan *et al.*, 2000), and are encountered particularly in granitic terrains in southern India. So these three contaminants (calcium, magnesium and fluoride) need to be treated after recovery. Among the numerous methods developed, lime softening is the most common and cheapest treatment method for fluoride-enriched groundwater (Crittenden *et al.*, 2005). Coagulation and precipitation and activated alumina are also used in certain regions depending on the cost and geological conditions (Meenakshi & Maheshwari, 2006). With respect to irrigation, the low salinity values show that the high KI and Na% can be discarded and no post-treatment of any kind is needed. Furthermore, as clay deposits in the tanks used for irrigation are reported, it is recommended to remove suspended solids present in the source water. There is no information about the microbial quality of the well water used for small-scale (household level) applications and irrigation. Microbial quality of the well water must be monitored and sufficient disinfection needs to be provided if necessary.

Raipur: In general and also for the parameters investigated, groundwater quality is suitable for drinking purposes except for a few samples which showed concentrations of sodium, chloride, nitrate, iron and manganese exceeding the guideline values. The study of Groeschke (2012) shows that nitrate, iron and manganese present in water probably need post-treatment before distributing the water through the public water supply system. Based on the experience at other sites in India, pre-treatment is required to remove suspended solids so as to improve the efficiency of infiltration. Also, disinfection of the recovered water will most probably be necessary before it can be distributed as drinking water.

12.3 PRE- AND POST-TREATMENT OF BF AND MAR IN INDIA IN THE FUTURE

Some of the future scenarios or global change pressures like (i) population increase and urbanization, (ii) climate change, (iii) emerging contaminants, (iv) energy crisis and (v) increasing environmental awareness and stricter regulations are likely to affect the performance of natural systems for water and water treatment as these factors have direct and indirect impact on the quantity and quality of different types of water (TECHNEAU, 2007; Lozàn *et al.*, 2007; Cooley *et al.*, 2013). Effects of these influencing factors on the performance of NTSs in India and requirements for pre- and post-treatment systems were studied in detail under the Saph Pani project (Saph Pani D4.3, 2014) which are summarized in the following sub-sections.

12.3.1 Post-treatment requirements for BF sites in India in the future

Inferences about the potential impact of future scenarios on pre- and post-treatment systems for BF systems in India can be drawn from various plausible sources/scenarios as well as from published literature on BF and climate change (Schoenheinz, 2004; Schoenheinz & Grischek, 2011; Sprenger *et al.*, 2011). The impact of future scenarios or influencing factors on the quantity and quality of water, the consequent impact on BF systems and the coping strategies as well as potential post-treatment requirements for BF sites in India are described in Table 12.4. It is clear from Table 12.4 that proper care should be taken in the planning and design of BF system (protection zones, well placement, construction of wells, flood proofing) and higher levels of post-treatment (including robust disinfection and specific treatment for emerging contaminants) will be required in the future, which is very much site specific.

12.3.2 Pre- and post-treatment requirements for MAR sites in India in the future

Environmental and socioeconomic developments such as increasing population, rapid urbanization and industrialization, along with climate change are important issues in the world's developing and newly industrialized countries. It is expected that these factors affect quality and quantity of water resources, thus directly or indirectly affecting MAR systems. Table 12.5 summarizes the possible impacts of population growth and urbanization on MAR systems and coping strategies including pre- and post-treatment requirements.

Table 12.4 Future scenarios and post-treatment requirements for BF in India.

Influencing Factors (Future Scenarios)	Effect on Water Quality	Effect on Water Quantity	Impact on BF	Coping Strategies
Population increase:				
Increased water demand for different uses Urbanization	– May increase pollution in river and wells – Clogging of riverbeds by organic matter and gas bubble formation	– May cause lower groundwater levels and surface recharge	– Limit on water abstraction from a single well	– Constructing more wells along the river Ensuring construction of well head & source protection zones
Climate change factors:				
Increase in average temperature	– Increased mineralization and dissolution; lower oxygen concentration may cause anoxic conditions	– Little higher portion of bank filtrate due to lower water viscosity may be compensated by higher clogging of riverbeds	– Floods may increase the portion of bank filtrate in the well; droughts may decrease the portion of bank filtrate and thus may cause water quality changes	– Construct wells closer to river – Adapt well operation
Increased flooding	– Pathogen breakthrough; Increasing dilution but higher input of pollutants from surface run-off	– Higher abstraction rates possible		– Flood-proofing of wells – Ensuring sufficient disinfection
Increased drought	– High mineralization and increase in contaminant concentrations due to lesser dilution	– Lower abstraction rates		– Construct wells closer to river – Intensify monitoring of well water quality
Variable precipitation patterns	– Fluctuating water quality	– May be variable		– Real-time water quality monitoring – Installing multiple wells at varying distances and switching operations of the wells as per the conditions
Emerging contaminants:				
Deteriorating source water quality (increasing pollution)	– Long-term deterioration will cause delayed deterioration of well water quality	– May not change	– Increasing removal rates but higher absolute concentrations, additional post-treatment measures may be required	– Increase travel time/distance of well to river – Additional post-treatment units according to the type of pollutant

The predicted population increase from 1.21 billion (2011) to 1.6 billion (2050) in India is likely to have severe effects on the already stressed water resources. The impacts of population growth on MAR could be increased sewage production, combined with probable groundwater over-exploitation, increased use of fertilizers etc. A total of 32% increase in water use is estimated for 2050 (Amarasinghe *et al.*, 2007). Higher total suspended solids (TSS) in source water from sewages may result in clogging of MAR systems. This may be addressed by installing sedimentation and sand filtration as additional pre-treatment. Higher salinity due to over-exploitation may be treated with reverse osmosis (RO). Increased use of fertilizers and improper sanitation facilities may result in elevated nitrogen in surface water. This can be treated by RO, ion exchange, and biological denitrification.

Table 12.5 Possible impacts of population growth and industries/urbanization on MAR.

Influencing Factors	Effect on Water Quality	Effect on Water Quantity	Impacts on MAR	Coping Strategies (Additional Pre- and Post-Treatment Required)
Population growth – Projected population in the year 2050:1.3 to 1.6 billion (PRB, 2007) – Increasing water demand (680 to 900[a] billion m^3) (Amarasinghe *et al.*, 2007; KPMG, 2010)	– Increasing sewage volume in Class I cities and Class II towns from 2003 to 2051: 26,254 to 83,300 MLD (Bhardwaj, 2005) – At the coastline groundwater over-exploitation and subsequent seawater intrusion – Inadequate sanitation systems cause increase in nutrients and pathogens in water – Algal growth due to nutrients	– Increasing pressure on water resources – Increasing water demand	– Higher sewage load makes surface water unfit for MAR – Higher sewage load may cause clogging and deterioration of recovered water's quality in MAR systems. Therefore pre-treatment becomes necessary – Anoxic condition through high organic carbon	– Reducing sewage load on water bodies by providing proper wastewater treatment systems – Sedimentation and sand filtration for total suspended solids (TSS) removal (pre-treatment) – Reverse Osmosis (RO) for salinity removal – Nitrate: RO, Ion exchange, biological denitrification etc. – Iron and manganese: Aeration and sand filtration
– Increasing industrialization,[b] urbanization[c] and intensification of agriculture: – Increasing pollution; deteriorating source water quality (emerging pollutants)	– Increased industrial effluents – Pollution from fertilizers and pesticides: (Pesticides levels of 13 µg/L against permissible limit of 1 µg/L were reported from the Ganga basin. Major threats were DDT and HCH (GRBMP, 2011), organochlorine pesticides higher than 1 µg/L observed in Delhi (Mutiyar *et al.*, 2011) – Higher CO_2 emission and acidic rainfall – Personal care products, pharmaceuticals, steroids and hormones and plasticizes levels will also increase in the future	– Stressed water resources	– Persistent pollutants (pesticides, PhACs, PCPs) may break through – Increased sulphur and nitrogen dioxides in the atmosphere may result in acidic rainfall. This lead to various hydrogeochemical reactions and subsequent source water contamination in the aquifer – Higher ammonium levels (>0.2 mg/L) in rain were reported from Ahmedabad (Rastogi and Sarin, 2005) – Source water rich in sodium (from industries) may trigger the release of geogenic contaminants such as fluoride and increase their concentration.	– Strict regulation on effluent disposal to maintain water quality of river and lakes – Along with the conventional treatment methods like activated carbon, oxidation-filtration or membrane filtration, process such as nanofiltration and reverse osmosis as post-treatment may be needed to remove the persistent compounds – Apart from first flush and metallic filters, activated carbon filters and biofilms may be implemented to remove ammonium. – pH regulation using Na_2CO_3, NaOH, $CaCO_3$ and $MgCO_3$ – Fluoride: electrocoagulation, ion exchange or membrane filtration as post-treatment

[a]forecasted to 2050; [b]Average industrial growth is reported as 8% per annum during 2007–2012 (Kaushal, 2012; PCI, 2012); [c]An average of 2% growth during past 3 decades (IUSSP, 2009).

The major impacts of urbanization and industrial growth were identified as increasing industrial effluents, higher sulphur dioxide and nitrogen dioxide emissions that may result in acid rain, increased use of pesticides and personal care products as well as an increase of other organic pollutants. For the emerging pollutants additional pre/post-treatment such as activated carbon, oxidation-filtration or membrane process such as nanofiltration (NF) and RO were suggested. For higher ammonium concentrations in rainwater, activated carbon filters may be used. In case of higher fluoride concentrations, electro-coagulation, ion exchange or membrane filtration may be adopted.

Different climate change scenarios were investigated based on scenarios of increasing average temperatures, floods and droughts. Table 12.6 summarizes the possible impacts of climate change scenarios on MAR systems.

Table 12.6 Possible impacts of climate change scenarios on MAR systems.

Influencing Factors	**Possible Impacts on Source Water**		**Impacts on MAR (Relies Mostly on SW)**	**Coping Strategies (Additional Pre- and Post-Treatment Required)**
Climate Change	**Surface Water (SW)**	**Groundwater**		
Temperature increase IPCC (2007) predicted 6 scenarios to 2090–2099 A1FI: 4°C A1T: 2.4°C A1B: 2.8°C A2: 3.4°C B1: 1.8°C B2: 2.4°C	– Increased evaporation – Decreased O_2 solubility in water – Increased algal and phytoplankton growth – Increased dissolved organic carbon (DOC) – Accelerated microbial regrowth in distribution systems (Zwolsman, 2008)	– Impact of temperature may not be so immediate in groundwater and surface water – Increased vegetation growth → CO_2 in the soil → groundwater may dissolve more calcium carbonate ($CaCO_3$) – Decreased viscosity and increased flow velocity	– Evaporation loss and concentration of salts – Higher temperature may result in anoxic conditions during underground passage which may result in the mobilization of redox sensitive species including Iron, Manganese and Arsenic	– Fluoride: Electro-coagulation, ion exchange or membrane filtration – Algal toxins: Activated carbon – Taste and odour: Oxidation, adsorption – Iron, manganese, arsenic: Aeration, Slow sand filtration – Salinity: Reverse Osmosis (RO) – Hardness: Ion Exchange
Flooding	– Surface run-off and inundation may mobilize point and non-point contaminants – Increase of total suspended solids (TSS) in flood water (Kale and Hire, 2004) – Increased river discharge and inundation	– Increased groundwater recharge – Possible introduction of contaminants by increased groundwater recharge	– Physical damage of the MAR structures – Frequent clogging of MAR structures by TSS and subsequent maintenance of CD, and CW	– TSS: Sedimentation, sand filtration – Heavy metals: Chemical precipitation, ion exchange, phytoremediation, biological treatment etc. – Pathogens: Chlorination, UV treatment or ozonation
Droughts	– Reduced/no discharge and reduced dilution potential – Algal blooms (TECHNEAU, 2009) – Increase in phosphate, pathogens and organic micropollutants (TECHNEAU, 2009; Hrdinka *et al.*, 2012) – Increasing concentration of salts due to reduced stream flow	– Reduced/no groundwater recharge Symbol Std declining water table – Lowering of freshwater table may cause intrusion of saltwater from the sea or inland salt lakes	– Complete/partial failure of the MAR due to lack of surface water – Increasing salinity	– Sulphate: Nanofiltration, ion exchange – Salinity: RO

Temperature has a gradual but continuous effect on water quality and quantity in terms of increased rates of algal growth, evaporation and concentration of salts, a change in the hydrological cycle and triggering of biogeochemical processes. For MAR, an increase in temperature may result in anoxic conditions during subsurface passage which may result in the mobilization of redox sensitive species like nitrate, manganese, iron oxide, hydroxide and sulphate. Loss of water by evaporation is an issue for MAR structures with large open surfaces (e.g. check dams and infiltration ponds). The pre-/post-treatment needs of water under increasing temperature are: (i) fluoride: with electro-coagulation, ion exchange or membrane filtration; (ii) algal toxins: with activated carbon, (iii) taste and odour: with oxidation and adsorption, (iv) iron, manganese and arsenic: with aeration + filtration, (v) salinity: with RO, and (vi) hardness: with lime softening or ion exchange.

If conditions are suitable flooding may increase groundwater recharge. However, uncontrolled flooding causes groundwater contamination through interaction with contaminant sources like agricultural fields, industrial and urban wastes. Increase of TSS loads in the groundwater is the most common problem anticipated. Higher TSS levels may cause clogging in NTS. In groundwater the flow length and/or travel times may be reduced and may affect the treatment efficiency. Physical damage of the MAR structures is also possible. TSS may also act as carriers for the heavy metals that potentially need treatment. Sedimentation tanks and Slow sand filters can be adopted to avoid higher TSS level.

Droughts reduce the discharge and dilution potential of the surface water which may cause more frequent algal blooms. Point source pollution and irrigation return flows are expected. Groundwater recharge will be reduced, decreasing freshwater flow which may result in saline intrusions. Lowering of water tables and subsequent atmospheric exposure may favour oxidation of sulphate minerals. Droughts may result in the failure of MAR structures. The potential coping strategies may be the treatment of sulphate with NF and ion exchange and salinity with RO.

MAR systems are generally robust, and flexible in regards to changing conditions. Failure to achieve required water quality objectives can be met by additional pre-or post-treatment.

12.4 CONCLUSIONS AND RECOMMENDATIONS

BF is being used in India for abstraction of water from rivers or lakes. However, traditionally these systems were designed mainly to improve the production of water from the quantitative perspective and did not aim to improve the water quality. In other words BF systems were not fully considered as a part of "water treatment systems". In India, chlorination is practiced at all the BF sites, whereas hardness, iron and manganese removal is also required at some sites. Nainital is the only place where hardness removal is practiced as a post-treatment of bank filtrate. At Mathura and some sites in Andhra Pradesh and Jharkhand, bank filtrate is aerated and filtered before chlorination. Limited data is available on the quality of bank filtrates and post-treatment being carried out at different BF sites in India.

In general, pathogens, hardness, iron and manganese, ammonium, bulk organic matter and OMPs (specifically in case of rivers with direct impact of wastewater and for future water quality considerations) are some of the critical water quality parameters of concern in India. There is a relatively high concentration of organic matter in some of the bank filtrates from wells with short travel times located at very polluted surface water bodies in India and chlorination is the only treatment applied. Because of this, formation of disinfection by-products is the main quality concern. Limited information is available on the concentrations of OMPs in the raw water and filtrates at the BF sites in India.

To date the majority of the MAR systems in India are designed and constructed with the aim of using rainwater or flood water for groundwater augmentation. Data available in the literature on pre- or post-treatment of MAR systems in India is scarce and covers systems that are mainly linked to rooftop RWH and few surface run-off recharge systems only. Pre-treatment is mostly limited to sedimentation basins and sand filters to avoid clogging and to maintain the infiltration rate. Pre-treatment, if any, often forms a part of the recharge structure.

The parameters that need treatment are high amounts of suspended materials and turbidity along with elevated fluoride and nitrate concentrations in certain locations. The water quality problems found in the literature on groundwater treatment in India (salinity, fluoride, nitrate, iron, manganese and pathogens) – which may indicate necessary post-treatment – are similar to those that have been encountered in the Saph Pani case study sites.

It was observed that in many of the RWH systems elimination of the first flush is used as pre-treatment. Pre-treatment is necessary during the monsoon season to remove TSS and turbidity in water collected as surface run-off. Sedimentation and sand filters are found to be the most suitable low cost treatment options for TSS and turbidity removal in MAR systems. Descriptions were found in the literature on using filters made up of metals and coir (coconut fibre) as pre-treatment for small scale systems. An alternative would be to remove the clogging mechanically at regular intervals. It is recommended to perform a cost analysis to select the viable one among the two options.

Analysis of the data obtained from the Saph Pani field sites showed that post-treatment is necessary to meet the guidelines for drinking water. Low salinity in all three MAR locations (except few wells in Chennai) suggests that groundwater is suitable for irrigation. The salinity in that region is mainly due to the seawater intrusion and it is expected to be pushed back after the implementation of MAR structures. Microbial quality of the groundwater has not been assessed in these cases. However, the most common disinfection method, chlorination, will probably be applicable here as well. The parameters that need treatment are salinity, magnesium hardness, iron, manganese, nitrate and fluoride.

It was found that BF systems in the northern part of India, specifically at Saph Pani project case study sites in Uttarakhand, where there are relatively clean rivers and lakes, treat water so effectively it nearly meets the quality requirements for drinking water supply. Chlorination is the only treatment necessary and applied to maintain residual chlorine in the distribution system. Some additional chlorination may be required for control of pathogenic organisms during flooding periods. However, BF systems along relatively polluted stretches of river like Yamuna (in Delhi and Mathura) will require extensive post-treatment to remove bulk organic matter as well as OMPs and some inorganic parameters. It is expected that post-treatment of BF with advanced treatment processes like ozonation, activated carbon filtration and membrane filtration will be required for such BF systems to meet the water quality guidelines. The post-treatment requirements for BF systems are likely to increase in the future unless comprehensive programmes are implemented to control of indiscriminate wastewater disposal to water bodies. Construction of wells at optimum distance from the river, increasing the number of wells as well as flood proofing the wells and event-adapted operation of well schemes are some of the coping strategies for the design of BF systems to meet future challenges with respect to quantity and quality of water.

Pre-treatment of MAR systems requirements will be more critical in the future if the quality of the rain or storm water changes significantly and if wastewater treatment plant effluents are used for MAR. Secondly, the direct and planned abstraction of recharged water after MAR for municipal and industrial uses is not practiced. Therefore, post-treatment requirements for MAR systems are not well known or documented. Post-treatment of "groundwater" in India is mainly limited to either disinfection to remove pathogens (in case of shallow and contaminated aquifers) or to removal of geogenic contaminants like iron, manganese, arsenic and fluoride.

The impacts of population growth on MAR could be increased sewage production, combined with probable groundwater over-exploitation, increased use of fertilizers and pesticides etc. Post-treatment methods like aeration followed by rapid sand filtration, coagulation or adsorption-based processes would be relevant for the removal of geogenic contaminants (iron, manganese, arsenic and fluoride) now and in the future due to several potential water quality impacts of global change pressures. Additional advanced post-treatment options would be required in the future to treat water contaminated with organics, nutrients and micropollutants from sewage pollution, fertilizers and pesticides. A potential increase in the incidences of floods and droughts in parts of India would further demand the increased use of MAR as an integrated water resources management option which will require suitable pre- and post-treatment systems.

12.5 REFERENCES

Amarasinghe U. A., Shah T., Turral H. and Anand B. K. (2007). India's Water Future to 2025–2050: Business-as-Usual Scenario and Deviations. IWMI Research Report 123, International Water Management Institute, Colombo, Sri Lanka.

Amy G. and Drewes J. (2007). Soil aquifer treatment (SAT) as a natural and sustainable wastewater reclamation/reuse technology: Fate of wastewater effluent organic matter (EfOM) and trace organic compounds. *Environmental Monitoring and Assessment*, **129**, 19–26.

Bhardwaj R. M. (2005). Status of Wastewater Generation and Treatment in India. IWG-Env, International Work Session on Water Statistics, Vienna, Austria, pp. 1–9.

Bureau of Indian Standards (BIS) 10500 (2012). Drinking Water – Specification. Indian Standard, 2nd Revision. Bureau of Indian Standards, New Delhi, India, pp. 1–5.

Chadha, D. K. (2003) State of the art of artificial recharge applied in village level schemes in India. In: Management of Aquifer Recharge and Subsurface Storage: Making Better Use of Our Largest Reservoir. A. Tuinhof and J. P. Heederik (eds). Netherlands National Committee for the IAH, The Netherlands, pp. 19–24.

Cooley H., Ajami N., Ha, M. L., Srinivasan V., Morrison J., Donnelly K. and Christian-Smith J. (2013). Global Water Governance in the 21st Century, Pacific Institute, USA, pp. 3–9.

Crittenden J. C., Trussell R. R., Hand D. W., Howe K. J. and Tchobanoglous G. (2005). Water Treatment: Principles and Design. MWH Publications, New Jersey, United States, pp. 165–224.

Dash R. R., Mehrotra I., Kumar I. and Grischek T. (2008). Lake bank filtration at Nainital, India: Water-quality evaluation. *Hydrogeology Journal*, **16**(6), 1089–1099.

Dash R. R., Bhanu Prakash E. V. P, Kumar P., Mehrotra I., Sandhu C. and Grischek T. (2010). River bank filtration in Haridwar, India: Removal of turbidity, organics and bacteria. *Hydrogeology Journal*, **18**(4), 973–983.

Deepa K. and Krishnaveni M. (2012). Water quality performance of Soil Aquifer Treatment (SAT) using municipal treated wastewater of Chennai city, India. *Journal of Environmental Hydrology*, **20**, 1–9.

Dillon P. (2005). Future management of aquifer recharge. *Hydrogeology Journal*, **13**(1), 313–316.

Dwarakanath K. (2006). Rain Water Harvesting in Urban Areas and its Impact. National Seminar on Rainwater Harvesting and Water Management Nagpur UNESCO 2006, Nagpur, India, pp. 231–235.

GRBMP (2011). Emerging Contaminants in Ganga River Basin with Special Emphasis on Pesticides. Ganga River Basin Management Plan. GBP_IIT_EQP_SOA_02_report _ver 01. p. 33.

Groeschke M. (2012). Hydrogeological Characterization of Raipur Urban Area, Chhattisgarh, India – Hydrochemistry and Isotope Investigations Hydrogeology Group, Department of Earth Sciences, Freie Universität Berlin, Berlin, Germany, pp. 1–20.

Holländer H. M., Mull R. and Panda S. N. (2009). A concept for managed aquifer recharge using ASR-wells for sustainable use of groundwater resources in an alluvial coastal aquifer in Eastern India. *Physics and Chemistry of the Earth*, **34**, 270–278.

Hrdinka T., Novicky O., Hanslik E. and Rieder M. (2012). Possible impacts of floods and droughts on water quality. *Journal of Hydro-Environment Research*, **6**, 145–150.

IPCC (2007). Fourth Assessment Report: Climate Change 2007: Impacts, Adaptation and Vulnerability Summary for Policy-Makers. Intergovernmental Panel on Climate Change, Geneva, Switzerland, pp. 7–22.

IUSSP (2009). Urbanization in India: Dynamics & Consequences. XXVI IUSSP International Population Conference. http://iussp2009.princeton.edu/papers/91026. (Accessed 1 August 2013)

Jamwal P. and Mittal A. K. (2010). Reuse of treated sewage in Delhi city: Microbial evaluation of STPs and reuse options. *Resources, Conservation and Recycling*, **54**, 211–221.

Jebamalar A. and Ravikumar G. (2006). Performance Evaluation of Rain Water Harvesting – A Case Study in Chennai City. National Seminar on Rainwater Harvesting and Water Management UNESCO 2006, Nagpur, India, pp. 267–274.

Kale V. S. and Hire P. S. (2004). Effectiveness of monsoon floods on the Tapi River, India: Role of channel geometry and hydrologic regime. *Geomorphology*, **57**, 275–291.

Kaledhonkar M. J., Singh O. P. Ambast S .K. Tyagi N. K. and Tyagi K.C (2003). Artificial groundwater recharge through recharge tube wells: A case study. *IE (I) Journal AG*, **84**, 28–32.

Kanhe N. M. and Bhole A. G. (2006). Study of Experiments on Roof Top Rainwater Harvesting and Surface Rainwater Harvesting for Recharging of Wells. National Seminar on Rainwater Harvesting and Water Management UNESCO 2006, Nagpur, India, pp. 288–291.

Kaushal, R. (2012) Industrial Growth During Five Year Plans. http://www.slideshare.net/crazykudi/industrial-growth-in-in. (Accessed 1 September 2013)

Kimothi P. C., Dimri D. D., Adlakha L. K., Kumar S., Rawat O. P., Patwal P. S., Grischek T., Sandhu C., Ebermann J., Ruppert M., Dobhal R., Ronghang M., Kumar P., Mehrotra I. and Uniyal H. P. (2012). Development of riverbank filtration in uttarakhand. *J. Indian Water Works Association*, Special Issue on River Bank Filtration, 13–18.

KPMG (2010). Water Sector in India: Overview and Focus Areas for the Future. http://kpmg.com/in. (Accessed 1 August 2013)

Kumar P., Mehrotra I., Boernick H., Schmalz V., Worch E., Schmidt W. and Grischek T. (2012). Riverbank filtration: An alternative to pre-chlorination. *J. Indian Water Works Association*, Special Issue on River Bank Filtration, 50–58.

Lorenzen G., Sprenger C., Taute T., Pekdeger A., Mittal A. and Massmann G. (2010). Assessment of the potential for bank filtration in a water-stressed megacity (Delhi, India). *Environ. Earth Sci.*, **61**(7), 1419–1434.

Lozán J. L., Grassl H., Hupfer P., Menzel L. and Schönwiese C. D. (2007). Global Change: Enough Water for All? University of Hamburg, Hamburg, Germany, pp. 13–18.

Mariappan P., Yegnaraman V. and Vasudevan T. (2000). Occurrence and removal possibilities of fluoride in ground waters of India. *Poll. Res*, **19**(2), 165–177.

Meenakshi and Maheshwari R. C. (2006). Fluoride in drinking water and its removal. *Journal of Hazardous Materials*, **137**(1), 456–463.

Mutiyar, P. K., Mittal, A. K. and Pekdeger A. (2011). Status of organochlorine pesticides in the drinking water well-field located in the Delhi region of the flood plains of river Yamuna. *Drink. Water Eng. Sci.*, **4**, 51–60.

PCI (2012). Report of the Working Group on "Effectively Integrating Industrial Growth and Environment Sustainability". Planning Commission of India. Twelfth Five Year Plan (2012–2017).

PRB (2007). The Future Population of India. Population Foundation of India, http://www.prb.org/pdf07/futurepopulationofindia.pdf. (Accessed 1 August 2013)

Rastogi N. and Sarin M. M. (2005). Chemical characteristics of individual rain events from a semi-arid region in India: Three-year study. *Atmospheric Environment*, **39**, 3313–3323.

Ray C. (2008).Worldwide potential of riverbank filtration. *Clean Technology Environmental Policy*, **10**(3), 223–225.

Ronghang M., Kumar P., Mehrotra I., Kimothi P .C., Adlakha L. K., Sandhu C., Grischek T. and Voltz T. J. (2012). Application of riverbank filtration for year-round drinking water production in a small town in the hills of Uttarakhand. *J. Indian Water Works Association*, Special Issue on River Bank Filtration, 19–24.

Sandhu C., Grischek T., Kumar P. and Ray C. (2011a). Potential for riverbank filtration in India. *Clean Techn Environ Policy*, **13**(2), 295–316.

Sandhu C., Grischek T., Schoenheinz D., Prasad T. and Thakur A. K. (2011b). Evaluation of bank filtration for drinking water supply in Patna by the Ganga River, India. In: Riverbank Filtration for Water Security in Desert Countries, C. Ray and M. Shamrukh (eds), NATO Science for Peace and Security Series, Springer, Dordrecht, The Netherlands, pp. 203–222. doi: 10.1007/978–94–007–0026-0_4

Sandhu C. and Grischek T. (2012). Riverbank filtration in India – using ecosystem services to safeguard human health. *Water Science & Technology: Water Supply*, **12**(6), 783–790.

Sandhu C., Grischek T., Ronghan, M., Mehrotra I., Kumar P., Ghosh N.C., Rao Y. R. S., Chakraborty B., Patwal P. S. and Kimothi P. C. (2015). Overview of bank filtration in India and the need for flood-proof RBF systems. In: Saph Pani – Enhancement of Natural Water Systems and Treatment Methods for Safe and Sustainable Water Supply in India. T. Wintgens and A. Nättorp (eds), IWA Publishing, London, United Kingdom.

Saph Pani D1.4 (2014). Report on Bank Filtration Economics and Cost-Estimates or Options of Additional treatment Measures. EU Saph Pani Project Deliverable D1.4. www.saphpani.eu.

Saph Pani D2.2 (2013). Documentation of Acquired Data and Conceptual Model of MAR Impact as Input for WP5 Modelling. EU Saph Pani Project Deliverable D2.2. www.saphpani.eu.

Saph Pani D4.1 (2013). Review of the Post-Treatment Applied to Natural Treatment Systems in India and Critical Water Quality Parameters. EU Saph Pani Project Deliverable. www.saphpani.eu.

Saph Pani D4.3 (2014). Post-Treatment Requirements of Different Natural Treatment Systems. EU Saph Pani Project Deliverable. www.saphpani.eu.

Schoenheinz D. (2004). DOC as Control Parameter for the Evaluation of Management of Aquifers with Anthropogenic Influenced Infiltration. PhD Thesis, Dresden University of Technology. Faculty of Forest-, Geo- and Hydro Sciences, Dresden, Germany.

Schoenheinz D. and Grischek, T. (2011). Behavior of dissolved organic carbon during bank filtration under extreme conditions. In: *Riverbank Filtration for Water Security in Desert Countries*, C. Ray and M. Shamrukh (eds), NATO Science for Peace and Security Series, Springer, Dordrecht, The Netherlands, pp. 51–67. doi: 10.1007/978–94–007–0026-0_4.

Sharma S. K. and Amy G. (2010). Natural treatment systems. In: *Water Quality and Treatment: Handbook of Community Water Supply*, 6th edn, American Water Works Association and McGraw Hill Inc., USA, Chapter 15, Sections 15.1–15.33.

Sharma S. K., Ernst M., Hein A., Jekel M., Jefferson B. and Amy G. (2012). Treatment trains utilising natural and hybrid processes. In: Water Reclamation Technologies for Safe Managed Aquifer Recharge. C. Kazner, T. Wintgens and P. Dillon (eds), IWA Publishing, United Kingdom, pp. 239–257. ISBN 978–184–339–3443.

Shivakumar B. L. (2006). Conservation of Ground Water Through Artificial Recharge Using Roof Top Rainwater in RV College of Engineering Campus Bangalore (a case study). National Seminar on Rainwater Harvesting and Water Management UNESCO 2006, Nagpur, India.

Singh P., Kumar P., Mehrotra I. and Grischek T. (2010). Impact of Riverbank Filtration on Treatment of Polluted River Water. *Journal of Environmental Management*, **91**(5), 1055–1062.

Sprenger C., Lorenzen G. and Pekdeger A. (2008). Occurrence and Fate of Microbial Pathogens and Organic trace Compounds at Riverbank Filtration Sites in Delhi, India. TECHNEAU Integrated Project, Deliverable D5.2.6., http://www.techneau.org (Accessed 5 September 2012)

Sprenger C., Lorenzen G., Hülshoff I., Grützmacher G., Ronghang M. and Pekdeger A. (2011). Vulnerability of bank filtration systems to climate change. *Sci. Total Environ.*, **409**(4), 655–663.

TECHNEAU (2007). Global Trends Affecting the Water Cycle. Wind of Change in the World of Water. TECHNEAU Integrated Project, Deliverable D1.1.7, EU Sixth Framework Programme, http://www.techneau.org

TECHNEAU (2009). Analysis of the Vulnerability of Bank Filtration Systems to Climate Change by Comparing their Effectiveness Under Varying Environmental Conditions. TECHNEAU Integrated Project, Deliverable D5.2.3, EU Sixth Framework Programme, http://www.techneau.org

Thakur A. K., Ojha C. S. P., Grischek T., Ray C. and Jha R. (2009). River Bank Filtration in Extreme Environment Conditions. Proc. Int. Conf. Water, Environment, Energy and Society (WEES-2009), New Delhi, India, pp. 1340–1347.

Vasudevan P. and Tandon M. (2006). Microbial quality of rainwater from roof surfaces. *Journal of Scientific and Industrial Research*, **67**, 432–35.

Zwolsman J. J. G. (2008). Impacts of Climate Change on River Water Quality and Drinking Water Production in the Netherlands. KWR Watercycle Research Institute, http://www.ipcc.ch/publications_and_data/ar4/wg1/en/spmsspm-projections-of.html (Accessed 10 February 2009).

Chapter 13

General framework and methodology for selection of pre- and post-treatment for soil aquifer-based natural treatment systems

Saroj Kumar Sharma, Richard Missa, Maria Kennedy, Cornelius Sandhu, Thomas Grischek and Anders Nättorp

13.1 INTRODUCTION

Soil aquifer-based natural treatment systems (NTSs) namely bank filtration (BF), artificial recharge and recovery (ARR) and soil aquifer treatment (SAT) have been employed for water and wastewater treatment and reuse in different parts of the world. BF (river or lake) has been practiced as a method of "abstraction" of water from surface water sources for more than 100 years (Eckert & Irmscher, 2006). ARR has been employed using many techniques (infiltration ponds, dug wells, trenches, vadose zone wells and direct injection wells) mainly for replenishment of groundwater resources. Sometimes they are also used as a "natural method of water treatment" for surface water sources when the source water quality and/or hydrogeological conditions are not suitable to employ BF. SAT is a specific term designated for methods employed to recharge wastewater treatment plant (WWTP) effluents aiming at subsequent reuse (Sharma & Amy, 2010).

In general, soil-based NTSs or managed aquifer recharge (MAR) systems will be feasible where the following three key areas are adequately addressed (Martin & Dillion, 2002):

i) hydrogeological and technical system design and operation to achieve benefits that exceed costs
ii) system compliance with regulations, within a progressive regulatory regime
iii) establishment of suitable consultative mechanisms to allow satisfactory stakeholder negotiations

Several factors influence the feasibility of a soil-based NTS at a particular site. These include among others, (i) source water quality, (ii) variation in available quantity, (iii) hydrogeological conditions at the site (soil type and permeability, depth of groundwater table, type of the aquifer available, storage capacity of the aquifer, mineralogy of the aquifer material) as well as (iv) treated water quality requirements for intended use.

Pre- and post-treatment systems are integral components of natural systems employed for water and wastewater treatment. These systems not only enhance the performance of NTSs but also help to meet the water quality requirements for different applications. Pre-treatment is relevant for MAR (ARR and SAT) systems. Sedimentation (using detention tanks, reservoirs, settling basin) and filtration (roughing or rapid sand) are common pre-treatments applied to ARR systems (CGWB, 2007; Holländer *et al.* 2009). Sometimes coagulation, adsorption, membrane filtration, advanced oxidation, disinfection and their combinations have been applied as pre-treatment in some NTSs (van der Hoek, 2000; van Houtte & Verbauwhede, 2005; Tielemans, 2007; Sharma *et al.* 2011) to reduce clogging and contamination of the aquifers.

Post-treatment is often required after NTSs to meet the local water quality standards and guidelines for subsequent (re) use. Commonly used post-treatment methods for NTSs include (i) disinfection/chlorination to ensure microbial safety and disinfectant residual in the water distribution system, (ii) aeration/chemical oxidation-rapid sand filtration to remove common groundwater contaminants like iron, manganese and ammonium, (iii) ozonation for oxidation of bulk organics and organic micropollutants (OMPs), (iv) activated carbon filtration (with or without pre-ozonation) to remove the OMPs and colour/taste and odour present in the water, (v) softening and pH correction to remove the hardness and to ensure that there is no scaling or corrosion of water distribution system.

Disinfection (by chlorination) is the most common post-treatment applied to bank filtrates mainly in northern India such as in Uttarakhand, where the critical surface water quality parameters are mainly pathogens and very high turbidity in monsoon. Other systems also use aeration followed by rapid sand filtration before chlorination (e.g. Mathura, Ahmedabad and various BF sites in Jharkhand and Andhra Pradesh; Saph Pani D1.1, 2012; D1.4, 2014). Suspended solids removal by sedimentation in settling basins, detention tanks/chambers or ponds followed by sand filtration is the most common pre-treatment applied to rainwater or stormwater or river water used for MAR in India. Sometimes both of these two pre-treatment processes (sedimentation and filtration) are achieved in a combined unit which forms a part of recharge structure (Saph Pani D4.1, 2013).

13.2 TYPICAL POLLUTANTS AND PRE- AND POST-TREATMENT FOR SOIL/AQUIFER-BASED NTSS

13.2.1 Removal of pollutants by NTSs and pre- and post-treatment systems

Soil-based NTSs are capable of removing several pollutants from water sources. Their removal efficiencies are highly dependent on source water quality and the hydrogeological conditions on site. The type of pre- and post-treatment systems required depend on the type of NTS employed, source water type and quality (rainwater, urban runoff, river or lake water, WWTP effluent), local hydrogeological conditions, process conditions (hydraulic loading rate or HLR, travel time/distance, abstraction rate) applied and intended use of the water after the NTSs (Figure 13.1). Furthermore, required pre- and post-treatment is influenced by national and local regulations regarding groundwater recharge, wastewater reuse and water quality standards and guidelines in place (Sharma & Amy, 2010). Inadequate pre-treatment may clog the NTSs, reduce their runtime and removal capability and consequently make additional post-treatment necessary. On the other hand, a well-designed NTS with proper pre-treatment will require minimal post-treatment. Sometimes, pre- or post-treatment is required to ensure that there is no detrimental effect on aquifers or other receiving water bodies.

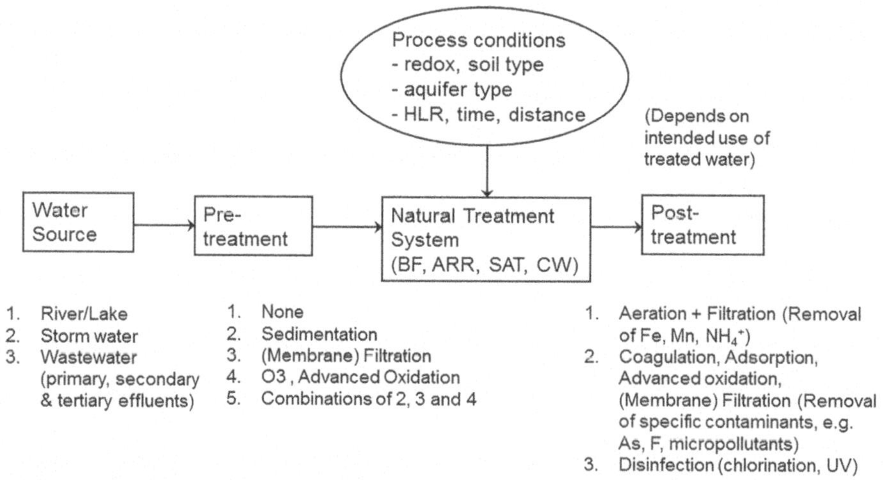

Figure 13.1 Natural treatment system components.

It is to be noted that the pollutant removal efficiencies of NTSs and conventional above-the-ground-treatment processes (pre- and post-treatment) are highly dependent on the raw water quality as well as process conditions applied locally. Some indicative values of the efficiencies of three different NTSs in removing different selected pollutants as collected from various literature sources are shown in Table 13.1. Additionally, based on the data collected from literature sources, lists of common pollutants to be removed by different possible pre- and post-treatment processes for BF, ARR and SAT and their typical removal efficiencies were compiled. These data are presented in detail in Missa (2014) and summarised in Tables 13.7, 13.8 and 13.9 which provide matrices for the selection of pre- and post-treatment for BF, ARR and SAT systems respectively.

Table 13.1 Indicative removal efficiency of typical pollutants by different NTSs.

Pollutant	BF	ARR	SAT*		References
Heavy metals	>90%	>90%	PE	100%	Idelovitch (2003)
			SE	100%	
			TE	100%	
Total Suspended Solids (TSS)	90–100%	90–100%	PE	86–100%	Goldschneider *et al.* (2007), Akber *et al.* (2003), Idelovitch (2003), Abel *et al.* (2014)
			SE	>90–100%	
			TE	>90–100%	
Turbidity	≤1 NTU (50–100%)	≤1 NTU (50–100%)	PE	≤1 NTU (50–100%)	Sharma (2013), BF[1]: Dash *et al.* (2008; 2010); see also Chapter 3
			SE	50–100%	
			TE	50–100%	
Colour	50–100%	50–100%	PE	50–100%	Saph Pani D4.2 (2013); BF[1]: Singh *et al.* (2010); Kumar *et al.* (2012)
			SE	50–100%	
			TE	50–100%	
Pathogens					
• Bacteria including indicators	2–6 Log	2–6 Log	PE	1.2–6.9 Log	WHO (2011); see specific references for BF site-investigations in India [2]
			SE	3–6.5 Log	
			TE	2.4–3 Log	
• Viruses	2.1–8.3 Log	2.1–8.3 Log	PE	4 Log	WHO (2011)
			SE	0≥4 Log	
			TE	0.4–4 Log	
• *Giardia*	1≥2 Log	1≥2 Log	PE	1≥2 Log	WHO (2011)
			SE	1≥2 Log	
			TE	1≥2 Log	
• *Cryptosporidium*	1≥2 Log	1≥2 Log	PE	1≥2 Log	WHO (2011)
			SE	1≥2 Log	
			TE	1≥2 Log	
Iron	0% Sometimes increase	0% Sometimes increase	PE	0%	Sharma (2013)
			SE	0%	
			TE	0%	
Manganese	0% Sometimes increase	0% Sometimes increase	PE	0%	Sharma (2013); de Vet *et al.* (2010)
			SE	0%	
			TE	0%	
Nitrate	50–100%	50–100%	PE	57–100%	Sharma (2013), Saph Pani D4.2 (2013), Essandoh *et al.* (2013), Akber *et al.* (2003), Idelovitch (2003), Al-Kubati (2013)
			SE	3≥90%	
			TE	0–22%	
Ammonium	53–90%	53–90%	PE	17–100%	Saph Pani D4.2 (2013), Sharma *et al.* (2012b), Essandoh *et al.* (2011), Akber *et al.* (2003), Abel *et al.* (2014)
			SE	0–99.2%	
			TE	17≥85%	
Phosphate	≥64%	≥64%	PE	4–100%	Cha *et al.* (2006), Akber *et al.* (2003)
			SE	30≥99%	
			TE	37≥80%	
Organic micropollutants (OMPs)**	≥50%	≥50%	PE	75–100%	Sharma (2013), BF[1]: Saph Pani D1.4 (2014)
			SE	20–100%	
			TE	10–100%	

(*Continued*)

Table 13.1 Indicative removal efficiency of typical pollutants by different NTSs (*Continued*).

Pollutant	BF	ARR	SAT*		References
Dissolved organic carbon (DOC)/ Total organic carbon (TOC)	>25%–≥50%	≥50%	PE SE TE	10–91% 10≥90 20≥80	Sharma (2013), Miehe *et al.* (2010), Quanrud *et al.* (2003), Abel *et al.* (2014); BF[3]
Salinity	Not removed	Not removed	Not removed		
Hardness	Not removed	Not removed	Not removed		

*SAT: PE = primary effluent; SE = secondary effluent; TE = tertiary effluent.
**Removal of OMPs is highly dependent on type of pollutant and redox conditions.
[1]specific reference for BF site-investigations in India.
[2]Dash *et al.* (2008, 2010); Sprenger *et al.* (2008, 2012); Singh *et al.* (2010); Sandhu and Grischek (2012); Saph Pani D1.2 (2013); Bartak *et al.* (2015); see also Chapter 3.
[3]Singh *et al.* (2010); Sandhu *et al.* (2011a, 2011b); Kumar *et al.* (2012); Saph Pani D1.4 (2014); see also Chapter 3.

The conventional physico-chemical treatment processes as pre- or post-treatment for NTS are capable of removing several main pollutants with varying removal efficiencies. This is obvious from the removal efficiencies data collected from different literature sources (Maeng, 2010; Abel, 2014; Missa, 2014) presented in Tables 13.7, 13.8 and 13.9 These tables also show that there is a wide range of options available for selecting conventional treatment processes for pre- and post-treatment of NTS depending on quality of water to be treated and final water quality requirements and costs. It is also to be noted that one treatment method may be able to remove several contaminants and often a combination of different treatment methods are employed to ensure that all pollutants are removed up to the desired level and to provide multiple barriers in the treatment system.

13.3 TYPICAL COSTS OF NTS AND PRE- AND POST-TREATMENT SYSTEMS

Estimation of total costs of treatment (capital costs as well as operation and maintenance (O&M) costs) is critical for assessing whether NTS (together with associated pre- and post-treatment) are competitive in terms of water quality and costs with the conventional surface water treatment options. Cost of water treatment depends on the size of the plant (treatment capacity) and varies from place to place depending upon the capital costs for installation of the facility (land costs, equipments and treatment units) and O&M costs (chemical, energy, manpower and routine maintenance).

Limited data are available on the costs of NTS in developing countries (some examples for India are Essl *et al.* 2014; Saph Pani D1.4; and D6.1) and most of the NTS in developed countries (where some cost data is available) are often of relatively large treatment capacities. These data often include the cost of pre-treatment as well as transmission and water distribution systems, and thus it is difficult to separate the cost of the NTS only. It has been estimated that the cost of the artificial recharge schemes varies from 7–100 USD/m^3 of daily infiltration capacity. The capital costs of artificial recharge schemes are comparable with those of treatment works for surface water for drinking water supply, but costs of operation and maintenance in recharge schemes are likely to be less. Estimates of operation and maintenance costs for artificial recharge schemes vary from of 0.05–0.30 USD/m^3 of water throughput (Hofkes & Visscher, 1986).

The following sub-sections present some estimated total costs (sum of capital and O&M costs) obtained from literature sources for NTS and conventional treatment systems per m^3 of water produced. These cost tables are indicative and can be used to make a relative comparison of costs of different pre- and post-treatment options with NTS combinations obtained from the matrices for feasibility study and preliminary decision making. Local capital and O&M costs should be calculated for each option at each site to obtain a realistic comparison with the alternatives.

13.3.1 Typical costs of NTS

Table 13.2 shows an example of the costs for NTS (BF, ARR, and SAT) based on literature review. The NTS costs vary from place to place and include construction costs, equipment costs (capital/investment costs). It also includes energy costs, chemical costs as well as other O&M costs.

Typical structures used for artificial recharge in India include percolation tanks (with or without recharge shafts), check dams, nala bunds, gabion structures, dug wells, injection wells, sub-surface dykes or underground bandhars, and roof top rainwater harvesting with recharge system. The sizes and costs of these recharge systems varies from state to state. Typical costs of different types of artificial recharge structures applicable in different states of India are presented in detail in "Master Plan for Artificial Recharge Ground Water in India" (MWR, 2013). Ranges of costs of recharge structures are summarized in Table 13.3.

Table 13.2 Indicative costs of soil-based NTS.

NTS	Costs		References
	Total Costs [INR/m³]	Relative Cost Class*	
BF	2.43–13.77	Low	Bosuben (2007), Sharma *et al.* (2012a), Saph Pani D1.4 (2014)
ARR	7.29–17.22	Low	Kumar and Aiyagari (2007), Osborn *et al.* (1997), Gale *et al.* (2002)
SAT	26.73–40.5	Low–Medium	Aharoni *et al.* (2011), Sharma *et al.* (2012b)

*Low = <0.40 EUR/m³, Medium = 0.40–.00 EUR/m³, High = 1.00–2.00 EUR/m³

Table 13.3 Typical investment costs of different artificial recharge systems in India.

Artificial Recharge Structure	Typical Cost Range [Million INR]
Percolation tank	0.5–6
Check dam	0.4–2
Nala bund	0.2–0.3
Recharge shaft/bore hole	0.2–0.35
Rooftop rainwater harvesting system	0.1–0.5

Source: Adapted from MWR, 2013.

Nema *et al.* (2001) based on the detailed cost analysis of a 55 MLD SAT system, revealed the cost competitiveness of the SAT system with the conventional aerobic and anaerobic wastewater treatment systems (Table 13.4). The SAT system was found to be economical, specifically in terms of recurring O&M costs. The capital costs of a SAT system mainly consist of land costs and the overall cost of a SAT system is lower if the land is available at a reasonable cost.

Table 13.4 Cost comparison of SAT system with other conventional wastewater treatment systems (system capacity: 55 MLD).

Treatment System	Capital Cost [Million INR]	Annualized Investment Cost [Million INR]	O&M Cost [Million INR]	Total Annualized Cost [Million INR]	Specific Treatment Cost [INR/m³]	Cost Ratio (Specific Treatment Cost Basis)
Activated sludge plant (conventional)	145.0	20.4	29.0	49.4	2.45	1.55
Activated sludge plant (extended aeration)	129.0	18.0	34.0	52.0	2.60	1.65
Trickling filter	139.7	19.3	35.0	54.3	2.70	1.70
Anaerobic filter	130.0	16.9	26.0	42.9	2.13	1.35
Up-flow anaerobic sludge blanket	110.0	17.5	20.0	37.5	1.86	1.17
SAT	90.0	12.6	19.2	31.8	1.58	1.00

Source: Nema *et al.* (2001).

13.3.2 Typical costs of surface water treatment

The surface water treatment costs vary considerably due to the type and size of treatment plant and location of the plant, construction costs, equipment costs and additional costs like licenses, taxes (capital/investment costs). Table 13.5 shows some examples of typical costs of conventional surface water treatment processes based on the literature review. These are the total costs per m^3 of water treated which include installation costs as well as O&M costs (including energy and chemical costs, but excluding the costs of waste/sludge disposal).

Table 13.5 Typical costs of conventional water treatment processes.

Treatment Process	Total Costs [INR/m^3]	Relative Cost Class*	References
Coagulation	8.1–20.25	Low	de Moel *et al.* (2006)
Sedimentation	4.05–20.25	Low	de Moel *et al.* (2006)
Aeration	8.1–44.55	Low–Medium	de Moel *et al.* (2006)
Rapid sand filtration	24.3–44.55	Low–Medium	de Moel *et al.* (2006)
Slow sand filtration	56.7–121.5	Medium–High	de Moel *et al.* (2006)
Cl_2	0.57–8.91	Low	Dore *et al.* (2014)
O_3	1.22–12.15	Low	Dore *et al.* (2014)
UV	0.89–3.65	Low	Dore *et al.* (2014)
AOP	6.48–365.31	High	Goi (2005)
GAC	40.50–72.90	Medium–High	de Moel *et al.* (2006)
Activated Alumina	36.45–59.13	Medium	USEPA (2000)
Lime softening	28.35–48.60	Low–Medium	de Moel *et al.* (2006)
Ion Exchange	4.86–12.96	Low	Kratochvil *et al.* (2009)
MF	4.05–16.20	Low	Kennedy *et al.* (2013)
UF	4.05–16.20	Low	Kennedy *et al.* (2013)
NF	12.15–162.00	Low–High	Kennedy *et al.* (2013), de Moel *et al.* (2006)
RO	20.25–162.00	Low–High	Kennedy *et al.* (2013), de Moel *et al.* (2006)

*Low = <0.40 EUR/m^3, Medium = 0.40–1.00 EUR/m^3, High = 1.00–2.00 EUR/m^3.

Costs of water treatment in India

The capital cost of conventional surface water treatment (with relatively clean source water except in Delhi and Agra where it is significantly polluted) in India currently ranges from 2 to 2.2 million INR/MLD with minimal operation costs of 0.01–0.10 INR/m^3. The most expensive water treatment plant in India is in Agra with capital costs of 10 million INR/MLD and O&M costs of 4–5 INR/m^3 (WG-UIWSS, 2011). This is attributed to the extreme pollution of the Yamuna river that is currently used as a raw water source. The new water treatment plant under construction in Agra (located in Sikandra), will source its raw water through a 130 km long pipeline from an irrigation canal that carries relatively clean water as it originates from the Ganga river. Table 13.6 presents the costs of some modern water treatment plants in India.

13.4 MATRICES FOR SELECTION OF PRE- AND POST-TREATMENT FOR NTS

This section presents the matrices to be used for selection of the appropriate pre- and post-treatment for NTSs (BF, ARR and SAT). The selection matrices are in the form of tables. Each matrix includes a list of pollutants to be removed, pre-treatment/ and post-treatment system to be selected for a NTS with their indicative removal efficiencies and guidelines for drinking water quality. Where available, WHO (2011) and Indian Standard (BIS, 10500, 2012) guideline values for drinking water quality have been included in the matrices as water quality requirements to be met. These guideline values vary for some parameters in India because they have been prepared based on the exposure, magnitude of concentration and occurrence or prevalence (spatial distribution) of a specific parameter of concern and consequent risk to human health (e.g. widespread, relatively high hardness and fluoride concentrations in ambient groundwater in India). Thus according to BIS 10500 (2012), if the required concentration of a certain parameter is exceeded, it may still be tolerated up to the specified tolerance limit in the absence of an alternative source.

Table 13.6 Cost of water treatment with modern plants in India.

Treatment Plant	Technology	Capacity [MLD]	Capital Cost [Million INR]	Capital Cost [Million INR/MLD]	O&M Costs [INR/m^3]	Power Costs [INR/m^3]	Total O&M Costs [INR/m^3]
Sonia Vihar, Delhi	Pre-settler-Pulsator + Aquazur (Degremont)	635	1,890	3	0.38	1.04	1.43
Chemba-rambakkam	Pulsator + Aquazur (Degremont)	530	1,350	2.5	0.39	0.82	1.21
TK-Halli-1	Pulsator + Aquazur (Degremont)	300	450	1.5	0.22	0.10	0.32
Nagpur	Pulsator + Aquazur (Degremont)	120	150	1.3	0.39	1.04	1.43
TK Halli-II	Aquadaf + Aquazur (Degremont)	550	1,900	3.4	0.32	0.10	0.42
Agra (Sikandra)	Conventional + MBBR	144	1,560	10.8	3–4	n.a.	4–5
Minjur, Chennai	Desalination	100	4,730	47.3	48.66	10–12	59–61
Nemmeli	Desalination	100	10,340	100	n.a.	n.a.	21

n.a.: not available.
Source: WG-UIWSS, 2011.

13.4.1 Matrix for selection of appropriate post-treatment for BF systems

Table 13.7 shows a matrix for the selection of post-treatment options for BF. The selection matrix of BF is different compared to the selection matrices of ARR and SAT because it includes only post-treatment and no pre-treatment. Post-treatment is required for BF systems when some water quality parameters of concern in bank filtrate or extracted water do not meet the drinking water guidelines and standards.

As shown in the above matrix developed for selection of post-treatment for BF, depending upon on the water quality and site conditions, typical examples of post-treatment combinations for the removal of key contaminants in bank filtrate could be:

- BF only (when there are no water quality problems with bank filtrate and water distribution systems are in very good conditions and well-maintained)
- BF + Disinfection (for removal of pathogens, and presence of low concentration of ammonium)
- BF + Aeration + Rapid Sand Filtration (RSF) + Disinfection (for removal of pathogens, ammonium, nitrate, iron and manganese)
- BF + Microfiltration (MF)/Nanofiltration (NF) + Reverse Osmosis (RO) (for pathogens, ammonium, nitrate, micropollutants, hardness and fluoride)
- BF + Aeration + RSF + Ozonation + Activated carbon filtration + Disinfection (for removal of pathogens, iron, mangancsc and OMPs)

13.4.2 Matrix for selection of appropriate pre- and post-treatment for ARR systems

The selection matrix for ARR (with their different possible pre- and post-treatment options together with their removal efficiencies) is presented in Table 13.8. From this table, possible combinations for ARR system can be: (i) pre-treatment + ARR, (ii) ARR + post-treatment or (iii) pre-treatment + ARR + post-treatment. ARR systems generally include pre-treatment because clogging is the critical problem in soil-based NTSs. Additionally, post-treatment may be necessary to meet the water quality standards and guidelines as some pollutants may not be removed adequately during the soil passage or because some other contaminants may be introduced into the water during the soil passage (depending on local hydrogeology/mineralogy and redox conditions).

Table 13.7 Matrix for selection of appropriate post-treatment options for BF.

Parameter	Removal Efficiency of BF	Post-treatment		Guideline Values (WHO, 2011; BIS 10500, 2012)
		Type*	Removal Efficiency	
Pathogens	1≤8.3 Log	Chlorination	1–4 Log	No pathogen in 100 mL sample
		UV	1–4 Log	
		MF/NF	3–7 Log	
		Ozonation	1–4 Log	
Hardness	–	Lime softening	60%	500 mg/L (WHO, 2011); 200 mg/L (required) & 600 mg/L in the absence of an alternate source (BIS, 10500, 2012)
		Ion exchange	35%	
		NF	85–99%	
		RO	>99%	
		Ion exchange + RO	35≥99%	
Iron/ Manganese	–	Aeration + RSF	Fe 92–97% Mn 17–79%	0.3 mg/L Fe Recommended value for aesthetic reason
		Aeration + RSF + Aeration + RSF	Fe 92≥99% Mn 17–96%	
		Aeration + Coagulation + RSF	Fe 92–99% Mn 17–92%	
		Aeration + Coagulation + Sedimentation + RSF	Fe 95≥99% Mn 38–87%	<0.1 mg/L Mn Recommended value for aesthetic reason; 0.3 mg/L in the absence of an alternate source (BIS, 10500, 2012)
		Aeration + Coagulation + RSF + MF/UF	Fe >60–100% Mn <20–90%	
		Aeration + Coagulation + RSF + Ion exchange	Fe >60–100% Mn <20–92%	
Fluoride	–	NF/RO	92%	1.0 mg/L (required by BIS 10500, 2012) and 1.5 mg/L (WHO, 2011; in the absence of an alternate source by BIS 10500, 2012)
		Activated Alumina	75%	
		Coagulation + NF/RO	20–97%	
		Coagulation+ Activated Alumina	20–90%	
		Ion exchange	95%	
Nitrate	50–100%	Ion exchange	90%	50 mg/L (WHO, 2011); 45 mg/L (BIS, 10500, 2012)
		NF/RO	65≥95%	
		Ion exchange + NF/RO	90–100%	
Ammonium	53–90%	Chlorination	100%	1.5 mg/L as threshold odour concentration (WHO, 2011); 0.5 mg/L (as total ammonia-N; BIS 10500, 2012)
		NF	90–98%	
		Aeration + RSF	40–50%	
		Aeration + RSF + RSF	50–75%	
		Ion exchange	97%	
OMPs**	≥50%	NF	>99%	For pesticides 0.01 (Alpha HCH) to 190 μg/L (Malathion) depending upon type (BIS, 10500, 2012)
		GAC	0–70%	
		AOP	20–99.9%	
		Ion exchange	40–100%	
		Ion exchange + NF	82–100%	
Salinity	–	NF	40–99%	50 mg/L
		RO	≥98.5%	
		NF + RO	>99%	

*Type: UV = Ultraviolet; MF = Microfiltration, NF = Nanofiltration, UF = Ultrafiltration, RO = Reverse Osmosis, RSF = Rapid sand filtration, GAC = Granular activate d carbon, AOP = Advanced oxidation process.
**Removal of OMPs is highly dependent on type of pollutant and redox conditions.

Table 13.8 Matrix for selection of appropriate pre- and post-treatment options for ARR.

Pollutants to be Removed	Pre-treatment		Removal Efficiency of ARR	Post-treatment		Guideline Values (WHO, 2011)
	Type*	Removal Efficiency		Type*	Removal Efficiency	
Pathogens	Ozonation	1–4 Log	1≤8.3 Log	Chlorination	1–4 Log	No pathogens in 100 mL sample
	UV	1–4 Log		Ozonation	1–4 Log	
	Chlorination	1–4 Log		UV	1–4 Log	
				NF	3–6 Log	
Hardness	Lime softening	60%	–	–	–	500 mg/L
	NF	85–99%				
Turbidity	Sedimentation + Aeration + RSF/SSF	>95–100%	50–100%	MF/UF	>98%	<5 NTU
				NF	70–86%	
	MF/UF	>98%				
TSS	Sedimentation + Aeration + RSF/SSF	100%	90–100%	–	–	<1,000 mg/L
	Coagulation/ Sedimentation	50 ≥ 85%				
	UF	85–99.9%				
Iron/ Manganese	Aeration + RSF	Fe 92–97%	–	Coagulation + Sedimentation	Fe 95–96%	0.3 mg/L Fe**
		Mn 17–79%			Mn 37–38%	
	Aeration + RSF + Aeration + RSF	Fe >99%		Aeration + RSF	Fe 92–97%	<0.1 mg/L Mn**
					Mn 17–79%	
		Mn 31–96%		(Coagulation + Sedimentation) + Aeration + RSF	Fe 95≥99%	
					Mn 34–84%	
				MF/UF	Fe 95–97%	
					Mn 37–43%	
Fluoride	–	–	–	Activated alumina	75%	1.5 mg/L
				Coagulation	71%	
				NF	92%	
				RO	92%	
				Ion exchange	95%	

(Continued)

Table 13.8 Matrix for selection of appropriate pre- and post-treatment options for ARR (*Continued*).

Pollutants to be Removed	Pre-treatment		Removal Efficiency of ARR	Post-treatment		Guideline Values (WHO, 2011)
	Type*	Removal Efficiency		Type*	Removal Efficiency	
Arsenic	–	–	90%	Coagulation and filtration	>20%	0.01 mg/L
				Activated alumina	96%	
				NF/RO	93%	
				Ion exchange	99.43%	
				Lime softening	91%	
Nitrate	Ion exchange	90%	50–100%	Ion exchange	90%	50 mg/L
				RO	65≥95%	
Ammonium	Chlorination	100%	53–90%	NF	90–98%	
	Aeration + RSF	40–50%		Chlorination	100%	–
				Aeration + RSF	40–50%	
OMPs***	Ozonation	50≥90%	≥50%	Ozonation	50≥90%	–
				AOP	20–99.9%	
	GAC	0–70%		GAC	0–70%	
				Ion exchange	40–100%	
				NF	>99%	
Colour	Aeration + Coagulation + RSF	>60–64%	50–100%	AOP	<48%	50 mg/L
	GAC	<55%		Coagulation + Sedimentation	>60%	
				NF	70–94%	
Salinity	–	–	–	NF	40–99%	50 mg/L
				RO	≥98.5%	

*UV = Ultraviolet Radiation; MF = Microfiltration, NF = Nanofiltration, UF = Ultrafiltration, RO = Reverse Osmosis, RSF = Rapid sand filtration, SSF = Slow Sand filtration, GAC = Granular activated carbon, AOP = Advanced oxidation process.

**Recommended Fe and Mn values for aesthetic reasons.

***Removal of OMPs is highly dependent on type of pollutant and redox conditions.

Based on the water quality and site conditions, the following are typical examples of key contaminants and relevant pre- and/or post-treatment systems for the ARR to handle these key contaminants:

- Pathogens and stabilization of temperature: ARR only
- Pathogens, bulk organic matter and OMPs: Ozonation + ARR
- Pathogens: ARR + Disinfection only
- Pathogens and arsenic or other metals at low concentrations: ARR + Lime softening or Coagulation + RSF
- Pathogens, TSS and turbidity: Sedimentation + RSF + ARR
- Pathogens, TSS, turbidity, ammonium, iron and manganese: Sedimentation + RSF + ARR + Aeration + RSF + Chlorination

13.4.3 Matrix for selection of appropriate pre- and post-treatment for SAT systems

Table 13.9 shows a matrix which can be used to select the pre-and post-treatment for SAT. It also includes the pollutants to be removed by each pre- and post-treatment together with SAT. Moreover the table contains the removal efficiencies for each treatment step and guidelines values.

Depending upon source of water, quality and site conditions, the following are typical examples of the key contaminants and pre- and/or post-treatment system for SAT to handle them:

- TSS, turbidity at low concentrations: Sedimentation + SAT
- TSS, turbidity at higher concentrations: Sedimentation + Coagulation + SAT
- TSS, turbidity: UF + SAT
- Pathogens, TSS, turbidity and ammonium: MF/UF + SAT + Chlorination
- Pathogens, ammonium, nitrate and salinity: SAT + NF/RO
- Pathogens,TSS,turbidity,ironandmanganese:Coagulation + Sedimentation + RSF + SAT + Aeration + RSF +Chlorination
- Pathogens, salinity, iron, manganese, ammonium, bulk organic matter and OMPs: Ozonation + SAT + Aeration + RSF + NF

13.4.4 Use of the matrices for selection of pre- and post-treatment options

The stepwise procedure to use the matrix tables for selection of pre- and post-treatment for different NTS is as follows (Figure 13.2). However, under certain circumstances NTS can also cause risks to ambient groundwater (environment) and human health. Thus, the implementation of NTS should be undertaken using a structured management approach to assess the risks. In this context and in addition to the use of the following matrix tables, approaches presented in well-established MAR Guidelines (NRMMC–EPHC–NHMRC, 2009; Page *et al.* 2010) and a risk-based assessment and management approach to BF in India under local conditions (Bartak *et al.* 2014), should be considered:

- Collect raw water quality and hydrological/hydrogeological data for the given site intended for NTS.
- Select the type of NTS to be used based on the water quality and hydrological/hydrogeological data.
- Check in the appropriate matrix table which contaminants require pre-treatment or post-treatment or both to meet the water quality guidelines.
- Make all possible combinations of pre- and post-treatment options for those contaminants from the matrix table.
- Estimate the final water quality with different combinations of pre- and post-treatment options. In order to assess the final quality, first calculate the removal of a given contaminant in pre-treatment, NTS and post-treatment. Also consider the effects of dilution if some natural groundwater is also abstracted together with the infiltrated or recharged water (Sharma et al., 2012a). When the dilution effect is taken into consideration separately, the final concentration of a pollutant can be computed using the following relation:

  $$C_{\text{final}} = C_{\text{source}} * (1\text{-}R_{\text{PRE}}) * (1\text{-}R_{\text{NTS}}) * (1\text{-}R_{\text{POST}}) \quad (13.1)$$

  where C_{source} = concentration of a pollutant in source water, C_{final} = final concentration of a pollutant after post-treatment, R_{PRE}, R_{NTS}, R_{POST} = removal efficiency of a pollutant in pre-treatment system, NTS and post-treatment system respectively.
- If there is more than one treatment step in pre-treatment or post-treatment, then R_{PRE} and R_{POST} refers to overall removal efficiencies of all the steps involved.
- Assess the final results by comparing them with the guidelines or standards.
- Select the options that meet the water quality requirements.
- For each alternative (that meet water quality requirements) estimate the total costs by adding the costs of pre-treatment, NTS and post-treatment.

Table 13.9 Matrix for selection of appropriate pre- and post-treatment options for SAT.

Pollutants to be Removed	Pre-treatment		Removal Efficiency of SAT**		Post-treatment			Guideline Values (WHO, 2011)
	Type*	Removal Efficiency			Type*	Removal Efficiency		
Pathogens	Chlorination	1–4 Log	PE	>1–6.9 Log	Chlorination	1–4 Log		No pathogens in 100 mL sample
	UV	1–4 Log	SE	0–6.5 Log	Aeration + RSF	0.4–4 Log		
	Ozonation	1–4 Log	TE	0.4–4.0 Log	Ozonation	1–4 Log		
	MF/UF	0–7 Log			UV	1–4 Log		
					NF/RO	3–7 Log		
Hardness	–	–	PE	–	NF	85–99%		500 mg/L
			SE	–	Ion exchange + NF	90–99%		
			TE	–	RO	>99%		
Turbidity	UF	>98%	PE	50–100%	–	–		<5 NTU
			SE	50–100%				
	Coagulation + Sedimentation	>95%	TE	50–100%				
TSS	UF	85–99.9%	PE	86–100%	–	–		<1,000 mg/L
	Coagulation + Sedimentation	60≥85%	SE	>90–100%	–	–		
	Aeration + RSF	70–80%	TE	>90–100%	–	–		
Iron/ Manganese	–	–			Coagulation + Sedimentation	Fe	95–96%	0.3 mg/L Fe***
						Mn	37–38%	
			PE	–	Aeration + RSF	Fe	92–97%	<0.1mg/L Mn***
						Mn	17–79%	
			SE	–	(Coagulation + Sedimentation)+ (Aeration + RSF)	Fe	95≥99%	
			TE	–		Mn	48–87%	
					Aeration + RSF + MF/UF	Fe	>99%	
						Mn	48–88%	

Nitrate	Ion exchange	90%	PE	57–100%	Ion exchange	90%	50 mg/L
	RO	65≥95%	SE	3≥90%	RO	65≥95%	
			TE	0–22%			
Ammonium	NF/RO	90–98%	PE	25–99%	Ion exchange	98%	–
			SE	0–99%	NF/RO	90–98%	
					Chlorination	100%	
			TE	17–100%	Aeration + RSF	40–50%	
OMPs****	UF	>90%	PE	75–100%	Ion exchange	40–100%	–
	RO	70–99.9%			Ozonation	50≥90%	
			SE	20–100%	AOP	20–99.9%	
					GAC + AOP	20–100%	
					Ion exchange	40–100%	
			TE	10–100%	NF	>99%	
					RO	70–99.9%	
Salinity	–	–	PE	–	NF	40–99%	50 mg/L
			SE	–	RO	≥98.5%	
			TE	–	NF + RO	>99%	

*Type: UV = Ultraviolet Radiation; MF = Microfiltration, NF = Nanofiltration, UF = Ultrafiltration, RO = Reverse Osmosis, RSF = Rapid sand filtration, GAC = Granular activated carbon, AOP = Advanced oxidation process.

**SAT: PE = primary effluent, SE = secondary effluent, TE = tertiary effluent.

***Recommended Fe and Mn values for aesthetic reasons.

****Removal of OMPs is highly dependent on type of pollutant and redox conditions.

- Rank different possible combinations of pre- and post-treatment for a given NTS based on the removal efficiencies and cost effectiveness for decision making.

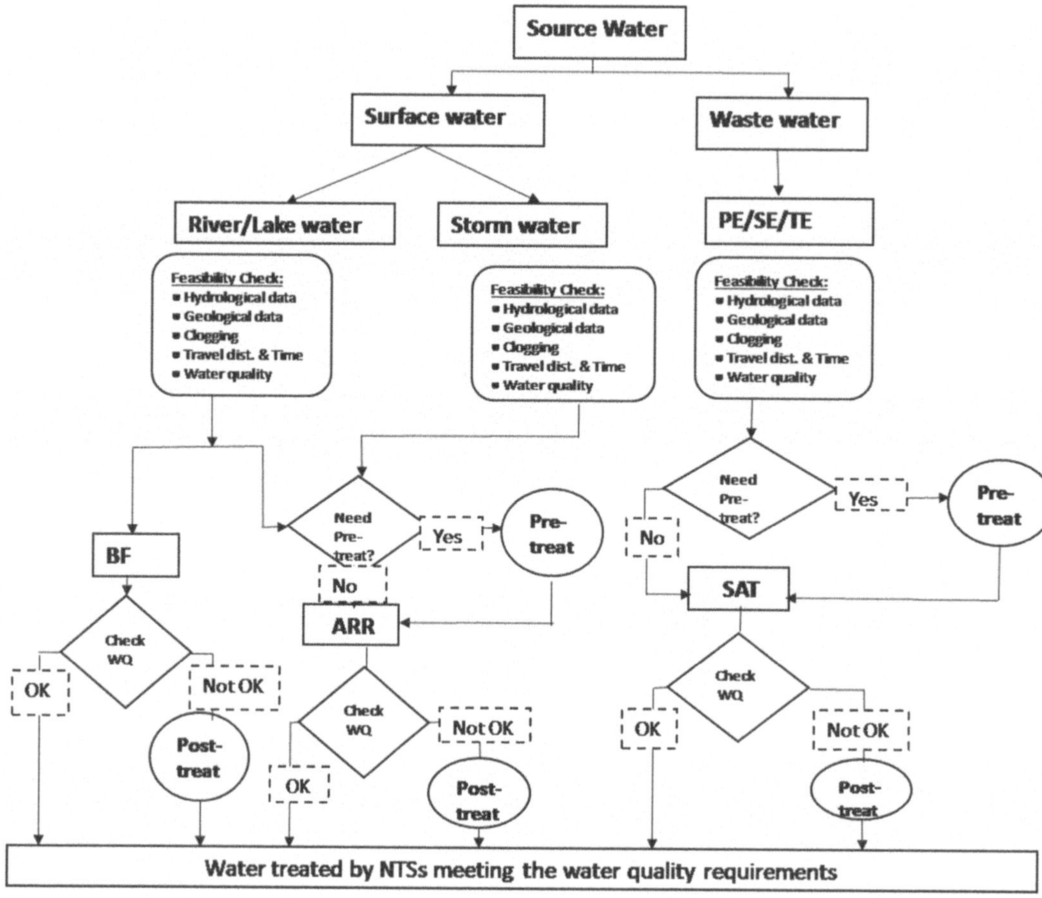

Figure 13.2 Framework for selection of pre- and/or post-treatment (PE = primary effluent, SE = secondary effluent, TE = tertiary effluent, WQ = Water quality).

Two examples of the use of the matrices developed for the selection of pre- and post-treatment of NTS under given conditions (BF and ARR respectively) are presented in the Annex of this chapter.

It is to be noted that engineering judgment in the selection of a proper treatment combination (pre-treatment + NTS + post-treatment) is required. For the correct selection of a treatment system availability of energy, chemicals and skilled manpower as well as cost of land play an important role.

13.5 CONCLUSION

Soil-based NTSs, namely BF, ARR and SAT, have been used in different parts of the world for water and wastewater treatment and reuse. While assessing the feasibility of NTSs at a given site, all the components of the NTS (including pre- and post-treatment) as well as local regulations, water quality guidelines and institutional capacities should be taken into consideration. Source water quality and local hydrological/hydrogeological conditions determine the type of NTS which is most favorable and feasible under given conditions. Furthermore, treated water quality requirements (local guidelines and standards) as well hydrogeological conditions at the intended site determine the pre- and post-treatment requirements.

Comprehensive literature data on cost of NTS as well as some common conventional treatment processes (used as pre- and post-treatment) were compiled. In general, when the source water quality and local hydrogeological conditions are favorable, BF is the cheapest and most effective method of water treatment for developing countries requiring no or minimal post-treatment. ARR is attractive when relatively cheap land is available nearby and BF is not feasible due to local hydrogeological conditions. SAT is an attractive option for polishing wastewater effluents with the aim of water reuse provided the local regulation permits such technology and if the clogging of the aquifer can be minimized by proper pre-treatment and operation of the system.

Also, a comprehensive compilation of removal efficiencies of NTS was made as well as of common conventional treatment processes for different pollutants generally present in water. They were presented in the form of matrices/tables to facilitate selection of appropriate treatment process to remove a particular pollutant. In addition, stepwise procedures for the selection of the most suitable pre- and post-treatment systems for any given NTS were developed.

Several combinations of pre- and post-treatment together with a NTS can meet the water quality guideline values and standards for the intended use. Determination of capital and O&M costs of each of the feasible options is required to rank them in terms of cost effectiveness. It is expected that the matrices and the developed selection procedure can be used by designers and planners to make a preliminary selection of NTS and associated pre- and post-treatment systems.

13.6 REFERENCES

Abel C. D. T. (2014). Soil Aquifer Treatment: Assessment and Applicability of Primary Effluent Reuse in Developing Countries. PhD thesis, UNESCO-IHE, Delft, The Netherlands, CRC Press/Balkema.

Abel C. D. T., Sharma S. K., Mersha S. A. and Kennedy M. D. (2014). Influence of intermittent infiltration of primary effluent on removal of suspended solids, bulk organic matter, nitrogen and pathogens indicators in a simulated managed aquifer recharge system. *Journal of Ecological Engineering*, **64**, 100–107.

Aharoni A., Guttman J., Cikurel H. and Sharma S. K. (2011). D3.2.1.f Guidelines for Design and Operation and Maintenance of SAT (and Hybrid SAT) System. Report EU SWITCH Project, MEKOROT and UNESCO-IHE.

Akber A., Al-Awadi E. and Rashid T. (2003). Assessment of the use soil aquifer treatment (SAT) technology in improving quality of tertiary treated wastewater in Kuwait. *Emirates Journal for Engineering Research*, **8**(2), 25–31.

Al-Kubati K. M. A. (2013). Development of Framework for Site Selection, Design, Operation and Maintenance for Soil Aquifer Treatment (SAT) Systems. MSc Thesis, MWI 2013-17, UNESCO-IHE, Delft, The Netherlands.

Bartak R., Page D., Sandhu C., Grischek T., Saini B., Mehrotra I., Jain C. K. and Ghosh N. C. (2015). Application of risk-based assessment and management to riverbank filtration sites in India. *Journal of Water and Health*, **13**(1), 174–189. doi:10.2166/wh.2014.075

BIS 10500 (2012). Drinking Water – Specification. Indian Standard (second revision), Bureau of Indian Standards, New Delhi, India, 1–5.

Bosuben N. K. (2007). Framework for Feasibility of Bank Filtration Technology for Water Treatment in Developing Countries. MSc Thesis, MWI 2007-17, UNESCO-IHE, Delft, The Netherlands.

CGWB (2007). Manual on Artificial Recharge of Groundwater. Central Ground Water Board, Government of India. http://www.cseindia.org/userfiles/Manual%20on%20Artificial%20Recharge%20of%20Ground%20Water.pdf. (Accessed 1 December 2012)

Cha W., Kim J. and Ckoi H. (2006). Evaluation of steel slag for organic and inorganic removal in soil aquifer treatment. *Water Research*, **40**, 1034–1042.

Dash R. R., Mehrotra I., Kumar P. and Grischek T. (2008). Lake bank filtration at Nainital, India: Water quality investigations. *Hydrogeology Journal*, **16**(6), 1089–1099.

Dash R. R., Bhanu Prakash, E. V. P., Kumar P., Mehrotra I., Sandhu C. and Grischek T. (2010). River bank filtration in Haridwar, India: Removal of turbidity, organics and bacteria. *Hydrogeology Journal*, **18**(4), 973–983.

De Moel P. J., Verberk J. Q. J. C. and van Rijk J. C. (2006). Drinking Water: Principles and Practices. TU Delft and Kiwa, The Netherlands, 154–184.

De Vet W. W. J. M., van Genuchten C. C. A., van Loosdrecht M. C. M. and van Dijk J. C. (2010). Water quality and treatment of river bank filtrate. *Drinking Water Engineering and Science*, **3**(1), 79–90.

Dore M. H. I., Moghadam A. K., Singh R. G. and Achari G. (2014). Costs and the Choice of Drinking Water Treatment Technology in Small and Rural Systems. Res'Eay-Waterment_Project4.2-powerpoint\presentation. http://www.env.gov.nl.ca/env/waterres/training/adww/decade/05_Mohammed_Dore.pdf. (Accessed 15 February 2014)

Eckert P. and Irmscher R. (2006). Over 130 years of experience with riverbank filtration in Düsseldorf, Germany. *Journal of Water Supply: Research and Technology – AQUA*, **55**(4), 283–291.

Essandoh H. M. K, Tizaoui C. and Mohammed M. H. A (2013). Removal of dissolved organic carbon and nitrogen during simulated soil aquifer treatment. *Water Research*, **47**, 3559–3577.

Essl L., Starkl M., Kimothi P. C., Sandhu C. and Grischek T. (2014). Riverbank filtration and managed aquifer recharge as alternative water supply technologies for India: Strengths – weaknesses – opportunities – threats analysis. *Water Science & Technology: Water Supply*, **14**(4), 690–697.

Gale I. M., Newmann I., Calow R. C. and Moench M. (2002). The Effectiveness of Artificial Recharge of Groundwater: Review. Groundwater Systems and Water Quality Programme, Phase 1 Final Report CR/02/108N, British Geological Survey, Keyworth, England.

Goi A. (2005). Advanced Oxidation Processes for Water Purification and Soil Remediation. PhD thesis, Tallinn University of Technology, Estonia.

Goldschneider A. A., Harzlampides K. A. and Marcquarine K. T. B. (2007). River sediment and flow characteristics near bank filtration water supply: Implication for riverbed clogging. *Journal of Hydrology*, **344**, 55–69.

Gupta A., Singh H., Mehrotra I., Kumar P., Kumar S., Grischek T. and Sandhu C. (2015). Lake bank filtration for water supply in Nainital. In: Natural Water Treatment Systems for Safe and Sustainable Water Supply in the Indian Context *Saph Pani*, Wintgens T., Nättorp A., Elango L. and Asolekar S. R. (eds), IWA Publishing, London, United Kingdom, 39–56.

Hofkes E. H. and Visscher J. T. (1986). Artificial Groundwater Recharge for Water Supply of Medium-Size Communities in Developing Countries. International Reference Centre, The Netherlands, 36–37.

Holländer H. M., Mull R. and Panda S. N. (2009). A concept for managed aquifer recharge using ASR-wells for sustainable use of groundwater resources in an alluvial coastal aquifer in Eastern India. *Physics and Chemistry of the Earth*, **34**, 270–78.

Idelovitch E. (2003). SAT (Soil Aquifer Treatment) – The long-term performance of the Dan Region Reclamation Project. PowerPoint presentation – The World Bank Water Week 2003, Washington D.C., USA.

Kennedy M. D., Rodriguez S. G. S. and Schippers J. C. (2013). Advanced Water Treatment: Low Pressure Membranes Technology. Lecture notes-LN0424/13/1, UNESCO-IHE, Delft, The Netherlands.

Kratochvil D., Bratty M., Sanguinetti D., Ye S. and Murray T. (2009). Innovative Ion Exchange Systems for Cost-Effective Increase in Water Re-Use in Power Plants. 70th Annual International Water Conference (IWC-09–40), Orlando, Florida, 521–533.

Kumar N. N. and Aiyagari N. (1997). Artificial Recharge of Groundwater. Civil Engineering Dept, Virginia Tech, USA, 1–9.

Kumar P., Mehrotra, I., Boernick H., Schmalz V., Worch E., Schmidt W. and Grischek T. (2012). Riverbank filtration: An alternative to pre-chlorination. *Journal of Indian Water Works Association*, Special Issue on River Bank Filtration, 50–58.

Maeng S. K. (2010). Multiple Objective Treatment Aspects of Bank Filtration. PhD thesis, UNESCO-IHE, Delft, The Netherlands, CRC Press/Balkema.

Martin R. and Dillion P. (2002). Aquifer Storage and Recovery: Future Directions for South Australia. Department of Water, Land and Biodiversity Conservation (DWLBC) and CSIRO Land and Water, Australia. Available: http://webdoc.sub.gwdg.de/ebook/serien/ud/DWLBC/DWLBC2002_04.pdf (Accessed 12 November 2014)

Miehe U., Hinz C., Hoa E. and Grutzmacher G. (2010). D1.1b. DOC and Trace Organic Removal via Ozonation Underground Passage - Expected Benefit and Limitations. Kompetenz Zentrum Wasser Berlin gGmbH 49(0), Berlin, Germany, pp. 2–21.

Missa R. (2014). Framework for Selection of Appropriate Pre- and Post-Treatment Options for Different Natural Treatment Systems. MSc thesis, MWI 2014–37, UNESCO-IHE, Delft, The Netherlands.

MWR (2013). Master Plan for Artificial Recharge to Ground Water in India. Ministry of Water Resources, Government of India, New Delhi, India.

Nema P., Ojha C. S., Kumar A. and Khanna P. (2001). Techno-economic evaluation of soil aquifer treatment using primary effluent at Ahmedabad, India. *Water Research*, **35**(9), 2179–2190.

NRMMC–EPHC–NHMRC (2009). Australian Guidelines for Water Recycling: Managed Aquifer Recharge (Phase 2). Report of the Natural Resource Management Ministerial Council, Environment Protection and Heritage Council and National Health and Medical Research Council, Canberra, Australia, pp. 23–52.

Osborn N. I., Eckenstein E. and Fabian R. S. (1997). Demonstration and Evaluation of Artificial Recharge to the Blaine Aquifer in Southwestern Oklahoma. Oklahoma Water Resource Board-OWRB. Oklahoma, United States, pp. 1–18.

Page D., Dillon P., Vanderzalm J., Toze S., Sidhu S., Barry K., Levett K., Kremer S. and Regel R. (2010). Risk assessment of aquifer storage transfer and recovery with urban stormwater for producing water of a potable quality. *J. Environ. Qual.*, **39**(6), 2029–2039.

Quanrud D. M., Hafer J., Karpiscak M. M., Zhang J., Lansey K. E. and Anold R. G. (2003). Fate of organics during soil aquifer treatment; sustainability of removal in the field. *Water Research*, **37**, 3401–3411.

Sandhu C., Grischek T., Schoenheinz D., Prasad T. and Thakur A. K. (2011a). Evaluation of bank filtration for drinking water supply in Patna by the Ganga River, India. In: Riverbank Filtration for Water Security in Desert Countries, C. Ray and M. Shamrukh (eds), NATO Science for Peace and Security Series, Springer, Dordrecht, The Netherlands, pp. 203–222.

Sandhu C., Grischek T., Kumar P. and Ray C. (2011b). Potential for riverbank filtration in India. *Clean Technologies and Environmental Policy*, **13**(2), 295–316.

Sandhu C. and Grischek T. (2012). Riverbank filtration in India – using ecosystem services to safeguard human health. *Water Science and Technology: Water Supply*, **12**(6), 783–790.

Saph Pani D1.1 (2012). Database of Relevant Pollutants in Urban Areas and Their Attenuation at RBF Sites. EU Saph Pani Project Deliverable, www.saphpani.eu.

Saph Pani D1.2 (2013). Guidelines for Flood-Risk Management of Bank Filtration Schemes during Monsoon in India. Saph Pani Project Deliverable, http://www.saphpani.eu/downloads (Accessed 19 August 2014)

Saph Pani D1.4 (2014). Report on Bank Filtration Economics and Cost-Estimates or Options of Additional Treatment Measures. EU Saph Pani Project Deliverable D1.4, www.saphpani.eu

Saph Pani D4.1 (2013). Review of the Post-treatment Applied to Natural Treatment Systems in India and Critical Water Quality Parameters. EU Saph Pani Project Deliverable, www.saphpani.eu

Saph Pani D4.2 (2013). Removal Efficiency of Conventional Post-Treatment after Natural Treatment Systems. EU Saph Pani Project Deliverable, www.saphpani.eu

Saph Pani D6.1 (2013). Report on Initial Sustainability Assessment. EU Saph Pani Project Deliverable, www.saphpani.eu

Sharma S.K. and Amy G. (2010). Natural Treatment Systems. In: Water Quality and Treatment: Handbook of Community Water Supply, 6th edn, American Water Works Association and McGraw Hill Inc., USA, Chapter 15, Sections 15.1–15.33.

Sharma S. K., Hussen M. and Amy G. (2011). Soil aquifer treatment using advanced primary effluent. *Water Science and Technology*, **64**(3), 640–646.

Sharma S. K., Chaweza D., Bosuben N., Holzbecher E. and Amy G. (2012a). Framework for feasibility assessment and performance analysis of riverbank filtration system for water treatment. *Journal of Water Supply; Research and Technology–AQUA*, **61**(2), 73–81.

Sharma S. K., Ernst M., Hein A., Jekel M., Jefferson B. and Amy G. (2012b). Treatment trains utilising natural and hybrid processes. In: Water Reclamation Technologies for Safe Managed Aquifer Recharge, C. Kazner, T. Wintgens and P. Dillon (eds), IWA Publishing, United Kingdom, pp. 239–257. ISBN 978–184–339–3443.

Sharma S. K. (2013). Natural Treatment Systems: Bank Filtration and Artificial Recharge. Lecture Notes, UNESCO – IHE, Delft, The Netherlands.

Singh P., Kumar P., Mehrotra I. and Grischek T. (2010). Impact of riverbank filtration on treatment of polluted river water. *Journal of Environmental Management*, **91**(5), 1055–1062.

Sprenger, C., Lorenzen, G. and Pekdeger, A. (2008) Occurrence and Fate of Microbial Pathogens and Organic Trace Compounds at Riverbank Filtration Sites in Delhi, India. TECHNEAU integrated project: Deliverable D 5.2.6. Available: http://www.techneau.org (Accessed November 2014).

Sprenger C., Lorenzen G. and Pekdeger A. (2012). Environmental tracer application and purification capacity at a riverbank filtration well in Delhi (India). *Journal of Indian Water Works Association*, Special Issue on River Bank Filtration (December, 2012), 25–32.

Tielemans M. W. M. (2007). Artificial Recharge of the Groundwater in the Netherlands. Proceedings of IWA Regional Conference on Groundwater Management in the Danube River Basin and Other Large River Basins, Belgrade, Serbia, pp. 1–12.

USEPA (2000). Technologies and Costs for Removal of Arsenic from Drinking Water. EPA 815-R-00–028, United States Environmental Protection Agency.

Van der Hoek J. P., Hofman J. A. M. H. and Graveland A. (2000). Benefits of ozone-activated carbon filtration in integrated treatment processes, including membrane systems. *Journal of Water Supply: Research and Technology, AQUA*, **49**(6), 341–356.

Van Houtte E. and Verbauwhede J. (2005). Artificial Recharge of Treated Wastewater Effluent Enables Sustainable Groundwater Management of a Dune Aquifer in Flanders, Belgium. Proceedings of ISMAR5 Conference, 10–16 June 2005, Berlin, Germany.

WHO (2011). Guidelines for Drinking-water Quality. 4th edn, World Health Organization, Geneva, Switzerland, pp. 149–201.

WG-UIWSS (2011). Report of the Working Group on Urban and Industrial Water Supply and Sanitation for the Twelfth Five-Year-Plan (2012–2017). Submitted to the Steering Group on Water Sector, Planning Commission, November 2011, New Delhi, India.

13.7 APPENDIX

Example of application of matrices developed for selection of pre- and post-treatment options

A.1 Example of selecting post-treatment for a BF system

River water is proposed to be used as a source; there is an alluvial aquifer of 60 m depth at site and travel distance and travel time are expected to be 150 m and 4 months respectively. Critical pollutants to be treated after BF are iron and manganese (due to local hydro-geological conditions) with estimated concentrations of 5 mg/L and 1 mg/L respectively in the bank filtrate. It is required to find the appropriate treatment train with or without post-treatment processes.

The selection of post-treatment alternatives for iron and manganese removal, calculations of removal efficiencies of each alternative and comparison of the costs of selected alternatives are presented in the following tables.

Table 13A.1 Treatment alternatives to remove iron and manganese (from the BF matrix table 13.7).

Water Type	NTS	Post-treatment	Output
River water	BF	Aeration + RSF	Treated water meeting guideline values
		Aeration + RSF + Aeration + RSF*	
		Aeration + Coagulation + RSF	
		Aeration + Coagulation + Sedimentation + RSF	

*Applied when iron and manganese concentrations are high or when iron, manganese and ammonium are present.

From the above selection table it is clear that both options meet the guidelines although option 2 is more efficient than option 1. In terms of costs, option 1 is cheaper than option 2. Consequently, option 1 is selected because it meets the guidelines and is cheaper than option 2. Furthermore, the water is generally chlorinated before supply to maintain disinfectant residual in the distribution system. Then, a schematic diagram of the proposed treatment system for given condition would be:

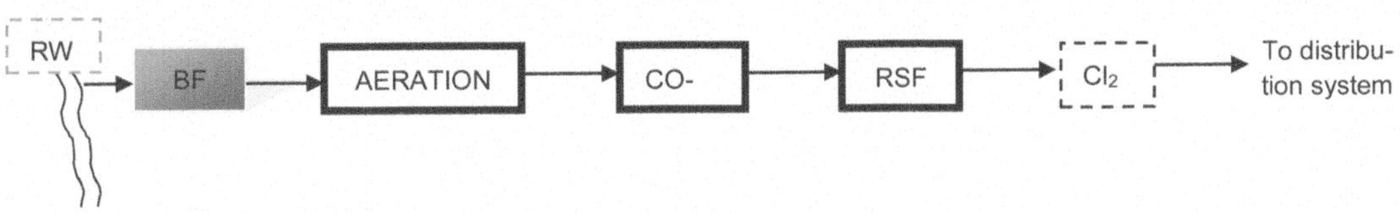

Table 13A.2 Calculation of removal efficiencies and comparison with guidelines.

River Water		BF		Post-treatment			Comparison	
Pollutant	**C_{source} mg/L**	**Eff_0 %**	**Conc. rem_{i+1} mg/L**	**Process**	**Eff_1 %**	**Conc. rem_{i+2} mg/L**	**Guideline Values mg/L**	**Remarks**
Fe	–	–	5	Aeration +	92–97	0.15–0.3	0.3	Yes
Mn	–	–	1	RSF	17–79	0.21–0.83	<0.1	No
Fe	–	–	5	**Aeration +**	92≥99	**<0.05–0.4**	**0.3**	**Yes**
Mn	–	–	1	**RSF + Aeration + RSF**	17–96	**0.04–0.83**	**<0.1**	**Yes**
Fe	–	–	5	**Aeration +**	92–99	**0.05–0.4**	**0.3**	**Yes**
Mn	–	–	1	**Coagulation + RSF**	17–92	**0.08–0.83**	**<0.1**	**Yes**
Fe	–	–	5	Aeration +	95≥99	>0.05–0.25	0.3	Yes
Mn	–	–	1	Coagulation + Sedimentation + RSF	38–87	0.13–0.62	<0.1	No
Fe	–	–	5	Aeration +	>60–100	0.00–2.00	0.3	Yes
Mn	–	–	1	Coagulation + RSF + MF/UF	<20–90	0.1≥0.8	<0.1	No

NB: Conc. rem. $_{i+(n+1)}$ = (100–eff_n/100) Conc.$_{\cdot i+n}$.
Yes: Means pollutant can be removed either with minimum removal efficiency or maximum removal efficiency.
No: Means pollutant cannot be removed at up to the required level with the proposed treatment process.

Table 13A.3 Cost comparison of selected alternatives.

NTS		Post-treatment		Comparison	
Type	**$Costs_1$ (Euro/m³)**	**Process**	**$Costs_2$ (Euro/m³)**	**Total $Costs_{1+2}$ (Euro/m³)**	**Rank**
BF	0.03–0.17	**Aeration + Coagulation + RSF**	**0.20–1.35**	**0.23–1.52**	**1**
		Aeration + RSF + Aeration + RSF	0.80–2.20	0.83–2.37	2

A.2 Example of selecting pre-and post-treatment options for an ARR system

Stormwater is available as the source of the water; the soil-aquifer system which can be used for NTS is made up of a phreatic aquifer; travel distance and travel time are expected to be 150 m and 4 months respectively. The depth of vadose zone was estimated to be 5.0 m. The major pollutants to be removed are iron, manganese and hardness with estimated concentrations of 3 mg/L, 0.8 mg/L and 240 mg/L respectively in the source water. It is required to determine the appropriate pre-and/ post-treatment processes for ARR system.

The selection of pre- and post-treatment alternatives for iron, manganese and hardness removal, calculations of removal efficiencies of each treatment alternative and comparison of the costs of selected alternatives are presented in the following tables.

Table 13A.4 Treatment alternatives to remove iron and manganese (from the ARR matrix table 13.8).

Water Type	Pre-treatment	NTS	Post-treatment	Output
Storm water	Aeration + RSF	ARR	ARR Aeration + RSF	Treated water meeting guideline values
			Aeration + RSF + RSF	
			Aeration + Coagulation + RSF	
			Aeration + Coagulation + Sedimentation + RSF	
			Aeration+ Coagulation + RSF + MF/UF	

Table 13A.5 Treatment alternatives to hardness (from the ARR matrix table 13.8).

Water Type	Pre-treatment	NTS	Post-treatment	Outcome
Storm water	–	ARR	–	Treated water meeting guideline values
	Lime softening	ARR	–	
	FN	ARR	–	

Table 13A.6 Calculation of removal efficiencies for iron and manganese and comparisons with guidelines.

River Water		Pre-treatment			ARR		Post-treatment			Comparison	
Pollutant	**C_{source} mg/L**	**Process**	**Eff_1 (%)**	**Conc. rem_{i+2} mg/L**	**Eff_0 (%)**	**Conc. rem_{i+1} mg/L**	**Process**	**Eff_1 (%)**	**Conc. rem_{i+2} mg/L**	**Guideline Values mg/L**	**Remarks**
Fe	3	Aeration + RSF	92–97	0.09–0.24	–	0.09–0.24	Aeration + RSF	92–97	0.003–0.020	0.3	Yes
Mn	0.8		17–79	0.17–0.66	–	0.17–0.66		17–79	0.04–0.55	<0.1	Yes
Fe	3	Aeration + RSF	92–97	0.09–0.24	–	0.09–0.24	Aeration + RSF + Aeration +	92≥99	0.001–0.02	0.3	Yes
Mn	0.8		17–79	0.17–0.66	–	0.17–0.66	RSF	17–96	0.007–0.55	<0.1	Yes
Fe	3	Aeration + RSF	92–97	0.09–0.24	–	0.09–0.24	Aeration + Coagulation +	92–99	<0.001–0.02	0.3	Yes
Mn	0.8		17–79	0.17–0.66	–	0.17–0.66	Aeration + RSF	17–92	0.01–0.55	<0.1	Yes
Fe	3	Aeration + RSF	92–97	0.09–0.24	–	0.09–0.24	Aeration + Coagulation +	95≥99	0.001–0.012	0.3	Yes
Mn	0.8		17–79	0.17–0.66	–	0.17–0.66	Sedimentation + RSF	38–87	0.02–0.41	<0.1	Yes
Fe	3	Aeration + RSF	92–97	0.09–0.24	–	0.09–0.24	Aeration + Coagulation +	>60–100	0≥0.096	0.3	Yes
Mn	0.8		17–79	0.17–0.66	–	0.17–0.66	RSF + MF/UF	<20–90	0.02–0.53	<0.1	Yes
Fe	3	Aeration + RSF	92–97	0.09–0.24	–	0.09–0.24	–	–	0.09–0.24	0.3	Yes
Mn	0.8		17–79	0.17–0.66	–	0.17–0.66		–	0.17–0.66	<0.1	No
Fe	3	–	–	3	–	3	Aeration + RSF	92–97	0.09–0.24	0.3	Yes
Mn	0.8		–	0.8	–	0.8		17–79	0.17–0.66	<0.1	No
Fe	3	–	–	3	–	3	Aeration + RSF +	92≥99	0.03–0.24	0.3	Yes
Mn	0.8		–	0.8	–	0.8	Aeration + RSF	17–96	0.03–0.66	<0.1	Yes
Fe	3	–	–	3	–	3	Aeration + Coagulation +	92–99	0.03–0.24	0.3	Yes
Mn	0.8		–	0.8	–	0.8	RSF	17–92	0.06–0.66	<0.1	Yes
Fe	3	–	–	3	–	3	Aeration + Coagulation +	95≥99	0.03–0.15	0.3	Yes
Mn	0.8		–	0.8	–	0.8	Sedimentation + RSF	38–87	0.10–0.5	<0.1	No
Fe	3	–	–	3	–	3	Aeration + Coagulation +	>60–100	0–1.2	0.3	Yes
Mn	0.8		–	0.8	–	0.8	RSF + MF/UF	<20–90	0.08–0.64	<0.1	Yes

Table 13A.7 Calculation of removal efficiencies for hardness and comparisons with guidelines.

River Water		Pre-treatment			ARR		Post-treatment			Comparison	
Pollutant	**C_{source} mg/L**	**Process**	**Eff_0 (%)**	**Conc. rem_{i+1} mg/L**	**Eff_1**	**Conc. rem_{i+2} mg/L**	**Process**	**Eff_2**	**Conc. rem_{i+3} mg/L**	**Guideline mg/L**	**Remarks**
Hardness	240	–	–	240	–	240	–	–	240	500	Yes
Hardness	240	Lime softening	60	96	–	96	–	–	96	500	Yes
Hardness	240	**NF**	**85–99**	**2.4–36**	–	**2.4–36**	–	–	**2.4–36**	**500**	**Yes**

NB: Conc. rem. $_{i+(n+1)}$ = (100-eff$_n$/100) Conc$_{i+n}$.
Yes: Means pollutant can be removed either from minimum removal efficiency or/to maximum removal efficiency.

Table 13A.8 Cost comparison of selected alternatives for iron and manganese removal.

Pre-treatment		NTS		Post-treatment		Comparison	
Process	**Costs$_2$ (Euro/m^3)**	**Type**	**Costs$_1$ (Euro/m^3)**	**Processes**	**Costs$_2$ (Euro/m^3)**	**Total Costs$_{1+2}$ (Euro/m^3)**	**Rank**
Aeration + RSF	0.40–1.10	ARR	0.093–0.206	Aeration + RSF	0.40–1.10	0.893–2.406	3
Aeration + RSF	0.40–1.10			Aeration + RSF + Aeration + RSF	0.80–2.20	1.293–3.506	7
Aeration + RSF	0.40–1.10			Aeration + Coagulation + RSF	0.50–1.35	0.993–2.656	4
Aeration + RSF	0.40–1.10			Aeration + Coagulation + Sedimentation + RSF	0.55–1.60	0.943–2.906	6
Aeration + RSF	0.40–1.10			Aeration + Coagulation + RSF + MF/UF	0.55–1.55	1.043–2.856	5
–	–			Aeration + RSF + Aeration + RSF	0.80–2.20	0.893–2.406	3
–	–			**Aeration + Coagulation + RSF**	**0.50–1.35**	**0.593–1.556**	**1**
–	–			Aeration + Coagulation + RSF + MF/UF	0.55–1.55	0.643–1.756	2

Table 13A.9 Cost comparison of selected alternatives for hardness removal.

Pre-treatment		NTS		Post-treatment		Comparison	
Process	**Costs$_2$ (Euro/m^3)**	**Type**	**Costs$_1$ (Euro/m^3)**	**Process**	**Costs$_2$ (Euro/m^3)**	**Total Costs$_{1+2}$ (Euro/m^3)**	**Rank**
–	–	ARR	0.093–0.206	–	–	0.093–0.206	1
Lime softening	0.35–0.60			–	–	0.443–0.806	2
NF	0.15–2.00			–	–	0.243–2.166	3

It is clear from the above table 13A.9 that option 1 is the cheapest and it will be selected for hardness removal. Table 13A.8 shows that, for iron and manganese removal, option 1 is the cheapest and will be selected. The two options will be combined to form the post-treatment system. Although ARR may not be effective for removal of iron, manganese and hardness but it will be used for removal of some turbidity and TSS and some pathogens that may exist in the source water. Iron and manganese in abstracted water will be removed by the post-treatment system. Furthermore, the water is generally chlorinated before supply to maintain disinfectant residual in the distribution system. The following is the schematic diagram of the treatment system proposed for removal of iron, manganese and hardness:

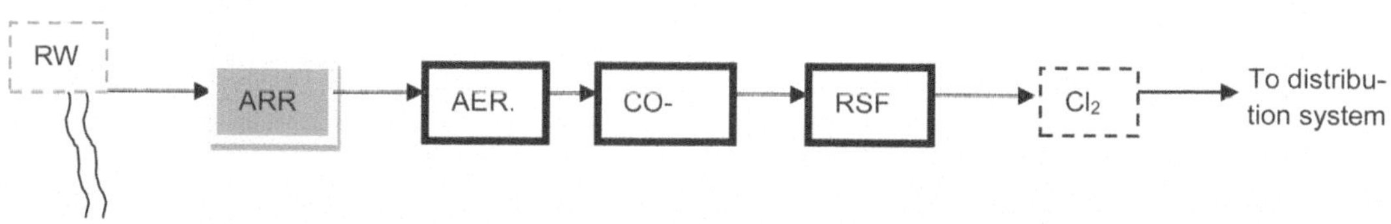

Chapter 14

Modelling of natural water treatment systems in India: Learning from the Saph Pani case studies

Wolfram Kloppmann, Cornelius Sandhu, Maike Groeschke, Rajaveni Sundara Pandian, Géraldine Picot-Colbeau, Mohammad Fahimuddin, Shakeel Ahmed, Marina Alazard, Priyanie Amerasinghe, Punit Bhola, Alexandre Boisson, Lakshmanan Elango, Ulrike Feistel, Stefanie Fischer, Narayan C. Ghosh, Thomas Grischek, Gesche Grützmacher, E. Hamann, Indu Sumadevi Nair, Mahesh Jampani, N. C. Mondal, Bertram Monninkhoff, Marie Pettenati, S. Rao, Sarah Sarah, Michael Schneider, Sebastian Sklorz, Dominique Thiéry and Anna Zabel

14.1 INTRODUCTION

Analytical or numerical models can be used at all stages of natural treatment system (NTS) implementation, from initial planning of individual systems, to upscaling at the watershed scale and to system optimisation so as to reach defined water quantity and quality targets. Such models enable water managers to test diverse scenarios so that they can optimise implementation of NTS's within a watershed (which type? where? how big?). At local scale, models also may be useful for improving any individual NTS by fine-tuning technical options. Generally spoken, they are management tools that help avoid costly, real-size trial-and-error testing of NTS's and also may avoid surprises with respect to the expected impact of NTSs on water quantity and quality of a watershed or aquifer.

Within the Saph Pani project, different types of NTSs (river bank filtration, constructed wetlands and managed aquifer recharge) have been modelled in a large variety of geological and hydro-climatic settings, representative of the Indian subcontinent, thus demonstrating the usability of state-of-the art integrated surface-groundwater flow and transport models as planning and management tools.

Numerical flow or flow and transport modelling has been applied to the following problems within the Saph Pani case studies:

River Bank Filtration (RBF)

- Optimisation of well technology and exploitation schemes assisted by flow modelling for RBF in Haridwar, Uttarakhand
- Contaminant transport/attenuation in an RBF scheme in the urban alluvial aquifer of Yamuna River, New Delhi

Managed Aquifer Recharge (MAR)

- Saline intrusion management and implementation of MAR check dams in the coastal Arani and Koratalaiyar watershed, Chennai, Tamil Nadu
- Behaviour of individual MAR-SAT percolation tanks, Maheshwaram, Telangana

Constructed Wetlands

- Wetland impacts on water balance in the Musi watershed, Hyderabad, Telangana

14.2 MODELLING OF RIVER BANK FILTRATION (RBF)

Direct surface water abstraction from river networks for drinking water and irrigation bears important sanitary risks, both microbial and chemical (pathogens, organic contaminants including emerging substance classes, inorganic major, minor and trace compounds). These risks can be considerably reduced through indirect river water abstraction from wells and boreholes within the accompanying alluvial aquifer. Water pumped from such wells will contain a variable fraction of river water (up to 100%) and recharge coming from direct infiltration of rainwater or from groundwater flow from the piedmont area. The river water fraction, called river bank filtrate, will be naturally purified during the passage through (1) the river bed, often rich in clay minerals and organic matter and (2) the alluvial aquifer. Main processes are (1) physical retention of suspended matter and microbes depending on the porosity of the filtering media, (2) chemical interaction of the migrating water with the aquifer material, notably sorption-desorption processes, ion exchange on clay minerals, dissolution-precipitation reactions and (3) microbiologically mediated processes mainly taking place in contact with biofilms on the aquifer material, in particular biodegradation of organic substances, transformation of organic matter, nitrification-denitrification processes. Those processes all have their proper kinetics and are therefore time-dependent. The purifying action of RBF will largely vary in function depending on the contact time of the migrating water with minerals and biofilms.

In the following sub-chapter we will outline some crucial aspects of RBF, including the determination of key parameters for purification capacity, that is, the mixing proportion of river bank filtrated in the pumped alluvial groundwater and the travel time from the river to the pumping wells (Haridwar case study) as well as the simulation of transport of nitrogen species within the aquifer material, both on lab scale and field scale (Delhi case study).

14.2.1 RBF at River Ganga, Haridwar, Uttarakhand: groundwater flow modelling

Site description

The importance of RBF as a sustainable year-round natural treatment technology for the provision of drinking water to the permanent residents of Haridwar (>225,000) and the highly variable number of pilgrims (at least 50,000 daily, with up to 8.2 million on specific days such as *Kumbh Mela*; Gangwar & Joshi, 2004) for the removal of bacteriological indicators (total coliforms and *E. coli*) and for meeting the dynamic drinking water demand has been highlighted in Chapters 1 and 2.

As of 2013, at least two-thirds, or 59,000 to 67,000 m^3/day (Bartak *et al.*, 2015), of the total raw water for drinking was abstracted from a total of 22 RBF wells with the remainder supplied by deep groundwater abstraction wells. The RBF wells are located on the west-bank of the Ganga River in the North and south part of the city, on Pant Dweep Island and on a narrow stretch of land between the Upper Ganga Canal (UGC) (Figure 14.1). Thus, by virtue of their proximity to the Ganga River and UGC that form natural recharge boundaries, the RBF wells abstract around 40 to 90% bank filtrate (Bartak *et al.*, 2015). The portion of bank filtrate abstracted by the wells located on Pant Dweep Island and further south is greater than those to the north. This is due to their location in an area influenced by the naturally occurring flow of bank filtrate between the UGC and Ganga River due to the difference in hydraulic gradient. The naturally pre-treated RBF water is abstracted from the upper unconfined alluvial aquifer, which is in hydraulic contact with the Ganga River and UGC. The aquifer comprises fluvial deposits of poorly graded sand (0.0075–4.75 mm) beneath which lies a lower layer of silty sand (Dash *et al.*, 2010). After abstraction, the water is disinfected with sodium hypochlorite at the well prior to being distributed to the consumer. Although these wells are relatively shallow (7–10 m deep), they have a large storage capacity due to their large diameter (~10 m). The abstraction from these wells is highly variable (790 to 7,530 m^3/day) and dependent on the season, with higher daily abstractions during monsoon as a result of longer operating hours due to a greater water demand, but also due to an increase in groundwater levels due to greater recharge from the surface water bodies.

The discharge of partially treated sewage and untreated storm water run-off into the Ganga River (and UGC in Haridwar) and its upstream tributaries, as well as large-scale ritualistic bathing, are a source of thermotolerant coliforms (TTC; *E. coli)* present in the surface water. In this context, mean TTC numbers measured in the 22 RBF wells, calculated from long term water quality data (2005–2013), were 18 TTC/100 mL during monsoon and 1 TTC/100 mL during non-monsoon compared with 10^4–10^5 TTC/100 mL in the Ganga, including UGC (Bartak *et al.*, 2015). This highlights the significant removal efficiency of 3.5 to 4.4 $\log_{10}$ of TTC by RBF (Dash *et al.*, 2010).

Problems to be solved

Despite the observed high TTC removal efficiency, TTC counts up to 93 MPN/100 mL were still observed in some RBF wells (Bartak *et al.*, 2015). In this context it has been observed that some RBF wells which are very close to the surface water body show the presence of coliforms, such as wells 27, 2, 16, 40, 42, 17, 21 and 24 (for details see Chapter 2) that are located at a distance of 6–36 m from the UGC and its escape channel.

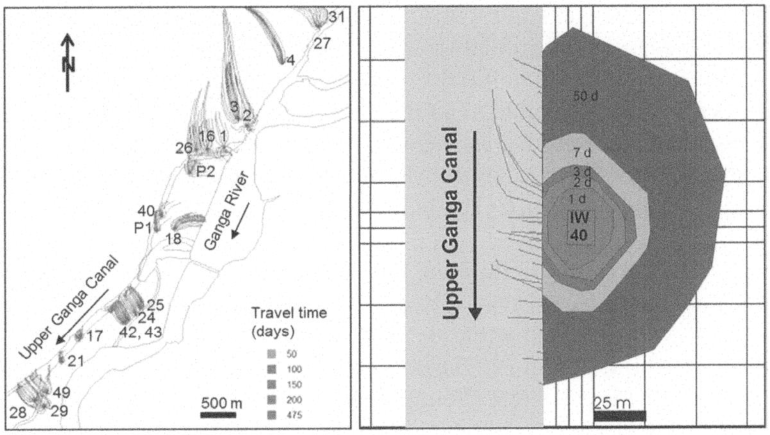

Figure 14.1 Travel time and flow paths of bank filtrate for RBF system in Haridwar (left) and travel time of bank filtrate during monsoon for RBF well 40.

However, it is also evident that RBF wells which are comparatively far away from the Ganga or UGC (48–190 m) and in the area where the Ganga enters Haridwar in the northern part of the city (thus low impact of upstream pollution), such as wells 3, 4, 26, 1 and 31, also have comparably high coliform counts of <2–93 MPN/100 mL (Bartak *et al.*, 2015). Normally one would not expect wells to be contaminated at such a relatively far distance. But considering that the lack of well head protection zones, social land use practices such as public bathing/washing at the well heads, well head housing, cattle in and around the RBF wells and unsanitary defecation practices near/around the wells were identified as high risks for the Haridwar RBF system (Bartak *et al.*, 2015), it is conceivable that the origin of coliforms in some RBF wells is not the bank filtrate from the Ganga River but rather ambient landside groundwater.

Thus, the objective of the numerical groundwater flow modelling study for Haridwar was to identify the flow paths of the bank filtrate to the RBF wells and the travel time in order to analyse the source of contamination of the wells. Additionally, an overall understanding of the hydrogeological system in response to the dynamic hydrological regime of the Ganga River (high monsoon and low non-monsoon water levels) was to be achieved. The degree of confidence in the numerically simulated portion of bank filtrate abstracted by the RBF wells was to be ascertained through comparisons with analytical calculations from mean electrical conductivity (EC) and Oxygen-18 isotope values.

Tools and modelling strategy

A three-dimensional, finite element, two–layered, numerical groundwater flow model was set up in Visual MODFLOW (version, 2011.1). The spatial extent of the entire model area is 5,000 m (East-West) × 6,000 m (North-South). However, the active model domain is assigned only to the area of the Ganga floodplain with the remaining cells inactivated. The model domain was discretised to obtain a cell area of 12.5 × 12.5 to 100 × 100 m. The upper model layer coincides in general with the upper sandy layer of the aquifer (thickness from 0 m to around 12 m below ground level (bgl) and with the bottom of the partially penetrating RBF wells. The lower layer represents the silty sand (thickness 12 m to around 21 m bgl). The Ganga River, UGC and its escape channel are represented by the river boundary condition. The hydraulic conductivity of the riverbed was determined from sieve analyses at various points and accordingly assigned to calculate the riverbed conductance. Reference day water level measurements were conducted on three specific days, one each in August 2012, October 2012 and January 2013 to represent monsoon, post and pre-monsoon conditions and a relatively good calibration was achieved for each of these conditions. The actual hourly abstraction rates of the RBF wells for the monsoon and non-monsoon operating hours were normalised for a continuous operation (24 hour period) and assigned using the well boundary condition for the 22 RBF wells in the model at their respective locations. Subsequently the particle tracking tool was used in MODPATH to visualize the flow paths and travel times of water to the RBF wells (Figure 14.1). The zone budget method in MODFLOW was used to determine the portion of bank filtrate abstracted by the RBF wells.

Outcome, added value and perspectives

The simulated flow paths of the water to the RBF wells (Figure 14.1, left) corroborate to the portion of bank filtrate abstracted by them that have been calculated from long-term mean EC values (Bartak *et al.*, 2015) and Oxygen-18 isotope values (Saph

Pani D1.2, 2013). The flow of water to the wells in Figure 14.1 indicate that the RBF wells located in the northern part of Haridwar also receive a considerable portion of groundwater in addition to some bank filtrate. For wells 3, 4, 26 and 1, the portion of ambient landside groundwater is between 40–60% with the remainder being bank filtrate. Consequently a greater portion of bank filtrate is abstracted in monsoon due to an increase in the Ganga River levels and thereby the water line of the river moves closer to the bank and the wells. But as the area that lies in the groundwater catchment of the RBF wells is densely populated and substantially large, there is a greater risk of contamination from decentralised sewage disposal (septic tanks) and leaky wastewater drains. This would also explain the high TTC counts in the RBF wells in relation to a relatively low portion of bank filtrate.

On Pant Dweep Island the shortest travel time of the bank filtrate to the wells 40 and P1, located only 15 m from the UGC, is around 3 days during the non-monsoon period that decreases to 2 days during monsoon (Figure 14.1, right). The mean portion of bank filtrate abstracted is 60–70% and while the TTC counts in well 40 are <2–93 MPN/100 mL, they are <3 MPN/100 mL or below the detectable limit in well P1 (Bartak *et al.*, 2015). Although both wells exhibit short travel times of bank filtrate, bathing and washing activities take place immediately next to well 40 by means of a tap attached to the main distribution pipe at the well. Thus the higher TTC count in well 40 and other wells located close to the UGC bank with similar short travel times, can be explained due to the preferential flow of water into the RBF wells from above ground and around the wells (not river/canal water) due to flooding, intensive rainfalls event or regular seepage / drainage of wastewater from bathing and washing activities (Saph Pani D1.2, 2013). This results in very short travel times (45 minutes to 4.5 hours) as demonstrated by a NaCl tracer experiment on well 40 in Chapter 2 (Sandhu *et al.*, 2014). For RBF wells 18 and P2, located between 110 m and 320 m from the UGC and Ganga River, the travel time of the bank filtrate is substantially longer (up to a year, Figure 14.1, left). Compared to well 40, the TTC count is lower with a maximum of only 15 MPN/100 mL. As the bank filtrate to these wells has considerably long travel times, the likely reason is above-ground contamination from wide spread defecation on the vast open spaces of the Pant Dweep Island that has an extremely large influx of pilgrims and tourists daily, especially during festivals like the *Kanwar Mela*. During longer festivals like the *Kumbh* and *Ardh Kumbh Melas*, pilgrims reside on the island for up to 4 months. Unlined pit-latrines are dug for such events that have been assessed as a risk to the wells (Bartak *et al.*, 2015).

On the other hand, the remaining wells that are located at a distance of 15 m and more from the UGC in the southern part of Haridwar abstract the highest portion of bank filtrate of all RBF wells in Haridwar (80–90%). The simulated portion of bank filtrate (using the zone budget tool in MODPATH) abstracted by these wells lies within a ±10% confidence limit of the analytically calculated portions using EC and Oxygen-18 isotope data. The maximum TTC count observed in some wells was up to 15 MPN/100 mL while in the others it was below the detectable limit of <3 MPN/100 mL, with the exception of one well having a maximum TTC count of 93 MPN/100 mL (Bartak *et al.*, 2015). As the area between these wells and the UGC, its escape channels and Ganga River is not residential, the impact from domestic sewage (septic tanks, pit-latrines) is low. However, occasional high TTC counts can be attributed to washing and bathing activities and inappropriate drainage of water (wastewater, rainfall and/or storm water runoff) near/around the wells.

Most importantly, the comparatively overall low TTC counts highlights the high removal efficiency of the RBF system, because most public bathing takes place daily in this stretch of the UGC from which the bank filtrate to these wells originates. Furthermore, the annual monsoon and the location of the wells in the area result in a natural recharge to the RBF wells thereby ensuring sustainable operation during periods of peak water demand.

Conclusions

The Haridwar RBF system is operating sustainably since 1965. The groundwater flow modelling study of the RBF system in Haridwar has identified the flow paths of bank filtrate and the groundwater catchment areas of the RBF wells. In conjunction with investigations on the risk of floods and health risk assessment to RBF wells using Haridwar as an example (Chapter 2; Sandhu *et al.*, 2015; Bartak *et al.*, 2015), the groundwater flow modelling investigation has helped to identify potential sources of contamination to the wells. Consequently the study has shown that the wells which abstract the highest portion of bank filtrate, have overall lower or at the most an equal magnitude (only in some cases) of TTC counts compared to RBF wells that abstract an equal or greater portion of ambient groundwater. On one hand the flow modelling study has helped to signify the effectiveness of the natural RBF system to remove pathogens, and on the other hand it illustrates the risk of contamination to unconfined aquifers from inhabited areas without appropriate collection, treatment and discharge of domestic sewage and wastewater. This highlights the need for the implementation of well-head and catchment protection zone measures. These measures have to be prioritised in lieu of the growing pressure on land use and conflicting interests. The flow modelling study has also shown the benefit of locating RBF wells on islands and in areas where a natural flow between surface water bodies occurs to ensure sustainable abstraction. The groundwater flow model of the Haridwar RBF system is a useful tool to compliment the water quality and isotope investigations and can be integrated into a regional hydrogeological assessment of the Haridwar urban area.

14.2.2 RBF at Yamuna River, New Delhi: Ammonium reactive transport modelling

Site description

A further RBF field site investigated as part of Saph Pani is located in New Delhi. The capital city is located in North India in the Indo Gangetic Plain along the banks of the Yamuna River. Within the city the river is dammed by two barrages, one in the north of the city and one in the south. In between both barrages, treated, partially treated, and untreated sewage water feed the river through 16 drains (Government of Delhi, 2006). Numerous production wells draw water from the floodplain aquifer, shallow sand and kankar aquifer made up of river deposits. Due to high groundwater abstraction in the city, losing stream conditions are dominating (Lorenzen *et al.*, 2010) and therefore some of the wells draw a high share of bank filtrate. Through the infiltration of river water, sewage-borne contaminants can enter the aquifer and, depending on their retention and degradation rates in the sediments, can eventually reach the production wells. In this context one parameter of concern is ammonium (drinking water limit: 0.5 mg/L, BIS 10500, 2012).

The Delhi study site comprises a transect of observation points across the flood plain on the East bank of the Yamuna River in central Delhi. It includes several hand pumps and observation wells as well as four Ranney wells (large horizontal collector wells) (Figure 14.2). Because the well field was not specifically designed for RBF, the production wells are not parallel to the river bank but are constructed across the complete width of the undeveloped flood plain. The main focus of the study lies on a Ranney well (P3) located at a distance of 500 m from the river bank. A detailed description of the field site is given in Chapter 4.

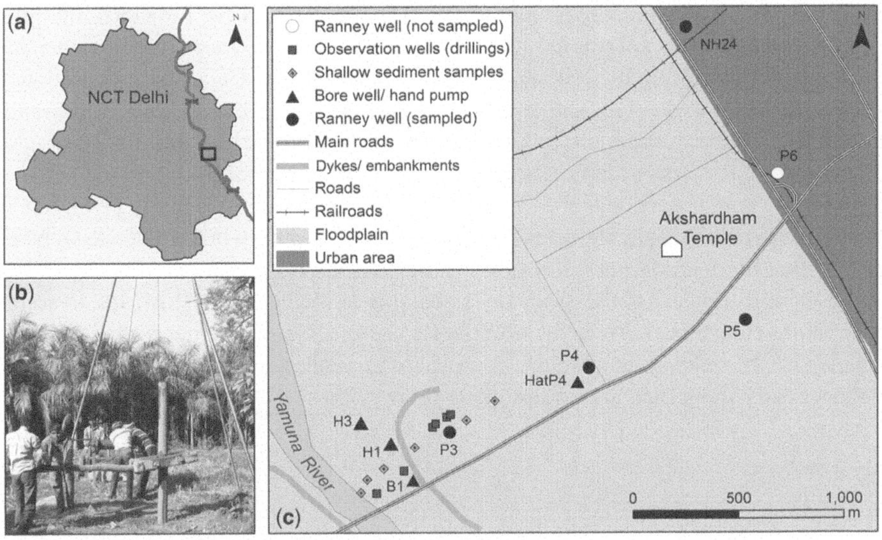

Figure 14.2 (a) Location of the study area (b) Drilling of observation well (c) Location of the hand pumps, Ranney wells and of the shallow and deep drillings conducted during the field work (modified after Groeschke, 2013). Detailed explanation can be found in Chapter 4.

Problems to be solved

The floodplain aquifer is the aquifer with the highest fresh water potential in Delhi (Kumar *et al.*, 2006). However, high ammonium concentrations were found in the river water and in the aquifer close to the river. In the river water ammonium concentrations up to 20 mg/L were measured during the Saph Pani sampling campaigns and concentrations up to 33.3 mg/L were reported by the Central Pollution Control Board (2006). In the sampling points B1, H1, and H3 high variations in ammonium concentrations between 4.5 mg/L and 35 mg/L were found. While 2012 the development of concentrations was similar in the three sampling points, this was not the case in 2013 (Groeschke *et al.*, *submitted a*). In the Ranney well P3, at a distance of 500 m from the river, an increase of ammonium concentrations has been observed for the past years. In 2012 and 2013, the concentrations varied between 5.5 and 8 mg/L and the well was not used for the production of drinking water.

In wells further away from the riverbank, ammonium concentrations remained below 1.7 mg/L in both years. Ammonium concentrations in December 2013 are shown in Figure 14.3. Due to the high ammonium concentrations, the aquifer might be only of limited use for the production of safe drinking water in the future if no appropriate post-treatment or remediation concept is installed. For this, the prediction of future ammonium concentrations is of utmost importance.

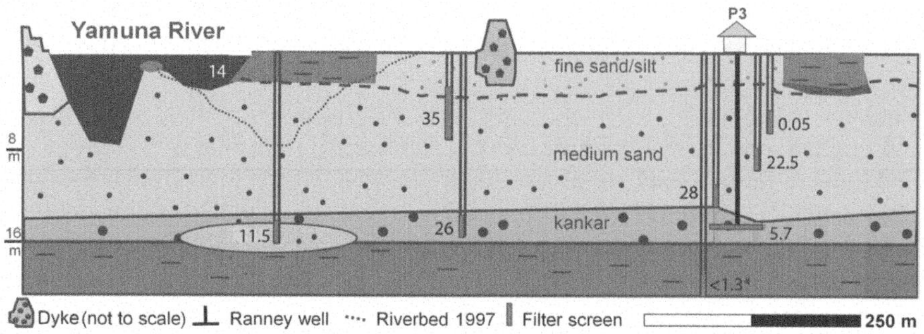

Figure 14.3 Ammonium concentrations in mg/L in December 2013. *Observation well not sampled in December 2013 but all previous concentrations were below 1.3 mg/L. Cross section modified after Groeschke *et al.* (*submitted a*).

Various processes influence the transport and fate of ammonium in an aquifer. Due to interactions with the sediment particle surfaces (cation exchange), ammonium does not move with linear groundwater flow velocity but is retarded. The retardation of ammonium strongly depends on site-specific sediment characteristics and therefore transport cannot be predicted using the retardation factors already published, which vary in magnitudes between 10^0 and 10^2 for sands and gravel, depending on the clay content and the feed concentrations of ammonium (Buss *et al.* 2003). Furthermore, the presence of the reduced nitrogen species ammonium is strongly dependent on the redox conditions in the aquifer. Under oxic conditions it can be biologically oxidised to nitrate in the process of nitrification. Under anoxic conditions it can be also oxidised in the anammox process if nitrate or nitrite are present as electron acceptors (van de Graaf *et al.*, 1995; Clark *et al.*, 2008). In addition, the irreversible fixation of ammonium in clay minerals could also occur as well as the mineralization of organic N as an additional source of ammonium. A detailed description of the redox states, occurrence and effects of nitrogen is given in Chapter 4.

Laboratory column experiments show that fixation or degradation of ammonium takes place to some extent in the sediments of the unsaturated zone and that no degradation takes place in the saturated zone under suboxic and anoxic conditions (see Chapter 4). These results give indications as to the processes occurring at the field site. However, to completely understand the developments of the ammonium concentrations, especially the strong variations, it is necessary to set up a 2D or 3D reactive transport model of the field site. In order to be able to set up such a model, several small scale modelling approaches were applied to determine the necessary input data and to test different hypotheses.

Tools and modelling strategy

The following modelling techniques were applied to gain a better understanding of the processes occurring in the columns and at the field site:

- Inverse geochemical modelling to determine precipitation and dissolution processes occurring during infiltration
- 2D flow and nonreactive transport modelling of column experiments to determine transport parameters of the different lithological units
- 2D and 1D reactive transport modelling of column experiments implementing cation exchange by adapting selectivity coefficients
- 2D and 1D reactive transport modelling of column experiments adding a nonreactive tracer to determine retardation factors
- 1D modelling of two 500 m flow paths

Inverse modelling was conducted with PHREEQC v3 (Parkhurst & Appelo, 2013) to identify reactions which can explain the evolution of the water composition from infiltration to the wells. A sample from the Yamuna River taken in October 2012 was used as the initial water composition and a sample from bore well B1, taken in June 2013, was used as the final solution. Calcite, clay minerals, iron bearing minerals (iron hydroxides, iron sulphides), organic matter, and the exchange species NH_4X, NaX, KX, MgX_2, CaX_2 were included as potential reacting phases (X is a cation exchanger like clay minerals).

A travel time of approximately eight months for the distance of 250 m is in accordance with the average linear groundwater velocity of 0.9 m/d published for this field site (Sprenger, 2011).

A *2D flow and nonreactive transport model* was developed to determine the effective porosities and the dispersivities of the different sediments. The flow simulations were carried out with MODFLOW and the advective-dispersive transport of the NaCl tracer was simulated with the transport simulator MT3DMS and additionally with PHT3D. Tracer breakthrough curves were fitted by adjusting dispersivities and effective porosities, taking into account measured total porosities and literature values (e.g. Johnson, 1967). To ensure that no numerical dispersion or oscillations occurred, the simulations were run with TVD and MMOC solver and selected models were furthermore rerun with smaller grid spacing.

Reactive transport modelling with adapted selectivity coefficients was carried out in 2D and 1D. Using the transport parameters determined with the non-reactive tracer modelling, flow and reactive transport models were developed with MODFLOW and PHT3D to simulate the adsorption and desorption experiments. Many investigations show that at contaminant sites, where the infiltrating water is strongly influenced by one contaminant, simple sorption isotherm models are insufficient to describe the ammonium behaviour at field scale (Buss *et al.*, 2003). Ion exchange models, which consider all species that compete for the exchange sites give better results (Hamann, 2009). Therefore, reactive transport models were developed which consider ion exchange of all the main cations present (Haerens *et al.*, 2002). The reactive transport was computed with PHT3D using the Amm.dat database provided with the software PHREEQC v2. The Amm.dat database decouples ammonium from the nitrogen system, which means that no oxidation of ammonium can occur in the model, which is in accordance with the experimental results showing no oxidation of ammonium to nitrite and nitrite at significant levels (Groeschke *et al., submitted b*). The cation exchange selectivity coefficient is the relative preference of an exchanger to adsorb different cations. It is not a thermodynamic constant, but varies with the exchanger composition (e.g. Tournassat *et al.*, 2007 after Jensen, 1973). For the exchanger phases of the three sediment types (sand, sand with kankar, kankar), equilibrium equations for Na/K, Na/Mg, Na/Ca, Na/NH_4 were set up using the Gaines Thomas convention (Gaines & Thomas, 1953) and measurements of the cation compositions on the exchanger as well as the corresponding activities in groundwater samples.

Retardation factors for ammonium were determined simulating a conservative tracer test by adding a non-reactive, conservative tracer to the reactive transport model (Groeschke *et al., submitted b*). The retardation factors of NH4 were then calculated from the modelled conservative tracer. Also,NH4 breakthrough curves from the time required for the ammonium to reach a relative concentration (C/C_0) of 0.5 at the outlet of the column were compared to the time required for the tracer to reach $C/C_0 = 0.5$ (Steefel *et al.*, 2003), whereby the conservative tracer represents the velocity of the water (Appelo & Postma, 2007 after Sillén, 1951).

1D modelling of flow paths was applied to determine how long it would take to flush out the ammonium from the 500 m wide strip near the river. Detailed description of the model set-up is given in Chapter 4.

Outcome, added value and perspectives

With the help of 2D models, the transport parameters of the two main lithological units of the aquifer (sand and kankar) as well as a transitional unit (sand with kankar) were determined. Reactive transport models were set-up using adapted selectivity coefficients for the different lithological units. This modelling approach gives good results with respect to the development of ammonium concentrations as well as the development of the concentrations of the main cations. In the sand it takes about 10–12 flushes until the 100% ammonium breakthrough is reached and in the kankar it takes about 30–35 flushed pore volumes (Groeschke *et al., submitted b*). The measured and modelled results of one sand and one kankar column experiment are shown in Figure 14.4. Retardation factors were determined by adding a non-reactive tracer to the models; resulting factors are higher than retardation factors published previously for sand and gravel aquifers- between 6.7 and 19.8 (Groeschke *et al., submitted b*) as opposed to between 2.8 and 6.4 (Böhlke *et al.*, 2006). Using the information from all modelling steps, two simplified 1D flow paths from the river were modelled, one in the gravel and one in the kankar layer. Under conditions where only cation exchange occurs and no oxidation of ammonium takes place, it would take 19 years to flush the ammonium from the sand layer and 61 years to flush the ammonium from the kankar layer, provided that the assumed linear flow velocity is accurate.

Conclusions

The enhancement of the river water quality by effective sewage treatment is essential for long term improvement of the groundwater quality in the floodplain aquifer. However, even if this would occur on short term, it would still be a long-term measure due to the strong retardation of ammonium in the aquifer. Therefore, remediation measures or adapted post treatment have to be installed as short and medium term measures, especially as the ammonium concentrations in well P3 will further increase. For this it is essential to know how the ammonium concentrations will develop in the future and if the ammonium plume will reach the drinking water production wells farther away from the river. To obtain more detailed predictions, the information obtained with the modelling techniques described above have to be used to set up a 2D or 3D hydrogeological flow and (reactive) transport models for the floodplain and different source water qualities and pumping scenarios have to be considered.

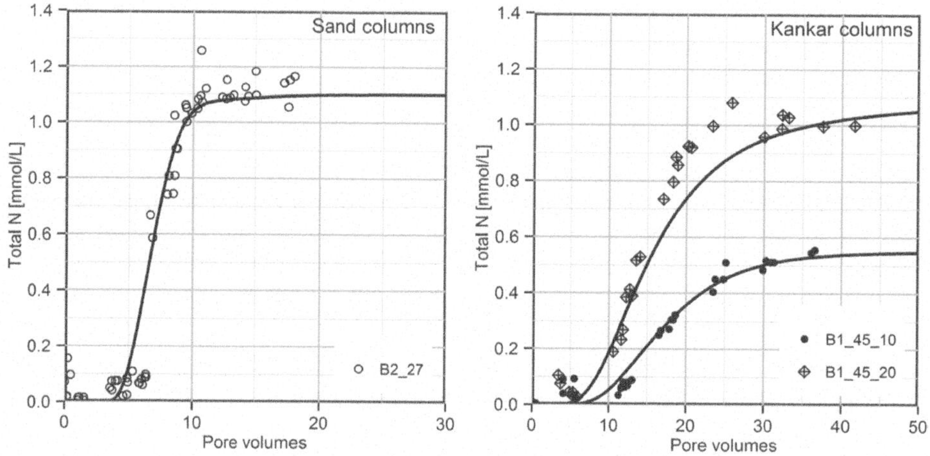

Figure 14.4 Measured (symbols) and modelled (lines) data of selected adsorption experiments. Note the different scales on the x-axes. The breakthrough of ammonium occurs after 12 pore volumes in the sand and after 30–35 pore volumes in the kankar. Modified after Groeschke *et al.* (*submitted b*).

14.3 MODELLING OF MANAGED AQUIFER RECHARGE (MAR)

Managed Aquifer Recharge (MAR) is a method to enhance groundwater quantity and, particularly when combined with Soil-Aquifer-Treatment (SAT), groundwater quality, through the implementation of different types of structures. MAR structures include aquifer storage and recovery (ASR), aquifer storage, transfer and recovery (ASTR), infiltration ponds, infiltration galleries, SAT, percolation tanks and check dams. One important aspect of water quality improvement is also the remediation of saline intrusion into coastal aquifers through increased groundwater recharge upstream in the watershed or through MAR systems (injection well galleries) at the very limit of the salt water wedge.

A thorough understanding of hydrodynamics, at local and watershed scale, is crucial for the selection of the type, dimension and location of MAR structures within a watershed. Flow and transport models, established for those different scales, are important tools to estimate the benefits of MAR-SAT systems for water quality and quantity before implementation and to optimise existing structures. This sub-chapter will look in more details on modelling of a coastal watershed in Tamil Nadu, impacted by over exploitation and saline intrusion (Chennai case study) and on MAR implementation in a watershed typical for Central India with crystalline fractured bedrock overlain by a more or less porous weathering zone (saprolite).

14.3.1 MAR in a coastal aquifer affected by seawater intrusion: Chennai, Tamil Nadu

Site description

The Arani and Koratalaiyar (A–K) basin is located around 40 km north of Chennai. Surface water from reservoirs and groundwater mostly from well fields of this watershed are one of the major sources for the Chennai city water supply. Excessive and heavy pumping of groundwater from the A–K basin, tidal water ingress, relatively low recharge, generally poor land and water management are the most obvious causes for seawater intrusion. Artificial recharge methods include rainwater harvesting, construction of infiltration wells, percolation tanks, recharge pits and shafts, managing runoff water and facilitating utmost recharge (Asano, 1985). Several check dams were constructed across the Arani and Koratalaiyar rivers to mitigate the problem of seawater intrusion by increasing the groundwater recharge.

The study area comprises two non-perennial river basins Arani-Koratalaiyar (A–K) which are flowing through north of Chennai. The rivers generally flow only for few days during the north east monsoon (November–January). A very dry period occurs in this region during April to May when the temperature rises above 45°C. A colder (winter) period occurs during November to January, experiencing an average temperature of 25°C. The average annual rainfall is around 1200 mm, 35% falling in the south west monsoon (June–September) and 60% during the north east monsoon (October–December). Modelling work has been carried out for an area of 1,455 km^2 in a part of A–K river basin. The Eastern model boundary is delimited by the Bay of Bengal and the south western side is bounded by the Palar River. The elevation in the model area ranges from sea level in the eastern side to 130 m above mean sea level in south west side as observed in the survey of India toposheet. Groundwater has being exploited for the purpose of agricultural and Chennai city water supply. Five well fields were constructed to withdraw groundwater to supply the city with water (Figure 14.5).

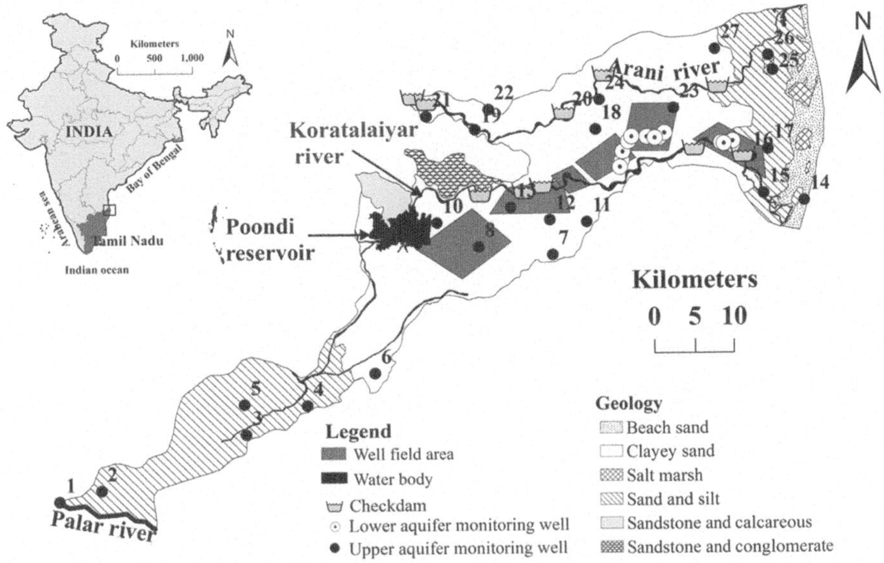

Figure 14.5 Geology of the study area (adapted from Rajaveni *et al.*, 2014a).

Geology and hydrogeology

In this area, the basement Archaean rocks are overlaid by boulders, clay, shale and sandstone of Mesozoic age, the stratigraphic succession of the geologic formations is given in Table 14.1 (UNDP, 1987). The geological outcrop of the A-K basin is shown in Figure 14.5.

Table 14.1 Stratigraphic succession of the geological formation (after UNDP, 1987).

Stratigraphic Age and Thickness	Geological Description
Quaternary (up to 40 m)	Fine to coarse sand, gravel, laterite
Tertiary (45–50 m)	Shale, clay and sand stone
Mesozoic	Gondwana shale and clay
Archaean	Crystalline rocks

The main aquifer in the area is the quaternary alluvium and predominantly consists of fine grained material, reflecting a buried channel system. The subsurface lithology has been characterized by boreholes with depths of 50 m thickness penetrating the coastal alluvium with thicknesses up to 35 m. Groundwater in the area occurs in shallow alluvial zone near the coast and the depth to groundwater level increases with the elevation of the area. The thick clay lenses form a semi confined aquifer system. The groundwater levels in the unconfined aquifer ranges from 2 m to 6 m bgl (below ground level) and in semi confined aquifer it ranges from 14 m to 20 m bgl. A west to east geological cross section is given in Figure 14.6 (A–A′).

Problems to be solved

The A–K basin is characterized by severe over-extraction of groundwater for agricultural activities and water supply to the Chennai metropolitan area, which has been identified to cause significant seawater intrusion (UNDP, 1987). Numerical modelling can help to analyse seawater intrusion by using models to simulate different pumping conditions and quantifying the effect of possible mitigation measures.

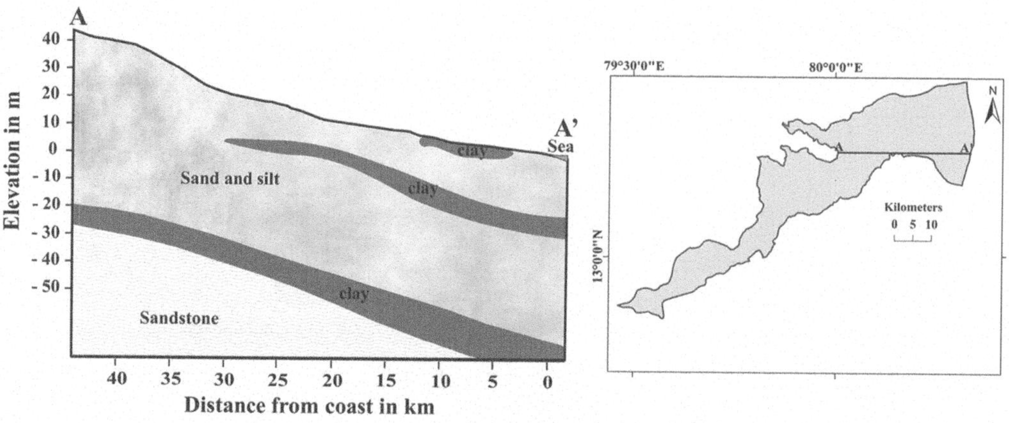

Figure 14.6 Geological cross section along A–A′ line (right figure).

The general objectives are:

- Simulate the current seawater intrusion.
- Representation of check dams and potential other artificial recharge structures in the model to predict future seawater intrusion and to analyse measures to push back the saltwater front.

Tools and modelling strategy

The methods and tools used to generate the coupled model are as follows (Bhola *et al.*, 2014):

1) A rainfall-runoff model (NAM) to produce surface water inflow at the sub-catchment scale as well as the infiltration into the subsoil, integrated in the 1D surface water model.
2) A 1D surface water model (MIKE 11) for the two rivers Arani and Koratalaiyar.
3) A 3D groundwater model (FEFLOW) for the alluvial aquifers of A–K basin which is coupled to the MIKE 11 model using the coupling interface IfmMIKE11 (Monninkhoff, 2011), to describe the interaction between the groundwater and surface water in detail.

Outcome, added value and perspectives

The NAM model parameters were calibrated and extended homogeneously over the entire A–K basin. Since the model was calibrated for an eight year time period, it covers a wide range of hydrologic and climate conditions, which builds confidence in the model's ability to predict stream flow conditions under a variety of scenarios. The model gives a satisfactory comparison with observed flow records with an R^2 value of 0.6. Main focus was given to achieving least volume and peak errors (Figure 14.7). The model over-predicts the total volume in eight years by 10.5%, and a peak error of almost 7% for a discharge greater than 300 m^3/day. The NAM model does not predict low flow accurately due to high surface and root zone storage coefficients. These coefficients define the water holding capacity of the soil, i.e. overland flow will occur once the rainfall is greater than the thresholds of these coefficients. In the observed discharge records, it was found that the response of a rainfall that results in runoff is relatively high and therefore it was implemented accordingly in the model.

Groundwater model The model was calibrated in two stages, steady and transient state condition. The steady state calibration was carried out to achieve an average match between the available observed and simulated groundwater heads and to define a suitable distribution of conductivities. The transient state was carried out for a period of 13 years from January 1996 to March 2009 (Rajaveni *et al.*, *submitted*). Basically transient state calibration was conducted by adapting local conductivities and porosities until the best fit curve was obtained for observed and simulated groundwater heads. A R-square value of 0.901 was obtained during steady state calibration. In the transient state calibration, the simulated groundwater heads were accurately describing the groundwater dynamics of the observed groundwater head in most of the wells. The observed and simulated groundwater head variations in the transient state calibration are exemplarily shown in Figure 14.8 for one of the observation wells.

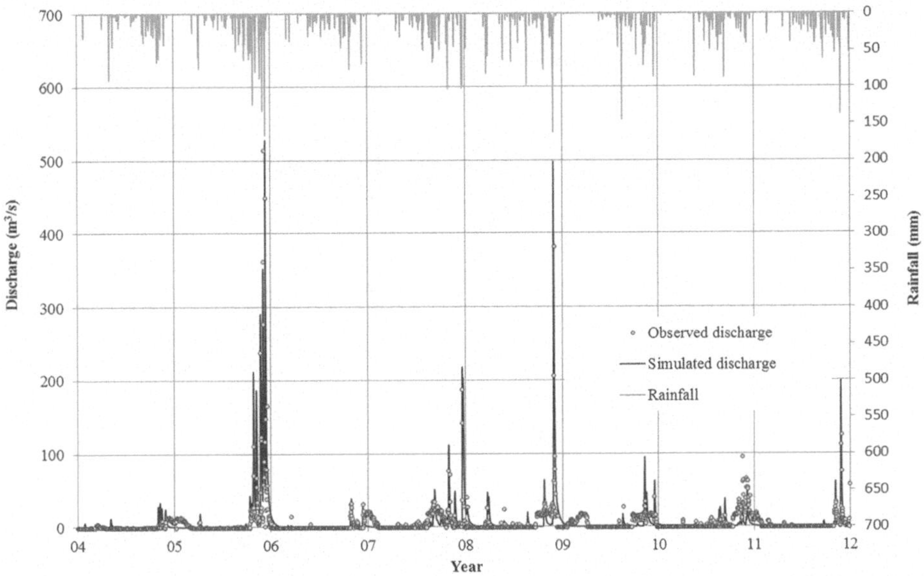

Figure 14.7 Comparison of observed and simulated discharge from 2004 to 2012 at the inlet of Poondi reservoir (Bhola *et al.*, 2014).

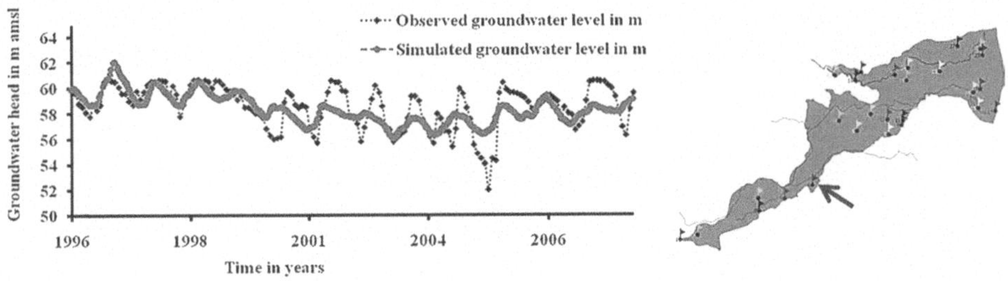

Figure 14.8 Observed and simulated groundwater head variations during transient state calibration in the single aquifer system.

Density dependent model Density dependent parameters were applied to the uncoupled 3D-groundwater model, to report fresh water-seawater interactions. The hydraulic head boundary condition (BC) at the Bay of Bengal was assigned as saltwater head BC. Mass concentration BC was assigned as 500 mg/L in the existing hydraulic head location and 35,000 mg/L in the eastern boundary (coast). An initial mass-concentration distribution was defined, according to the range used in the boundary conditions. To avoid numerical instabilities, the mesh was refined in the coastal region (high density gradient area), which increased the total number of elements from 1 to 1.5 million (Rajaveni *et al.*, *submitted*). An uncoupled density dependent seawater intrusion was simulated and the result shows seawater has intruded from 3.5 km in the year May 1997 to 7 km during May 2003 (Figure 14.9).

Principle simulation of the effect of MAR The general aim of this study is to improve groundwater quantity and quality through MAR structures. As a first step the calibrated 3D groundwater model was used to evaluate the effect of recharge from MAR structures on groundwater heads in the basin. A total of 9 check dams, 4 in the Arani River and 5 in the Koratalaiyar River, existed during 1996 in the study area and were implemented in the uncoupled model. The effect of check dams was computed and predicted by representing the check dams as a fluid transfer BC with different realistic time series (Rajaveni *et al.*, *submitted*). Groundwater head variations were simulated under 2 scenarios (i) with and (ii) without check dams. Observation well 10 has been chosen to explain this study since this observation well is located at the centre of the modelled area. Figure 14.10 shows a maximum of 2 m increase in groundwater heads with the implementation of check dams in the model at this location. The highest differences can be observed during monsoon seasons. During non-monsoon seasons the groundwater head at this location will eventually reach the level representing the situation without check dams.

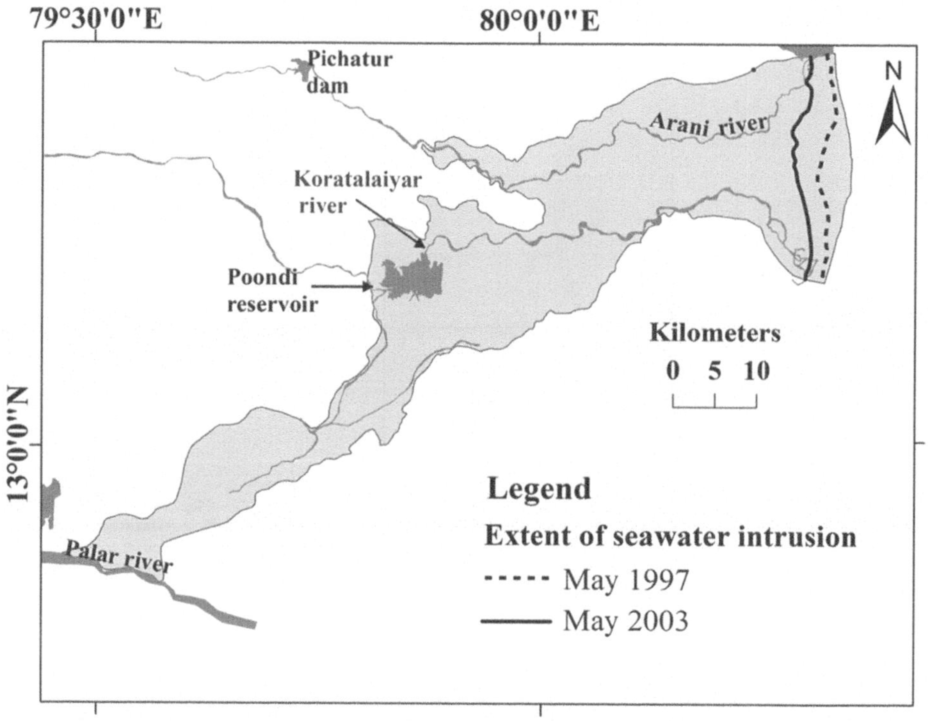

Figure 14.9 Extent of seawater intrusion for two time periods; May 1997 and May 2003 (Rajaveni *et al., submitted*).

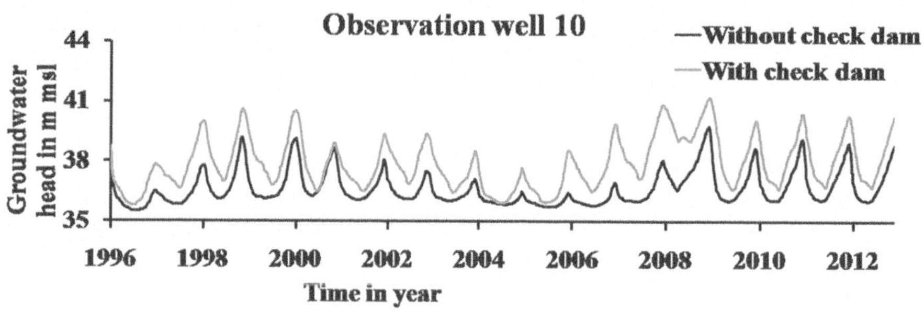

Figure 14.10 Groundwater head variations with and without check dam at the observation well 10 which is located near Poondi.

Coupled surface and groundwater model An uncoupled groundwater flow model was coupled with the surface water model MIKE11 and simulations were performed for the period 2004 to 2009. Three scenarios were simulated: one scenario without check dams, scenario 1 considering most of the existing check dams until present day (9 in total) and scenario 2 with three additional check dams (12 in total) as well as an increased dam crest of about1 m at the already existing check dams (Rajaveni *et al., submitted*).

Figure 14.11 compares the simulated groundwater head along the Arani River, at locations close to the implemented check dams in scenario 2. The last figure (bottom) also includes the simulated water level in the check dam, calculated by the coupled surface water model MIKE11.The following table gives an overview of the situation at 4 selected locations in Figure 14.11. The scenario without check dams represents a situation with approximated natural river courses.

At location A1 there is an existing check dam in scenario 1 and in scenario 2 the same check dam has been raised by 1 m (Rajaveni *et al., submitted*). As expected, the results show an increase in groundwater levels through the implementation of the check dam in scenario 1 and a further, though less significant increase in groundwater heads by raising the dam wall in scenario 2.

Table 14.2 Overview of the scenario definitions at 4 selected locations at the Arani River.

Location	Scenario without Check Dams	Scenario 1	Scenario 2
A1	No check dam	Check dam	Check dam raised
A2	No check dam	No check dam	Check dam implemented
A3	No check dam	Check dam	Check dam raised
A4	No check dam	Check dam	Check dam raised

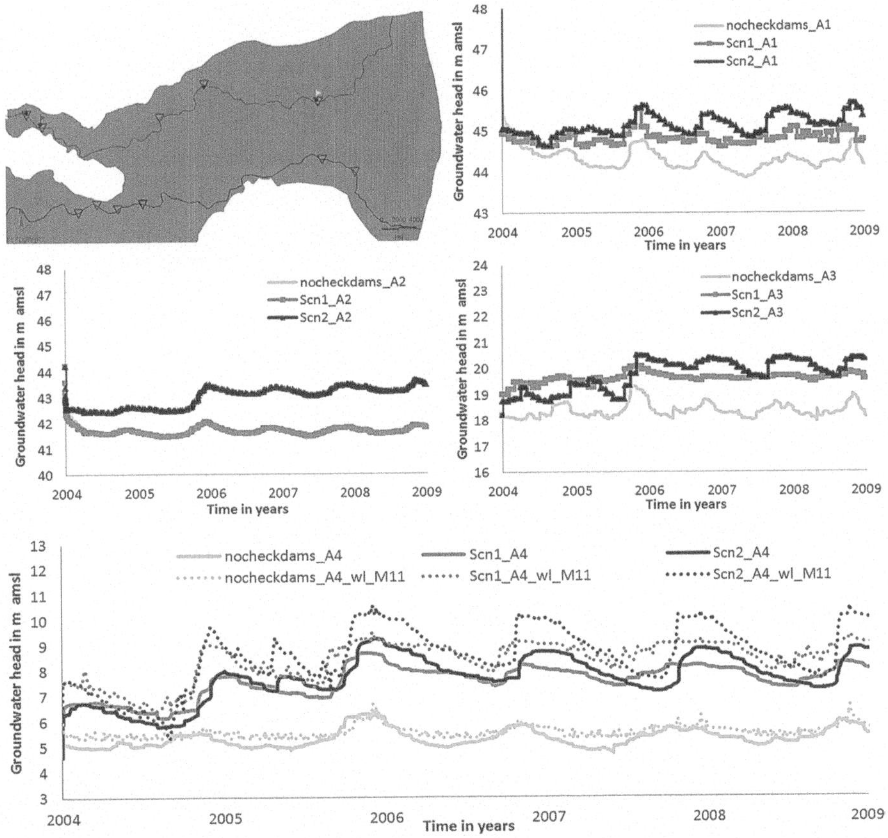

Figure 14.11 Comparison of groundwater heads in the vicinity of existing and plannedcheck damsalongthe Araniyar River for different scenarios (Rajaveni *et al.*, *submitted*).

Scenario 1 has no check dam at location A2, leading to no difference compared to the scenario without check dams. Higher groundwater heads were obtained through the additional check dam in scenario 2.

For location A3 both scenario 1 and 2 have a check dam implemented, while the check dam in scenario 2 has been raised (similar to location A1). The simulation results show that the check dam in scenario 1 increased the groundwater heads and an additional increase in levels can be obtained by constructing a higher check dam wall. At the beginning of the simulation, scenario 1 has slightly higher groundwater heads at this location which is related to the fact that the water is not retained upstream in this scenario since there is no check dam between location A1 and A3 in this simulation. In scenario 2, however, there are two check dams which can retain the water along this river stretch, preventing the check dam at location A3 from being continuously refilled (Rajaveni *et al.*, *submitted*). After 2006, the results show that during wet periods the check dam at A3 can also be filled with release water from the newly implemented check dams upstream. The higher crest level in scenario 2 causes higher groundwater heads in that scenario at this location during that period.

At location A4, the situation is identical to A3; a check dam has been implemented in scenario 1 and the dam level has been raised in scenario 2, leading to the highest groundwater heads in scenario 2. A lag was identified between scenario 1 and 2 due to the retention effect of the upstream check dams in scenario 2. This can also be seen in the additionally displayed water levels directly at the check dam (Rajaveni *et al.*, *submitted*).

In summary, the results indicate that additional check dams have a positive (local) effect on the groundwater heads, just as the raising of the dams, though the effect is considerably smaller. The results also show that the implementation of additional check dams can retain water further upstream, possibly leading to a delay or even a lack of groundwater recharge in the downstream part of the catchment.

Conclusions and outlook

An integrated surface and groundwater model using MIKE11 and FEFLOW has been setup and was successfully calibrated. With the model it was possible to display salt water intrusion processes. Using the scenarios presented in this chapter, which show a significant local effect of the MAR structures on groundwater levels, the model is ready to be used to analyse the benefits of MAR structures on the saltwater intrusion process. For this, long-term analyses will be necessary. These simulations could be set up using yearly returning seasonal cycles for climatological conditions as well as natural groundwater recharge conditions. The simulations should cover a period of at least 50 years to analyse the effect of MAR structures on the long-term perspective. Furthermore, the model could also be used to predict the effect of long-term climatological changes.

14.3.2 MAR in a weathered crystalline hardrock aquifer: Maheshwaram, Telangana

Site description

One of the main experimental watersheds relevant for MAR studies in Saph Pani is located around the town of Maheshwaram (Figure 14.12) near Hyderabad, Telangana. With a total area of 53 km^2 and a semi-arid climate, it is situated on a weathered crystalline rock substratum, a geological and climatic context typical for the entire region where the saprolite weathering layer (10–20 m thick) is usually unsaturated. It is a watershed with a high density of groundwater production wells (>700) mostly for paddy irrigation. Changes in land use have occurred since 2006, the new Hyderabad international airport being located less than 10 km away. It is expected to become a peri-urban area in the coming years as significant housing projects are planned. MAR systems exist throughout the watershed in the form of percolation tanks, check dams, defunct dug wells, etc.

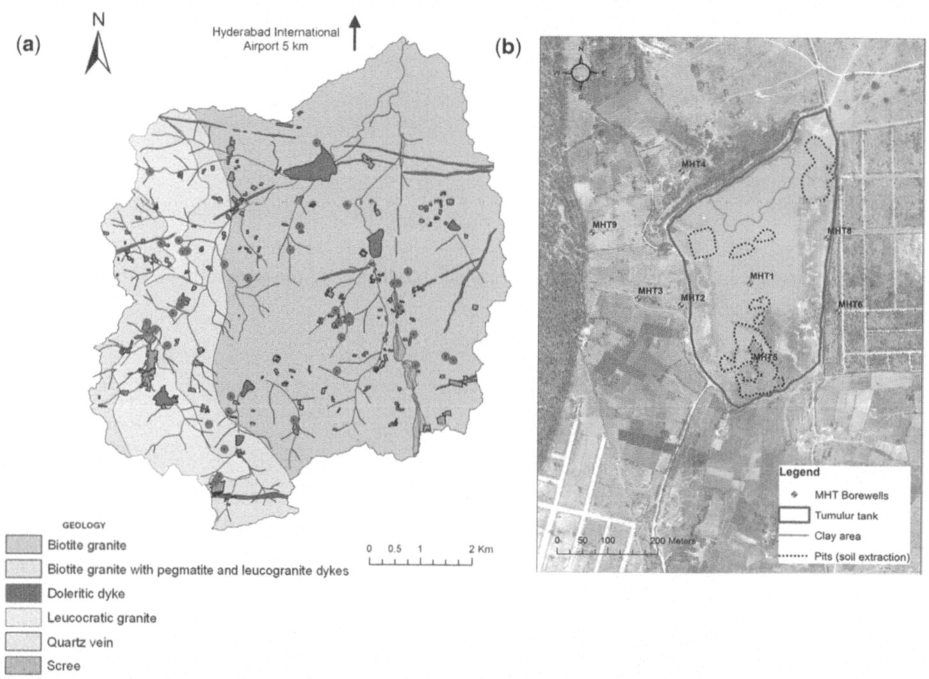

Figure 14.12 (a) Geological map of Maheshwaram watershed. Main MAR (percolation tanks and defunct dug wells) structures are indicated in dark grey. Middle grey areas are irrigated paddy fields. (b) Tumulur tank structure with pumping and observation wells (open circles).

Intensive groundwater exploitation for irrigation has resulted in aquifer over-exploitation and deterioration of groundwater quality (fluoride above maximum permissible limit of 1.2 mg/L (BIS, 10500:2012), salinisation and agricultural inputs). MAR is an attractive concept for groundwater augmentation and enhanced groundwater quality near wells exploited for domestic use.

Problems to be solved

The modelling objective for the case study in Maheshwaram is triple:

1) Develop tools to take into account the highly variable geometry of percolation tanks on weathered crystalline basement rocks under the specific Indian climate (dry season vs. wet season) through a specifically developed module for the 3D finite difference transient groundwater flow model MARTHE (Thiéry, 2010).
2) To assess the influence of percolation tanks on water quantity at local and regional scale.
3) To assess the influence of percolation tanks on crystalline basement rocks on water quality, in particular on fluoride concentrations, triggered by water-rock interactions with fluoride-containing minerals and evaporation together with agricultural backflow on paddy fields.

Tools and modelling strategy

Modelling infiltration from percolation tanks of variable geometry via a partially saturated weathering zone To assess the performance of percolation tanks the three-dimensional, finite difference transient state numerical groundwater code MARTHE was optimized by implementing three-dimensional, non-perennial surface water bodies in continuity with groundwater via an unsaturated zone. Implementation included the spatiotemporal evolution of the natural percolation tanks (i.e. changes in volume and geometry) linked to topography, taking into account heavy rainfalls during monsoon, evapotranspiration, infiltration, runoff, and groundwater dynamics. Part of the rain water stored in such tanks during the monsoon season infiltrates into the soil (variably-saturated media) and reaches the aquifer, while the rest evaporates. Theoretical simulations show that the new developed module "LAC" is able to simulate the relation between surface water and groundwater while respecting the water balance and to assess the highly variable geometry of infiltration tanks over the dry and wet season (Figure 14.13).

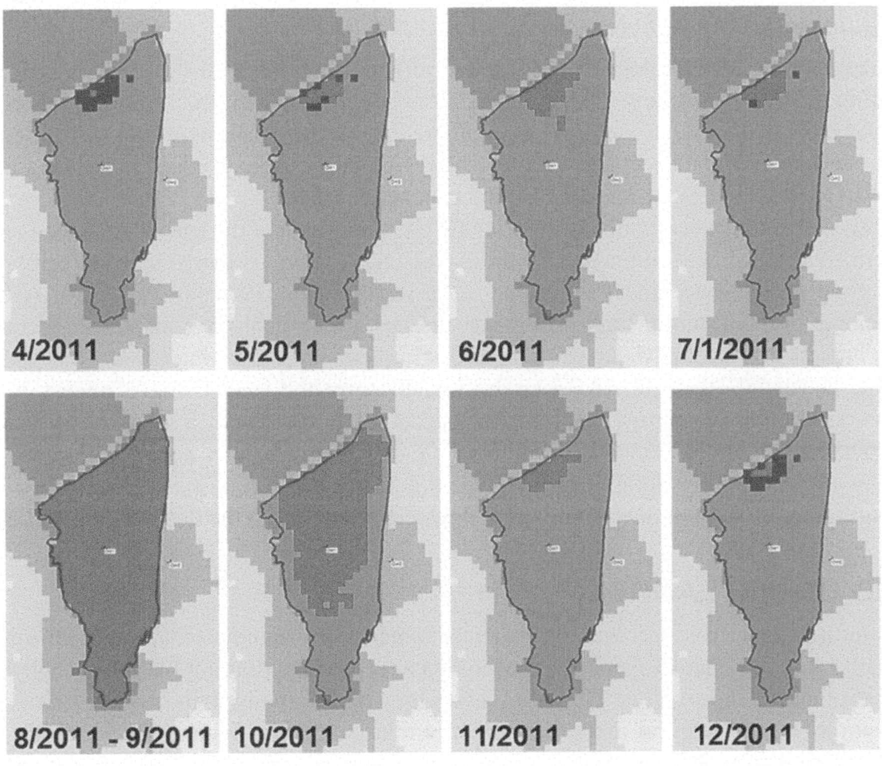

Figure 14.13 Simulated variable extension of the Tumulur tank filling (dark grey) over a monsoon season in 2011, Maheshwaram study site near Hyderabad, Telangana, India.

Modelling influence of percolation tank systems on fluoride concentrations: A geochemical model of solute recycling had been developed previously (Pettenati *et al.*, 2013) for paddy field irrigation using a 1D PHREEQC reactive-transport column (Parkhurst & Appello, 1999). This model was further developed and adapted to the percolation tank problem, on the basis of new monitoring data in order to test the conceptual geochemical model of MAR (Figure 14.14).

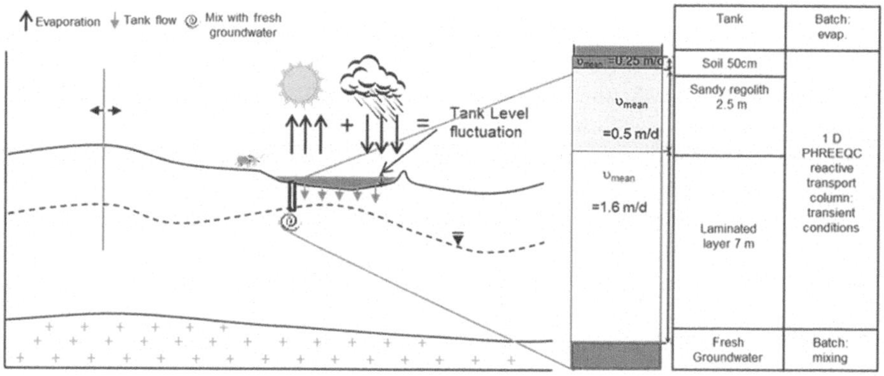

Figure 14.14 Conceptual model of hard-rock aquifer in southern India with Managed Aquifer Recharge (MAR) through an infiltration tank used for the development of a1D Phreeqc reactive column model. v_{mean} is the mean pore flow velocity (Pettenati *et al.*, 2014).

Reactive transport column modelling was performed over a period of 110 days with the calculated pore flow velocity, taking into account the mineral composition of the 3 distinct layers of altered biotite granite (determined by XRD analysis) the Cation Exchange Capacity (CEC) of the weathering profile (determined by cobalt hexamine chloride solution) and the measured initial groundwater composition.

Outcome, added value and perspectives

The 3D MARTHE software was first developed in 1990, and already integrated surface-groundwater flow under varying saturation states including density driven flow (Thiéry, 2010). It is now ready, with the implementation of a specific module for percolation tanks, to be applied to MAR systems on weathered crystalline basement rocks in India and elsewhere. Such massively integrated models are still the exception and will be increasingly used as decision-making tools for assessing the quantitative effects of MAR on groundwater resources at the watershed scale.

The geochemical 1D reactive transport model, using PHREEQC, investigated the role of managed aquifer recharge under variable climatic conditions and its impact on groundwater chemistry. A previous model satisfactorily reproduced the solute behaviour in Maheshwaram groundwater under the influence of paddy fields (Pettenati *et al.*, 2013). Based on that model, the reactive transport model of the Tumulur tank infiltration through the critical zone helps to understand the evolution of fluoride enrichment or depletion in groundwater when MAR is implemented in a watershed. Results of the first scenarios simulation show that the beneficial effect of MAR may be variable over the year, being strongest during monsoon where significant dilution occurs, whereas during the dry period F- accumulation occurs. In sum, the beneficial effects observed during monsoon are countered by the adverse effects during the dry period so that no overall water quality improvement related to the MAR system can be expected at neither the local nor, most likely, at the regional scale. Extrapolating to the regional scale would require integration of 3D groundwater flow approaches with the developed geochemical model.

14.4 MODELLING OF WETLANDS

Natural wetlands play an important role in regulating surface water and groundwater flows within a watershed and also possess a purifying quality through intensive and diverse biological processes, ranging from macrophyte uptake of nutrients and contaminants to microbiological processes. Those processes are voluntarily used and optimised in constructed wetlands. In an intermediate position between natural and engineered systems, there are man-made wetlands used for agriculture, notably paddy fields, with important effects through (1) supplementary water abstraction from the watershed, both from surface water and groundwater, (2) enhanced evaporation, (3) nutrient and trace element uptake by crops and (4) agricultural return flow towards the aquifer. Effects on groundwater quality may be either beneficial (through filtration, water-rock interaction, biological processes in soil, the underlying variably saturated zone and the aquifer, in an analogous way to SAT systems) or, on the contrary, adverse (mainly through evaporation and return flow causing enhanced salinity, trace elements and wastewater-related contaminants and pathogens). In this sub-chapter, we investigate the impact of indirect wastewater recycling for irrigation in a peri-urban watershed in Telangana through an integrated modelling approach.

14.4.1 Integrated modelling of the Musi River Wetlands: Hyderabad, Telangana

Site description

The Musi River is a major tributary of Krishna River, originates in the north west of Hyderabad in Rangareddy district and flows down in a south east direction, passes through Hyderabad city and then joins Krishna River at Wazirabad in Nalgonda District. The Musi River has been intercepted by two major reservoirs, Himayat sagar and Osman sagar, upstream of the city. Below these two reservoirs, the Musi River receives only the city's wastewater and storm water. It receives water from its upper catchment only when excess flood water is released from these reservoirs. The study area lies between coordinates 17° 15′ N, 17° 30′N and 78° 30′E, 78° 45.0′E and includes the villages Peerzadaguda, Kachiwani singram and Mutialguda, situated in peri-urban Hyderabad and on the northern side of the Musi River (Figure 14.15).

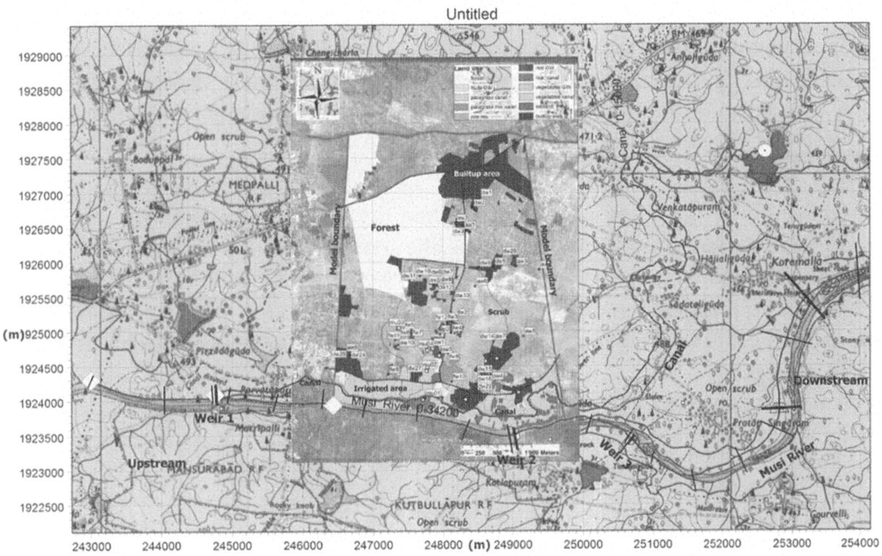

Figure 14.15 The study area east of Hyderabad, showing the Musi River, and the irrigation canal.

The Musi River downstream of Hyderabad has a cascade of overflowing weirs/ponds at which water is diverted into irrigation channels on both sides of the river. The wastewater flow in the river has made it a perennial river which is a significant resource in this semi-arid peri-urban environment where the cultivation of fodder grass, paddy and vegetables has provided economic benefits to many peri-urban inhabitants. Year round cultivation, water in the irrigation canal, overflowing diversion structures (weirs), storage ponds etc. have resulted in arise of the water table and converted the riparian zone along the river into a wetland.

Problems to be solved

Wastewater irrigation vs natural treatment systems Wastewater reuse has become a major area of interest to engineers, biologists, chemists, agronomists, water supply authorities, industries, water resources authorities, etc. Different agencies and stakeholders have different concerns such as prevention of surface water pollution, conservation and recycling of soil nutrients and development of additional water sources for agriculture, industries or non-potable supplies. Irrigation practice with wastewater is one of the reuse options for wastewater. The livelihood and economic activities of peri-urban farmers are the key drivers of wastewater reuse, especially irrigation for agriculture production and, as a secondary advantage, by-products and indirect benefits, for example, cheap water, perennial supply, reduction in surface water pollution, increase in soil nutrient and groundwater recharge (increase in specific yield of underlying aquifer increases). Some obvious downside aspects are soil degradation, degradation of ambient groundwater, cropping pattern change, aesthetics, and health risk for consumers and farmers.

Wetlands "Wetlands are areas where water plays an important role, creating a suitable environment for the associated plant and animal life. They occur where the water table is at or near the surface of the land, or where the land is covered by shallow water" (Ramsar, 2013). Wetlands, whether human made, (i.e. constructed) or purely natural, are also considered to be a cheaper and low-cost alternative technology for wastewater natural treatment. Distribution and differences in the types of natural wetlands are caused by topography, soil, drainage, vegetation, geology, climate, land use, as well as infrastructures like canals, controlled and impounded natural drainages or human-induced disturbances. Water table depth and its temporal variability and movement of water from one level to another through the wetland are the key parameters in characterizing the types and behaviour of a wetland.

Groundwater surface water interactions Surface water and groundwater have often been managed separately by completely different branches of the government. It is now recognized that water resources problems cannot be treated in isolation. Problems like wetland protection or the conjunctive use of surface water and groundwater resources require the integrated management of surface water and groundwater including water chemistry and ecology. Increasingly, water resources are managed on a watershed basis while addressing problems at the local scale. Watershed-based water management systems require new and more sophisticated tools. Traditional groundwater and surface water models were not designed to answer questions related to conjunctive use of groundwater and surface water, water quality impacts of surface water on groundwater, impact of land-use changes and urban development on water resources, and floodplain and wetland management. Instead, fully integrated hydrologic models of the watershed behaviour are required.

Objectives The main objective of the study was to understand the hydrodynamic behaviour of the groundwater-surface water systems under the influence of anthropogenic activities like irrigation, canal construction (seepage), weirs/ponding in the natural drainage of a riverine wastewater impacted (agriculture) wetland. The better understanding of the surface and sub-surface hydrologic processes in an integrated manner will help in assessing the positive and negatives impacts of wastewater irrigation practice on the groundwater and surface water systems. Considering the overall objective of the study dealing with models, it was necessary to understand the movement and exchange of water among the various zones of the system like the overland surface, unsaturated zone (sub surface), aquifer, vegetation and the exchange with surface water bodies (rivers/canals).Therefore, a distributed hydrologic tool, MIKE SHE was selected for carrying out the study.

Tools and modelling strategy

Integrated catchment modelling: application of MIKE SHE

MIKE SHE has been widely used for integrated hydrologic modeling. MIKE SHE's process based framework allows each hydrologic process to be represented according to the problem needs at different spatial and temporal scales. The water movement module of the software has a modular structure which includes six process-oriented components of the hydrological cycle. These are interception/evapotranspiration, overland/channel flow, unsaturated zone, saturated zone, snow melt and the exchange between aquifers and rivers (Figure 14.16) (DHI, 2014). MIKE SHE uses MIKE 11 to simulate channel flow and interact with surface water. MIKE SHE 's strength lies in its feature to provide a simulation of coupled, unsaturated-saturated zones, interaction between evaporation and shallow water tables and a better evapotranspiration module with root zone exchange apart from efficient coupling with open channels.

Modelling strategy In the present case, the area of interest was wastewater irrigated area along the Musi River which includes the river, weirs, cultivation practices, pumping etc. However, in this instance also, it appeared that the model domain needed to be suitably up-scaled to have a realistic groundwater boundary. For this reason, the model catchment was up scaled towards upland on the northern side up to near village Narapalli (Figure 14.16). The main input parameters for the model setup include topography, soils, land use and land cover, natural and canal drainage networks, locations of weirs and their hydraulic parameters, well numbers and locations, agriculture and irrigation data, rainfall, potential evapotranspiration, aquifer parameters etc. These parameters were gathered from field visits, primary survey, monitoring and also taken from secondary sources of research reports conducted in the area. The model domain (12.68 km^2) was divided into 60 m x 60 m cells. The irrigated area inside the model domain is about 1.73 km^2. In the present model, the study area is very small and highly vegetative and in the catchment no stream or ditch of significant size which carries significant surface runoff during dry or even rainfall period is present. However, there was a good number of observed groundwater table data across the model domain; hence the model was calibrated with groundwater depth only. All the processes like overland flow, unsaturated zone flow, saturated zone, evapotranspiration and exchange with surface water were included in the model setup as well as simulations considering their roles in the wastewater irrigation practice as a natural treatment system, i.e., soil-aquifer-treatment. MIKE 11 was setup and simulated as stand-alone including the Musi River, canal and Weir 2 and later on was integrated/ coupled with MIKE SHE. The coupled length of the Musi and the canal with MIKE SHE are 2.28 km and 4.05 km respectively.

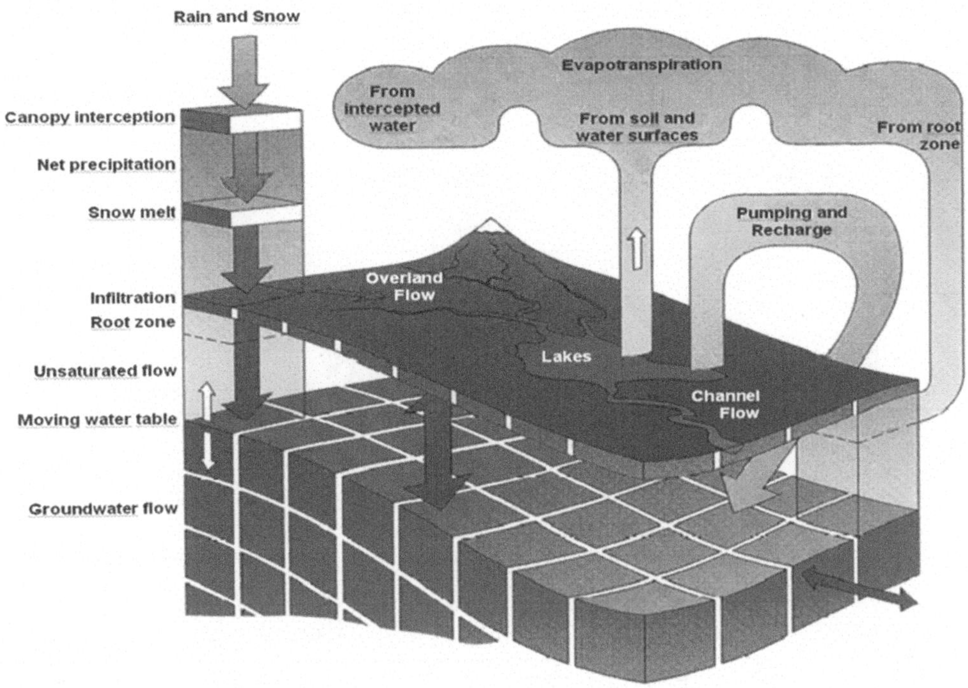

Figure 14.16 Hydrological processes in MIKE SHE (DHI, 2014).

Outcome, added value and perspectives

Two basic scenarios were simulated. The pristine scenario, assuming there are no canals and weirs/ponds across the Musi River and no additional irrigation except rainfall, and the baseline scenario, i.e. the existing condition. Irrigation, weirs and canal seepage have changed the hydrodynamic characteristic of the area; it is functioning like a wetland where there is direct exchange of groundwater with overland flow and the water table is close to the surface. The results shows that the area with a groundwater table within 6 m from the surface, i.e. Wetland, (Ramsar, 2013) has increased from 0.74 km^2 (Pristine condition) to 1.47 km^2 (Baseline) over the years (Figure 14.17).

In order to evaluate wastewater irrigation practices through the purifying action of agricultural return flow through soil, the variably saturated zone above the groundwater level and the aquifer itself as a natural treatment system, the first requirement is to know the movement of water through various zones and to quantify the exchange of water among them through water balance analysis. The MIKE SHE water balance tool provides a detailed account of the water balance. The water balance in terms of mean annual flow (Million cubic meters, Mm^3) including the losses and the return flows from different components of the system is presented in Table 14.3.

Overall the groundwater flow gradient is towards the Musi River and the gradient inverses locally due to pumping. Overland zone and saturated zone are interacting and exchanging water directly which is a typical feature of a wetland. In addition to salinity due to wastewater application salinity occurs because of soil and saturated zone evaporation. Farmers apply water when the deficit reaches around 50% (Maximum allowable deficit = 0.5). Even though wastewater supply is continuous and free, and farmers are conscious of the benefits related to the free nutrients wastewater contains, water application in the area is limited mainly by two factors, (1) the energy needed to lift water from the canal to the upland area and interruptions of power supply and (2) the farmers' fear of unnecessary contact with poor quality water. Thus over-irrigation and pumping occur, especially in the paragrass and vegetable growing areas. In the irrigated area, the consumptive loss is about 25% of total inflow and the return flow is therefore 75%. The modelling confirms the infiltration from Musi to the aquifer in the upstream of the weir where the ponding level is above the groundwater level. Seepage from the canal contributes to a rising water table and return flows. The stretch just downstream of the weir receives water from the aquifer (base flow) and is in gaining state.

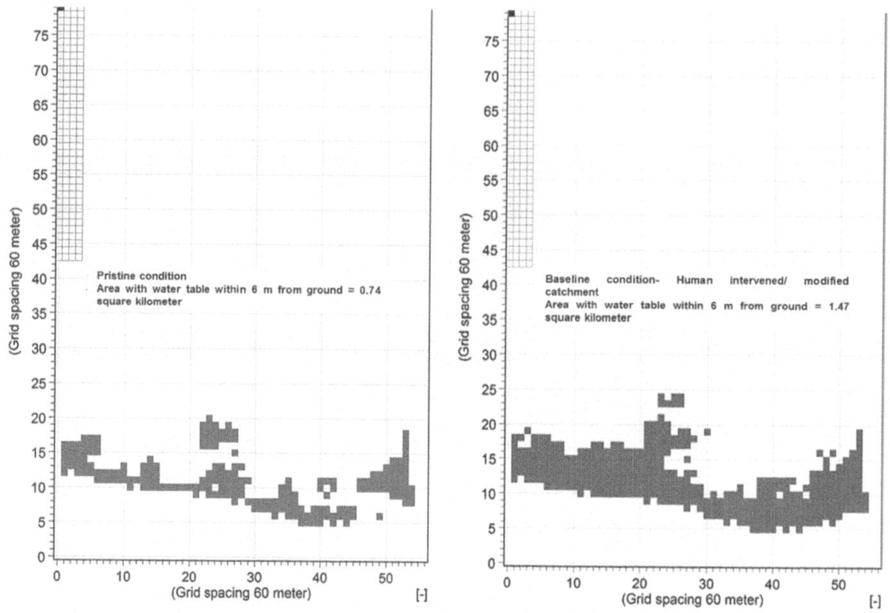

Figure 14.17 Impact of irrigation, canal and weirs on water table: **a.** pristine condition and **b.** baseline condition.

Table 14.3 Detailed mean annual inflows, losses and return flows in million m^3.

Components	Pristine (model Domain –12.68 km^2)	Baseline (Model Domain –12.68 km^2)	Baseline (Irrigated Area – 1.73 km^2)
Inflow			
Precipitation	10.22	10.22	2.14
Irrigation	0.00	10.29	10.17
Infiltration from canal & Musi river	0.00	2.57	2.56
Total	10.22	23.08	14.87
Loss			
Canopy evaporation	0.55	0.62	0.15
Overland evaporation	2.44	5.26	2.95
Soil evaporation	0.71	0.64	0.13
Saturated zone evaporation	0.02	0.03	0.03
Plant transpiration	2.40	2.11	0.48
Total	6.12	8.67	3.73
Return flow			
Direct surface return flow	1.56	7.35	5.73
Sub surface(interflow)	0.16	5.11	4.27
Base flow-through aquifer	1.40	0.92	0.95
Groundwater recovery	0.97	0.97	0.17
Storage	0.00	0.04	–0.01
Total	4.09	14.39	11.10
Error	–0.01	–0.02	–0.03

In addition to other parameters like soil, geology, vegetation, irrigation practice etc. the key system parameters for wastewater irrigation and for the natural treatment related to agricultural return flows through different milieus (soil, variably saturated zone and the groundwater body), are suitable topography and boundary conditions. Human interference in terms of irrigation infrastructures has increased the area of the riverine wetland. Canal seepage and ponding at weirs have made a significant contribution to the rise in water tables. Wastewater application on land has increased salinity (Biggs & Jiang, 2009; Hofstedt, 2005; McCartney *et al.*, 2008; Ensink *et al.*, 2009) but high water tables might also have contributed to soil salinity. However, the positive outcome is that the specific capacities of wells in and around the irrigated area have increased. To protect native groundwater resources outside the wastewater irrigated system (soil aquifer treatment system), the movement of wastewater in the wastewater irrigated area could be managed with well-planned recovery wells (appropriate locations, capacity, types and depths pumping schedules, etc.) and artificial or natural collector drains (appropriate depth, size, locations etc.) Therefore the share of groundwater and (wastewater-containing) surface water used for irrigation should be optimised through watershed–wide, integrated modelling to maximize the benefit and minimize the negative impact of wastewater irrigation practices. Several important agencies need to play an active role in encouraging and regulating wastewater irrigation practice in this area for example the Department of Irrigation and Agriculture and The State Pollution Control Board.

The distributed hydrologic modelling of the Musi wetland using MIKE SHE has demonstrated MIKE SHE's ability to represent complex hydrological systems found within many wetland environments where groundwater and surface water interactions are common hydrological processes. The detailed water balance analysis helps us to understand the movement and quantity of water from one level to other.

14.5 GENERAL CONCLUSIONS

In the light of the modelling exercises applied to the different NTS case studies, the biggest challenge for modelling NTS is model integration. When looking at NTS's like constructed wetlands or percolation tanks (soil-aquifer treatment) we need to take into account surface runoff, the unsaturated soil zone (complex but crucial for water purification), the saturated groundwater flow and even the density-driven saltwater flow in coastal aquifers. Water flow is a continuum but most currently available models are not yet able to treat it as such. One of the major advances in Saph Pani was the establishment of integrated models that take into account the whole water cycle at the watershed scale from surface flows, unsaturated and saturated flows to density driven flows. The project studies also integrated scales: Modelling NTS's needs both, a close look on their behaviour at a very local scale but also upscaling to a watershed scale to simulate effects if a large number of them were implemented. A typical example is percolation tanks. Our observations at the Maheshwaram site showed that their extension in all three dimensions varies widely with rainfall from close-to-nil during the dry season to maximum extension during monsoon. Treating their geometry as constant over time is an oversimplification that can lead to erroneous results if we want to estimate their real impact on groundwater recharge. For this reason, a specific module simulating infiltration was developed for the MARTHE software, already massively integrated with respect to all flow types (surface flow, unsaturated, saturated and density driven flow), able to simulate realistically the behaviour of infiltration tanks from rainfall, evaporation data and surface topography.

Another type of integration that revealed to be a crucial factor was the effects of water flow on water quality changes. Here, the most instructive example from the Saph Pani project is the simulation of ammonium transport from the heavily polluted Yamuna River, across the alluvial aquifer before reaching the wells that pump the river's bank filtrate. Ammonium breakthrough was first measured and modelled at the laboratory scale through percolation experiments in sediment columns and then up scaled to the aquifer scale through reactive transport modelling. An important result is the considerable residence time of several decades of ammonium in the aquifer due to sorption onto the aquifer material.

Models have been developed for all three types of NTS's studied in Saph Pani; managed aquifer recharge combined with soil-aquifer treatment, constructed wetlands and river bank filtration. This has demonstrated how these approaches can be used for understanding, planning and optimising NTSs. The modelling tools used are widespread and accessible (e.g. MODFLOW, FEFLOW, MIKE11, MARTHE, and MIKE-SHE). Even though the application of those tools to the specific problems of NTS implementation in the Indian context requires specialists trained in integrated modelling of complex systems like NTS's on different scales (up to basin scale), the knowledge and knowhow created in the project needs to be transmitted widely to young scientists and engineers through training programmes organised by the Indian institutions that were involved in the development of those methods within Saph Pani.

14.6 REFERENCES

Appelo C. A. J. and Postma D. (2007). Geochemistry, Groundwater and Pollution. 2nd edn, A. A. Balkema Publishers, Leiden.
Asano T. (1985). Artificial Recharge of Groundwater. Butterworth Publishers, Boston.

Bartak R., Page D., Sandhu C., Grischek T., Saini B., Mehrotra I., Jain C. K. and Ghosh N. C. (2014). Application of risk-based assessment and management to riverbank filtration sites in India. *Journal of Water and Health*, doi: 10.2166/wh.2014.075 (available online 31 May 2014).

Bhola P. K., Zabel A. K., Rajaveni S. P., Indu S. N., Monninkhoff B. and Elango L. (2014). Integrated Surface Water and Groundwater Modeling for Optimizing MAR Structures in the Chennai Region. ISMAR 8 conference Managed Aquifer Recharge, Beijing, China.

Biggs T. W. and Jiang B. B. (2009). Soil Salinity and Exchangeable Cations in a Wastewater Irrigated Area, India. *Journal of Environmental Quality*, **38**, May–June.

BIS 10500 (2012). Drinking Water – Specification. Indian standard, 2nd Revision. Bureau of Indian Standards, New Delhi, India.

Böhlke J. K., Smith R. L. and Miller D. N. (2006). Ammonium transport and reaction in contaminated groundwater: Application of isotope tracers and isotope fractionation studies. *Water Resources Research*, **42**(5), W05411.

Buss S. R., Herbert A. W., Morgan P. and Thornton S. F. (2003). Review of Ammonium Attenuation in Soil and Groundwater. Environ. Agency, Almondsbury, United Kingdom.

Central Pollution Control Board (2006). Water Quality Status of Yamuna River 1999–2005, Assessment and Development of River Basin Series: ADSORBS/41/2006–07.

Clark I., Timlin R., Bourbonnais A., Jones K., Lafleur D. and Wickens K. (2008). Origin and Fate of Industrial Ammonium in Anoxic Ground Water – 15N Evidence for Anaerobic Oxidation (Anammox). *Ground Water Monitoring & Remediation*, **28**(3), 73–82.

Dash R. R., Bhanu Prakash E. V. P., Kumar P., Mehrotra I., Sandhu C. and Grischek T. (2010). River bank filtration in Haridwar, India: removal of turbidity, organics and bacteria. *Hydrogeology Journal*, **18**(4), 973–983.

DHI (2014). MIKE SHE Technical Reference, version 2014. DHI Water and Environment, Denmark.

Hofstedt C. (2005). Wastewater use in Agriculture in Andhra Pradesh, India. An Evaluation of Irrigation Water Quality in Reference to associated Health Risks and Agricultural Suitability. Master thesis, Department of Soil Sciences, Swedish University of Agricultural Sciences, Ulls väg 17, SE 756 51 UPPSALA.

Gaines G. L. and Thomas H. C. (1953). Adsorption Studies on Clay Minerals. II. A Formulation of the Thermodynamics of Exchange Adsorption. *The Journal of Chemical Physics*, **21**(4), 714–718.

Gangwar K. K. and Joshi B. D. (2004). A preliminary study on solid waste generation at Har Ki Pauri, Haridwar, around the Ardh-Kumbh period of sacred bathing in the river Ganga in 2004. *Environmentalist*, **28**(3), 297–300.

Government of Delhi (2006). City Development Plan Delhi – chapter 8: Water Supply. Prepared by IL&FS Ecosmart Limited, New Delhi for the Department of Urban Development. http://jnnurm.nic.in/wp-content/uploads/2010/12/CDP_Delhi.pdf (accessed 8 July 2014).

Groeschke M. (2013). Challenges to riverbank filtration in Delhi (India): Elevated ammonium concentrations in the groundwater of an alluvial aquifer. *Zbl. Geol. Paläont*, **1**(1/2), 1–9.

Groeschke, M., Taute, T., Eybing, M. and Schneider, M. (*submitted a*). Assessment of an Ammonium Contamination at a Riverbank Filtration Site in Central Delhi, India.

Groeschke M., Hamann, E., Frommen T., Grützmacher G. and Schneider M. (*submitted b*). Transport of Ammonium in Alluvial Sediments – Column Experiments and Reactive Transport Modelling.

Haerens B., Dassargues A. and Lerner D. N. (2002). Reactive Transport Modelling of Ammonium: 1D conceptual modelling and comparison of reactive transport codes. Acta Universitatis Carolinae. *Geologica*, **46**(2–3), 27–31.

Hamann E. (2009). Reaktive Stofftransportmodellierung Einer Urbanen Grundwasserkontamination Aus Einem ehemaligen Rieselfeld (Reactive Transport Modeling of an Urban Groundwater Contamination Originating from a Former Sewage Farm). PhD thesis, Faculty of Mathematics and Science II, Humboldt Universität zu Berlin, Germany.

Jensen H. E. (1973). Potassium–calcium exchange equilibria on a montmorillonite and a kaolinite clay. *Agrochimica*, **17**, 181–190.

Ensink J. H. J., Scott C. A., Brooker S. and Cairncross S. (2010). Sewage disposal in the Musi-River, India: water quality remediation through irrigation infrastructure. *Irrigation and Drainage Systems*, **24**, 65–77.

Johnson A. I. (1967). Specific Yield – Compilation of Specific Yields for Various Materials. U.S. Geological Survey, Geological Survey Water-Supply Paper 1662.

Kumar M., Ramanathan A. L., Rao M. and Kumar B. (2006). Identification and evaluation of hydrogeochemical processes in the groundwater environment of Delhi, India. *Environmental Geology*, **50**(7), 1025–1039.

Lorenzen G., Sprenger C., Taute T., Pekdeger A., Mittal A. and Massmann G. (2010). Assessment of the potential for bank filtration in a water-stressed megacity (Delhi, India). *Environmental Earth Sciences*, **61**(7), 1419–1434.

Monninkhoff B. (2011). Coupling the Groundwater Model FEFLOW and the Surface Water, IfmMIKE11 2.0 User Manual. DHI-WASY Software.

McCartney M., Scott C. Ensink J., Jiang B. and Biggs T. (2008). Salinity Implications of Wastewater Irrigation in the Musi River Catchment in India. *Cey. J. Sci. (Bio. Sci.)*, **37**(1), 49–59.

Parkhurst D. L. and Appelo C. A. J. (2013). Description of Input and Examples for PHREEQC Version 3–a Computer Program for Speciation, Batch-reaction, One-dimensional Transport, and Inverse Geochemical Calculations.

Parkhurst D. L. and Appelo C. A. J. (1999). User's Guide to PHREEQC (Version 2)- A Computer Program for Speciation, Batch-Reaction, One Dimensional Transport, and Inverse Geochemical Calculation. USGS Water Res. Invest. Rept. 99–4259.

Pettenati M., Picot-Colbeaux G., Thiéry D., Boisson A., Alazard M., Perrin J., Dewandel B., Maréchal J. C., Ahmed S. and Kloppmann W. (2014). Water quality evolution during managed aquifer recharge (MAR) in Indian crystalline basement aquifers: reactive transport modeling in the critical zone. *Procedia Earth and Planetary Science*, **10**, 82–87.

Pettenati M., Perrin J., Pauwels H. and Ahmed S. (2013). Simulating fluoride evolution in groundwater using a reactive multicomponent transient transport model: application to a crystalline aquifer of Southern India. *Appl. Geochem.*, **29**, 102–116.

Rajaveni S. P., Indu S. N. and Elango L. (2014a). Application of remote sensing and GIS techniques for estimation of seasonal groundwater abstraction at Arani-Koratalaiyar river basin, Chennai, Tamil Nadu, India. *International Journal of Earth Sciences and Engineering*, **7(1)**, 248–251.

Rajaveni S. P., Indu S. N., Zabel A. K., Sklorz S., Bhola P., Monninkhoff B. and Elango L. (*submitted*). Coupled Surface Water and Groundwater Model to Identify Methods for Mitigation of Seawater Intrusion. *Submitted* to Water Resources Research.

Ramsar (2013). The Ramsar Convention Manual, 6th edn, Ramsar Convention Secretariat, Gland, Switzerland.

Saph Pani D1.2 (2013). Guidelines for Flood-Risk Management of Bank Filtration Schemes During Monsoon in India. Saph Pani Project Deliverable. http://www.saphpani.eu/downloads (accessed 19 August 2014).

Sillén L. G. (1951). On filtration through a sorbent layer. IV. The ψ-Condition, a Simple Approach to the Theory of Sorption Columns. *Ark. Kemi*, **2**, 477–498.

Sprenger C. (2011). Surface-/Groundwater Interactions Associated with River Bank Filtration in Delhi (India) – Investigation and Modelling of Hydraulic and Hydrochemical Processes. PhD thesis, Hydrogeology Group, Freie Universität Berlin, Germany.

Steefel C. I., Carroll S., Zhao P. and Roberts S. (2003). Cesium migration in Hanford sediment: a multisite cation exchange model based on laboratory transport experiments. *Journal of Contaminant Hydrology*, **67**(1–4), 219–246.

Thiéry D. (2010). Groundwater flow modeling in porous media using MARTHE. In: Modeling Software Volume 5, Tanguy J. M. (ed.), Environmental Hydraulics Series, Editions Wiley/ISTE London, pp. 45–60. ISBN: 978-1-84821-157-5

Tournassat C., Gailhanou H., Crouzet C., Braibant G., Gautier A., Lassin A., Blanc P. and Gaucher E. C. (2007). Two cation exchange models for direct and inverse modelling of solution major cation composition in equilibrium with illite surfaces. *Geochimica et Cosmochimica Acta*, **71**(5), 1098–1114.

UNDP (1987). Hydrogeological and Artificial Recharge Studies, Madras. Technical report, United Nations Department of technical co-operation for development, New York, USA.

Van de Graaf A. A., Mulder A., de Bruijn P., Jetten M. S., Robertson L. A. and Kuenen J. G. (1995). Anaerobic oxidation of ammonium is a biologically mediated process. *Applied and Environmental Microbiology*, **61**(4), 1246–1251.

Chapter 15

Developing integrated management plans for natural treatment systems in urbanised areas – case studies from Hyderabad and Chennai

Priyanie Amerasinghe,Sahebrao Sonkamble, Mahesh Jampani, Md. Wajihuddin, Lakshmanan Elango, Markus Starkl, Sarah Sarah, Md. Fahimuddin and Shakeel Ahmed

15.1 INTRODUCTION

Many natural water systems in India are contaminated due to uncontrolled discharges from cities and industries (CPCB, 2009). As a consequence, hydrological cycles are impacted to varying degrees, and the biggest impact is on potable water sources as well as irrigation waters for food production. The problems are most acute in the urban and peri-urban settings where development is rapid, and combined with water scarcity issues there is a need to look at how to prevent as well as treat and reuse used water for potable as well as non-potable uses. The projected water demand for a rising population will exert more pressure on the natural water supplies, therefore, it is important to treat contaminated waters and explore its safe reuse, especially where water scarcity is experienced. While multiple treatment technologies exist, not all are suitable for a given setting. The nature of treatment will depend on the source and reuse options, therefore, assessment of levels of contamination and the treatment strategy is an important consideration. Of the many technologies that are used for treatment of contaminants, natural treatment systems (NTS) are quite common and cost-effective and are being used world over with considerable success (Arceivala & Asolekar, 2006; Reed *et al.*, 1995; Shipin *et al.*, 2005; Crites *et al.*, 2014).

Existing NTS can play an important role in contaminant attenuation, enhancing its potential use and address the water scarcity issues in a country like India. The natural treatment processes are simple, and offer treatment of polluted waters through a combination of natural soil aquifer processes and plant-root systems (Nema *et al.*, 2001; Kaldec & Wallace, 2009). NTS like soil aquifer treatment (SAT), Managed Aquifer Recharge (MAR) or constructed wetlands (CW) are robust barriers, that can remove multiple contaminants, minimise the use of chemicals, use relatively low energy and have a small carbon footprint (Sharma *et al.*, 2008; Zurita *et al.*, 2009; Missimer *et al.*, 2014), and utilise locally available resources. These systems have been applied for wastewater treatment (as pre-treatment, main treatment or post-treatment) and reused in multiple ways (Sharma *et al.*, 2011).

Thus, the NTS can offer cost-effective means of treating contaminated water and can be integrated into the urban water management cycle to cope with different needs in and around the cities. However, when treated wastewater is integrated into an urban water management cycle, a number of issues have to be considered. Human health, environment, economic and social considerations are some of them. City sanitation plans (CTPs) are geared to address the sanitation related issues and can shed some light on how existing integrated management plans work within the cities. This paper describes a framework for introducing NTS as part of urban water management cycle, addresses some of key issues through the sanitation lens, and discusses how integrated management plans can help relieve some of the pressures of freshwater needs for the cities. The concept is introduced using two case studies from Hyderabad and Chennai, respectively, where NTS have proved to be a useful method for treating water in urban and peri-urban settings, and where the treated water has reuse potential in agriculture and aquifer recharge.

15.2 NATURAL TREATMENT SYSTEMS IN INDIA

A national survey of NTS identified over 108 sites that have operational systems for treatment of water to varying degrees (see Chapter 8). The study showed that most of the operating systems of the NTS were enhanced by the addition of mechanical pre-treatment for the removal of gross solids, especially where sufficient land suitable for the purpose is available. Further, it was observed that these options were cost effective in terms of both construction and operation especially in the urban areas. The operation and maintenance of these systems were managed by communities who were users or agencies, and the involvement was either direct or indirect and included collection, treatment and disposal. The primary aim of the agencies was to improve the sanitation facilities as well as to safeguard human health. Some of the types of examples are hyacinth and duckweed ponds, lemna ponds, fish ponds, waste stabilisation ponds, oxidation ponds and lagoons, algal bacterial ponds, and polishing ponds of Wastewater Treatment Plants (WWTPs). Of these, the most common system was the waste stabilisation pond which accounted for nearly 73% of the cases.

In India, only 37% of the wastewater generated in Class I and II cities and towns is treated (CPCB, 2009). Where the inadequate network coverage is attributed to the poor state of sanitation, there is ample opportunity to set up NTS as decentralised systems to lessen the burden on the larger network systems. It is known that natural water bodies will vary in response to environmental conditions. Therefore, it is important to understand how the systems function in nature, which in turn helps to identify the treatment potential and the fate of contaminants. Once the mode of contaminant removal is identified, it can be engineered to enhance its functions, and become part of a large scale wastewater treatment system. However, this requires a good assessment of the source water, and a well-defined management plan, for collection, treatment and final disposal to the environment. This will help free up the fresh water supplies needed for domestic purposes, and support the reuse of treated water for non-potable uses. Such a plan can also help network the small-scale businesses and illegal connections, and reduce the indiscriminate discharge to the environment that leads to pollution. While the construction aspects are relatively easy to deal with, its sustainable management requires a good plan that involves the communities and the responsible stakeholders. When such an alternative water supply is added to a total water management plan of a city, the largest beneficiaries of the treated water will be the industry, agriculture and city landscapers. NTS, can be cost effective and can be modelled to suit a setting, which is manageable and involves communities or public-private partnerships for sustainable solutions. Further, it helps environmental experts and policy makers to define legislation with the intention that the water is supplied and maintained at an appropriate quality for its identified use.

15.3 POLLUTION REDUCTION – CITY SANITATION PLANS

Sanitation policies, guidelines and regulatory processes are aimed at improving public health and environmental outcomes. In particular, these help to reduce pollution of water sources and indiscriminate dumping of waste in unauthorised places. The water safety plan, wastewater discharge standards and National Urban Sanitation Policy, which are designed to achieve sanitation targets, provide the necessary backdrop to develop the CSPs. Country-wide coverage has been effected by a directive given by the Ministry of Urban Development in India (MoUD, 2008). The requisite guidelines are given in its policy and sets out the overall strategy to address the proper sanitary disposal of all types of waste and behavioural attitudes of people. Thus, a CSP outlines the strategy of a city to achieve total sanitation status, which encompasses many elements as given in Figure 15.1. For example, a CSP demands multistakeholder participation, public consultation, baseline data collection and sanitation mapping, awareness raising, suitable technology selection and adoption, capacity building and monitoring and evaluation. Institutions like the Administrative Staff College of India (ASCI, n.d) has helped to develop a number of CSPs, where cities have lacked the capacity. It is expected that such a plan would transform urban India to meet international standards of healthy living, enhance tourism for revenue generation, and also improve the health and hygiene standards of people. Cities and towns have been requested to develop CSPs, addressing the technical and non-technical aspects of sanitation services delivery. However, the development of CSPs have not taken off as expected, reflecting in some ways the lack of capacity and training in the relevant institutions that are expected to take on the responsibility. The framework for CSPs demands a good knowledge on urban planning, sanitation, technical infrastructure and financing as well as to cater to the needs of local communities. Participatory development of CSPs is recommended, as planning for sanitation infrastructure alone cannot meet the requirement of achieving a high sanitation status. Further, in the implementation, the cities have to incur capital investments, adjustments of bylaws, strengthen administrative structures and develop capacity for sustainable management.

Thus, the CSPs address both technical and non-technical strategies, which include, the domestic water supply, solid waste management and drainage systems for disposal (MoUD, 2008; WSP, 2010). A gamut of non-technical aspects include policies and regulations, enhancement of institutional capacity, finances, community awareness and participation, private sector engagement, NGO engagement, and monitoring and evaluation. While Hyderabad is yet to develop a CSP, at the state level some guidelines have been established after discussion. The target areas for consideration are water supply (all types of sources, treatment to drinking water standards and distribution), solid waste management, wastewater collection and disposal, and storm water drainage. Further, an overview of the state level sanitation – current situation and challenges – achieving sanitation rankings for the cities in the state (National Award Scheme for Sanitation for Indian Cities, *The Nirmal Shahar Puraskar* – Press Information Bureau, Government of India, 2010), securing funds for sustainable sanitation activities has also been considered.

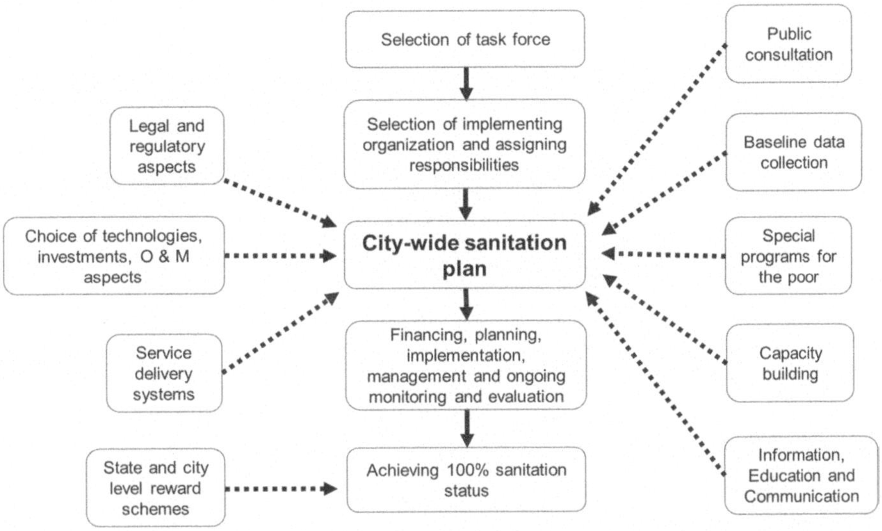

Figure 15.1 Key elements of a City Sanitation Plan (adapted from MoUD, 2008).

It appears that cities and Urban Local Bodies (ULBs) can be well endowed with grants under the 13th Finance Commission recommendations and further enhanced by schemes such as the Jawaharlal Nehru National Urban Renewal Mission, Urban Infrastructure Scheme for Small and Medium Towns, Infrastructure Development Scheme for Satellite Towns, North Eastern Region Urban Development Programme, Backward Region Grant Fund, multilateral and bilateral funds and significant initiatives by States themselves. However, such programs have not been fully utilised to achieve 100% sanitation status, and many cities still lag behind not being able to start up leave alone maintain. While the sanitation ratings were seen as a stimulus to engender change, its intended impact has not been fully achieved. The review on the CSPs show that adoption requires a good action plan, a dedicated set of staff, capacity building, community support, incentives for workers, and innovative ways to deal with each setting. In 2014, yet another program on sanitation was launched by the MoUD titled *Swachh Bharat Abhyan*, targeting 4,041 statutory towns, with objectives similar to that found in the CSPs (MoUD, n.d.). These are elimination of open defecation, eradication of manual scavenging, modern and scientific municipal solid waste management, bring about behavioural change to embrace healthy sanitation practices, awareness building on links between sanitation and public health, capacity augmentation for ULB's, and create an enabling environment for private sector participation in CAPEX (capital expenditure) and OPEX (operation and maintenance). In each state, a designated officer from the MoUD is appointed as the responsible officer to manage the program.

15.4 WATER SUPPLY AND SEWERAGE MANAGEMENT IN HYDERABAD AND CHENNAI

15.4.1 Hyderabad

City water supply

Hyderabad is the fourth most populous city in India with its 6.8 million people. With the recent bifurcation of the former state of Andhra Pradesh, it will serve as the capital for both states for a period of 10 years, after which it will be the capital city of Telangana (Figure 15.2). The city administration is governed by the Greater Hyderabad Municipal Corporation (GHMC), covering an area of 650 km^2. The Hyderabad Metropolitan Water Supply and Sewerage Board (HMWSSB) is responsible for the city water supply and wastewater management. Once dependant on lakes and reservoirs closer to the city, Hyderabad now lifts water from over 114 km to bring Krishna water to the city. This increasing demand has placed pressure on freshwater resources more than ever, and draws attention to the fact the freshwater sources should be protected as much as possible, while treating wastewater for alternative uses. The division of the former state has brought in a separate issue to the forefront, which is sharing the source water from the Krishna river (TToI, 2014a). This places pressures on the city of Hyderabad to clean up some of the lakes and rivers within the state to provide drinking water for its citizens and also water for food production. However, most natural water bodies are polluted including the Musi River, which runs through the city carrying large volumes of wastewater (1.25 million m^3/d – both domestic and industrial) discharged from the city.

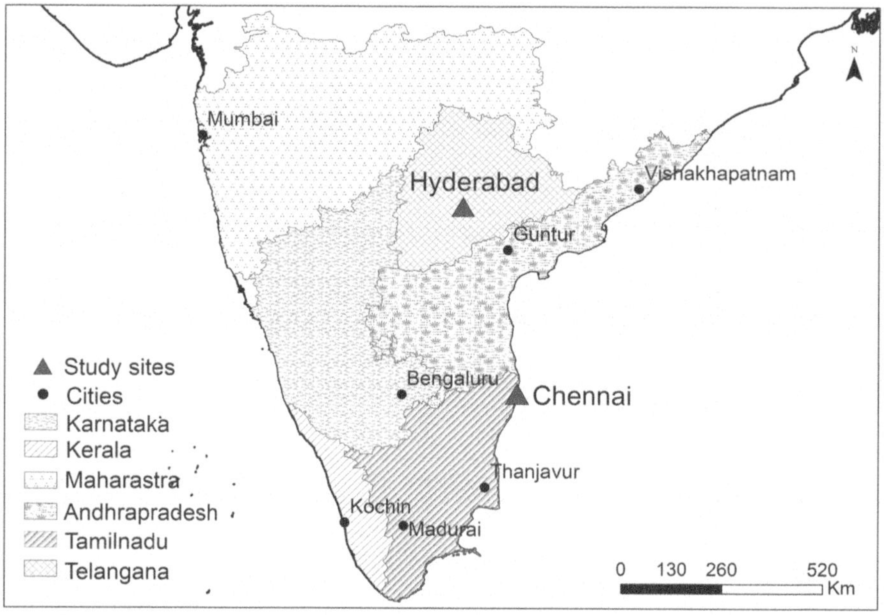

Figure 15.2 Study sites for natural treatment systems in the cities of Hyderabad (Telangana) and Chennai (Tamil Nadu) (Source: IWMI).

Hyderabad receives water from five sources (HMWSSB, 2013), which are Osmansagar on Musi River, Himayatsagar on Esau River, Manjira Barrage on Manjira River, Singur Dam on Manjira River and Krishna River Water (Figure 15.3). The chronology of abstractions from these sources span over the period 1920 to 2008. As against the demand of 2.18–2.27 million m^3/d, HMWSSB is presently supplying water to their full potential at 1.50–1.55 million m^3/d from all the five sources, and in some instances lifting water from a distance of 110 km (Krishna River). With around 0.68 million m^3/d water shortage and an additional demand of 20 per cent, water supply status is far from adequate. While the groundwater does cover the gap to a certain extent (an estimated 40% of the supply is said to be from groundwater), in the drier months the bore wells dry up, which increases the demand for fresh water supplies from HMWSSB (The New Times of India, 2014). The plans for lifting water from Krishna (phase III) and Godavari (Phase I), are underway at a cost of INR 33 per kL and INR 38 per kL respectively (1USD = INR 60). The highly subsidised domestic tariff (INR 26 per kL) indicates the huge gap between the expenditure and revenue of HMWSSB. The loss of revenue is attributed to leakages in old network systems, lack of metering for over 78% of the consumers, illegal connections and inability to issue new connections due to the defict in supply.

Wastewater management and treatment

The wastewater is collected via a sewerage network system and treated at three treatment plants (Table 15.1). The treatment plant at Attapur is yet to be commissioned. The network coverage for conveyance is estimated at only 60–70% therefore, only 52% of the wastewater is treated at these plants that are operational. The rest is discharged into waterways via 19 nalas which ultimately reach the Musi River. Thus, the Musi River, which runs through the city, receives both partially treated and untreated wastewater and immediately downstream it is used for agriculture. The safety concerns for food production are not addressed as farmers lift the water for irrigation and use it as a nutrient rich water supply. This water supply cannot only be easily treated via NTS to a level suitable for irrigation, it can also be used for other uses like gardening, washing cars to relieve the pressure on the domestic water supply. It can also help in improving the sanitation conditions in the city, if it is planned as part of the water supply and wastewater treatment systems.

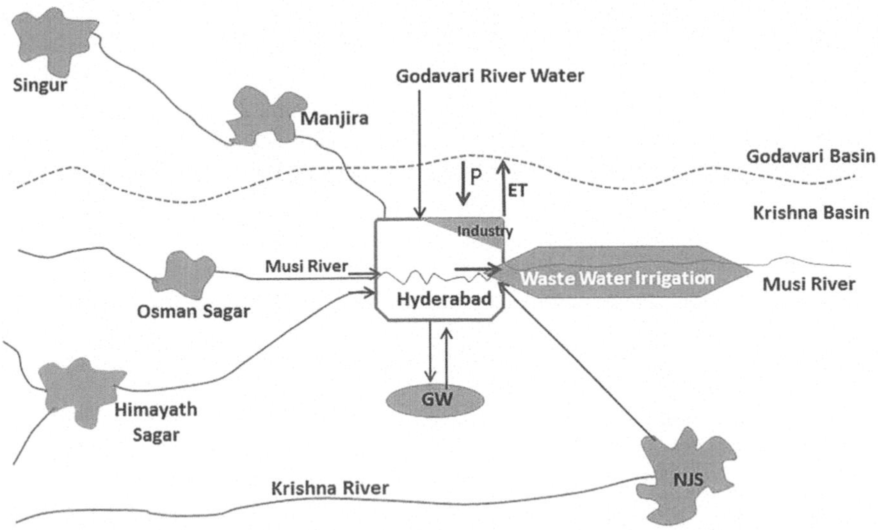

Figure 15.3 Drinking water supply to the city of Hyderabad. GW = groundwater, NJS = Nagarjuna Sagar Reservoir. (Source: Adapted from Van Rooijen *et al.*, 2005).

Table 15.1 Current wastewater treatment capacity of wastewater treatment plants and systems.

Description of Type and Site	Treatment Capacity [1000* m³/d]
Wastewater (HMWSSB)[1]	
Amberpet	339
Nagole	172
Nallacheruvu	30
Attapur	51
Lakes (GHMC)[2]	
Hussain Sagar	20
Lakes (HMDA)[3]	
Patel Cheruvu	2.5
Pedda Cheruvu	10
Durgam Cheruvu	5
Mir Alam Cheruvu	10
Saroor Nagar Lake	2.5
Safil Guda Lake	0.6
LangarHouz Lake	1.2
Noor Mohammad Kunta	4
Ranghadhamini Lake	5
Treated wastewater	652.8
Untreated wastewater[4]	599.2

[1]HMWSSB – Hyderabad Metropolitan Water Supply and Sewerage Board.
[2]GHMC – Greater Hyderabad Municipal Corporation (2014).
[3]HMDA – Hyderabad Metropolitan Development Authority (2014).
[4]Wastewater generated is 80% of water supply to the city.

Wastewater drainage network system coverage in the core city area is estimated to be 80% and also over 30–40 years old (HMWSSB, 2013). Therefore, it is not sufficient to cater to the current wastewater flows. The peripheral city network coverage is about 30%, but new connections get added on periodically. The city needs about 1.3 million m³/d wastewater treatment capacity, but the existing treatment capacity is about 0.7 million m^3/d. The WWTPs depend on a water cess for its maintenance activities which is far from adequate. The high cost of urban land makes new construction very expensive.

Currently, the conventional WWTPs are capable of treating only 52% of the wastewater generated in Hyderabad city. The existing wastewater disposal systems include an underground sewerage system in the urban areas, septic tanks in suburban areas and pit systems in the peri-urban settings. HMWSSB is responsible for providing the services to the city people. Rehabilitation and new network connections are made with loan assistance (JICA) or state government funds. The National River Conservation Plan of the National River Conservation Directorate, Government of India, has been by far the largest contributor. Detailed project reports are prepared for specific activities to secure funding from the government. Models are being worked out to recycle the WWTP water, though at present the treated water is released back to the river. Continuous monitoring of water quality is now being carried out by the Central Pollution Control Board (CPCB, 2014).

Municipal solid waste management

The management and handling of the municipal solid waste in cities are governed by the Environment (Protection) Act, 1986 (29 of 1986), as prescribed in sections 3, 6 and 25, and detailed out as a set of rules, namely, "Municipal Solid Waste Rules 2000" which was gazetted in 2000. It stipulates that municipal authorities are responsible for the collection, storage, segregation, processing and disposal of Municipal Solid Waste (MoEF, 2000; CPCB, 2000). The regulatory authority for monitoring and evaluation is the Central Pollution Control Board (CPCB, n.d. *a).* It has been estimated that a person living in Hyderabad produces, on average, 0.57 kg (2005) of waste per day (CPCB, n.d. *b*). For a projected population of nine million, the GHMC may have to deal with 5,181 tons of waste per day. The GHMC has introduced an integrated Solid Waste Management system through private sector participation, direct community involvement and effective public participation in segregation of recyclable waste. Local welfare associations and self-help groups play a major role in collecting waste, however, the bulk of the waste is collected and disposed of by a private company, Ramky Infrastructure Ltd. (REE, n.d.; TToI, 2013; TToI, 2014b). The landfill sites are 40 km away from the city, and the GHMC operates on BOOT/PPP (Build-own-operate-transfer/Public-private partnership) mode to dispose of the waste. A study that assessed the status of Municipal Solid Waste management in the country found that composting was the most popular method (Annepu, 2012). While the municipal solid waste collection has improved, it does not cover all parts of the city yet. Dumping in waterways still continues, a problem that might be seen for the foreseeable future.

City sanitation plan for Hyderabad

Currently, there is no CSP for Hyderabad, the reason for this is not clear. However, a state level plan has been outlined during a state level workshop. In general, a three tier approach has been suggested for better exchange of knowledge and experiences in sanitation issues. These are National Urban Sanitation Plan, State Sanitation Plan (SSP) and Urban Local Body (CSP-municipality/ULB) levels. The different components of the SSP are highlighted in Table 15.2.

Table 15.2 The activity plan for the state level sanitation plan.

Component 1	Component 2	Component 3
Support to central and state level urban sanitation program	Preparation and implementation of city sanitation plans (CSPs)	Knowledge management, communication and awareness raising
Management and planning instruments for the urban sanitation sector improved at central and state level	Support to develop CSPs, and enable municipalities to improve management skills, technical and financial aspects related to urban sanitation planning	All relevant stakeholders, (city authorities, community members, schools etc.) gain knowledge, and skills, improve communication related to urban sanitation

Source: APSSS & CSP (2013).

15.4.2 Chennai

City water supply

Chennai is one of the metro cities of India (Figure 15.2) in the State of Tamil Nadu, which relies on a system of reservoirs and lakes for its water supply. It is heavily dependent on the monsoon rains for recharging the water bodies, as three of the largest rivers within the state are heavily polluted. With a population of 4,681,087 (Census, 2011) the demand for domestic water supply has increased over time and as a consequence ground abstraction has increased, which has resulted in the decrease of water levels in aquifers. Currently, the water needs of Chennai city are met by desalination plants at Nemelli and Minjur, aquifers in Neyveli, Minjur and Panchetty, Cauvery water from Veeranam Lake, Krishna River from Andhra Pradesh, Poondi reservoir, and Red Hills, Chembarambakkam and Cholavaram lakes (CMW, 2015; Mariappan, 2014). Figure 15.4 presents an overview of the different sources of water that the City Metro Water relies on to meet the increasing water demand. Since 2002 rain water harvesting has been mandatory, but the challenges have impeded the successful adoption of the practice. The lack of adequate awareness raising, education and guidance of the entire process were seen as the key reasons for poor uptake. As means to augment the water supply the government investment on desalination plants has also increased, which is not very well received by the environmentalists. The current water supply from all sources is 0.98 million m^3/d against a demand of 1.2 million m^3/d. The water supply to the city is managed by the Chennai Metro Water (CMW), which covers an area of 426 km^2. The state 12th Five Year Plan proposes an integrated urban water management plan, which takes into account a holistic approach of economic efficiency, social equity and evironmental sustainability, however, the city of Chennai is yet to implement such an integrated program. The key areas for consideration are explotation for alternative sources of water, alignment of formal and informal institutions and their practices, wastewater management and recycling, enhancement of water use efficiency and conservation practices, water sensitive urban planning, prevention of water source pollution, and utilisation of ecological sollutions for pollution control (Bahri, 2012).

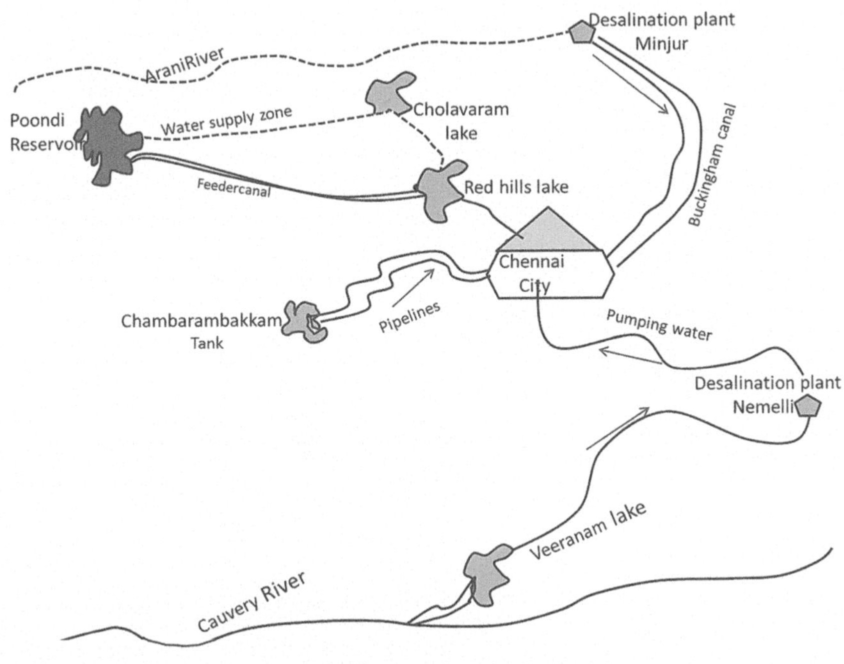

Figure 15.4 Urban water supply system in the city of Chennai. *Source*: IWMI.

Wastewater treatment and management

The Chennai City Wastewater System has been expanded over time to cater to the population increase. However, the sewerage network system is unable to handle the full flow, especially, when the storm water is released. The excess is drained into the natural waters close to the city viz. Cooum river, Adyar river, Buckingham canal and Otteri Nalla, which contributes to the large pollution loads in natural waterways and the drains. The following are the WWTPs of Chennai Metro Water Supply and Sewerage Board (CMWSSB) as at February 2014: Kodungaiyur 270 thousand m^3/d (110 + 80 + 80); Villivakkam 5 thousand m^3/d; Koyambedu 94 thousand m^3/d (34 + 60); Nesapakkam 117 thousand m^3/d (23 + 40 + 54); Perungudi 126 thousand m^3/d (54 + 60 + 12), totaling 612 thousand m^3/d of which 378 thousand m^3/d is powered by biogas (produced during the wastewater treatment) engines. To cope with the increasing population, the municipalities have been increasing the treatment capacity of existing treatment plants and the additions for each of the plants are stated in brackets. However, there is still a large treatment gap yet to be filled, which contributes to the overall pollution load in the waterways. It is felt that the city's inability to cope with treatment poses a great threat to the drinking water supply, especially in an area where the overflow pipes of septic tanks are connected to open drains and waterways.

City sanitation plan for Chennai

Like for Hyderabad, there is no sanitation plan for Chennai. While the national and state level guidelines are available, the activities have been planned in an ad hoc manner. In a recent report it was stated that on the national sanitation ranking Chennai received a score of 53% out of 100%. A study carried out on the sanitation services showed that the public sanitation services in the city were poor (IMRF-CDF, 2011). The key findings were that city public toilets were not adequate for the size of the population moving in the city, existing toilets were poorly maintained and underutilised, since city authorities did not have adequate funds for maintenance and governance. The 12th plan for the state of Tamil Nadu reports that sanitation coverage in the core areas of the city of Chennai is 99%. The CMWSSB is rensponsible for managing an estimated 610,000 network connections, 2,600 km of sewer lines and 180 pumping stations. The treatment capacity projected for 2020, is 1.49 million m^3/d, and highlights the capacity gap that has to be met in the coming years. The plan also highlights special programs for schools, solid waste management, and recycling of water in apartment buildings in the coming years (TNSPC, 2012).

15.5 CASE STUDIES

15.5.1 Natural wetland in the Musi River micro-watershed

A small wetland of the size of 4.5 ha, with a volume of standing water of 16,295 m^3, was studied for its function. The wetland was predominantly of *Typha capensis* grass with a submerged biomass of 18 kg/m^2 and a discharge rate of 1,812 m^3/d. The hydrogeological, geophysical and bio-geochemical investigations revealed that this wetland had potential to remove selected contaminants in its natural state (see Chapter 11). Natural systems both transform and capture many of the common pollutants that occur in domestic wastewater. These pollutants include nutrients like nitrogen (N) and sulphur (S), heavy metals, trace organic chemicals as well as disease causing pathogens, suspended sediments, and organic matter. The major natural processes occurring in natural systems include chemical adsorption (fixation), sedimentation (settling of solids), plant uptake, and bacterial degradation. This wetland, which is situated in the midst of rice, paragrass and vegetable farms has received wastewater for over 20 years, due to the agriculture practices. Over the years, the farmers have been collecting wastewater from the irrigation canal (gravity flows) to a large well and lifting the water by pumps to irrigate the farms farther afield. Thus, the wetland receives water that is lifted and also the run-off from the adjacent fields (Figure 15.5). It is envisaged that in a functional wetland the outlet water can be stored in a well (TWW) for irrigation purposes.

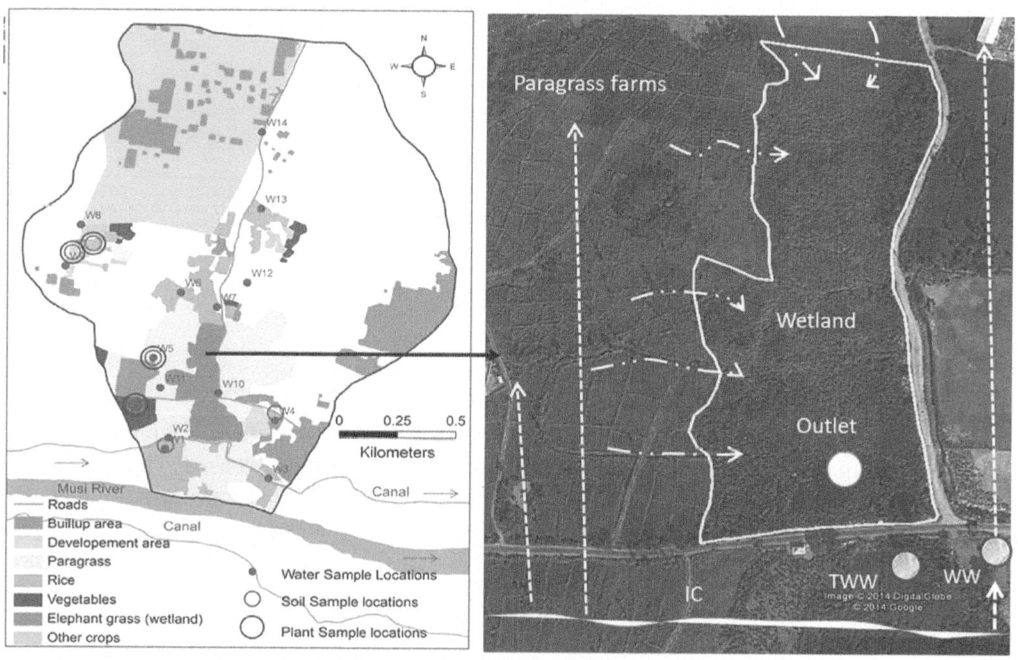

Figure 15.5 Map of the study area (see Chapter 11) and proposed community-managed wetland in Kachiwani Singaram, Hyderabad. IC = Irrigation channel; TWW = Proposed treated wastewater storage well; WW = Wastewater refill well (from irrigation channel). White straight arrows = wastewater lifted for irrigation; curved white arrows = waterflow direction towards the wetland (Google Maps, 2014).

The chemical analysis of water at the inlet and outlet points of the wetland showed that reduction in nitrates and sulphates were significant, during both pre- and post-monsoon seasons (Figure 15.6). Phophate reduction was not consistent through the seasons. It is well known that the removal of nitrates is by plant uptake, and usually occurs in the submerged vegetative parts of the plant and close to the soil, where the oxygen is less. Phosphates usually gets bound to the soil and is removed from water, contributing to the reductions. More detailed investigations are needed to fully understand the dynamics of pollutant removal, however, these results indicate that the hydraulic residence time and the discharge rates are conducive for contaminant removal, for a column of water that passes through this wetland. As such, there is potential for the wetland to be engineered to treat the wastewater for irrigation use.

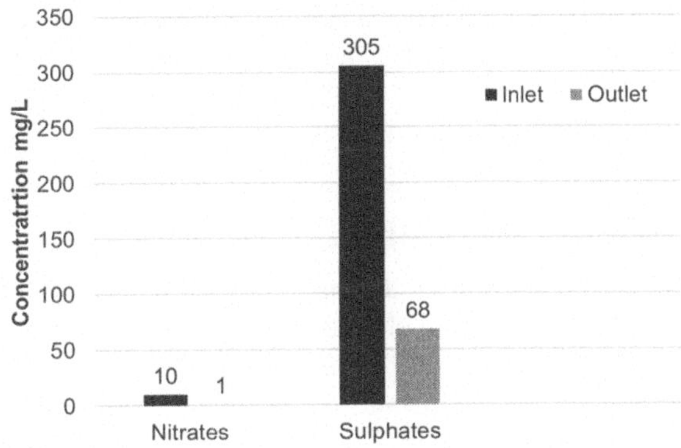

Figure 15.6 The nitrate and sulphate concentrations during the pre-monsoon season (2013) at the inlet and outlet points for the natural wetland.

15.5.2 Percolation pond and check dam in Chennai

Percolation pond

Percolation ponds are effective recharge structures that also offer contaminant removal as the water is passaged through the soils. Ultimately, they help to recharge the aquifer and reduce salinity. A constructed farm pond (8 × 8 × 1.75 m) with a capacity of 112 m^3 was able to dilute highly salinized groundwater and also augment the groundwater levels in the surrounding areas (Deliverable D 2.3, Saph Pani). This was constructed in Andarmadam, Thiruvallur district of Tamil Nadu, to collect the run-off water from the surrounding areas during the rainy season. These ponds can be constructed in farms with clear management instructions so that groundwater storage can be augmented through aquifer recharge. This type of recharge strucuture can be positioned where the drinking water supply sources are, especially close to townships and can be placed under the CMW integrated urban water management plan.

Check Dam

Check dams across drains, ditches and rivers enhance infiltration and help to recharge aquifers. It also affords the removal of contaminants, during the infiltration process and can help augment the water supply for the city of Chennai through aquifer recharge. A check dam located at a distance of 50 km northwest of Chennai across Arani River at Paleswaram was investigated for its treatment capacity. The study revealed that about 50% of the water harvested is recharged every year. Approximately 1.1 million m^3 of water is recharged if the annual rainfall is about 1,200 mm. A total of 19 check dams have been constructed in the Arani as well as Korattalaiyar River, which are expected to improve the groundwater storage in the area.

15.6 AN INTEGRATED MANAGEMENT PLAN FOR NATURAL TREATMENT SYSTEMS

The major natural processes that take place in the NTS are sedimentation, bacterial degradation, chemical adsorption and plant uptake. An understading of these processes that take place in any given setting is important for its successful adoption. The treated water can be used for an array of different purposes within a city, based on the level of treatment. Careful monitoring is thus required if it is to be used for potable or non-potable use. The integration of an NTS into the current water supply and sanitation plans is a new idea for these two cities. While both cities are facing water scarcity issues, novel ideas such as these have to be discussed among stakeholders that have a stake in it. The reasons for this are, firstly, that NTS have not been considered as part of an urban water management cycle, therefore, their acceptability for any proposed use has to be endorsed by the relevant authorities; secondly, their management framework has to be developed and integrated into the exisiting water supply systems for sustainability. Finally, the personnel and capacity building of its staff have to be identified and training methods have to be developed.

The two NTS described here are to be used for two different purposes. The one in Hyderabad is for agriculture use and the one in Chennai is for MAR to augment the groundwater supply. The studies showed that periodic monitoring of NTS performance is a must, as it is important to see if contaminant removal takes place continuously, in a satisfactory manner. Thus, the water quality monitoring should be carried out systematically, to see that water is cleaned to a desired standard for its reuse. While both systems described here are compared in their natural states, further fine tuning and adaptation to the local settings is recommended. Each system should have its own management plan, linked to the overall water management plan, which can be monitored by the relevant municipality. Since, NTS have not been considered as a treatment method in both cities, it was viewed as important to discuss the relative merits of NTS to raise awareness among different stakeholders and to obtain concurrence. Here we discuss the acceptance of the NTS as treatment systems by local stakeholders and how they could fit into an integrated water management plan of a city.

15.6.1 Stakeholder concurrence – Hyderabad

The stakeholder perceptions and concurrence were sought during a one day workshop titled "Wetlands as Natural Treatment Systems for Wastewater Treatement and Reuse". The stakeholders as well as the project partners were of the view that NTS could be a viable option for cities where drinking water supply becomes an issue during the summer months. The government stakeholders from 13 departments discussed aspects relevant to implementation, cost benefits and health concerns (WHO, 2006) as the proposed plan was for a community-managed system that was producing food for human consumption. The Hyderabad NTS was perceived as a viable option for farmers using marginal quality water without treatment in the peri-urban areas, which was a major concern especially in the production of green leafy vegetables. It was also felt that further elaboration on how farmers can establish such systems in the field, field testing with multiple field crops and appropriate water/food quality tests as per the recommendations of the WHO guidelines should be carried out. Thus, its potential to be used across agricultural and horticulture crops as well as non-edible crops, were well received. The proposal for two other possible wetland scenarios were also accepted, and these were, i) constructed wetlands and ii) mini wetlands for individual farmers. A few of the important considerations were that gravity flow was adequate to run the systems (no energy use) and required low maintenance, however, quality control has to be a key component of the integrated management plan. A separate stakeholder study revealed that constructed wetlands could be a viable option if the costs of establishment could be supported by the government, and will have important policy implications (Starkl *et al.*, 2015).

Table 15.3 List of agencies that participated in the stakeholder meeting held in Hyderabad.

	Department/Institute
1	Groundwater Department, Government of Telangana, India
2	Central Ground Water Board, Ministry of Water Resources, Government of India
3	National Remote Sensing Centre, ISRO, Hyderabad, India
4	Hyderabad Metropolitan Development Authority, India
5	Central Research Institute for Dryland Agriculture, Hyderabad, India
6	Commissioner, Agriculture, Telangana, India
7	Irrigation & Command Area Development, Water and Land Management Training and Research Institute, Telangana, India
8	Irrigation & Command Area Development, Andhra Pradesh, India
9	Special Commissioner, Department of Rural Development, Andhra Pradesh, India
10	Dept. of Geochemistry, Osmania University, India
11	International Water Management Institute, Patancheru, Telangana, India
12	CSIR-National Geophysical Research Institute, Hyderabad, India
13	Centre for Environment Management and Decision Support, Austria

Source: Saph Pani, News Letter 6 (2014).

The following is a synopsis of the outcomes of the workshop:

- The wastewater in irrigation canals should not be applied directly to crops, unless the water quality is tested prior to application, and is found to be suitable – following the WHO guidelines.
- Based on the contaminants the NTS should be further designed for contaminant removal.
- NTS treated wastewater can be used conjunctively with groundwater.
- Since it is a community-based NTS, the government involvement is a must for quality control.
- Finance schemes should be aligned to provide financial support communities as an incentive to use NTS as part of their agriculture practices.

Figure 15.7 depicts a possible integrated urban water management plan for Hyderabad. The city managers accept that urbanisation has influenced the development of large peri-urban spaces which provide many types of services that are needed for city dwellers. As such, there is a constant demand for the utility services to be extended to these spaces and the GHMC has been heeding to some of the requests over the years. However, the administration is beginning to realise that it is a timely proposition to explore some of the alternative sources to meet the potable as well as non-potable demands and also see that both quality and quantity are maintained for these sources through natural processes that are more cost-effective. For example the Musi River, which provided clean water for peri-urban agriculture in the past, has become a highly polluted river, rendering it unsuitable for irrigating food crops. The NTS that were studied offer new opportunities for municipalities to explore new methods of treatment of wastewater for food crops as well as non-potable uses like landscaping and industrial cooling. The municipality can play a key role in the design and management and has access to more water treated to a level that can be used. In this chapter we propose two types of constructed wetlands, one at the point of the irrigation channels and the other on farmer plots where the non-point source pollution contamination can be addressed before it is used for food production. Both types of constructed wetlands can be part of the urban water cycle where the municipality is responsible for the overall management and quality control. The consutruction of the larger wetlands can be introduced as part of the overall sanitation program and the farmers can be incentivised to build and manage the small NTS on farmer plots. In this way, the urban water supply, city sanitation plans and natural treatment systems can be part of an integrated plan, where the quantity and quality of water supply can be addressed at the same time.

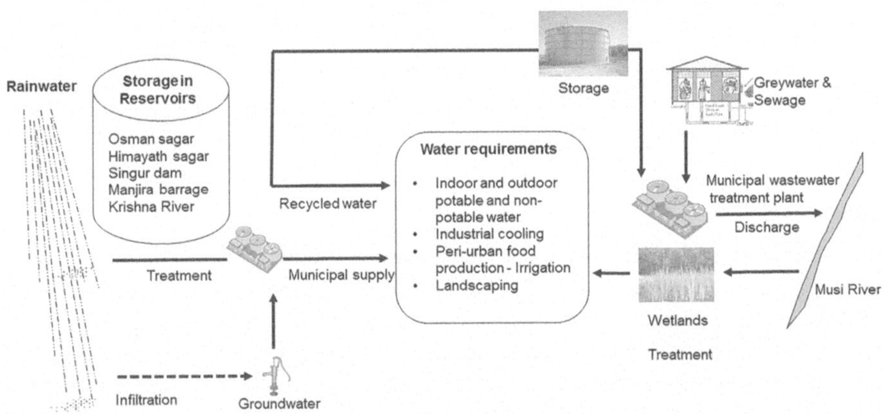

Figure 15.7 A schematic diagram of an integrated water management plan for Hyderabad that includes a natural water treatment system (Wetland).

15.6.2 Stakeholder concurrence – Chennai

The gap between the supply and demand for water for the city of Chennai has been increasing over the years. The current thinking of the water supply agencies is to plan for additional desalination plants to reduce the gap. However, the cost of desalination plants work out to be INR 8,712 million (Deliverable D6.2, Saph Pani). To improve the groundwater storage in the urban part of the city, rainwater harvesting has been in practice since 2003. However, the efficiency of the rainwater harvesting structures is questionable. There is a need for a regulatory authority to monitor the installation and functioning of these recharge structures and to assess the efficiency annually. In the urban and peri- urban as well as rural areas the groundwater recharge needs to be increased by the construction of small percolation ponds, and check dams to harvest the surface run-off, which is currently about 95 million m^3 from the Arani River basin to the sea, north of the city of Chennai. If

10,000 percolation ponds (8 × 8 × 1.75 m), are constructed, in the rural areas, north of the city of Chennai, it will result in the harvest of about 11 million m^3, which is about 12% of the existing run-off to the sea. The suggested recharge initiatives may lead to an increase in groundwater availability by 10 million m^3/year. Hence, the contribution of groundwater to meet the city water requirements can be increased rather than increasing the number of desalination plants.

The percolation ponds and the check dams can be viewed as part of both urban and rural watersheds and integrated into the city water supply system if a good monitoring system is established to test the water quality at these water harvesting structures. In both cases maintaining a good hydraulic flow is important for contaminant removal. A possible schematic framework for management of these strucutures is given in figure 15.8. The design and management of these structures can be part of the mandate of the municipality and/or the watershed development department depending on the positioning of strucutures. Both the watershed department and the municipality can monitor these strucutres together, ensuring that the desirable standards for drinking water is maintained. One important component of such a system is capacity building and must be inlcuded as part of the regular management plans.

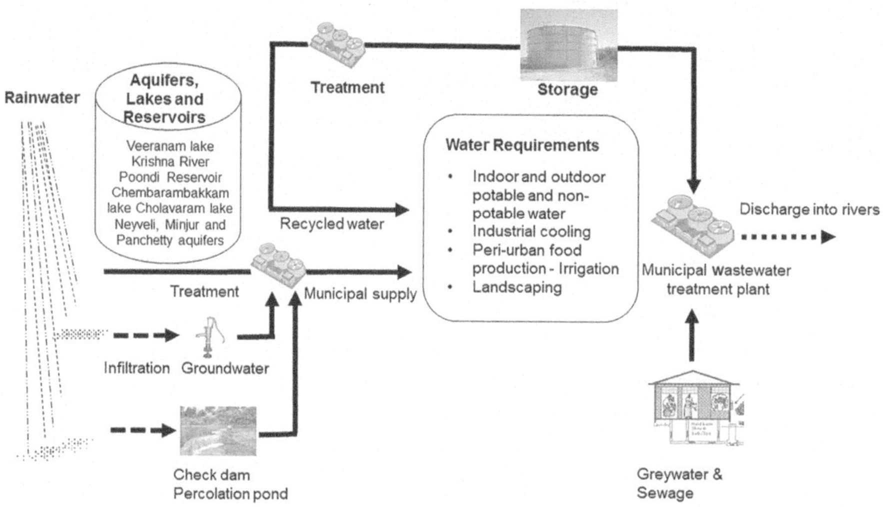

Figure 15.8 A schematic sketch of an integrated urban water management plan for Chennai that includes two natural water treatment systems (Check dam and percolation pond).

15.7 CONCLUSION

This chapter explores the integration of two NTSs into the urban water supply systems, in each of the cities. These systems can be cost-effective means of treating alternative sources of water that is available in and around cities, for food production and to augment the drinking water supplies. In Hyderabad a natural wetland, situated about 15 km downstream of the city, was capable of reducing contaminants in its natural state and displays the potential to be further engineered to achieve the desired water quality levels for reuse in agriculture. If such engineered wetlands can be utilised within urban and peri-urban settings, it can help to treat contaminated free flowing water to be used for alternative purposes like agriculture, landscaping, industrial cooling and other non-potable uses in the city.

The rapid and heavy rains in Chennai result in quick run-off with poor recharge via infiltration. The run-off water could be, to an extent, harvested by different structures developed for MAR. In Chennai, the check dam and pilot percolation pond that were studied for their treatment potentials indicated that both can be successful methods to lower salinity and unwanted contaminants. It was a good example of a successful MAR, which can be used for augmenting freshwater supplies, especially in coastal areas.

Urban water management is facing unprecedented challenges in both cities and the exploration of alternative sources is well timed. With source water contamination and diminishing aquifer sources, India needs to look at some of the natural systems for treatment and integrate it as part of its water resource management plans. Integrated water management plans can bring together all sectors linked with water to look at issues holistically. Involvement of key stakeholders in the discussion enhanced the awareness on NTS, and the importance having alternative solutions to treatment of urban run-off and other contaminated water sources in and around the cities, especially in peri-urban spaces. The government stakeholders were willing to pilot these ideas furthers in another phase. The Figures 15.7 and 15.8 illustrate how NTS can be positioned in a total urban water management plan, which can then be considered as part of a larger planning and management process. The need for running pilots for up-scaling can be a part of a short and long term goal setting process of the municipalities, where cities can look for alternative measures to augment the water supply to meet different demands. It is clear that testing and validating is a time consuming process. The need for key government institutions to play an active role in identifying the relative roles for effective and sustainable management will herald a change in the current thinking of using only expensive methods of treatment. Once the NTS are included as part of the urban water supply system, its overall management and quality control can ensure sustainability. Integrated water resource management allows coordinated, responsive actions across water users and service providers to plan for innovative sustainable options for augmentation. However, success also depends on securing significant funds for operation and maintenance of systems, proper capacity building of its staff and supportive policies.

15.8 REFERENCES

Annepu R. K. (2012). Solid Waste Management in India. M.Sc. Thesis. Master of Science in Earth Resources Engineering, Department of Earth and Environmental Engineering, Fu Foundation School of Engineering and Applied Science, Columbia University in the City of New York.

APSSS & CSP (Andhra Pradesh State Sanitation Strategy and City Sanitation Plan) (2013). Workshop on Urban School Sanitation and Urban Local Bodies Consultation. July 5, 2013, Hyderabad, United Andhra Pradesh.

Arceivala S. J. and Asolekar S. R. (2006). Wastewater Treatment for Pollution Control and Reuse. 3rd edn, Tata McGraw Hill, New Delhi, pp. 132–231.

ASCI (n.d.) http://www.asci.org.in/ (accessed 2 February 2015).

IMRF-CDF (Institute for Financial Management Research-Centre for Development Finance) (2011). Public Toilets in Chennai. IFMR Research-Issue Brief. http://www.indiasanitationportal.org/1148 (accessed 4 May 2015).

Bahri A. (2012). Integrated Urban Water Management, TEC Background Papers no. 16, Global Water Partnership (GWP), Drottninggatan, Sweden, pp. 1–89.

CMW (Chennai Metro Water) (2015). Chenni Water Supply. http://www.chennaimetrowater.tn.nic.in/index.htm (accessed 12 April 2015).

CPCB (Central Pollution Control Body) (2009). Status of water supply, wastewater generation and treatment in class-I cities and class-I towns of India. Control of urban pollution series: CUPS/70/2009–10.

CPCB (n.d. *a*). http://cpcb.nic.in/ (accessed 8 May 2015).

CPCB, (n.d. *b*). http://www.cpcb.nic.in/wast/municipalwast/Waste_generation_Composition.pdf (accessed 8 October 2014).

CPCB (2000). http://www.cpcb.nic.in/divisionsofheadoffice/pcp/MSW_Report.pdf (accessed 8 October 2014).

CPCB (2014). Guidelines for Real-Time Effluent Quality Monitoring System. Central Pollution Control Board, Delhi, India, pp. 21–32.

Crites R. W., Middlebrooks E. J. and Bastian R. K. (2014). Natural Wastewater Treatment Systems. CRC Press, Taylor & Francis Group, Florida.

HMWSSB (Hyderabad Metropolitan Water Supply and Sewarage Board) (2013). Hyderabad. http://www.icrier.org/pdf/hyd_short_presentation.pdf (accessed 6 October 2014).

HMWSSB (2014). Hyderabad Metropolitan Water Supply & Sewerage Board. http://www.hyderabadwater.gov.in/wwo/UI/soruce_info.aspx. (accessed 6 October 2014).

Kadlec R. H. and Wallace S. D. (2009) Treatment wetlands, 2nd edn, CRC Press, Boca Raton, Florida.

Mariappan J. (2014). As population surges, Chennai's water needs to double in 3 years. The Times of India (Chennai: The Times Group). http://en.wikipedia.org/wiki/Water_management_in_Chennai (accessed 9 October 2014).

Missimer T. M., Maliva R. G., Ghaffour N., Leiknes T. and Amy G. L. (2014). Managed aquifer recharge (MAR) economics for wastewater reuse in low population Wadi communities, kingdom of Saudi Arabia. *Water*, **6**, 2322–2338; doi:10.3390/w6082322

MoUD, Ministry of Urban Development (Editor) (2008). National Urban Sanitation Policy, Ministry of Urban Development, New Delhi. http://moud.gov.in/sites/upload_files/moud/files/NUSP_0.pdf (accessed 7 October 2010).

MoEF (2000). http://www.moef.nic.in/legis/hsm/mswmhr.html (accessed 8 October 2014).

MoUD (n.d.). http://moud.gov.in/SwachchBharat (accessed 2 February 2015).

National Urban Sanitation Policy (2014). http://www.wsp.org/sites/wsp.org/files/rank_sanitation_rating.pdf. (accessed 28 January 2015).

Nema P., Ojha C. S. P., Kumar A. and Khanna P. (2001). Techno-economic evaluation of soil aquifer treatment using primary effluent at Ahmedabad, India. *Water Research*, **35**(9), 2179–2190.

Press Information Bureau, Government of India (2010). http://pib.nic.in/newsite/erelease.aspx?relid=61761 (accessed 4 February 2015).

REE (n.d). Ramky Enviro Engineers, Hyderabad. http://www.ramkyenviroengineers.com/images/enviro/enviro%20engeneering%20brochure_latest.pdf. (accessed 5 May 2014).

Reed S. C., Crites R. W. and Middlebrooks E. J. (1995). Natural systems for waste management and treatment, 2nd edn, McGraw-Hill Inc., New York.

Saph Pani D6.2 (2014). Deliverable D 6.2. Report on case study assessments and feasibility studies.

Saph Pani, Biannual News Letter 6 (2014). Enhancement of natural water systems and treatment methods for safe and sustainable water supply in India. Saph Pani, By annual News Letter 6, pp. 4 (www.saphpani.eu).

Sharma S. K., Harun C. M. and Amy G. (2008). Framework for assessment of performance of soil aquifer treatment systems. *Water Science & Technology*, **57**, 6941–6946.

Sharma S. K., Hussen M. and Amy G. (2011). Soil aquifer treatment using advanced primary effluent. *Water Science & Technology*, **64**(3), 640–646.

Shipin O., Koottatep T., Khanh N. T. T. and Polprasert C. (2005). Integrated natural treatment systems for developing communities: Low-tech N-removal through the fluctuating microbial pathways. *Water Science & Technology*, **51**(12), 299–306.

Starkl M., Brunner N., Amerasinghe P., Mahesh J., Kumar D., Asolekar Shyam R., Sonkamble S., Ahmed S., Mohammed W., Pratyusha A. and Sarah S. (2015). Stakeholder Views, financing and policy implications for reuse of wastewater for irrigation: A case from Hyderabad, *India Water*, **7**, 300–328; doi:10.3390/w7010300

TToI (The Times of India) (2014a). http://timesofindia.indiatimes.com/city/hyderabad/Telangana-Andhra-Pradesh-water-conflict-may-leave-city-thirsty/articleshow/44965332.cms. (accessed 3 November 2014).

TToI (The Times of India) (2014b). http://timesofindia.indiatimes.com/city/hyderabad/Waste-management-project-Greater-Hyderabad-civic-body-in-dilemma-as-workers-oppose-move/articleshow/30299989.cms (accessed 8 December 2014).

TToI (The Times of India) (2013). http://timesofindia.indiatimes.com/city/hyderabad/Greater-Hyderabad-Municipal-Corporation-to-hand-over-garbage-collection-work-to-Ramky/articleshow/20638884.cms (accessed 8 December 2014).

The New Indian Express (2014). http://www.newindianexpress.com/states/andhra_pradesh/Water-Woes-Rising-Demand-Poor-Supply-Plague-Hyderabad/2014/03/03/article2087706.ece (accessed 21 April 2014).

TNSPC (2012). Tamil Nadu State Planning Committee, Twelfth Five Year Plan (2012–2017). http://www.ap.gov.in/Other%20Docs/Socio%20Economic%20Survey%202012-13/Twelfth%20Five%20Year%20Plan (2012–17)%20A%20Way%20Forward.pdf. (accessed 2 May 2015).

Van Rooijen D., Turral H. and Biggs T. W. (2005). Sponge city: Water balance of mega-city water use and wastewater use in Hyderabad, India. *Irrigation and Drainage*, **54**, S81–S91.

WSP (2010). Marching Together with a Citywide Sanitation Strategy. http://www.wsp.org/sites/wsp.org/files/publications/citywide_sanitation.pdf (accessed 7 January 2015).

World Health Organization (2006). Guidelines for the safe use of wastewater, excreta and greywater. Volume 2: Wastewater use in agriculture. Geneva, Switzerland: WHO, pp. 1–50.

Zurita F., De-Anda J. and Belmont M. A. (2009). Treatment of domestic wastewater and production of commercial flowers in vertical and horizontal subsurface-flow constructed wetlands. *Ecological Engineering*, **35**, 861–869.

Chapter 16

Application of a water quality guide to managed aquifer recharge in India

Peter Dillon, Declan Page, Joanne Vanderzalm, Jatinder Sidhu, Cornelius Sandhu, Alexandre Boisson and Lakshmanan Elango

16.1 INTRODUCTION

India is an international leader in the volume of enhanced recharge (4 km^3/year, CGWB, 2005) and the number of recharge structures employed in national, state and local level programs to help sustain groundwater storages and irrigation livelihoods. Potential recharge of 85 km^3/year of surplus run-off has been identified to augment groundwater recharge in the revised Master Plan for Artificial Recharge to Groundwater in India (CGWB, 2013).

While increasing the quantity of groundwater, artificial recharge also impacts the quality of groundwater. In many cases the recharge water is of comparable quality to water already in the aquifer. Enhancing recharge may improve the quality of groundwater, for example by freshening brackish groundwater. However, recharge may also introduce microbiological or chemical pollutants to aquifers, or mobilise minerals from the aquifer matrix, such as arsenic. Although these hazards are generally of relatively low concern for irrigation, they may potentially cause harm where groundwater is used as a drinking water supply. Hence when the same aquifer being recharged is also used as a drinking water source, it should be an obligation of those enhancing recharge to protect the health of those whose drinking water is affected by their operations.

Artificial recharge can be undertaken in such a way as to avoid these adverse effects on groundwater quality. When such control is done intentionally to protect human health and the environment this is called 'managed aquifer recharge' (MAR). In 2009, the Chairman of the Central Ground Water Board (CGWB) raised the opportunity to specifically account for groundwater quality protection in the Indian Guidelines for Artificial Recharge (CGWB, 2000) and Manual for Artificial Recharge of Groundwater (CGWB, 2007). This initiated a proposal to the Australian Department of Foreign Affairs and Trade (DFAT) to draw from Australia's experience in developing risk-based guidelines for managed aquifer recharge (NRMMC-EPHC-NHMRC, 2009) and adapt this to India's needs for groundwater protection. The revised Master Plan for Artificial Recharge to Groundwater in India (CGWB, 2013) also references the purposes and types of water quality monitoring necessary to protect groundwater quality.

The Australian MAR Guidelines (NRMMC-EPHC-NHMRC, 2009) form part of the Australian National Water Quality Management Strategy and adopts a risk assessment and management framework consistent with World Health Organisation (WHO) Drinking Water Guidelines (WHO, 2011) and other Australian guideline documents including the Australian Drinking Water Guidelines (NHMRC-NRMMC, 2011) and Australian Guidelines for Water Recycling (NRMMC–EPHC–AHMC, 2006). However, in applying the Australian MAR Guidelines to assess risks in artificial recharge projects in India and several other countries in transition, it was found that water quality data, especially on the microbial pathogens, were not available to enable quantitative risk assessment (Dillon *et al.*, 2010). Consequently, the DFAT funded project 'A Water Quality Guide to Managed Aquifer Recharge in India' (Dillon *et al.*, 2014) combined elements of the 'entry level assessment' from the Australian MAR Guidelines (NRMMC-EPHC-NHMRC, 2009) with the WHO sanitary survey approach (WHO, 2011, 2014) to produce a water safety plan commensurate with the WHO (2011) drinking water guidelines as applied to small scale systems (Davison *et al.*, 2005; WHO, 2012). The guide relies only on visual observations and not on water quality sampling and analysis in order to be capable of application at village level throughout India by people with elementary training in water quality protection, without access to reliable laboratory resources.

The resulting guidance document (Dillon *et al.*, 2014) aims to improve water safety based on visual observations and actions in response. It is intended to inform guidance at national level, for application and capacity building at state and local level. It aims to ensure that safety of drinking water is considered as an integral part of recharge enhancement planning and practice, and that this is integrated within water safety plans for drinking water wells accounting for natural and enhanced recharge. However, it currently has no formal status that requires compliance.

The purpose of this chapter is to document the application of the approach that was developed within the DFAT project to three recharge operations within Saph Pani;

- Bank filtration at Haridwar on the Ganga River, Uttarakhand
- Percolation tanks at Maheshwaram, Telangana, and
- Check dam near Chennai, Tamil Nadu.

For two of these sites where public drinking water supplies are potentially impacted by recharge enhancement this guidance is considered necessary to improve the safety of supplies but insufficient to assure safety. A recommended subsequent step would involve sampling and analysis of water quality for use in quantitative risk assessment for these sites.

16.1.1 Scope of the water quality guidance

Existing CGWB Guidelines for Artificial Recharge (CGWB, 2000) and Manual for Artificial Recharge of Groundwater (CGWB, 2007) provide advice on location, design and operation of groundwater replenishment projects with the current emphasis on the hydraulic performance of such systems. The current master plan for artificial recharge (CGWB, 2013) however references the need to account for water quality.

The water quality guidance for MAR (Dillon *et al.*, 2014) refers to a number of existing specifications and guidelines, including the Bureau of Indian Standards (BIS) specifications for drinking water (BIS, 2012) and guidelines for the quality of irrigation water (BIS, 1986). Related guidance is available from Central Pollution Control Board (CPCB) on water quality monitoring protocols (CPCB, 2007) and on well head protection and from BIS on construction of drinking water wells. Ministry of Environment & Forests (1992) policy on pollution prevention has stimulated programs such as the CPCB's National River Conservation Plan largely focused on building sewage treatment plants, which is currently being strengthened and broadened by the Government of India. Plans to end open defecation in India on health and safety grounds are currently in development, and would also have benefits for run-off water quality.

For public drinking water supplies the WHO Guidelines (WHO, 2011) provide a comprehensive approach to managing risks to human health. This requires substantial effort for investigations, monitoring, analysis and evaluation. The Australian MAR Guidelines (NRRMC-EPHC-NHMRC, 2009) follow the same principles for risk management for public health for all types of uses of the water and a similar approach for managing environmental risks. They therefore also require substantial water quality data acquisition to support quantitative risk assessments. Based on experience reported earlier (Dillon *et al.*, 2010) these approaches were unlikely to be adopted in India and many other countries in transition, due to the unavailability of the necessary data to support those assessments. In light of this, the current water quality guide for MAR in India (Dillon *et al.*, 2014) provides a transitional pathway whereby basic information readily observable at an existing or proposed artificial recharge site is used to make water safer for its intended uses. In the absence of water quality data or further water treatment this approach cannot guarantee effective protection of groundwater used for drinking water supplies. However it highlights the need to consider water quality as standard practice for new and existing projects in India and other countries in transition. It may also serve as a screening tool to identify sites where more rigorous investigations are required, that would then support assessments that accord with WHO or Australian guidelines.

16.1.2 Sources of water, types of aquifers and purposes

The Indian guide to water quality in MAR is relevant to natural source waters in rural and peri-urban catchments. It is recommended not to apply this to other recharge sources, such as urban stormwater, sewage effluent, and industrial effluents that are expected to contain significantly higher contaminant loads. The current guidelines are considered applicable in unconfined alluvial aquifers. They should not be relied on for application in aquifers used as public drinking water supplies, confined aquifers (where changed reduction-oxidation status is likely) and fractured rock aquifers (where there is less exposure to sorption or net attachment sites than in primary porosity media, and contaminants may migrate over longer distances).

Hence, the proposed Indian guidelines (Dillon *et al.*, 2014) cover a more limited range of source water types, aquifer types and existing groundwater uses than the WHO (WHO, 2011) and Australian MAR Guidelines (NRRMC-EPHC-NHMRC, 2009). With the possible exception of Haridwar bank filtration, insufficient data on microbial pathogens were available from any of these sites where these proposed guidelines were applied to provide confident advice, particularly on microbial risks to public health and on release of toxic metals from aquifers.

16.1.3 Water governance issues

The National Water Policy (Government of India, 2012) represents a comprehensive approach to integrated water management of surface water and groundwater and considers quantity and quality with objectives of equity, social justice and sustainability. It declares that whereas groundwater is currently "still perceived as an individual property and exploited inequitably and unsustainably in places", water needs to be "managed as a community resource, held by the state under public trust doctrine to achieve food security, livelihood, and equitable and sustainable development for all."

The water quality guide for MAR in India presumes the adoption of the National Water Policy, to assure that taking surface water for recharge does not impoverish communities downstream, and that enhanced recharge will in fact make groundwater more plentiful enabling equitable use for the highest valued uses, such as to satisfy basic human needs. With allocation plans in place, the government can then optimise water efficiency improvement and recharge enhancement programs, including activities supported by the National Rural Employment Guarantee Act.

16.2 METHODOLOGY

The Australian MAR Guidelines (NRMMC-EPHC-NHMRC, 2009) use a risk assessment and management framework to ensure protection of public health and the environment. They adopt a staged approach to risk assessment, beginning with an 'entry level risk assessment' based on existing knowledge of the recharge scheme, followed by a semi-quantitative risk assessment which may require additional information to be obtained through experiment or investigation (Figure 16.1).

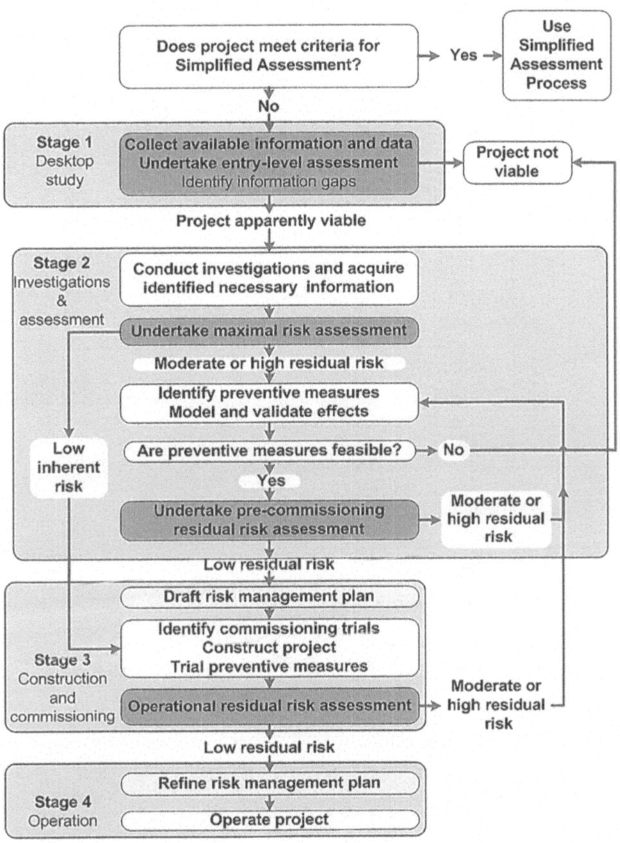

Figure 16.1 Australian MAR Guidelines risk assessment stages in managed aquifer recharge project development (NRMMC-EPHC-NHMRC, 2009).

Application of the Australian MAR Guidelines to assess risk in artificial recharge projects in India, China, Mexico, South Africa, and Jordan by experienced practitioners revealed that the initial entry level assessment was useful in providing a systematic pathway to identify any issues to be addressed (Dillon *et al.*, 2010). However, the water quality analyses of pathogens (viruses, protozoa and bacteria) required to complete the subsequent quantitative pathogen risk assessment were not available at any site. However, at the bank filtration system in Haridwar (Uttarakhand), bacteriological indicators (thermotolerant coliforms) were measured enabling a microbial risk assessment to be inferred (Bartak *et al.*, 2014). In the general absence of pathogen data 'A Water Quality Guide to Managed Aquifer Recharge in India' adapts elements of the Australian entry level assessment and incorporates the WHO sanitary survey approach (WHO, 2011). Together these identify hazards and hazardous events that are then addressed in a water safety plan.

A summary of the steps in the proposed Indian Guidelines, 'A Water Quality Guide to Managed Aquifer Recharge in India' is presented in Figure 16.2. This firstly eliminates from further consideration sites where risk to human health is inherently low (simple assessment). Then it addresses the viability of a project new project. It then assesses whether the nature of the project is within the constrained scope of applicability of the guidelines. Then it advances to a sanitary survey and aquifer assessment before compiling a water safety plan. An example of a water safety plan derived from this process is given in Table 16.1.

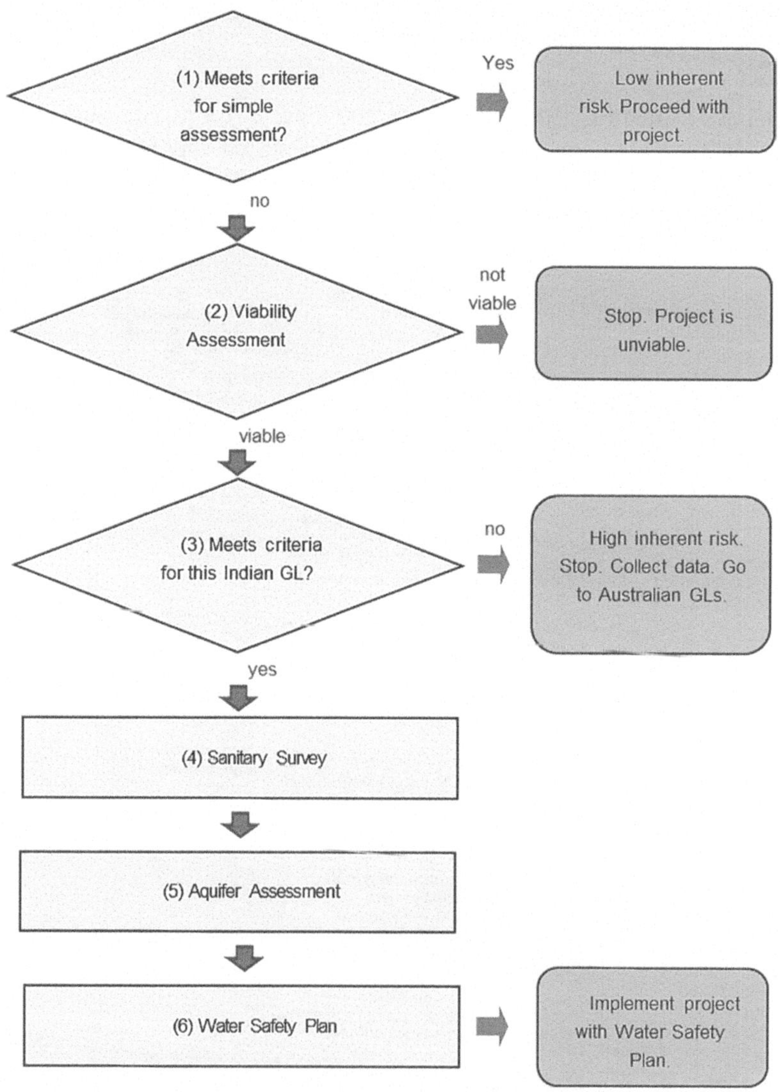

Figure 16.2 Outline of the steps involved in applying the Indian Guideline for managed aquifer recharge including implementation of a water safety plan (from Dillon *et al.*, 2014).

Table 16.1 Example of a managed aquifer recharge water safety plan (from Dillon *et al.*, 2014).

Hazardous Event	Cause	Control Measure	Critical Limits		Monitoring			Corrective Action	Verification
			Targets	Action	What	When	Who		
1. Human sewage entrainment in source water	Latrine leakage, open sewers, sewer pipe leaks, open defecation in catchment, and close to recharge facilities and recovery wells	More latrines with improved design, install separate sewage system from stormwater drains, or only harvest high wet weather flows, improve sewer capacity and response to chokes and leaks	Control sewage leaks, regulate sewage discharge points in catchment	Identify sewage leaks sewage discharge points. Repair, rebuild, or implement overflow diversion. Boil drinking water.	Sanitary inspection	Monthly	Op*	Remove pollutant sources, improve sanitation design, reduce sewer leakage or sewage discharge	Microbiological examination of water
2. Animal faecal matter entrainment in source water	Animal manure accumulation or cess pits in the catchment and close to recharge area and recovery well	Exclude livestock from water harvesting and recovery structures, collect animal faeces and store in dry areas with setback distance from water infrastructure	No overstocking in catchment, set back distances honoured, dry storage of animal manures	Controls on animal husbandry in catchment. Repair fences, exclusion zones. Boil drinking water.	Sanitary inspection	Monthly	Op	Removing livestock out of catchment, repair or erect fencing, arrange collection and removal of faeces.	Microbiological examination of water
3. Leaching of microbial contaminants into aquifer	Infiltration of water that has been in contact with human and animal wastes	Provide adequate setback distances to drinking water wells or springs	No sources of faecal material within setback distance	Close any latrines, and enclose or seal open sewers within setback distance. Boil drinking water.	Sanitary inspection	Monthly	Op	Remove sources of faecal material within setback distance, repair or erect fencing, improve sewerage.	Microbiological examination of water
4. Entrainment of chemicals in source water for recharge	Industry, transport and agricultural activities generating stockpiles, wastes, spills, and emissions reaching the catchment surface	Regulate industrial and agricultural activities in the catchment	No unauthorised sources of chemical contamination in catchment. All pollutants in wise use and management. Minimise spills through industry standards.	Remove wastes from catchment. Install bunding around industrial sites to prevent run-off to recharge system. Traffic loading regulations enforced.	Sanitary inspection	Monthly	Op	Move or bund polluting industries, regulate industrial discharge and agricultural use of chemicals	Sanitary inspection Analysis of source water and groundwater quality for pollutants.

(*Continued*)

Table 16.1 Example of a managed aquifer recharge water safety plan (from Dillon *et al.*, 2014) (*Continued*).

Hazardous Eevent	Cause	Control Measure	Critical Limits		Monitoring			Corrective Action	Verification
			Targets	Action	What	When	Who		
5. Leaching of chemicals into groundwater	Leaching from landfill, waste dumps and industrial discharge	Provide adequate set back distance, regulate industrial discharge	No source of chemicals within the set-back distance	Prevent pollutant discharge within set-back distance	Sanitary inspection	Monthly	Op	Improve containment and move or control pollution sources	Sanitary inspection Analysis of groundwater quality for pollutants.
6. Bypassing or failure of pre-treatment in recharge facility	Short-circuit of recharge flow. Clogging of filters, power and mechanical failures, treatment chemicals run out.	Design treatment to avoid admitting untreated water into the well for each of these hazardous events	No recharge of untreated water	Maintain treatment system regularly. Install system to shut-down recharge when alarm activated Boil drinking water.	Sanitary inspection	Monthly	Op	Maintain filter and any other treatment. Check alarm system operates correctly.	Sanitary inspection Analysis of source water and groundwater quality for pollutants.
7. Bypassing or failure of post treatment at recovery well	Clogging of filters, power and mechanical failures, treatment chemicals run out	Design treatment to avoid admitting untreated water into the water supply for each of these hazardous events	No distribution of untreated recovered water	Maintain treatment system regularly. Install system to shut-down recovery when alarm activated. Boil drinking water.	Sanitary inspection	Monthly	Op	Maintain filter and any other treatment. Check alarm system operates correctly.	Sanitary inspection Analysis of groundwater and recovered water for pollutants.

*Op = MAR scheme operator.

16.3 RESULTS AND DISCUSSION

'A Water Quality Guide to Managed Aquifer Recharge in India' was applied to three Saph Pani case study sites to determine the applicability and usefulness of a simplified approach to reduce health risk in the absence of data on pathogens of concern:

1) Bank filtration at Haridwar on Ganga River, Uttarakhand (see Chapters 2 and 12)
2) Percolation tanks in crystalline aquifers at Maheshwaram, Telangana, (see Chapter 7) and
3) Check dam near Chennai, Tamil Nadu (see Chapter 6).

A brief summary of each site is provided in sections 16.3.1–16.3.3 and the short form assessment in Table 16.2.

16.3.1 Bank filtration at Haridwar on Ganga River, Uttarakhand

Bank filtration has been used in Haridwar mainly since the 1980s as an alternative to surface water abstraction and to supplement groundwater resources (Sandhu *et al.*, 2011; Bartak *et al.*, 2014). The alluvium is comprised of unconsolidated to semi-consolidated, coarse to fine sand, silt and clay. The Haridwar district aquifer system consists of four water bearing layers, separated by confining clay layers in the western part of the district. The hydraulic conductivity of the aquifer is typically 16 to 50 m/d and the aquifer is hydraulically connected to the adjacent Ganga River (Dash *et al.*, 2010). Average discharge in the Ganga River at Haridwar is 200 m^3/s in non-monsoon months, rising to an average of 1,500 m^3/s during the monsoon (Das Gupta, 1975), with observed extreme peak flows up to 12,400 m^3/s in September 2010 (Saph Pani D1.2, 2013). During non-monsoon months, the average depth to the water table is 6.3 (2.8–9.4) m below ground level, while during the monsoon the depth to the water table rises by an average of 1.3 m (Saph Pani D1.2, 2013).

The source water for the BF system in Haridwar is the Ganga River and Upper Ganga Canal (UGC) that originate primarily from rural run-off. The presence of faecal indicators in groundwater wells was linked to areas where unsealed, temporary pit latrines are commonly used, social land use practices such as public bathing/washing at the well heads, presence of cattle in and around the RBF wells and unsanitary defecation practices (Bartak *et al.*, 2014). Haridwar is one of the most important Hindu pilgrimage sites in the world. As a result, ~50,000 visitors come to Haridwar daily with up to 8.2 million people during specific religious days such as Kumbh (Gangwar & Joshi, 2004) to wash themselves in the Ganga in close proximity (<50 m) to the bank filtration wells. The source water for recharge can be turbid, especially in high flows (monsoon). Filtration through the river bed provides water quality improvement, shown as a reduction in turbidity and coliform bacteria (Bartak *et al.*, 2014).

Faecal contamination sources are also present close to the extraction wells. Groundwater is used for drinking water supply; extraction is adjacent to the river (Figure 2.3; Figure 16.3). As of 2013, 59,000 to 67,000 m^3/d is sourced from 22 large diameter, bottom entry caisson wells (Figure 16.3), receiving a mixture of bank filtrate from the Ganga River, UGC and groundwater from the upper unconfined aquifer and around 50 tube wells, which target deeper confined aquifers. Average minimum travel times between the Ganga River and the production wells range from 30 to >88 d corresponding to flow distances of 4 to >490 m. During the monsoon the travel time to production wells can be as low as 2 d (Bartak *et al.*, 2014). Drinking water is treated by chlorination at the well using sodium hypochlorite (NaClO), but this can be interrupted by power failures or defective disinfectant dosage pumps. Recovery wells are housed to exclude livestock, but anthropogenic activities take place near/around the well houses. Flood water can directly enter wells during high flow. Private wells are located further from the recharge site.

16.3.2 Percolation tanks in crystalline aquifers at Maheshwaram, Telangana

Investigations have been carried out at the Tummulur tank in the Maheshwaram watershed, which is located 35 km south of Hyderabad (Telangana State, India). This watershed covers an area of 53 km^2 and due to the rapid growth in urbanisation this watershed is now in transition from a rural to suburban area. The area has a relatively flat topography ranging from 590 to 670 m above mean sea level with no perennial streams.

The region has a semi-arid climate with annual monsoon rains (rainy or "Kharif" season from June to October), producing more than 90% of the mean annual precipitation of about 750 mm. The mean annual temperature is approximately 26°C although during summer ("Rabi" season from October to May), the maximum temperature can reach 45°C (Maréchal *et al.*, 2006). The resulting potential evaporation from soil plus transpiration by plants is 1,800 mm/year. The watershed is overexploited with more than 700 boreholes used for agriculture dominated by rice paddy fields. As a result, the water table is currently 15 to 25 m below ground level and has become hydraulically disconnected from surface water.

The geology of the watershed is relatively homogeneous and is composed of the weathered Archean granite. Dewandel *et al.* (2006) describe a typical weathering profile from top to bottom:

- Saprolite (also called regolith or alterite) which, when saturated, constitutes a large part of the storage capacity of the aquifer.
- The fissured layer, which mainly assumes the transmissive function of the aquifer and is tapped by most of the wells drilled in the watershed. Hydraulic conductivity is found to range from 1.9×10^{-7} to 3.9×10^{-5} m/s (0.02 to 3.4 m/d). The effective porosity of the fissured layer is relatively low and is mainly ensured by small fissures affecting the blocks (Maréchal *et al.*, 2004).
- Impermeable basement, except where tectonic fractures are present.

Figure 16.3 Example of a large diameter caisson well adjacent to Ganga River (Photo: HTWD, 2011).

The monitored Tummulur tank is located in the downstream part of the watershed and has been used for more than 10 years for water storage. An earth bund in its northern part dams the natural stream outlet, and consequently, run-off water is stored over an estimated maximum area of 130,000 m^2 and a maximum water depth of 4 m. Before the first significant rainfall events, the soil is withered due to the dry period which creates important shrinkage cracks observable in the entire tank area. The soils specificities of the tank are the clayey zone, mainly in the lower northern part of the tank (flooded in Figure 16.4b). Most of the rest of the tank area is covered by silt loam soil on the surface underlain by sandy loam at a depth of 40 to 80 cm. Inflow to the tank can be highly variable (Figure 16.4). Monitoring wells were installed to measure piezometric levels and one staff gauge records the surface water level within the tank.

Figure 16.4 Views of the Tummulur tank from the dam, (a) 01/05/2012 and (b) 28/08/2013 (Photo: M.Viossanges, BRGM, 2012, 2013).

The water stored in the tank is used by the surrounding farmers (15) through individual borehole pumping. Irrigation duration and times are controlled by the availability of electricity (7 h per day). The percolation tank constitutes a drinking water supply source for livestock (goats and buffaloes in limited number) as well as for buffalo washing and hence may induce microbial contamination. The area is also known to be prone to geogenic fluoride contamination (Pettenati *et al.*, 2013, 2014) with concentration up to 5 mg/L. Tank functioning has been described in detail in Boisson *et al.* (2014).

16.3.3 Check dam at Chennai, Tamil Nadu

Check dams are constructed across the two non-perennial rivers flowing just north of Chennai, to improve the groundwater recharge and mitigate seawater intrusion. As a part of Saph Pani a detailed investigation was carried out in a check dam constructed across Arani River at Paleshwaram village, which is located 35 km from the Bay of Bengal. This check dam is 260 m long and 3.5 m height with the storage capacity of 0.8 Million m^3 (Figure 16.5).

Figure 16.5 Paleshwaram check dam across Arani River, north of Chennai, Tamil Nadu, India (Photo: S. Parimala Renganayaki, Anna University, Chennai).

The atmospheric temperature of this area ranges from 38° C to 42° C during May-June and from 18° C to 36° C during December-January. The average annual rainfall is around 1,200 mm of which 35% falls during the southwest monsoon (June-September) and 60% falls during the northeast monsoon (October-December). Rainfall is the major source of groundwater recharge in this area. This check dam receives water during monsoon rains (October to December) in the catchment and release of water from Pichature reservoir located at a distance of about 40 km on the upstream side of this check dam (Parimala Renganayaki & Elango, 2015).

Intensive agricultural activity takes place throughout the year, which mainly depends on groundwater. The major crops cultivated in this area are paddy, watermelon, spinach and cucumber. Flowers like jasmine and rose are also grown. The Tamil Nadu Water Supply and Drainage Board tap groundwater to supply water to the houses in nearby villages (Parimala Renganayaki & Elango, 2014).

This region comprises of alluvial deposits overlying the Gondwana clay and functions as an unconfined aquifer. As the clay lenses present in the alluvial formation of this area are only localised and do not extend over a large region, the groundwater occurs under unconfined conditions. Groundwater level measurements made during the study indicate that the water table occurs at a depth from 3 m to 12 m below the ground level (Parimala Renganayaki & Elango, 2015).

The quality of water in the check dam is generally very good with regard to the major ion composition and the total dissolved solids value, which was generally less than 720 mg/L. The groundwater was found suitable for domestic and irrigation purposes based on ionic concentrations in wells located 1.5 km from the dam due to the recharge from good quality water from the check dam (Parimala Renganayaki & Elango, 2014). *Escherichia spp., Salmonella spp., Shigella spp.* and *Clostridium spp.* were present in the water collected from the check dam and wells. Hence, the groundwater is not suitable for drinking purposes (Parimala Renganayaki *et al.*, 2015). As there were no spore producing organisms, the water from check dam and groundwater could be used for domestic purposes after boiling the water at house hold level, whereas municipal water supply requires the proper level of chlorination (Parimala Renganayaki, 2014).

Table 16.2 Application of 'A Water Quality Guide to Managed Aquifer Recharge in India' to three Saph Pani case study sites.

Case Study	Bank Filtration at Haridwar on Ganga River, Uttarakhand	Percolation Tanks at Maheshwaram, Andhra Pradesh	Check Dam near Chennai, Tamil Nadu
Source water and treatment	Mainly rural run-off, Filtration by recharge through river bed alluvium	Peri-urban run-off, infiltration through weathered sediments	River flow from rural run-off, infiltration through river bed
Aquifer type	alluvium	hard rock	alluvium
1. Simple Assessment	**Yes/No**	**Yes/No**	**Yes/No**
1. Is the aquifer being recharged used as a drinking water supply?	Y	Unknown	Y
2. Is the scale of recharge larger than domestic rainwater harvesting?	Y	Y	Y
3. Does the water being recharged contain sewage effluent, industrial wastewater, or urban stormwater?	Y	Y Animal excreta	Y Domestic solid waste disposal, pilgrim activity
4. Is the area around the recharge area ever waterlogged?	Y occasionally when river in flood	Y during monsoon	Y only during excess rains
Simple assessment is satisfied if all answers are No. No need to continue assessment. However if any answer is Yes proceed to Viability assessment.	*Proceed to Viability assessment*	*Proceed to Viability assessment*	*Proceed to Viability assessment*
2. Viability Assessment	**Yes/No**	**Yes/No**	**Yes/No**
1. Is there a sufficient demand for water?	Y	Y	Y
2. Is there an adequate source of water available for allocation to recharge?	Y	Y	Y but volume depends on rainfall in the upper catchment
3. Is there a suitable aquifer for storage and recovery of the required volume?	Y	Y Limited capacity	Y
4. Is there sufficient space available for capture and treatment of the water?	Y	Y	Y
If the answer to any question is No, then the project is not viable or has a major constraint. If answers are Yes, proceed to Guidelines applicability assessment.	*Viable, proceed to Guidelines applicability assessment*	*Viable, proceed to Guidelines applicability assessment*	*Viable, proceed to Guidelines applicability assessment*
3. Guideline Applicability Assessment	**Yes/No**	**Yes/No**	**Yes/No**
1. Is the source of water for recharge from a rooftop or a natural catchment? (i.e. not sewage effluent, industrial wastewater, or urban stormwater)	Y River is mostly rural run-off, but affected by towns and bathing	Y Rural run-off	Y Rural run-off, but some domestic solid waste disposal, pilgrim activity
2. Is the aquifer unconfined and not polluted?	Y Unconfined	Y Unconfined Some water quality issues (fluoride, pesticides)	Y Unconfined Some salinity issues, leading to use of check dams to freshen aquifer

3. Is the proposed recharge area remote from public drinking water supply systems?	N This is for public water supply.	Y There is not a public drinking water supply in the direct area.	N There are 2 public drinking water production wells, and many irrigation wells sometimes used for drinking.
These Guidelines are applicable if all answers above are Yes. Otherwise not applicable use alternate Guidelines e.g. Australian MAR Guidelines	*Not Applicable***	*Proceed to Viability assessment*	*Not Applicable***
4. Sanitary Survey	**Yes/No**	**Yes/No**	**Yes/No**
1. Is there a latrine, open sewer or leaky sewer or human or animal faeces within the catchment area of the recharge facility?	Y Unsealed & temporary pit latrines, public bathing/washing at the well heads, presence of cattle in and around the RBF wells, unsanitary defecation practices	Y Animals wash in percolation tanks	Y Latrines, animal waste, domestic waste and sewer
2. Is there a latrine, open sewer, leaky sewer or animal faeces in close proximity to the recharge structure or to the wells from which water will be recovered?	Y As above	Y Recovery is via farm wells.	Y Latrines, animal waste, domestic waste and sewer
3. Are there industrial, transport or agricultural activities generating stockpiles, wastes, spills, or emissions reaching the surface of the catchment area of the recharge facility?	Y Some use of agricultural chemicals and fertilisers and transport of fuels through the catchment	Y Agricultural activities	Y only agricultural stockpiles for short time periods in a year
4. Are there industrial, transport or agricultural activities generating stockpiles, wastes, spills, or emissions in close proximity to the recharge structure or the wells from which water will be recovered?	Y As above	Y Agricultural activities	Y only agricultural stockpiles for short time periods in a year
5. Is there pre-treatment or means of preventing contaminated water to be recharged? If so describe its design and resilience to power and mechanical failure, and any alarm systems.	Y Natural treatment by passage through aquifer before recovery	N	N
6. Is there post-treatment of water to be recovered? If so describe its design and resilience to power and mechanical failure, and any alarm systems.	Y Chlorination. Pumps shut down when power fails.	N	Y Chlorination of both public drinking water supply wells. No alarm system.
7. Does the existence and condition of any barriers around of the recharge structure and recovery wells prevent short circuit of contaminated water?	Y Concrete base around wells of limited effectiveness (see Chap. 2)	N	N

(*Continued*)

Table 16.2 Application of 'A Water Quality Guide to Managed Aquifer Recharge in India' to three Saph Pani case study sites (*Continued*).

Case study	**Bank filtration at Haridwar on Ganga River, Uttarakhand**	**Percolation tanks at Maheshwaram, Andhra Pradesh**	**Check dam near Chennai, Tamil Nadu**
*Any question answered by **Yes** needs to be taken into specific account in the Water Safety Plan below. Even if not observed, the possibility of these hazards occurring or barriers being breached also needs to be taken into account.* Proceed to Aquifer assessment	*Account for these in MAR water safety plan*	*Account for these in MAR water safety plan*	*Account for these in MAR water safety plan*
5. Aquifer Assessment	**Yes/No**	**Yes/No**	**Yes/No**
1. Does source water have low quality; is water turbid, coloured, contains algae, has a surface slick or does it smell?	Y Turbid especially in high flows	Y Turbid	N
2. Does the unconfined aquifer have a shallow water table, say <8m in urban area and say <4m in rural area?	Y Shallow water table. Increase in groundwater levels in monsoon due to elevated surface water levels (Ganga River and UGC)	Y Water table before monsoon ~25 m below ground level (bgl), water table after monsoon ~1 m bgl	Y Water table before monsoon ~15 m bgl after monsoon ~4 m bgl
3. Are there other groundwater users, groundwater-connected ecosystems or a property boundary within 100m of the recharge site?	Y Private wells located further from river than the bank filtration well	Y No management plan, other groundwater users are within 100m	Y Private wells for irrigation purposes and a few wells for domestic use
4. Is the aquifer known to contain reactive minerals (e.g. pyrite) or is groundwater in this area known to contain arsenic? Does the aquifer contain soluble minerals such as calcite and dolomite?	N Alluvial aquifer sourced by natural recharge of same source. No new geochemical reactions expected	Y Fluoride	N Alluvium comprising of sand, silt and clay.
5. Is the aquifer composed of fractured rock or karstic (fissured or cavernous) limestone or dolomite?	N	Y Fractured rock	N
6. Is the proposed project of such a scale that it requires development approval? Is it in a built up area; built on public, flood-prone or steep land; or close to a property boundary? Does it contain open water storages or engineering structures; or is it likely to cause public health or safety issues (e.g. falling or drowning), nuisance from noise, dust, odour or insects (during construction or operation), or adverse environmental impacts (e.g. from waste products of treatment processes)?	Y Existing project primarily since 1980 (Bartak *et al.*, 2014). Built on flood prone land.	Y Existing project	Y Existing project. In rural area and water is stored within the river. No issues during construction or operation regarding noise, dust, insects. No known adverse environmental impact. Prevents saline intrusion.

(In these three cases the projects already exist, and answered as though approaching the project prior to construction.)			
Any question answered by **Yes** *needs to be taken into specific account in the Water Safety Plan below. Even if not observed, the possibility of these hazards occurring or barriers being breached also needs to be taken into account.*	*Account for possibility of these in MAR water safety plan*	*Account for possibility of these in MAR water safety plan*	*Account for possibility of these in MAR water safety plan*
Proceed to Water Safety Plan			
6. Managed Aquifer Recharge Water Safety Plan	**Yes/No**	**Yes/No**	**Yes/No**
Human sewage entrainment in source water			
1. Do latrine leakage, open sewers, sewer pipe leaks, open defecation occur in the catchment or close to recharge facilities or recovery wells?	Y Latrines, leaky sewers, open defecation	Y Open defecation	Y Latrines, animal waste, domestic waste and sewer
Animal faecal matter entrainment in source water			
2. Are there any animal manure accumulations or cess pits in the catchment or close to the recharge area or recovery well?	Y Stockpiles of animal manure, fresh animal faecal material close to extraction wells	Y	Y Animal waste from nearby pilgrim centre, domestic waste and sewer draining into the river
Leaching of microbial contaminants into aquifer			
3. Can water infiltrate that has been in contact with human and animal wastes?	Y Floodplain and ambient ground-water are likely to contribute more contaminants to drinking water than bank-filtered river water	Y Animals have access to percolation tank	Y Bathing of humans and animals in the dam
Entrainment of chemicals in source water for recharge			
4. Are there industry, transport and agricultural activities generating stockpiles, wastes, spills, or emissions reaching the catchment surface?	Y Manures and agricultural chemicals	Y	Y but only agricultural stockpiles for short time periods in a year
Leaching of chemicals into groundwater			
5. Can water infiltrate from landfills, waste dumps or industrial discharge?	Y from manure stockpiles, possible spills of transport fuels, pesticides	N	N

(Continued)

Table 16.2 Application of 'A Water Quality Guide to Managed Aquifer Recharge in India' to three Saph Pani case study sites (*Continued*).

Case study	Bank filtration at Haridwar on Ganga River, Uttarakhand	Percolation tanks at Maheshwaram, Andhra Pradesh	Check dam near Chennai, Tamil Nadu
Bypassing or failure of pre-treatment in recharge facility			
6. Can recharging water short-circuit existing barriers to protect groundwater quality? (for example caused by clogging of filters, power or mechanical failures, or treatment chemicals running out)	Y Bank filtration is robust, but anthropogenic activities near/around well head houses may defeat their protective purpose	N Infiltration	N
Bypassing or failure of post treatment at recovery well			
7. Can water used for drinking bypass existing barriers or treatments (for example caused by clogging of filters, power or mechanical failures, lack of fuel for boiling water, or treatment chemicals running out)?	Y Chlorination, pumps shut down when power fails, but leaky mains allow ingress of contaminated shallow groundwater when pipes are not pressurised. Plan should include verifying chlorination level, procedures during floods and repairing leaky pipes	N No treatment	Y Treatment by chlorination of public water supply wells, but no alarm system in event of chlorination failure.
See Table 16.1 to identify suitable control measures, critical limits, monitoring, corrective actions and verification. Insert site-specific measures in consultation with stakeholders and agree on whom to take each action is required.	*Stakeholders capable of implementation to complete the MAR Water Safety Plan*	*Stakeholders capable of implementation to complete the MAR Water Safety Plan*	*Stakeholders capable of implementation to complete the MAR Water Safety Plan*
Extra work is warranted to monitor at these sites (**) with higher risks to better assess risks to public water supplies and implement remedies.	**		**

**Two of these aquifer recharge sites directly contribute to public drinking water supplies. Therefore this approach alone is considered unable to confidently protect human health (i.e. not applicable because of the high potential risk to drinking water supplies). However, the guidelines were followed through regardless to help form a water safety plan to encourage immediate actions to improve water quality and to identify issues needing to be considered in a more rigorous assessment.

16.4 CONCLUSION

For many years India has been the international leader in the volume of water intentionally recharged and the number of recharge structures employed to sustain groundwater resources. In many cases the source of water for recharge is the same source that recharges the aquifer naturally, and little change in groundwater quality is expected. However, enhancing recharge can shorten the residence time of water in the unsaturated zone or accelerate the movement of water from streams to wells, potentially diminishing the opportunity for contaminant removal. Recharge enhancement can result in groundwater quality improvement for agricultural irrigation supplies while at the same time deteriorating the quality for drinking water supplies. Where groundwater is used as a drinking water source, it should be an obligation of those enhancing recharge to protect the health of those whose drinking water is affected by their operations.

Application of 'A Water Guide to Managed Aquifer Recharge in India' (Dillon *et al.*, 2014), to three Indian case studies within Saph Pani identified practical protective measures that can be applied to improve the protection of groundwater from contamination arising from recharge operations. It was found that the example form of a water safety plan in Table 16.2 was relevant to the three sites evaluated. The next step would be to identify local stakeholders who would take responsibility to undertake and implement the water safety plan.

The Haridwar and Chennai case studies involve public drinking water supplies. These are outside of the recommended scope of application of this approach as it is not feasible to confidently protect human health in public water supplies in the absence of water quality data, particularly for microbial pathogens. However, the approach was followed to help identify the issues needing to be considered in a water safety plan to improve the safety of supplies and to also to inform on the types of data that would support a more rigorous assessment.

For the Haridwar bank filtration system, based on monitored water quality data for the period 2005–2013, a comprehensive staged approach to assess the risks from 12 hazards to human health and the environment was undertaken by Bartak *et al.* (2014). Accordingly, it was determined that the risks from inorganic chemicals, salinity, nutrients and turbidity were acceptable. A quantitative microbial risk assessment (QMRA) indicated that the risks to human health from bacterial pathogens (Escherichia coli O157:H7) were below the reference risk used in the study (Bartak *et al.*, 2014). This QMRA was limited by inadequate characterization of viral and protozoan pathogen numbers in source water. Pathogen removal capabilities for bank filtration at several sites reported in literature indicate high removal capabilities even for viral and protozoan pathogens. Given the potential for high pathogen numbers in source waters at Haridwar, the site-specific pathogen risks warrant better characterization.

In the absence of water quality data it cannot be claimed that the approach was validated, and that its application will assure safety. It can however be confidently claimed that the application of this approach, which requires only basic information that is readily observable, will make recharge enhancement safer. The next stage would be to apply, test and refine this document for artificial recharge in rural and urban conditions, for a wider range of waters, aquifer types, recharge methods, end uses of recovered water, and capabilities of implementing institutions, to facilitate adoption by state and federal authorities. Sites where a water safety plan is developed and implemented can then be claimed to have progressed from artificial recharge to managed aquifer recharge.

16.5 REFERENCES

Bartak R., Page D., Sandhu S., Grischek T., Saini B., Mehrotra I., Jain C. K. and Ghosh N. (2014). Application of risk-based assessment and management to riverbank filtration sites in India. *Journal of Water and Health*, **13**(1), 174–189.

Boisson A., Baisset M., Alazard M., Perrin J., Villesseche D., Dewandel B., Kloppmann W., Chandra S., Picot-Colbeaux G., Sarah S., Ahmed S. and Maréchal J.-C. (2014). Comparison of surface and groundwater balance approaches in the evaluation of managed aquifer recharge structures: Case of a percolation tank in a crystalline aquifer in India. *Journal of Hydrology*, **519**, Part B, 1620–1633.

Bureau of Indian Standards (BIS) (1986). Guidelines for the Quality of Irrigation Water. IS 11624 (1986, Reaffirmed 2009): UDC 631.671.03: 626.810 (026) https://archive.org/details/gov.law.is.11624.1986 (accessed 8 April 2015).

Bureau of Indian Standards (BIS) 10500 (2012). Drinking Water – Specification. Indian Standard, 2nd Revision. Bureau of Indian Standards, New Delhi, India. https://law.resource.org/pub/in/bis/S06/is.10500.2012.pdf (accessed 8 April 2015).

CGWB (2000). Guide on Artificial Recharge to Ground Water. Central Ground Water Board. Ministry of Water Resources, Government of India. New Delhi, May, 2000. http://www.cgwb.gov.in/documents/Guide_on_ArtificialRecharge.pdf (accessed 8 April 2015).

CGWB (2005). Master plan for artificial recharge to groundwater in India. Central Ground Water Board, Ministry of Water Resources, Government of India. http://cgwb.gov.in/documents/MASTER%20PLAN%20Final-2002.pdf (accessed 8 April 2015).

CGWB (2007). Manual on artificial recharge of groundwater. Central Ground Water Board, Ministry of Water Resources, Government of India, 185p. http://www.cseindia.org/userfiles/Manual%20on%20Artificial%20Recharge%20of%20Ground%20Water.pdf (accessed 8 April 2015).

CGWB (2013). Master Plan for Artificial Recharge to Groundwater in India. Central Ground Water Board, Ministry of Water Resources, Government of India. http://cgwb.gov.in/documents/MasterPlan-2013.pdf (accessed 8 April 2015).

CPCB (2007). Guidelines for Water Quality Monitoring MINARS/27/2007–08. Central Pollution Control Board, Government of India. http://cpcb.nic.in/upload/NewItems/NewItem_116_Guidelinesof%20waterqualitymonitoring_31.07.08.pdf (accessed 8 April 2015).

Das Gupta S. P. (1975). The flood plain of Ganga. In: Upper Gangetic Flood Plain. S. P. Das Gupta (ed.), National Atlas Organization Monograph, Govt. of India, Dept. of Sci. and Technol., Calcutta, pp. 1–7.

Dash R. R., Bhanu Prakash E. V. P., Kumar P., Mehrotra I., Sandhu C. and Grischek T. (2010). River bank filtration in Haridwar, India: Removal of turbidity, organics and bacteria. *Hydrogeology Journal*, **18**(4), 973–983.

Davison A., Howard G., Stevens M., Callan P., Fewtrell L., Deere D. and Bartram J. (2005). Water Safety Plans: Managing Drinking-Water Quality from Catchment to Consumer. World Health Organisation, Geneva. WHO/SDE/WSH/05.06. 235p. http://www.who.int/water_sanitation_health/dwq/wsp0506/en/ (accessed 8 April 2015).

Dewandel B., Lachassagne P., Wyns R., Maréchal J. C. and Krishnamurthy N. S. (2006). A generalized 3-D geological and hydrogeological conceptual model of granite aquifers controlled by single or multiphase weathering. *Journal of Hydrology*, **330**, 260–284. doi: 10.1016/j.jhydrol.2006.03.026

Dillon P., Page D., Vanderzalm J. and Chadha, D. (2013). Water quality considerations in managed aquifer recharge: From artificial recharge to managed aquifer recharge in India. *Journal of Groundwater Research*, **2**(2), 8–15.

Dillon P., Vanderzalm J., Page D., Toze S., Wolf L., Pavelic P., Cunliffe D., Wang W., Willardson B., Tredoux G., Jain R. and Raj R. (2010). Australian Guidelines for Managed Aquifer Recharge and their International Relevance. Proc. ISMAR7, Abu Dhabi 9–13 Oct 2010. www.ismar7.org (accessed 8 April 2015).

Dillon P., Vanderzalm J., Sidhu J., Page D. and Chadha D. (2014). A Water Quality Guide to Managed Aquifer Recharge in India. CSIRO Land and Water Flagship and UNESCO. https://publications.csiro.au/rpr/pub?pid=csiro:EP149116 (accessed 8 April 2015).

Gangwar K. K. and Joshi B. D. (2004). A preliminary study on solid waste generation at Har Ki Pauri, Haridwar, around the Ardh-Kumbh period of sacred bathing in the river Ganga in 2004. *Environmentalist*, **28**(3), 297–300.

Government of India Ministry of Water Resources (2012). National Water Policy. http://wrmin.nic.in/writereaddata/NationalWaterPolicy/NWP2012Eng6495132651.pdf (accessed 8 April 2015).

Maréchal J. C., Dewandel B. and Subrahmanyam K. (2004). Use of hydraulic tests at different scales to characterize fracture network properties in the weathered-fractured layer of a hard rock aquifer: Fracture network properties in hard rock. *Water Resources Research*, **40**. doi: 10.1029/2004WR003137.

Maréchal J. C., Dewandel B., Ahmed S., Galeazzi L. and Zaidi F. K. (2006). Combined estimation of specific yield and natural recharge in a semi-arid groundwater basin with irrigated agriculture. *Journal of Hydrology*, **329**, 281–293. doi:10.1016/j.jhydrol.2006.02.022

Ministry of Environment & Forests (1992). Policy Statement for Abatement of Pollution. Government of India, New Delhi http://envfor.nic.in/sites/default/files/annual_report/psap.pdf (accessed 8 April 2015).

NRMMC–EPHC–AHMC (2006). Australian Guidelines for Water Recycling: Managing Health and Environmental Risks (Phase 1), Natural Resources Ministerial Management Council, Environment Protection and Heritage Council and Australian Health Ministers' Conference, Canberra. http://www.environment.gov.au/resource/national-water-quality-management-strategy-australian-guidelines-water-recycling-managing-0 (accessed 8 April 2015).

NRMMC–EPHC–NHMRC (2009). Australian Guidelines for Water Recycling (Phase 2): Managed Aquifer Recharge. Natural Resource Ministerial Management Council, Environment Protection and Heritage Council and National Health and Medical Research Council, Canberra, http://www.environment.gov.au/resource/national-water-quality-management-strategy-australian-guidelines-water-recycling-managing-1 (accessed 8 April 2015).

NHMRC-NRMMC (2011). Australian Drinking Water Guidelines. National Health and Medical Research Council and Natural Resource Ministerial Management Council, Canberra https://www.nhmrc.gov.au/guidelines/publications/eh52 (accessed 8 April 2015).

Pettenati M., Perrin J., Pauwels H. and Ahmed S. (2013). Simulating fluoride evolution in groundwater using a reactive multicomponent transient transport model: Application to a crystalline aquifer of Southern India. *Applied Geochemistry*, **29**, 102–116. doi: 10.1016/j.apgeochem.2012.11.001.

Pettenati M., Picot-Colbeaux G., Thiéry D., Boisson A., Alazard M., Perrin J., Dewandel B., Maréchal J.-C., Ahmed S. and Kloppmann W. (2014). Water quality evolution during managed aquifer recharge (MAR) in Indian crystalline basement aquifers: Reactive transport modeling in the critical zone. *Procedia Earth and Planetary Science*, **10**, 82–87. doi: 10.1016/j.proeps.2014.08.016.

Parimala Renganayaki S. and Elango L. (2014). Impact of recharge from a check dam on groundwater quality and assessment of suitability for drinking and irrigation purposes. *Arabian Journal of Geosciences*, **7**(8), 3119–3129.

Parimala Renganayaki S. and Elango L. (2015). Assessment of recharge from a check dam as a method of Managed Aquifer Recharge by hydrogeological investigations. *Environmental Earth Sciences*, **73**(9), 5349–5361.

Parimala Renganayaki S. (2014). Managed Aquifer Recharge: An Integrated Approach to Assess the Impact of a Check Dam. PhD thesis, Anna University, Chennai, India.

Parimala Renganayaki S., Brindha K., Sankaran K. and Elango L. (2015). Effect of recharge from a check dam and river bank filtration on geochemical and microbial composition of groundwater. *Arabian Journal of Geosciences*, DOI: 10.1007/s12517-015-1825-4

Sandhu C., Grischek T., Kumar P. and Ray D. (2011). Potential for riverbank filtration in India. *Clean Technologies and Environmental Policy*, **13**(2), 295–316.

Saph Pani D1.2 (2013). Guidelines for flood-risk management of bank filtration schemes during monsoon in India. Saph Pani Project Deliverable. http://www.saphpani.eu/downloads (accessed 8 April 2015).

WHO (2011). Guidelines for Drinking Water quality 4th edition. World Health Organisation, Geneva. http://whqlibdoc.who.int/publications/2011/9789241548151_eng.pdf (accessed 8 April 2015).

WHO (2012). Water safety planning for small community water supplies: Step-by-step risk management guidance for drinking-water supplies in small communities. World Health Organisation, Geneva. http://www.rural-water-supply.net/en/resources/details/396 (accessed 8 April 2015).

WHO (2014). Water safety plan: a field guide to improving drinking water safety in small communities. (B. Rickert, O. Schmoll, A. Rinehold, E. Barrenberg) World Health Organisation, Regional Office for Europe, Copenhagen. http://www.euro.who.int/en/publications/abstracts/water-safety-plan-a-field-guide-to-improving-drinking-water-safety-in-small-communities (accessed 8 April 2015).

Chapter 17

Rapid assessment and SWOT analysis of non-technical aspects of natural wastewater treatment systems

Markus Starkl, Priyanie Amerasinghe, Laura Essl, Mahesh Jampani, Dinesh Kumar and Shyam R. Asolekar

17.1 INTRODUCTION

A general overview and technical details of natural treatment systems (NTS) including constructed wetlands (CWs), waste stabilization ponds (WSPs), duckweed ponds (DPs), water hyacinth ponds and polishing ponds have been provided in Chapters 8 and 10. As outlined in Starkl *et al.* (2013), often assessment studies focus on technical aspects only, with no or little consideration of the non-technical aspects. It has been argued that the non-technical aspects do influence the long-term sustainability of technologies and therefore their critical assessment is of importance. This chapter compliments the previous information through investigations on environmental, health and safety as well as economic, social and institutional aspects of those technologies. The work presented here encompasses an initial sustainability appraisal of currently existing NTSs followed by a strength-weaknesses-opportunities-threats (SWOT) analysis.

NTS utilise natural processes such as attenuation and buffering capacity of natural soil-aquifer and plant-root systems and as such, the process of contaminant removal is not aided by the input of significant amounts of energy and/or chemicals (Sharma & Amy, 2010). NTS can be classified as soil-based and aquatic treatment systems. Examples for soil-based systems are horizontal sub-surface flow constructed wetlands (HSSF-CWs), soil aquifer treatment systems or planted filters. Aquatic systems are DPs or WSPs. They can be used as secondary or tertiary treatment systems and in combination with conventional and other NTS (hybrids) or be solely based on the influent water quality and intended reuse of the treated water. It has also been reported that a combination of different treatment technologies allows for improved water quality of the effluent (Alvarez *et al.*, 2008; Mbuligwe, 2004; Kaseva, 2003).

The survey of existing NTS across India showed that the NTS for wastewater treatment are WSPs and DPs; other technologies such as modified CWs and floating wetlands have been implemented only at pilot scale so far. A detailed overview of NTS in India can be found in Chapter 8.

17.2 METHODOLOGY

The main intended benefit of all Saph Pani case studies is the provision of and access to safe water for human consumption or agricultural use. Thus, the rapid assessment evaluated selected case studies to see if intended benefits of the NTS (technologies) were achieved. Further, other relevant expected and unexpected benefits were also studied, for example income generation and employment for those communities that are associated with the systems and risks that could jeopardize the successful functioning of the systems. Based on the intended and unintended benefits, current risks and future risks, the case studies were classified as "success" or "failure" cases. During the rapid assessment the underlying reasons for success or failure of the cases were also studied and a SWOT analysis was carried out for a robust assessment. The methodology for the rapid assessment was based on previous studies conducted (Starkl *et al.*, 2010) and is comprised of the following four steps (Figure 17.1):

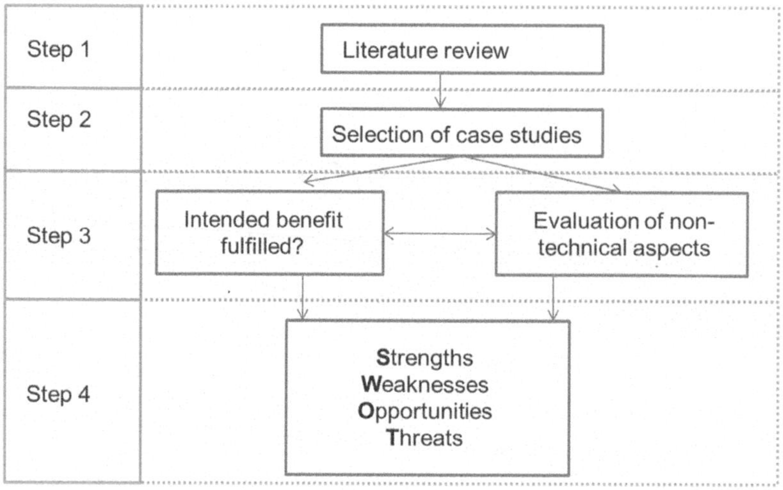

Figure 17.1 Methodology adopted for Assessment.

17.2.1 Step 1: Survey and review of existing information on Indian case studies

First, a survey and a review of natural treatment technologies that exist in India were carried out based on a literature search. Existing information on non-technical (environmental, health and safety, economic, social and institutional) aspects was summarised and relevant issues and knowledge gaps were highlighted.

17.2.2 Step 2: Identification of suitable case studies for the rapid assessment

After the survey and review of existing case studies, suitable case studies were selected for the rapid assessment. They were selected based on certain criteria such as existing knowledge gaps, its current use, accessibility or application under real life conditions.

17.2.3 Step 3: Rapid assessment

The rapid assessment was primarily based on questionnaires: a general questionnaire for all case studies and tailor-made, additional questionnaires were used for the different technology groups, considering the already available information and technology specifications. The general questionnaire was targeted at collecting basic background information, especially non-technical information. The specific questionnaires focused on aspects important for each of the technology groups (e.g. certain risks that are only relevant for a certain technology, such as e.g. health risks and safety of wastewater reuse in food production). Expert visits and initial interviews with targeted stakeholders and users were conducted to fill in the questionnaires and get an overall impression of the functioning of the NTSs.

17.2.4 Step 4: SWOT analysis

To assess the potential of the technologies in India, a SWOT-analysis was conducted. SWOT analysis was initially developed for business management, but has also been used in natural resource management (e.g. Srivastava *et al.*, 2005; Terrados *et al.*, 2007).

The SWOT analysis provides a framework for analyzing a situation by identifying strengths and weaknesses, but also recognises challenges and develops strategies for the future (Srivastava *et al.*, 2005). Thus, in this analysis, the strengths are viewed as advantages that support the decision to implement a system; weaknesses show what can be improved or what needs to be investigated before implementation. Opportunities refer to possible chances and positive improvements, whereas threats show risks and obstacles for the future.

17.3 RESULTS AND DISCUSSION

17.3.1 WSP in the city of Mathura, state of Uttar Pradesh in northern India: Case study 1

The intended benefit of WSP is treatment of wastewater according to Indian standards and reuse of the effluent if possibilities for reuse exist within the nearby surrounding of the treatment plant. WSPs work without energy input and operation and maintenance (O&M) is limited to the removal of solids from the pre-treatment unit. The WSP in the city of Mathura consists of a pre-treatment unit with rack and grit chamber and two treatment chains consisting of four ponds (Figures 17.2 and 17.3). The first pond is an anaerobic pond, followed by two facultative anaerobic ponds (FAP) and a maturation pond (MP). However, at the time of assessment, only one set of ponds is functioning and the other is being dried out for repairs. Currently, water is not reused, but it had been attempted to cultivate fish in the FAP and MP. Due to problems (see below) this practice was stopped.

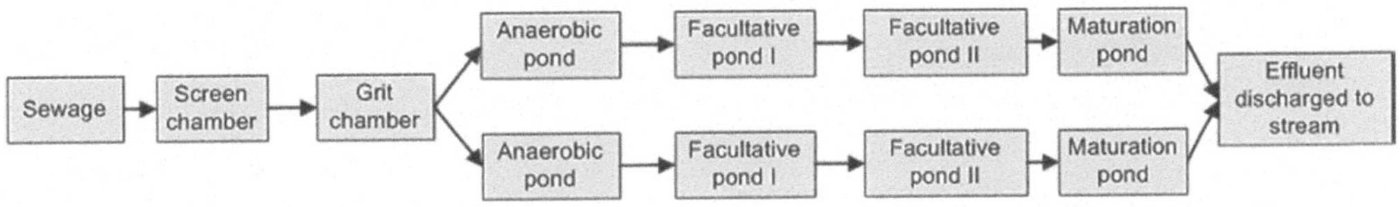

Figure 17.2 Schematic flow chart of WSP Mathura.

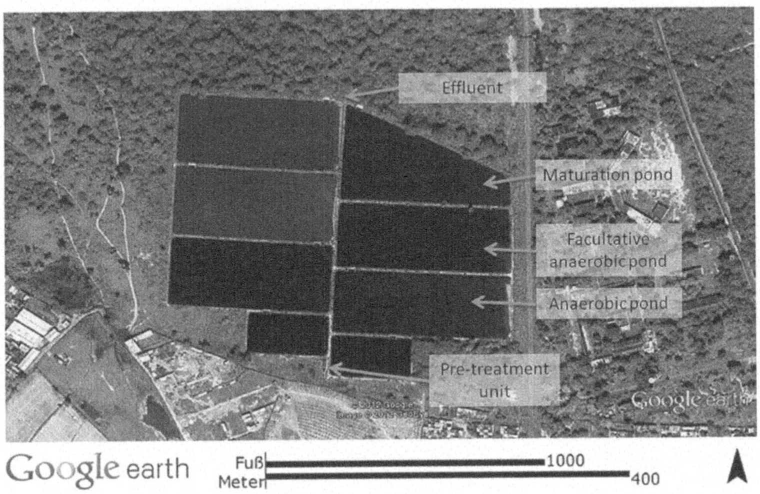

Figure 17.3 WSP in Mathura, picture from 2009 (*Source*: Google Earth).

Domestic wastewater (13.59 MLD) from the city of Mathura is conveyed to this treatment plant, but the community living close to it is not connected to the sewer network. The WSP was constructed and commissioned by the National River Conservation Directorate and operated and maintained by Mathura Jal Board (Water Board).

Health aspects

There could be risks related to faecal transmission of pathogenic agents to operators as they have no special equipment for the handling of primary and secondary sludge; they use a shovel and have no gloves. However, the degree of risk could not be assessed as the quality of the incoming wastewater was not available for study.

Communities living close to the treatment system appeared to have been affected by the WSP as commented by some members. According to these comments some members became sick due to the nearby drinking water pumps having become contaminated. While there is no water quality data available to support this, possible contamination of the groundwater could have occurred due to a hole in the cement lining in the FAP (Figure 17.4). At present, the WSPs are undergoing repairs and are therefore not functioning. Treated wastewater is not reused in agriculture; so there are no risks to farmers and consumers.

Figure 17.4 Hole in lining of facultative anearobic pond.

Social aspects

The field visit has revealed a problem of acceptance: according to three local people who were interviewed during the site visit, the last of the four ponds was used for cultivation of fish. One local user has informed that after a community member fell sick, the practice of rearing fish was stopped. Local people believed that the reason for the illness of the community member was due to eating contaminated fish from the pond.

As mentioned above, due to the problem related to contamination of groundwater, communities are unhappy with the placement of the WSP as the system is not even serving their community by collecting the sewage. These tensions could become a problem in the future.

Institutional aspects

The main institutions involved are the Mathura Jal Board, the Central Pollution Control Board (CPCB) and a private company that is contracted for one year by the Mathura Jal Board. One technical supervisor of the Mathura Jal Board was responsible for supervising all wastewater treatment plants (WWTPs) in Mathura.

The treatment performance is monitored every month by the Mathura Jal Board and the CPCB, but the information on the performance is not available to the public. The actual O&M is handled by the private company. Two operators have been selected from the local community. The operators are responsible for cleaning the rack and guiding the plant. They did not receive specific training. The site visit showed that the institutional arrangements worked well since technical problems such as infiltrating wastewater were being tackled immediately.

Economic aspects

The construction costs of the WSP are not known. The O&M of the WSP has been outsourced to a private company at a cost of 400,000 INR per year (≈5,000 EUR)[1] according to the operators. The salary of the operators was reported as 32,000 INR per year (≈400 EUR) and free housing was provided by the company close to the plant.

According to the operators, there is no revenue from selling any by-products. The treated water is discharged to the nearby stream and sludge is stacked around the premises of the treatment plant. The maturation pond was successfully used for rearing, but this practice was stopped due to acceptance problems.

Summary of evaluation results

The intended benefits of the treatment plant are mainly fulfilled (Table 17.1). Effluent quality could not be assessed as the monitoring results are not available public, but it seems that the system is working well based on the visual impression during the field visit. Reuse of the treated water for irrigation is not possible within the near surroundings as there is no farmland adjacent to the treatment plant (Figure 17.5). No energy is required to operate the treatment plant, but nevertheless power cuts appear to affect the system as the wastewater is pumped to the pond system. The operators who were selected from the local community have no special skills. Their main task is removing solids from the pre-treatment unit.

[1]Average currency exchange rate of year 2014: INR EUR = 0.0123 (Online Currency Converter, 2015). All amounts indicated in EUR are calculated with this currency exchange rate.

Table 17.1 WSP Mathura – Fulfilment of intended benefits.

Intended Benefit/Purpose	Fulfilled (yes/no)	Comments
Treatment of wastewater according to Indian standard	not available	Monitoring results not available. Based on the impression during the field visit, the system seems to be working well, even though only 50% of the treatment plant were operational, the effluent was clear (visual impression).
Reuse of treated wastewater	not applicable	Reuse within the surrounding not possible as no agriculture is practiced.
No energy requirements	yes	The treatment plant itself requires no energy to be functional. Pumping is required to transport wastewater to the treatment plant.
No skills for O&M required	yes	The operators are from the local community and need no special training; their main task is removal of solids from pre-treatment unit.

Figure 17.5 Maturation ponds: left side operational, right side not functional.

17.3.2 WSP in the city of Agra, state of Uttar Pradesh in northern India: Case study 2

This waste stabilization pond (WSP) was built 17 years ago in Agra by the local water board. It has a capacity of 10 MLD and treats mixed domestic and industrial wastewater. The water is passing two treatment chains with an anaerobic pond; two facultative ponds and one maturation pond on each side (Figures 17.6–17.8). Treated effluent is being discharged into a stream.

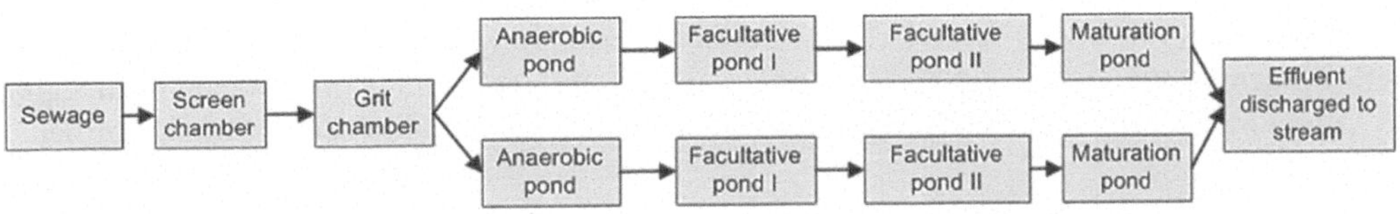

Figure 17.6 Schematic flow chart of WSP Agra.

Health aspects

The risks to operators are similar as in the WSP Mathura case study (Chapter 17.3.1). The operator reported that when the system was built 18 years ago, the ponds were not lined and water could infiltrate which caused groundwater pollution. Then the system was lined with concrete. Now there is a monitoring well to ensure that groundwater is not contaminated. Treated wastewater is not reused in agriculture, therefore no risks to farmers and consumers occur.

Social aspects

According to the operator, farmers do not want to use the water for irrigation due to the high salinity levels caused by the wastewater coming from the textile industry. It was reported that farmers also do not want to use the sludge because they believe that it is harmful to the plants due to an overdose of nutrients. According to the operator, there is no problem of acceptance of the treatment plant by the nearby residing local community.

Figure 17.7 Pre-treatment unit.

Figure 17.8 Maturation pond.

Institutional aspects

The main institutions involved are the local water board, the CPCB and a private company contracted for O&M. The local water board is responsible for monitoring of effluent quality and groundwater wells. The CPCB is conducting additional monitoring. Neither institution publishes the monitoring results.

The private company is responsible for O&M of the treatment plant. There were three operators and one supervisor working in the treatment plant. Their main tasks were cleaning the racks and the surrounding of the pond. Once a year each side of the treatment chain is being cleaned.

Economic aspects

The construction costs are not known. According to the operator the contracted company receives 700,000 INR annually (≈8,500 EUR) for the O&M of the treatment plant. The staffs receive a salary of 2,500 INR/month (≈30 EUR/month) (operator) and 5,000 INR/month (≈60 EUR/month) (supervisor). There are no benefits from the reuse of side products.

Summary of evaluation results

The intended benefits are partly fulfilled (Table 17.2). Based on the visual impression during the site visit, the WWTP seems to be working well. Effluent is not reused even though reuse was initially intended. No monitoring results are available to the public. No energy is required for the treatment process and O&M works well. No further risk was identified.

Table 17.2 WSP Agra – Fulfilment of intended benefit.

Intended Benefit/Purpose	Fulfilled (yes/no)	Comments
Treatment of wastewater according to Indian standard	not available	No monitoring results available. System seems to be working well.
Reuse of treated wastewater	no	The treated wastewater is not used for irrigation as farmers have reservations against the water quality.
No energy requirements	yes	The treatment plant itself requires no energy to be functional. Pumping is required to transport wastewater to the treatment plant.
No skills for O&M required	yes	The operators are from the local community and need no special training. Their main task is removal of solids from pre-treatment unit.

17.3.3 HSSF-CW in Katchpura slum, city of Agra, state of Uttar Pradesh in northern India: Case study 3

The horizontal sub-surface flow constructed wetland (HSSF-CW) is the last part of a treatment system consisting of a baffled septic tank, anaerobic baffled reactor and/or anaerobic filter. The intended benefit is treatment of wastewater according to Indian norms and reuse of treated wastewater. The system is easy to operate and maintain and requires no energy. The system consists of a pre-treatment unit, a baffled septic tank, a baffled HSSF-CW planted with reed beds (Figures 17.9 and 17.10). The assessment was made for the whole treatment plant, which receives 0.05 MLD of domestic wastewater per day.

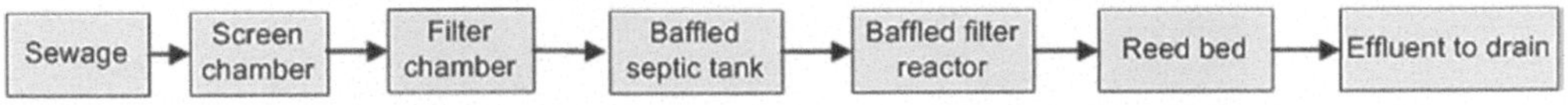

Figure 17.9 Schematic flow chart of HSSF-CW in Katchpura slum.

Figure 17.10 Treatment unit planted with Canna indica.

The treated wastewater is being reused for gardening. The remaining wastewater at the site enters into the storm water drainage (as the system is very small) and is conveyed to the Yamuna River. An additional intended benefit of the system is the improvement of the environmental situation in the area. It is planned to up-scale the system in the near future. For the larger system, reuse of water for irrigation purposes is intended. The location of the treatment system within the slum area can be seen in Figure 17.11.

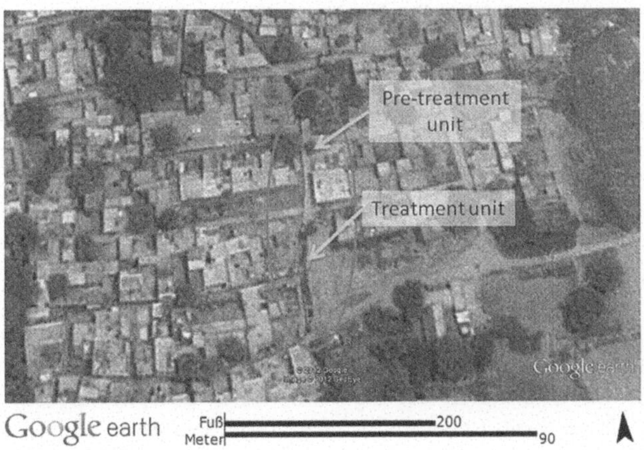

Figure 17.11 Planted gravel filter in Agra (Source: Google Earth).

Health aspects

Operators are exposed to a possible risk of faecal transmission when emptying the grit chamber. It is located under the pavement (Figure 17.12) and even more difficult to empty than conventional pre-treatment units. However, the risk cannot be quantified as the quality of the incoming wastewater is not known.

Figure 17.12 Primary treatment unit (baffled septic tank).

The system is located in a residential area. According to the operator, there were no complaints about mosquitoes or other nuisances affecting the health so far. There is no health risk emanating from the treatment system for farmers.

Social aspects

Treated wastewater is being well accepted in the nearby community for irrigation of gardens. There were no problems related to acceptance of the treatment plant so far. With the implementation of the system, employment opportunities during construction and for O&M were created and the environmental situation is improving. The community was involved in the construction and the newly created pavement on the top of the pre-treatment unit is a meeting place for the villagers. Community participation was an integral part of implementation and the community is also involved in the operation of the treatment plants.

Institutional aspects

The main institutions involved in the O&M of the treatment system is the Centre for Urban and Regional Excellence (CURE) which has already assisted in construction, with additional support provided by the Agra Nagar Nigam, USAID FIRE (D), Cities Alliance and financial assistance from Water Trust, United Kingdom and London Metropolitan University.

Two operators from the community are operating the treatment plant. They were trained and in case of problems the implementing NGO (CURE) can be contacted. Their main task is the cleaning of the rack, while all other task e.g. cleaning of filter material or removing of solids from grit chamber are done when necessary. Every three months, the effluent quality is monitored by the local NGO, but the results are not public.

The current institutional arrangement works well as in case of problems the local NGO provides support to the operators. Until now, only one problem occurred due to flooding of sewers, which could be solved by the operators in cooperation with the NGO: the system was blocked in March 2012 and the operators had to remove the filter material, wash it manually and put it back in the system.

Economic aspects

The overall O&M of the system was found to be satisfactory. Reportedly, the capital cost incurred for establishing this WWTP was ≈15,000 EUR and the O&M costs are currently of the order of ≈3,500 EUR per year. Cost recovery is done in an unconventional way: the revenues from the "Mughal Heritage Trail", which was initiated by the same NGO that implemented the WWTP (see institutional aspects), are used to pay the salary of the operators. The revenues from the trail are sufficient to pay five guides on the trail and two operators in the treatment plant (Table 17.3). The operators receive a salary of 3,500 INR/month (≈50 EUR/month) each.

Table 17.3 Costs and revenue for O&M of HSSF-CW in Katchpura slum, City of Agra.

Components	Value (INR/yr)	Value (EUR/yr)*
Revenue per visitor (=700 INR) × Number of visitors per year (=450)	315,000	3,900
Costs per operator/tour guide: 12 × 3,500 INR	42,000	520
Total revenues	**357,000**	**4,400**
Salary 5 guides	210,000	2,600
Salary 2 operators	84,000	1,000
Total costs	**294,000**	**3,600**

*Average exchange rate of year 2014: INR EUR = 0.0123 (Online Currency Converter, 2015).

Summary of evaluation results

The intended benefits are mainly fulfilled (Table 17.4). The monitoring results were not available, but based on the visual impression the treatment system seemed to work well. Water is being reused in gardening.

Apart from the treatment of domestic wastewater, another intended benefit was to improve the quality of the environment of the poor families in Katchpura. As reported by the National Institute of Urban Affairs (2011), the environmental situation has improved. The open channel that conveyed the wastewater to the Yamuna River is now covered and can be crossed easily even during monsoon.

Table 17.4 Planted gravel filter Agra – fulfilment of intended benefits.

Intended Benefit/Purpose	Fulfilled (yes/no)	Comments
Treatment of wastewater according to Indian standard	yes	Monitoring is conducted, but information about performance is not available in public.
Reuse of treatedwastewater	yes	The treated wastewater is being reused in gardening.
No energy requirements	yes	No energy is required for the treatment process. Wastewater is collected by a gravity sewer.
No skills for O&M required	partly	The operators have no special skills, but they received a short training from the supporting NGO.
Improvement of environmental situation	yes	Compared to the situation before, the environmental situation has improved. This was confirmed during the site visit by local people.

The treatment process requires no energy and only basic skills are required. However, continuous support from the NGO proved to be a reason for success. No additional risks could be detected.

17.3.4 HSSF-CW in Ekant Park, city of Bhopal, state of Madhya Pradesh in central India: Case study 4

The intended benefit of CWs are treatment of wastewater according to Indian norms and low energy requirements. Additional expected benefits may depend on local circumstances. Figure 17.13 shows the components of the treatment system: around 25% of the total wastewater is entering the HSSF-CW, the remaining 75% are directly entering the natural wetland with *canna indica* (Figure 17.15) as the capacity of the CW is not enough to treat the entire wastewater stream.

Figure 17.13 Flow chart of HSSF-CW in Ekant Park.

Treated effluent from the natural wetland is released to a small pond from where it is pumped off for irrigation. Nearby an open stream of untreated wastewater is crossing the park without being treated or used for any purpose. Additional intended benefits are the avoidance of problems with mosquitoes and the reuse of water.

Health aspects

There are no formally assigned operators. The park operators employed for the maintenance of the park have no special equipment and come directly into contact with the treated wastewater. The results of the monitoring in 2003 showed *E.Coli* of 8×10^3 MPN/100 mL in the effluent which is lower than the recommended standard of 10^4 MPN/100 mL, but higher than the desirable 10^3 MPN/100 mL. As no recent monitoring results are available, the risk cannot be quantified.

Visitors of the park are the main stakeholder group that is impacted by the CWs. As the systems are not well maintained, accumulated wastewater can become a breeding ground for mosquitoes. In the park area raw wastewater is conveyed to the treatment plant in an open unlined channel. Visitors, especially children who are not aware of the water quality, can easily come into contact with the untreated wastewater. Besides, a large stream of untreated wastewater is crossing the park.

The treated wastewater is used for gardening within the park. As mentioned above, there is a risk for the park staff that is getting into contact with the water. As water is not produced for agricultural purposes, there are no consumers who could get into contact with products irrigated with treated wastewater.

Social aspects

Park staff is handling the treated wastewater for gardening. Three operators were interviewed. They think that the use of treated wastewater contributes to water conservation. There are no communities around the treatment plant, however visitors to the park can be affected due to the quality of the water used for irrigating the lawns. Eleven visitors were interviewed and those who come regularly to the park were aware of the treatment plant located within the park area. There is also a stream of untreated wastewater crossing the park and 50% of the respondents reported odour emanating from this stream.

Also the appearance of mosquitoes was mentioned as problem, but the respondents think that the mosquitoes are not originating from the treatment plant, but from the untreated wastewater. One respondent reported that he had seen children playing at the outlet of the treatment plant where water accumulates to be later used for irrigation. All respondents think that wastewater is a safe water source for the irrigation of the park. Table 17.5 shows the sample characteristics and the results of the small survey.

Table 17.5 Sample characteristics and results, HSSF-CW in Ekant Park.

Descriptive Statistics	Sample Description	n = 11 Respondents
Gender	Male	64%
	Female	36%
Age	20–30	27%
	31–40	18%
	41–50	18%
	51–60	37%
Question	**Answers**	**Percentage**
How often are you visiting the park?	First time	18%
	Two times per week	27%
	Everyday	55%
Do you know which water is used in this park for irrigation?	Treated wastewater	45%
	No	55%
Did you experience any problems/risks related to the (treated) wastewater? (multiple answers possible)	Mosquitoes	45%
	Children playing with the treated water	9%
	Bad smell of untreated wastewater*	45%
	No	6%
Do you think that treated wastewater is safe to be used for irrigation?	Yes	100%
	No	0%

*note: not related to treatment plant.

Institutional aspects

The main stakeholder involved in O&M of the treatment plant is the Bhopal Municipal Corporation that is not continously operating the treatment plant, but reduces the activities to annual cleaning of the pre-treatment unit. Wild growth of plants (Figure 17.14) in the treatment unit was observed, but the initially planted *Phragmites karka* still prevails. The Madhya Pradesh Pollution Control Board monitored the effluent quality every month, but the results were not available for the case study.

Figure 17.14 HSSF-CW at Ekant Park.

Figure 17.15 Natural wetland at Ekant Park.

Economic aspects

The system was constructed 20 years ago and the costs were 1.4 million INR (≈17,000 EUR) according to a sign in the park. No O&M costs occur as there are no operators assigned and no electricity and spare material are required. The pre-treatment unit is cleaned when necessary by the park staff. Water is used for irrigating the entire park which has a size of 65 acres (≈26 ha).

Summary of evaluation results

The intended benefits are mainly fulfilled (Table 17.6), but odour and mosquitoes were evident in the stream of untreated wastewater crossing the park. Evaluation results from the year 2003 showed good performance of the treatment plant, no energy is required for the treatment process and water is completely reused. At present, the system appears to have degraded, and there could be health risks to the staff who come into contact with the treated water and children who play with the treated water that is collected for irrigation.

Table 17.6 Results of evaluation – Fulfilment of intended benefits.

Intended Benefit/Purpose	Fulfilled (yes/no)	Comments
Treatment of wastewater according to Indian standard	yes	Published evaluation results are 15 years old, data from central and local pollution control board not available.
Reuse of treatedwastewater	yes	The treated wastewater is used for gardening within the park (Figure 17.16).
No energy requirements	yes	The treatment unit requires no energy.
No odour and mosquito problem	partly	Odour and mosquitoes are a problem; they do not emanate from the treatment unit but from the stream of untreated wastewater crossing the park (Figure 17.17).

Figure 17.16 Treated water used for irrigation (without using gloves).

Figure 17.17 Wastewater stream crossing park.

17.3.5 Duckweed pond in village Saidpur, District Ludhiana, State of Punjab, northern India: Case study 5

The intended benefits of duckweed ponds (DPs) are treatment of wastewater according to Indian norms, reuse of treated wastewater and use of by-products such as duckweed and fish.

The DP was established in year 2004 and receives 0.35 MLD domestic wastewater from the village community. As depicted in Figure 17.18, wastewater enters the DP (Figure 17.19) via an open sewer (Figure 17.20) and then flows into the fish pond (Figure 17.21) after which it is extracted for irrigation (Figure 17.22). There is no outlet and only surplus water is used for irrigation to keep the level in the fish pond high.

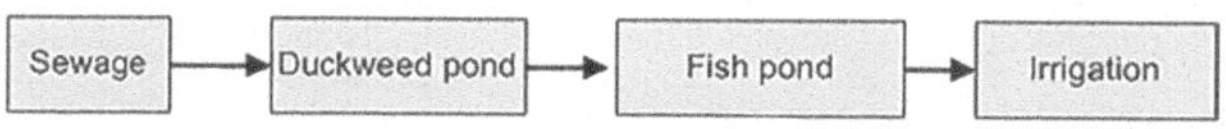

Figure 17.18 Flow chart of DP in village Saidpur.

Figure 17.19 Duckweed pond.

Figure 17.20 Inlet of duckweed pond.

Figure 17.21 Fish pond with orange trees planted around.

Figure 17.22 Agriculture around treatment plant.

Health aspects

There is a possible risk to the operators who remove the duckweed from the water surface. Currently, removal is carried out without any special equipment, and the cleaners are exposed to the contaminants. The treatment plant is not located in a residential area; therefore, no risks to the communities were detected.

There is a possible risk to farmers who come into contact with treated wastewater, which is used to irrigate wheat, sorghum and cotton. However, the risk cannot be quantified as the quality of the treated wastewater was not known at the time of assessment.

The quality of the fish was tested twice by the Food Corporation of India for heavy metals and pathogens and showed that fish was suitable for eating. The risk for the consumer of irrigated plants is low, as the types of crops grown are not eaten raw.

Social aspects

The treated effluent is being well accepted for irrigation and the two interviewed farmers even prefer it over groundwater as it contains more nutrients. According to the two farmers, there are no problems with the community as this treatment system is located outside the residential area.

Institutional aspects

The main institutions involved are the local government and the Punjab Pollution Control Board. One person was assigned by the local government to remove the duckweed from the pond every week. The quality of the effluent is monitored twice a year by the Punjab Pollution Control Board, but the information about the performance is not public.

Economic aspects

The construction costs are not known. Labour for construction was provided within the Mahatma Gandhi National Rural Employment Guarantee Act (MGNREGA) which is a guarantee scheme for hundred days of work at minimum wage for every adult living in rural households.

According to two local people who were interviewed, O&M costs are recovered with the revenues from auctioning of the fish 25% of the annual earnings of 50,000 INR (≈600 EUR) are used to pay the operator. The remaining money goes to the village fund. As an additional benefit, orange trees are planted around the treatment unit and villagers can pick oranges whenever they want.

Summary of evaluation results

The intended benefits are mainly fulfilled (Table 17.7) and no further risks were detected. As no monitoring results are available, no statement about the performance can be made, but the treatment plant seemed to work well based on the visual impression during the field visit. Treated water is used for irrigation. Operation costs can be recovered by the earnings made from selling the fish.

Table 17.7 Results of evaluation – Fulfilment of intended benefits.

Intended Benefit/Purpose	Fulfilled (yes/no)	Comments
Treatment of wastewater according to Indian standard	not available	Monitoring is conducted, but results are not public. Treatment plant seemed to work well.
Reuse of treatedwastewater	yes	Treated wastewater is used for irrigation.
Use of side-products	yes	Duckweed is used as fish fodder. Fish are cultivated in fish pond. Orange trees around ponds generate additional benefit.
No energy requirements	yes	No energy is required for the treatment process. Wastewater is collected by a gravity sewer.

17.3.6 Water hyacinth pond in village community in district Bathinda, state of Punjab, northern India: Case study 6

The intended benefit of the water hyacinth pond is treatment of wastewater from the local community and reuse of water for irrigation. The water hyacinth pond in Bathinda Distrcit is not a constructed but a natural system. It receives 0.25 MLD wastewater from a village community and has no outlet (Figure 17.23). The wastewater is conveyed in an open sewer (Figure 17.24) to a pond where water hyacinths (Figure 17.25) were planted around five years ago.

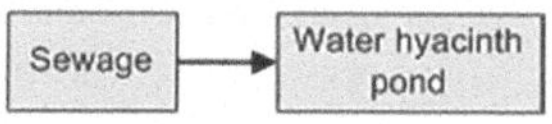

Figure 17.23 Flow chart of water hyacinth pond in village community in District Bathinda.

Figure 17.24 Inlet of water hyacinth pond.

Figure 17.25 Village pond covered with water hyacinths.

Health aspects

There is no formal arrangement for O&M. Therefore, the operating staff is not exposed to any risk. When attempting to remove the plants from the pond, people can come into contact with the wastewater and can become exposed to possible health risks.

Wastewater in the water hyacinth pond is hardly accessible as it is covered by a thick layer of water hyacinths. Neither smell nor problems with mosquitoes were reported by the community members. Water from the pond is not used for any purpose, therefore no health risk emanate from treated water.

Social aspects

Ten persons from the village expressed their opinion about the pond system. Treated water cannot be used due to the thick layer of plants covering the surface. Before the plants grew in the pond, villagers used to irrigate their fields with the water from the pond, but now the water is no longer accessible which is considered as problem by local people.

According to villagers, bad odour from the wastewater has disappeared, but the excessive grow of water hyacinths in the pond makes its use for cattle watering and irrigation impossible as the water disappears under a thick layer of plants. Local people are not aware of the beneficial effect of water hyacinths on wastewater but instead think that this plant is deterring them from using the pond water.

Institutional aspects

The local government is the only institution involved in O&M of the system. There is no arrangement for operation. An attempt was made to remove the water hyacinths with the help of 30 workers employed under the MGNREGA scheme, but after one month the water hyacinths once again covered the surface of the village pond. The villagers did not try to remove them again.

Economic aspects

This system was not constructed; it is a village pond that has already existed since a long time. There is no formal arrangement for O&M. Water hyacinths are not used for any purpose even though examples for their use in India are reported in literature.

Summary of evaluation results

The intended benefit is not fulfilled as the water in the pond is now not available for irrigation or cattle rearing (Table 17.8). This is due to the uncontrolled growth of water hyacinths. The people have thus lost a source of irrigation water. The treatment performance of the pond is not known as it is not monitored. Local people are not satisfied with the treatment system as they perceive the plants as an obstacle. However, according to local people, the bad smell has disappeared.

Table 17.8 Results of evaluation – Fulfilment of intended benefits.

Intended Benefit/Purpose	Fulfilled (yes/no)	Comments
Treatment of wastewater according to Indian standard	not available	No monitoring resultsavailable.
Reuse of treatedwastewater	no	Not possible due to thick layer of plants.
Use of sideproducts	no	No use of water hyacinths, no cultivation of fish.
No energy requirements	yes	No energy is required for the treatment process. Wastewater is collected by a gravity sewer.

17.4 SWOT ANALYSIS

A SWOT analysis has resulted in the following strengths, weaknesses, opportunities and threats:

17.4.1 Strengths

An important aspect in all of the cases studied was that the system required low or even no energy input. The study showed that economic benefits from by-products of wastewater treatment are numerous and not only limited to the use of treated wastewater for irrigation, but also for rearing fish. The planting of fruit trees in the area of the treatment plant as done in Punjab appears to have good acceptance.

In all case studies where the treated wastewater was reused, users were satisfied with the quality. When used for irrigation, farmers appreciated the content of nutrients in the water, which replaced chemical fertilizer. In Ekant Park in Bhopal, the State of Madhya Pardesh operating staff agreed that the reuse of the treated wastewater contributes to water conservation.

17.4.2 Weaknesses

Land requirement of NTS is higher than for mechanised treatment systems and varies between 1.5 m^2 per person for CWs and 6 m^2 per person for a pond system (e.g. Arceivala and Asolekar, 2006). Lower space requirements can be achieved by combining mechanised and NTS. For systems located near communities, problems with odour and mosquitoes were reported due to inadequate O&M of the systems.

17.4.3 Opportunities

Due to high land prices in peri-urban and urban areas NTS are mainly suitable for rural areas. However, NTS may be used for green zones in urban areas and be integrated in urban landscaping as in the example of the CW in Ekant Park in Bhopal. In rural areas, space can be saved if the existing village ponds are integrated with the wastewater treatment systems as seen in the examples from Punjab, northern India. An even higher use of by-products, than shown in the case studies, can be achieved, if e.g. sludge treatment becomes more popular.

17.4.4 Threats

Institutional and organisational issues are considered to be of high importance, similar to studies reported from CWs in Mexico (Starkl *et al.*, 2010) and Thailand (Brix, 2010). It has been clearly demonstrated that this aspect is relevant for the long-term sustainability as systems where no formal arrangement for O&M exists are prone to clogging and flooding and problems cannot be tackled immediately.

As mentioned above, the NTSs require low or even no energy input. However, as water is usually pumped from pumping stations to the treatment plants, power cuts can affect their functionality.

Potential risks for affected stakeholder groups (operators, neighbours, farmers, consumers) need to be further investigated. Municipalities should take particular care if the NTS are close to human habitations and groundwater aquifers, to anticipate health-related issues and be ready to address them. For the water users health risk assessments should be mandatory, and for the products food safety measures and testing should be part of the agriculture production process. For consumers it is advisable to follow a multi-barrier-approach and take measures to reduce contamination even at household level by washing and disinfecting vegetables before consumption.

17.5 CONCLUSIONS

The assessment provided insights into challenges and the potential of NTS. A summary of the main SWOTs can be seen in Figure 17.26.

STRENGTHS	WEAKNESSES
• low energy requirements • benefit of side products • high acceptance for reuse of treated wastewater	• high land requirements • odour and mosquitoes (if located near human settlements)
OPPORTUNITIES	**THREATS**
• potential for rural areas • increased use of side products possible	• improper O&M arrangement can endanger functioning • power cuts of pumping system affect treatment system • possible health risks to operators, neighbours, farmers and consumers

Figure 17.26 SWOT analysis based on existing evaluation results and rapid assessment.

One of the main problems was that monitoring results, even where they existed, were not accessible. An environmental information system providing information about monitoring results and updated information about WWTPs would be a desirable tool to ensure accessibility and increase transparency.

17.6 REFERENCES

Alvarez J. A., Armstrong E., Gómez M. and Soto M. (2008). Anaerobic treatment of low-strength municipal wastewater by a two-stage pilot plant under psychrophilic conditions. *Bioresource Technology*, **99**, 7051–7062.

Arceivala S. J. and Asolekar S. R. (2006). Wastewater treatment for pollution control and reuse, 3rd edn, Tata McGraw Hill, New Delhi.

Brix H., Koottatep T., Fryd O. and Laugesen C. H. (2010). Appropriate and Sustainable Wastewater Management in Developing countries by the use of Constructed Wetlands. Proceedings 686, Water Resource Management, Gaborone, Botswana.

Kaseva M. E. (2003). Performance of a sub-surface flow constructed wetland in polishing pre-treated wastewater—a tropical case study. *Water Research*, **38**, 681–687.

Mbuligwe S. E. (2004). Comparative effectiveness of engineered wetland systems in the treatment of anaerobically pre-treated domestic wastewater. *Ecological Engineering*, **23**, 269–284.

National Institute of Urban Affairs (2011). Decentralized Waste Water Treatment (DEWAT) System in Kachhpura Peer Experience and Reflective Learning (PEARL), Urban Initiatives, **5**, New Delhi, India.

Online Currency Converter (2015). http://www.freecurrencyrates.com/exchange-rate-history/EUR-INR/2014 (accessed 28 April 2015)

Sharma S. K. and Amy G. (2010). Natural treatment systems. In: Water Quality and Treatment: Handbook of Community Water Supply, J. Edzwald (ed.), 6th edn, American Water Works Association and McGraw Hill Inc., USA, Chapter 15. ISBN 0071630112/9780071630115.

Srivastava P. K., Kulshreshtha K., Mohanty C. S, Pushpangadan P. and Singh A. (2005). Stakeholder-based SWOT analysis for successful municipal solid waste management in Lucknow, India. *Waste Manage*, **25**(5), 531–537.

Starkl M., Essl L., Martinez J. L. and Lopez E. (2010). Water quality improvements through constructed wetlands: A case study from Mexico. Conference Proceedings of the 3rd African Conference on Water Resource Management of the International Association of Science and Technology for Development, Gaborone, Botswana.

Starkl M., Amerasinghe P., Essl L., Jampani M., Kumar D. and Asolekar S. R. (2013). Potential of natural treatment technologies for wastewater management in India. *Journal of Water, Sanitation and Hygiene for Development*, **3**(4), 500–511.

Terrados J., Almonacid G. and Hontoria L. (2007). Regional energy planning through SWOT analysis and strategic planning tools. Impact on renewable development. *Renew.Sustain Energy Rev 2007*, **11**(6), 1275–1287.

Chapter 18

Viewing sub-surface for an effective managed aquifer recharge from a geophysical perspective

Shakeel Ahmed, Tanvi Arora, Sarah Sarah, Farooq Ahmad Dar, Tarun Kumar Gaur, Taufique Warsi and Pasupunoori Raghuvendar

18.1 INTRODUCTION

Managed Aquifer Recharge (MAR) is a technique of intentional storage and treatment of water in aquifers where natural recharge processes have been affected (Dillon, 2005; Dillon *et al.*, 2009). MAR is a cost-effective process that has solved groundwater quantity and quality problems in many of the world's stressed aquifer systems (Gale & Dillon, 2005).

Natural processes of groundwater recharge have been highly influenced and affected by climate change, urbanization and other unmanaged anthropogenic practices. Groundwater is drafted at a rate which cannot be replenished through natural recharge conditions alone. Groundwater quality is constantly deteriorating due to uncontrolled dumping of waste directly in groundwater interacting zones. Therefore, artificial recharge and MAR is of great interest to replenish the heavy losses of groundwater for sustainable management of the aquifers by enhancing natural infiltration and recharge processes through purposeful and managed activities (Dillon *et al.*, 2009).

The process of aquifer recharge involves and is influenced by both surface as well as sub-surface characteristics. However, in the estimation and evaluation of aquifer recharge, often only surface features and not the sub-surface features are considered, as the latter are much more complex and hidden compared to the surface features that are visible and easily mapped. The sub-surface contains both unsaturated and saturated zones and the two have very contrasting properties. Thus, to investigate the sub-surface, geophysical methods are promising tools as they are mostly non-invasive and reveal the physical properties of the sub-surface. In this chapter relevant geophysical techniques applicable for MAR investigations and their application are briefly described.

18.2 UNIQUE CONTRIBUTIONS AND CHALLENGES OF HYDRO-GEOPHYSICS

Geophysical methods are merely based on the physical properties of sub-surfaces. Hydro-geophysics is an evolving but useful field for understanding aquifer systems, their geometry, the extents and characteristics. Due to non-uniformity in such systems, it is important to identify a suitable geophysical method to develop an approach for identifying water-bearing structures or specific points of recharge to aquifers. Audio Frequency telluric methods (AFTM) have been used to detect water in karst settings (Chen, 1988; CGS, 2005; Gan *et al.*, 2011) and also in combination with induced polarization to identify existing fissures (Li & Wang, 2009). Time domain electromagnetic is of limited use in carbonate terrenes (Chalikakis, 2006; Ezersky *et al.*, 2006). Very low frequency (VLF) methods have helped detect epikarst and also locate near surface heterogeneities, but with a limited resolution (Turberg & Barker, 1996; Bosch & Müller, 2001, 2005). Integrated with electrical resistivity tomography (ERT), VLF provides good results to map productive aquifers for precise location of wells (Alexopoulos *et al.*, 2011; Vargemezis *et al.*, 2011). Other methods like gravimetry (based on density variations), magnetic resonance sounding (to locate water content), mise-à-la-masse (to trace water filled cavities), self-potential (to map preferential pathways) are employed with specific success (Jacob *et al.*, 2009; Guérin *et al.*, 2009; Jardani *et al.*, 2006; Meyerhoff *et al.*, 2012). Very few studies focused on the use of geophysical microgravimetric and gravity gradient techniques to target sub-surface cavities and karst features (Colley, 1963; Neumann, 1965; Butler, 1984; Blizkovsky, 1979). Kaspar and Pecen (1975) used an electromagnetic (EM) method for tracing karst features in eastern Slovakia by means of differences in electrical properties between limestone and karst features. Moore and Stewart (1983) used seismic refraction, ERT and microgravity to delineate zones of increased fracture density.

In the past decade Electrical sounding and Resistivity tomography (VES & ERT) have evolved, potential geophysical technologies to monitor the natural changes in shallow sub-surface resistivity (Binley *et al.*, 2002; Dutta *et al.*, 2006, Arora & Ahmed, 2010, 2011; Loke *et al.*, 2014; Singha *et al.*, 2014). ERT can be applied to define the water table (Zaidi, 2012), to delineate aquifers (Dutta *et al.*, 2006), to search for karst geological structures (Leuccim, 2005) and find karst water (Metwaly *et al.*, 2012; Vlahović & Munda, 2011; Gan *et al.*, 2013) and also to study the unsaturated zone (Arora, 2013; Carriere *et al.*, 2013).

We present three case studies: one from the Chandi limestone of Raipur city, a second from a dugwell site in crystalline aquifers and a percolation tank in granitic terrene of Maheshwaram watershed, all of which are poorly understood in terms of hydrogeology which demands detailed research before taking any MAR program in the area. The purpose of MAR in these areas is to infiltrate excess surface run-off to the aquifer for a smooth supply and demand of groundwater for future uses. The implementation of MAR and its success depend highly on the knowledge of the hydrogeological setup of the aquifer.

The purpose of this chapter is to demonstrate, how conceptual models of the three aquifers are developed through hydro-geophysical approaches. The results will help locate the best suitable sites for implementing artificial recharge schemes and to assess their feasibility. This work will have great impact on the operation, management, maintenance and expansion of MAR in areas which are still little understood in the Indian context.

18.3 GEOPHYSICAL METHODS

18.3.1 Hydro-geophysical electrical methods

The relationship between electrical resistivity, current and the electrical potential is governed by Ohm's law. The Poisson's equation is used to calculate the potential in a continuous medium, the form of Ohm's Law, combined with the conservation of current. Loke (2013) defined the potential due to a point current source located at x is given by

$$\nabla\left[\frac{1}{\rho(x,y,z)}\right]\nabla\phi(x,y,z) = \frac{\partial j}{\partial t}\,\delta(xs) \tag{18.1}$$

where ρ is the resistivity, ϕ is the potential and j is the charge density.

Knowledge of the resistivity distribution can aid in calculating the potential at any point either on surface or in the medium. Here, the purpose of the resistivity method is to calculate the unknown distribution of electrical resistivity in the sub-surface. The measurements for the resistivity survey are made by passing a current into the ground through two current electrodes (usually metal rods), and measuring the difference between the voltage at two potential electrodes. Basically, the resistivity meter has a current source and voltage measuring circuitry that are connected by cables to the electrodes (Figure 18.1).

There exist different configurations as drawn in Figure 18.1. Many workers like Dahlin and Zhow (2004), Saydam and Duckworth (1978), Szalai and Szarka (2008) and Zhou *et al.* (2002) have clearly discussed the pro's and con's of various configurations. We need to accept and adopt a particular configuration depending on the need of the interest of the field objective. The multi-channel system utilises the multiple gradient configuration (shown above) and was specially described by Loke *et al.* (2013) and Dahlin and Zhou (2004).

We consider that potential difference is directly proportional to the current injected. By assuming a constant, called geometrical factor (k), we calculate the apparent resistivity (ρa) value by the formula given in equation (18.2).

$$\rho a = k\frac{\Delta V}{I} \tag{18.2}$$

k is the geometric factor which depends upon the array configuration adopted in field (Koefoed, 1979).

The equation (18.2) represents the inverse problem with the assumption of homogenous and isotropic sub-surface to carry out resistivity measurements.

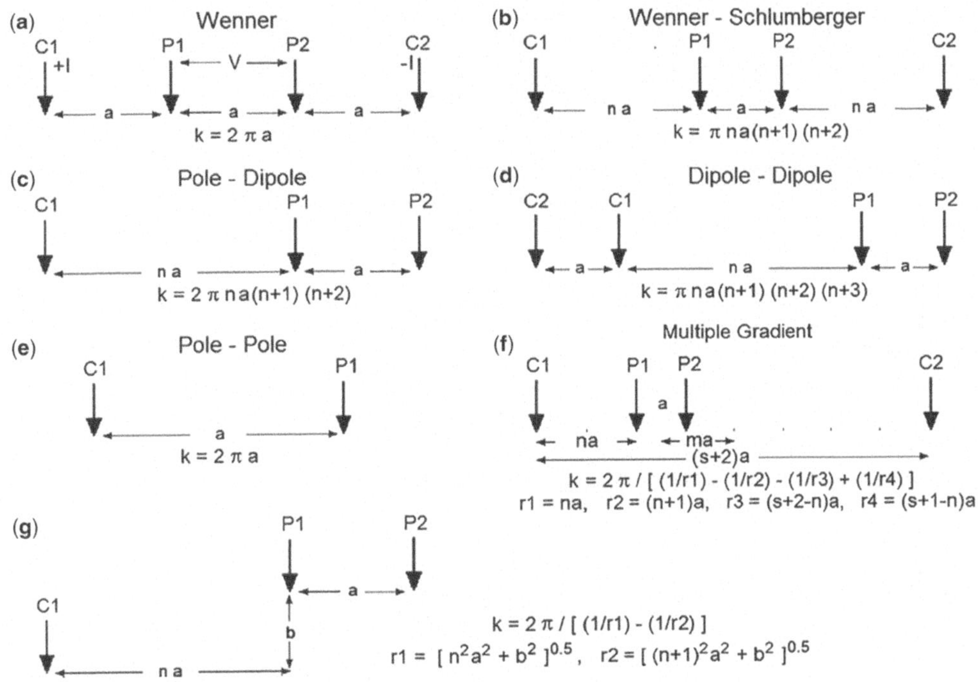

Figure 18.1 Some commonly used electrode arrays and their geometric factors. Note that for the multiple gradient array, the total array length is '(s + 2)a' and the distance between the centre of the potential dipole pair P1–P2 and the centre of the current pair C1–C2 is given by 'ma'. British Geological Survey National Environmental Research Council 2013 (modified after Loke *et al.*, 2014).

Let us consider the Schlumberger configuration (Figure 18.2). The distance between AB is 2s and between MN is 2b.

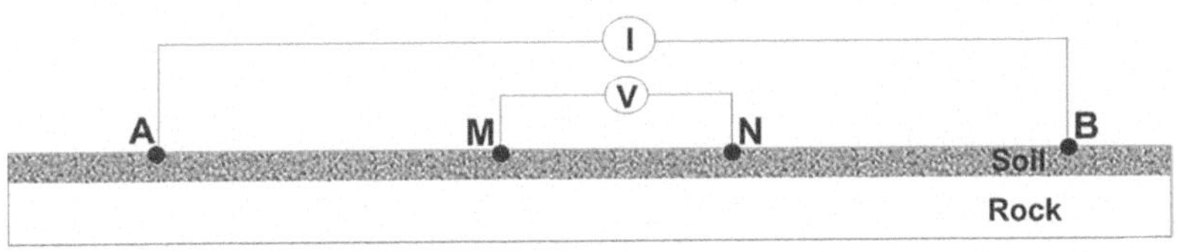

Figure 18.2 An electrode layout for the Schlumberger configuration.

The current I is flowing between AB, the potential at M due to current electrodes A and B is given by:

$$V_M = \frac{\rho_a I}{2\pi}\left[\frac{1}{\mathrm{AM}} - \frac{1}{\mathrm{MB}}\right] \tag{18.3}$$

when ρ_a is known as apparent resistivity.

Similarly, the potential at N due to current electrodes A and B is:

$$V_N = \frac{\rho_a \mathrm{I}}{2\pi}\left[\frac{1}{\mathrm{AN}} - \frac{1}{\mathrm{NB}}\right] \tag{18.4}$$

The potential difference will be

$$\Delta V = V_M - V_N \tag{18.5}$$

$$\Delta V = \frac{\rho_a I}{2\pi}\left[\frac{1}{\mathrm{AM}} - \frac{1}{\mathrm{MB}} - \frac{1}{\mathrm{AN}} + \frac{1}{\mathrm{NB}}\right] \tag{18.6}$$

After writing AM, MB, AN and NB in terms of s and b, the expression of apparent resistivity will be given by:

$$\rho_a = \frac{\pi(s^2 - b^2)}{2b} * \frac{\Delta V}{I} \tag{18.7}$$

Here $\frac{\pi(s^2 - b^2)}{2b}$ is known as geometrical factor for the Schlumberger configuration.

Vertical electrical soundings (VES)

Electrical resistivity surveys have been used for many decades in hydrogeological, mining and geotechnical investigations. The direct-current electrical resistivity method for conducting a vertical electrical sounding (VES) has proved very popular in groundwater studies due to the simplicity of the technique and the ruggedness of the instrumentation. There are a number of configurations available for electrical surveys of which either the Schlumberger or Wenner configuration are most useful for sounding, since all commonly available interpretation methods and interpretation aids for sounding are based on these two configurations. When using either method, the centre point of the configuration should be at a fixed location, while the electrode location varies around it. The apparent resistivity values and layer depths interpreted from them are referred to the centre point.

Multi-electrode systems and 2-D imaging surveys

The typical sounding and profiling resistivity methods makes use of four electrodes connected to four individual cables. On the one hand, these methods provide only 1-D section of the medium, the use of multi-electrode systems help in capturing 2-D vertical as well as horizontal pictures of the medium. Normally there are more than 24 electrodes connected through a multi-core cable (Figure 18.3). The current and potential electrodes are maintained at regular fixed interval from each other and are progressively moved in a line on the ground. At each step, one set of measurement is recorded. The set of all measurements at the first electrode gives a profile of resistivity values. Then the electrode spacing is increased further and second set of measurements are completed. This process of increasing spacing (by a factor of n = 2) is repeated until the maximum spacing between the electrodes is reached. It should be taken into consideration that the depth of investigation depends on the large values of "n".

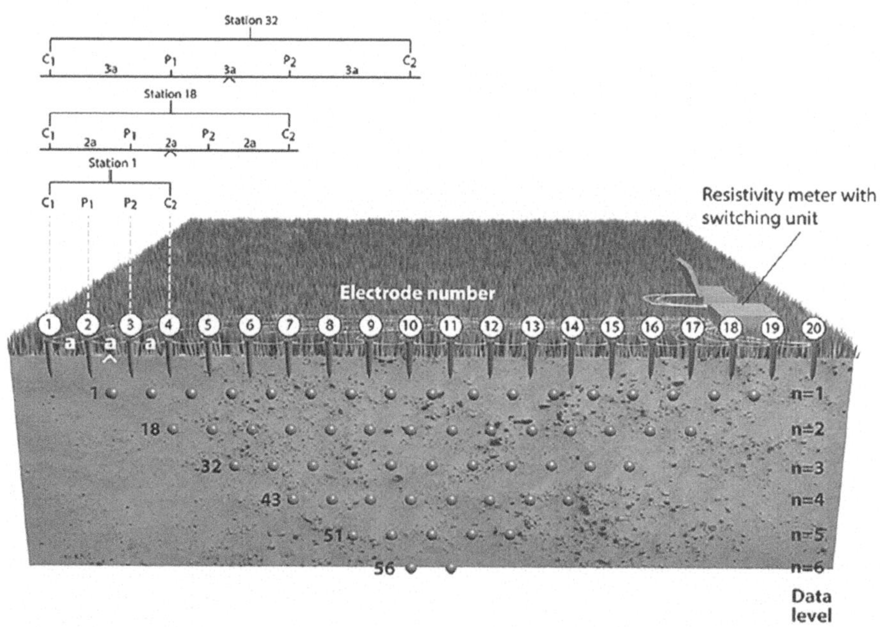

Figure 18.3 Schematic diagram of a multi-electrode system, and a possible sequence of measurements to create a 2-D pseudosection. British Geological Survey, National Environmental Research Council 2013 (modified after Loke *et al.*, 2014).

As suggested by Loke and Barker (1996a), there should be large number of rectangular cells in the 2-D section to infer the maximum information out of it. There are both horizontal and vertical variations in the resistivity values of the cells. Inversion of these measured resistivities is very important before interpretation because the raw resistivity measurements rarely give the true section of the surface.

In the presented study resistivity measurements were carried out with the resistivity meter Syscal R1 Junior (IRIS Instruments, Orléans, France) equipped with 48 electrodes in different spacing (4, 6 and 10 m depending on spatial constraints) and ABEM Terrameter equipped with 120 electrodes with various spacing. The intensity and voltage accuracy is 0.3%, which is consistent with the measurements carried out under constant hydro-geologic conditions for about ten hours persistent within a tenth of an ohm-meter, i.e. one thousandth of the measured resistivity could reach several ohm-meters between two sets of measurements 1 hour apart. Inversion was made with the RES2DINV software (Loke & Barker, 1996a). This technique, based on the smoothness-constrained least squares method, produced a 2-D sub-surface model of the resistivity section.

18.3.2 Time domain electromagnetic methods (TDEM)

Time Domain Electromagnetic Methods (TDEM) are also primarily used for resistivity sounding. TDEM measurements are carried out by sending an electrical current through a transmitter coil. When the transmitter current is shut off, induction creates a decaying primary magnetic field, which in turn induces secondary electric currents (that are essentially a shallow "image" of the former transmitter current) with their accompanying magnetic fields. The decay of the induced field over time (which depends on the electric properties of the ground) is monitored with a separate receiver coil. TDEM provides a wide range of effective sounding depths from approximately 6 to 900 meters. In karst hydrogeology, TDEM is useful for determining the depth and thickness of overburden, depth of the water table, depth and degree of saltwater (or ion-contaminated water) intrusion, and detection of fracture zones (e.g. Farrell *et al.*, 2003). TDEM provides good lateral as well as vertical resolution, and a wide range of effective survey depths. The method is sometimes limited by the requirement for impractically large (e.g. 300 × 300 m or larger) transmitter coils to achieve deep measurements. Though we adopted this technique in the Karst terrene the results of TDEM are not consistent with the ERT due to conductivity variations.

18.3.3 Borehole resistivity logging

Well logging, also known as borehole logging, is the practice of making a detailed record (a well log) of the geologic formations penetrated by a borehole. The log may be based either on visual inspection of samples brought to the surface (geological logs) or on physical measurements made by instruments lowered into the hole (geophysical logs). Resistivity logging is a method of well logging that works by characterizing the rock or sediment in a borehole by measuring its electrical resistivity.

This survey presented in this study was done to identify the contact zone between the weathered layer and the bedrock. It is also useful to investigate the zones supplying water to the wells. This is based on the principle of Ohm's law and is used to calculate the resistivity of the sub-surface formation. There are several electrode configuration methods like one-electrode, two-electrode (short normal, long normal) three-electrode methods etc. In the present case, a resistivity probe with two electrodes was used, which was fabricated at IFCGR, NGRI (Figure 18.4).

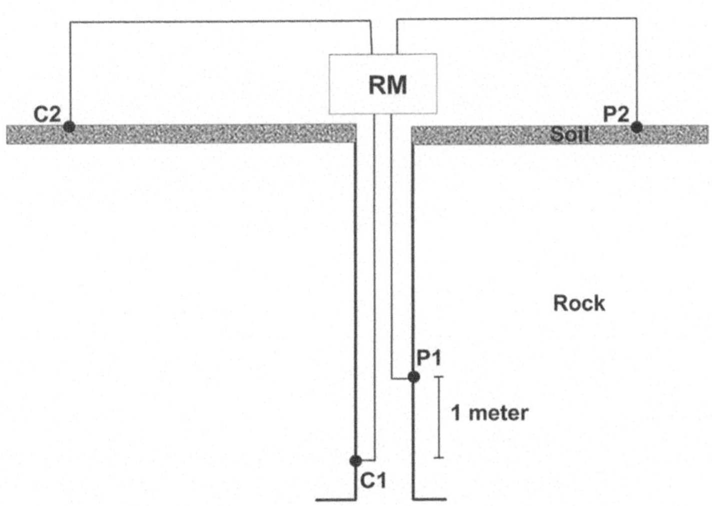

Figure 18.4 Logging circuit diagram of the geophysical borehole logger fabricated at IFCGR, NGRI.

The probe consists of two electrodes: one is used as current electrode and the second is used as potential electrode and the spacing between the two electrodes is 1m. The other two (current and potential) electrodes are kept far away from the bore hole. The probe is then lowered into the bore hole and every 0.5 m the measurement is recorded till the bottom of the well. The instrument used for this technique was "SYSCAL switch (Junior)" which measures current and potential. The resistivity values were plotted against corresponding depths which gave the resistivity log of the particular bore well. Boundaries of formations having different resistivities are detected most readily with short electrode spacing; whereas information on fluids in thick permeable formations can be best obtained with long spacing.

18.4 CASE STUDIES PERTAINING TO MAR

18.4.1 Finding conducting zones in Karst aquifer systems and analysing the efficacy of proposed recharge structure

The study area was Raipur, an example for a water stressed city where urbanization poses a threat by altering the hydrological scenarios. This work highlights possibilities for assessing the efficiency of MAR by using hydro-geophysical techniques to study the structure of the unsaturated zone of karst terrene of Central India, where soil cover is very thin and followed by shale formation that are not suitable for recharge.

Electrical resistivity sounding was carried out at 12 sites to detect conductive zones along the local surface drainage in the area. The Chokhra nalah is the main local drainage along which MAR structures need to be proposed in the Telibandha area of Raipur (Figure 18.5). We did close soundings from VES method thereby capturing the local resistivity variation in the area. After the dense VES, ERT with 4–10 m spacing using Wenner-Schlumberger and gradient methods was carried out.

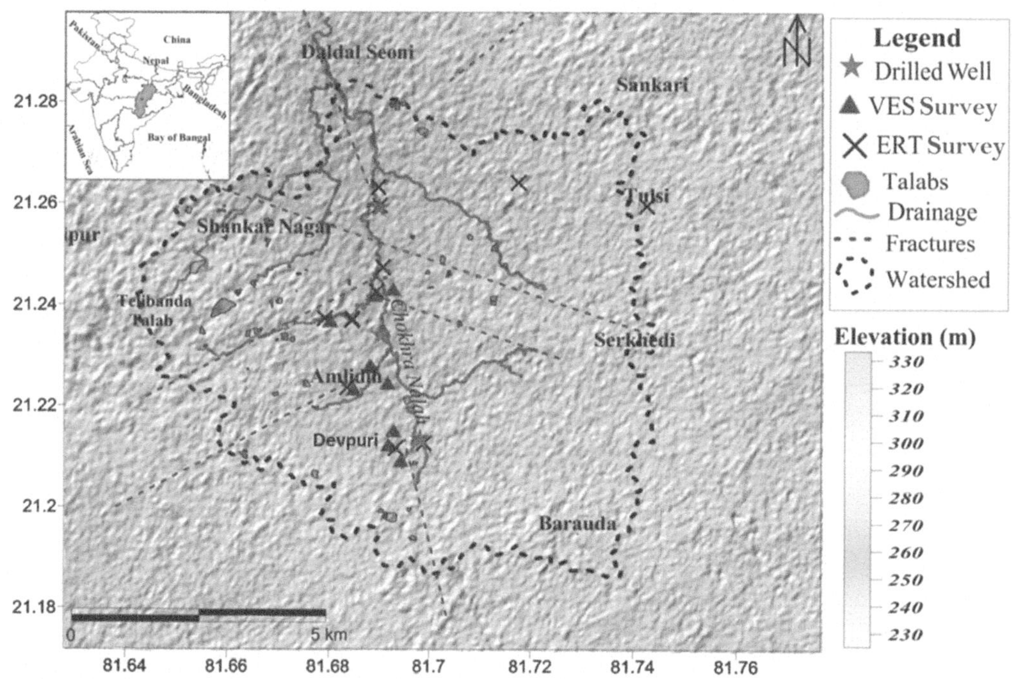

Figure 18.5 Digital Elevation map of the Telibandha area at the outskirts of Raipur city. The map also shows the location of various measurement points.

Digital elevation model (DEM)

A Digital Elevation Model (DEM) from ASTER (Advanced Spaceborne Thermal Emission and Reflection Radiometer) of 30×30 m resolution was used for topographic interpretation. Watershed and drainage were delineated using the Spatial Analysis Toolbox of ArcGIS software. The lineaments were digitized from the shaded relief maps of DEM with different sun angle (Figure 18.5).

Vertical electrical sounding (VES)

Figure 18.6 shows the resistivity values for layers at different depths interpreted by using the 1X-1D software (www.interprix.com). The resistivity values range from as low as 5–200 ohm m to as high as 2,000–3,000 ohm m.

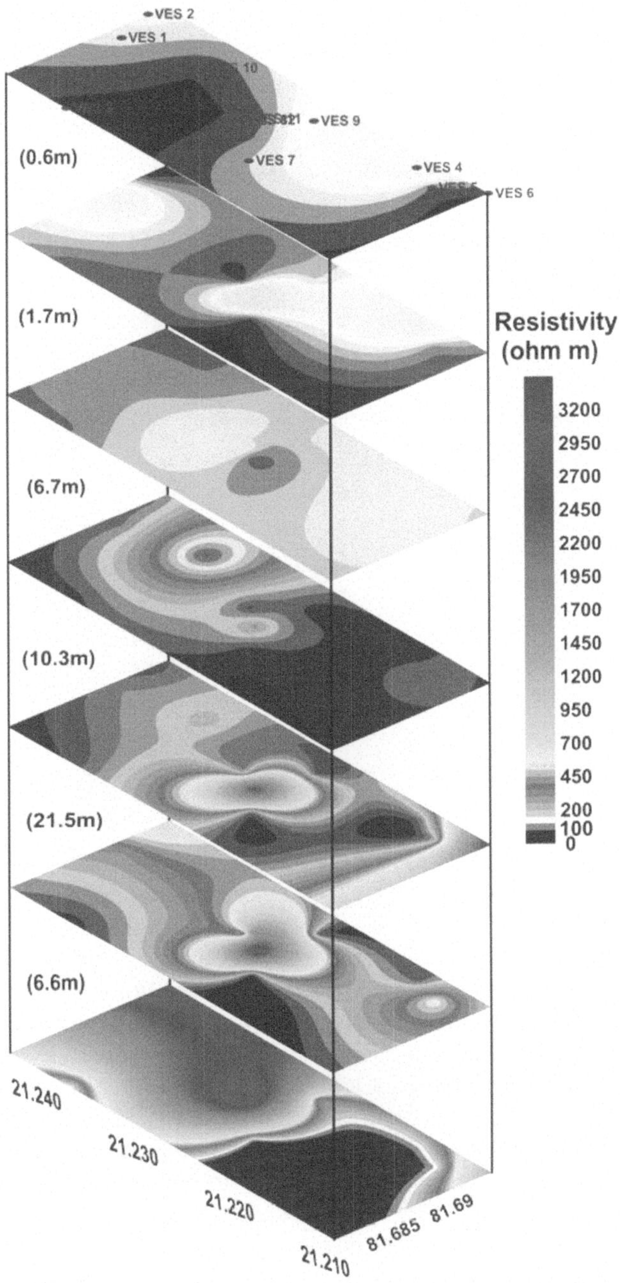

Figure 18.6 Results of Vertical Electrical Sounding (VES) at different depths. The black dots are the location points of the VES as shown in Figure 18.5.

The profiles show a sharp transition zone at the depth of 2.3 m (layer 3). As we interpret the deeper layers, the high resistivity value concentrates towards point VES 8 (at the centre). This shift gives an indication of an existing paleochannel between points VES8 and VES10or some weak structure, as it would be typical for a karst aquifer system. Accordingly, the area around VES 8 may be a suitable site for recharge.

Electrical resistivity tomography (ERT)

Electrical Resistivity Tomography (ERT) was carried out along 12 selected areas in and around Telibandha area. Exemplarily, the results for Station Devpuri (Figure 18.5) are given. The profile was laid down in NW-SE direction along 480 m with electrode spacing of 10 m. The Wenner-Schlumberger configuration was used to survey the profile. A total number of 529 datum points were measured at 23 data levels, allowing to plot a pseudo cross-section (Figure 18.7). The results clearly show the layered structure of the limestone in the sub-surface. Resistivity varied with the water content showing a thick layer of 15–20 m thickness of damp limestone with resistivity varying between 50–150 ohm m at a depth of 5 m and up to 90 m with high resistivity of more than 300 ohm m of dry limestone below.

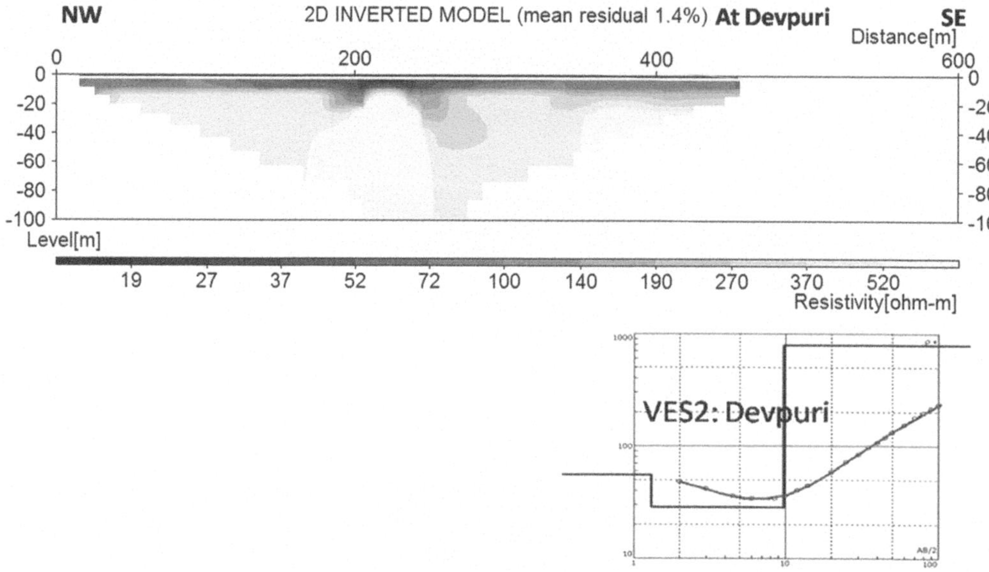

Figure 18.7 Electrical Resistivity Tomogram at station Devpuri.

The area of interest remains limited to 35–40 m below the surface, however many strong lateral and vertical anomalies were found in the ERT surveys. The low resistivity in the unsaturated zone, as compared to the high resistivity of limestone bedrock, is the target for preferred pathways. The low resistivity anomalies in the 2D inverted sections could be water-filled conduits or solution channels. These were later confirmed with the lithology of drilled wells at 3 sites. The borehole logging results correlated very well with the resistivity of ERT survey and lithology. Lithological information confirmed that the solution activity has modified the hydraulic properties of the aquifer by widening fractures, bedding planes and developing solution voids. Local heterogeneity was mapped based on the resistivity variations thereby proposing the favourable sites for construction of MAR structures. Along with it the surface implications of the hiding mafic dykes were inferred in the area.

After the complete detailed investigations, a combined map showing the conductive and resistive zones was generated (Figure 18.8). Then sites for 3 borewells were proposed and drilled to correlate the hydro-geophysical parameters to the lithology. The lithological data of three bore holes gave a clear understanding of the sub-surface geology of the limestone aquifer. The aquifer is overlain by a top black soil layer of 1–6 m thickness which possesses a high clay content, nodules (kanker) and is rich in iron matter. Below this layer flaggy limestone with a high clay content occurs down to the fresh limestone. This indicates that the weathered zone overlying the limestone is generally of ~6 m depth. Below this zone fresh limestone is encountered where two fracture zones and solution activity is observed mostly in the central part of the watershed. The solution channels are filled with sediments/soils and indicate their collapsed nature which can also be related to the presence of solution type features on the surface and presence of more surface ponds/talabs. The fresh black limestone up to about 40m depth has two fracture zones with high groundwater yield and little solution activity indicating the effect of karstification processes in the limestone. The solution weathering in other layers was found to be insignificant. The results were used to analyse the secondary porosity at a particular site to assess the quantity pore spaces available for recharging and storing water.

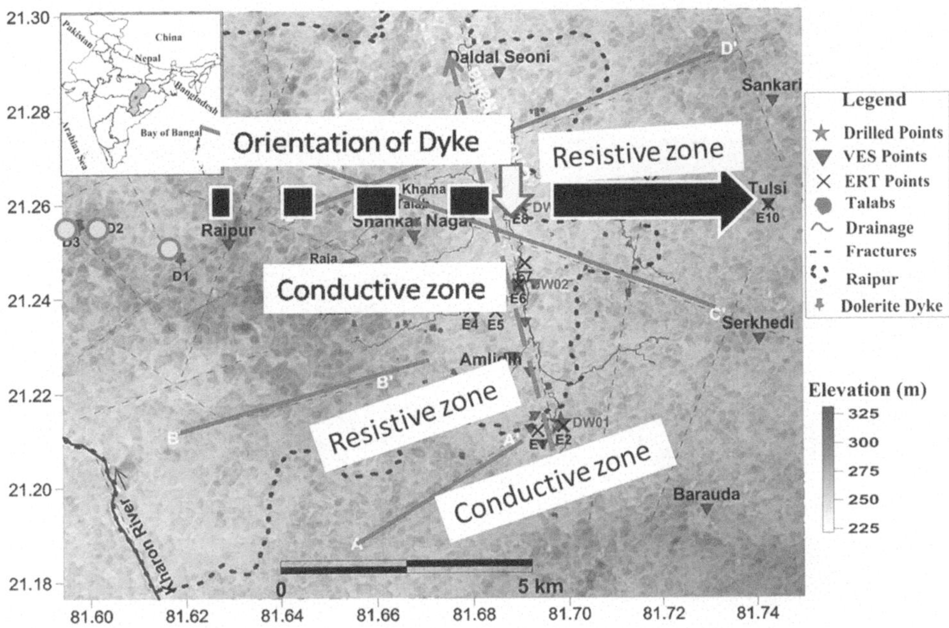

Figure 18.8 Results of Surface Electrical Structure mapping of the area suggesting the favourable zones for MAR. Black arrow indicates the possible trace of hidden sub-surface dyke. Black lines are drawn to demarcate the resistive and conductive zones on the surface based on resistivity data.

18.4.2 Recharge through intervention in dugwells in a crystalline aquifer: Assessment using time lapse electrical resistivity tomography (TLERT)

Natural recharge due to infiltration is variable and also influences the correlation between the structures deducted from geophysical data with respect to time. This spatial variation of the electrical resistivity has been documented in this study by the tomogram analysis of 2D electrical resistivity imaging data. The tomograms added precision to the analysis of the moisture movement, as it is more accurate and rigorous at spatial imaging of electrical resistivity data.

The long duration imaging surveys with a frequency of 14 days were carried out by Arora & Ahmed (2011) at a site along the profile with a length 96 m with electrode spacing of 2 m. The unsaturated zone had been mapped up to a depth of 12 meters. These tomograms of the data acquired between 15th October 2004 (post monsoon of 2004) and 15th October 2005 (during the monsoon period of 2005) at one of the sites were re-analysed and re-interpreted as a part of the project work as shown in corresponding Figure 18.9.

The resistivity data were processed and interpreted with the help of RES2DINV software developed by Loke and Barker (1996a) following an iterative optimization approach. The resulting tomograms of the data acquired between 15th October 2004 (post monsoon of 2004) and 15th October 2005 (during the monsoon period of 2005) at site S1 were further analysed in regards to soil water content in the scope of this study and interpreted with regard to the movement of the water infiltration front. Figure 18.9 shows the tomograms of the time period 12th July 2005 to 11th September 2005. Thereby, low electrical resistivity corresponds to increased soil water content and the subsequent plots (A to L) show the change in resistivity from one time-step to another. The observed changes indicated both, a vertical and horizontal movement of infiltrating water.

Published TLERT data from Arora & Ahmed (2011) for the monsoon cycle of 2005 were taken and further improved for analysis. In this study we concentrated on the resective zones where the recharge could possibly take place. Quantitatively the percent differences in electrical resistivities associated with the arrival of the recharge match the recharge measurement fairly well, both in time and space. Between 20 m and 42 m along Zone 1 (Figure 18.9), a non-uniform distribution of moisture in the vertical profile following the infiltration after a rainfall event indicates a zone of preferential natural recharge. Between 68 m and 76 m along profile the Zone 3, the variation in resistivity is most presumably due to different geological materials exhibiting different porosities.

After the time lapse experiment and observing the water level fluctuations in one of the monitoring dug wells, it was recommended to construct rain gardens (large recharge pits with plants) in the surrounding areas. In total, three rain gardens were constructed up to a depth of 1.2 m from the surface enabling maximum recharge to the aquifer. A positive effect of the recharge was observed in all the bore wells and dug wells within the study area as evident from 0.5m to 1.0 m additional rise in groundwater levels.

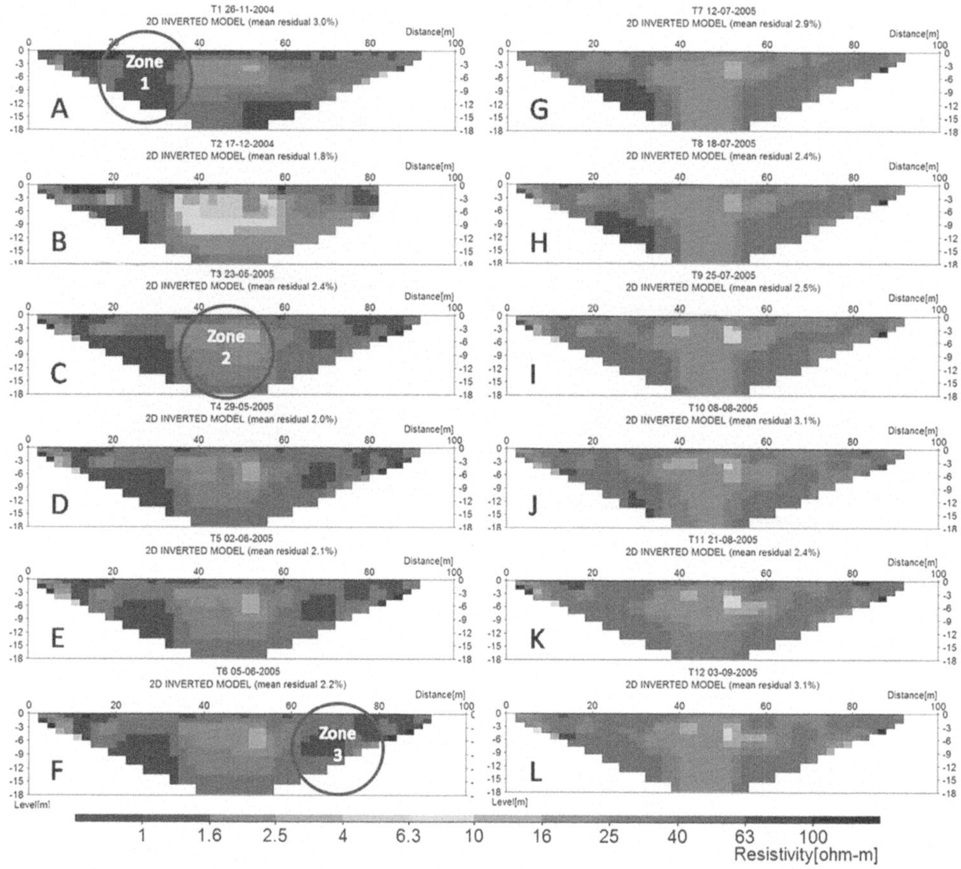

Figure 18.9 Inverted resistivity models for the time-lapse experiment carried out in the study area for the monsoon cycle of 2005 (labelled from A–L) (modified after Arora and Ahmed, 2011).

18.4.3 Infiltration in a percolation tank and effectiveness of check dams

Geophysical investigations around a percolation tank

The investigated area was the Tumulur percolation tank located in the north-eastern part of the Maheshwaram watershed, which is underlain by weathered Archaen biotite granite. Its weathering profile leads to a stratified aquifer with two distinct layers (Dewandel *et al.*, 2006): The saprolite (10–15 m thick layer from the surface) is characterized by sandy-clay material. Its total porosity is relatively high (up to 10%), but due to the clay content the effective porosity is low. In total five ERT profiles were acquired in different orientations to investigate the tank in and around area. The same methodology of ERT was undertaken (as mentioned in above sections) using the IRIS instrument Syscal Junior with 48 electrodes. The variable electrode spacing from 4–10 m was used for different profiles. After the data acquisition, the raw resistivity data was inverted and modelled to acquire the pseudo section using the RES2DINV software by Loke and Barker (1996a).

The ERT profile A (Figure 18.10) shows an increase of the saprolite layer thickness from west to east, at a depth of 10–25 m. In the profile B, the saprolite thickness trends around 10–15 m all through the N–S profile. But the fractured media is encountered towards the south of the profile, intersecting the profile A. Towards the northern end, there is basement rock dipping in N–S direction. While considering the profile C, there is a homogeneous saprolitic layer all through the NE-SW trending profile, but the fresh basement is quite shallow. This B and C profiles intersects at a point and confirms the results sections obtained individually. The information obtained is in agreement with the observations of the bore cutting (Boisson *et al.*, 2014). These profiles were used to complement previous data obtained on watershed scale to produce maps of saprolite thickness and alternated zone thickness (Boisson *et al.*, 2014) on tank scale. ERT data was integrated to the previous datasets through interpolation using the kriging method.

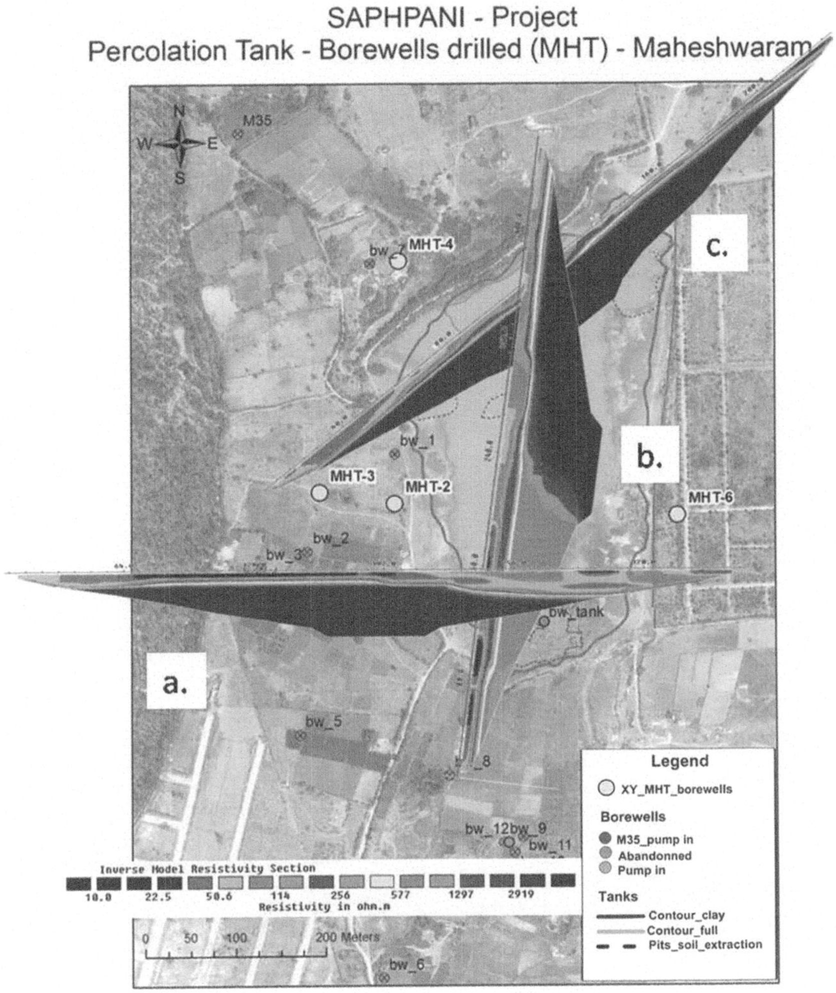

Figure 18.10 ERT profiles carried out around the percolation tank in different orientations. A total 5 profiles were taken, of which 3 are shown here.

Verification of artificial recharge effectiveness through simulation studies

The numerical model of aquifer flow in the area of Maheshwaram, including the simulation of tank recharge, was available in the same project (Chapter 14). It was therefore, useful to study and validate a new scenario for the Artificial Recharge through the tank at a favourable point provided by the geophysical investigations.

The effect of the Tumlur tank, with an area of 0.14 km^2, was initially simulated in steady state and then in transient condition within the regional model. The tank area was divided into meshes of 50 m by 50 m, giving about 57 active meshes for micro scale modelling. The steady state simulation had been performed for averaged rainfall and abstraction values in and around Tumlur tank for the period January 2001 and was then calibrated under transient conditions for the periods 2001 to 2006. This model was calibrated against the general recharge and hydraulic conductivity.

The meshes in the tank area (Figure 18.11) were divided into two parts through a hypothetical barrier, simulating a check-dam. Thus extra recharge (additional as Artificial Recharge) was assigned once to the northern side of the check dam as if the check dam was absent, simulating the logging of water under natural condition and on another run, the same extra recharge were assigned to the meshes at the southern side of the check dam only as the proposed check dam would not have allowed water to flow to the lowest point. The model was run in both conditions for a period of 6 years keeping all other inputs the same. The respective rises in groundwater levels at a representative observation point in both the cases were plotted for 6 years or 12 seasons (Figure 18.12b). It could clearly be seen that if the run-off water is applied to the southern portion of the tank, by creating an engineered structure, the benefit to the system will be manifold. To determine the exact location of such a check dam geophysical investigations are recommended.

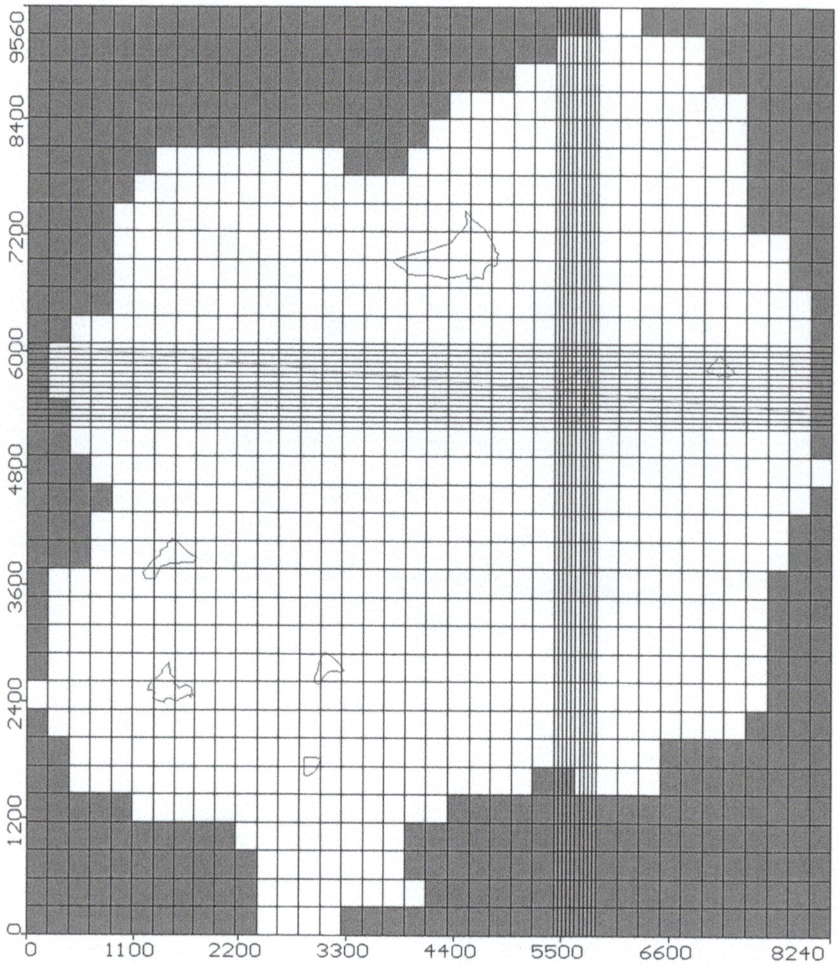

Figure 18.11 Discretized watershed with tanks.

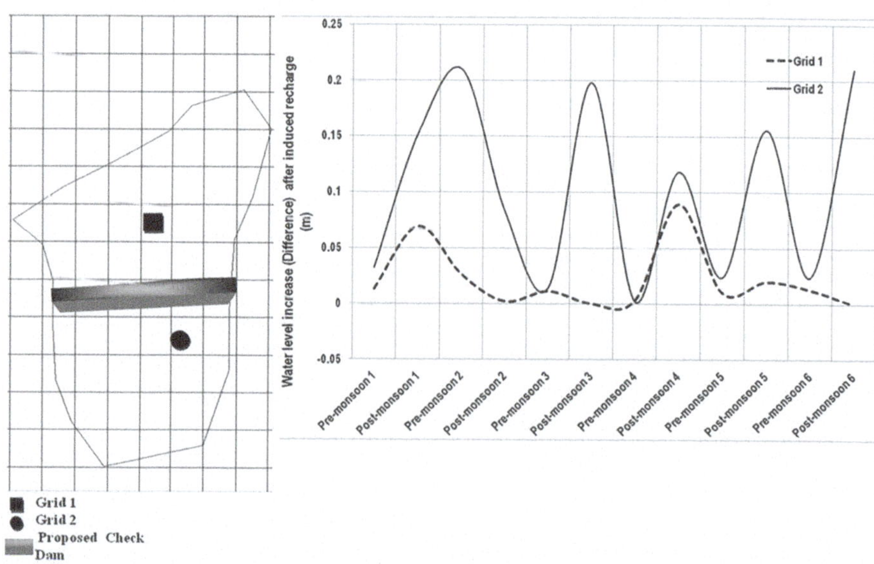

Figure 18.12 (a) Discretised Tumlur tank and (b) Impact of induced recharge on tank grid water level.

It is very clear that in spite of all other conditions being same, the space available in the northern part (that is lowest by natural way) was not enough for the artificial recharge to happen and the water would have been lost due to evaporation. The space available in the sub-surface has been comparatively more and hence with same amount of run-off water applied to the southern part using a check dam at an appropriate has provided more recharge to groundwater. This has been possible to acquire the prior knowledge using geophysical investigations.

18.5 CONCLUSIONS

The investigations documented above show that geophysical investigations along with geological knowledge provides insight to the sub-surface with considerable depth and have tremendous benefits for designing and performing MAR. MAR success could be badly affected in absence of such knowledge. In addition, we conclude that geophysical experiments have to be specially designed for MAR studies e.g., very close electrode spacing in case of VES as well as ERT and that application of TEM in alluvial formation etc. will enhance the applicability.

Artificial Recharge that is to enhance rainfall recharge by artificial means is a common and essential practice in MAR. The water available on the surface is put by a number of artificial means into the sub-surface and it is very important that the sub-surface and its characteristics are very well known. As MAR, check dams, dug wells and infiltration ponds are more prevalent, there is a need to understand the factors that control the movement of water in the unsaturated zone sub-surface beneath these ponds. Monitoring the hydrologic processes in the unsaturated zone is of great importance for improving MAR and geophysical instrumentation and monitoring is a potential method of identifying flow and transport processes in variably saturated media.

The geophysical information, to a great extent, provides knowledge and guidelines to determine the quantity and quality of the recharged water to the aquifer system and ensures both success and sustainability.

ACKNOWLEDGMENT

Authors are grateful to Director CSIR-NGRI, BRGM and the project team of Saph Pani. The work has been carried out with grant No. 281911 from the European Commission.

18.6 REFERENCES

Alexopoulos J. D., Dilalos S. and Vassilakis E. (2011). Adumbration of amvrakia's spring water pathways, based on detailed geophysical data(Kastraki – Meteora). *Advances in the Research of Aquatic Environment*, **2**, 105–122. doi:10.1007/978-3-642-24076-8.

Arora, T. and Shakeel A. (2010). Electrical structure of an unsaturated zone – a case study related to hard rock aquifer. *Current Science*, **99**(2), 216–220.

Arora T. and Ahmed S. (2011). Characterization of recharge through complex vadose zone of a granitic aquifer by time-lapse electrical resistivity tomography. *Journal of Applied Geophysics*, **73**, 35–44.

Arora T., Linde N., Revil A. and Castermant J. (2007). Non-intrusive characterization of the redox potential of landfill leachate plumes from self-potential data. *Journal of Contaminant Hydrology*, **92**, 274–292, doi:10.1016/j.jconhyd.2007.01.018.

Arora T. (2013). Hydrogeophysical characterization of unsaturated zone. Lambert Academic Publications, ISBN: 978–659–43870–7.

Binley A., Cassiani G., Middleton R. and Winship P. (2002). Vadose zone flow model parameterisation using cross-borehole radar and resistivity imaging. *Journal of Hydrology*, **267**(3–4), 147–159. doi:10.1016/S0022–1694.

Blizkovsky M. (1979). Processing and application in microgravity surveys. *Geophys. Prospect.*, **27**, 848–861.

Boisson A., Baisset M., Alazard M., Perrin J., Villesseche D., Dewandel B., Kloppmann W., Chandra S., Picot-Colbeaux G., Sarah S., Ahmed S. and Maréchal J. C. (2014). Comparison of surface and groundwater balance approaches in the evaluation of managed aquifer recharge structures: Case of a percolation tank in a crystalline aquifer in India, *Journal of Hydrology*, **519**, 1620–1633.

Bosch F. P. and Müller I. (2001). Continuous gradient VLF measurements: A new possibility for high resolution mapping of karst structures. *First Break*, **19**, 343–350.

Bosch F. P. and Müller I. (2005). Improved karst exploration by VLF-EM gradient survey: Comparison with other geophysical methods. *Near Surf Geophys*, **3**, 299–310.

Butler K. (1984). Microgravimetric and gravity gradient techniques for detection of subsurface cavities. *Geophysics*, **49**, 1084–1096.

Carriere S. D., Chalikakis K., Sénéchal G., Danquigny C.and Emblanch C. (2013) Combining electrical resistivity tomography and ground penetrating radar to study geological structuring of karst unsaturated zone. *J. Appl. Geophysm.*, **94**, 31–40.

China Geological Survey (CGS). (2005). Groundwater exploration in the arid and semiarid areas in China (Episode 2). Geological Publishing House, Beijing.

Chalikakis K. (2006). Application de méthodes géophysiques pour la reconnaissance et la protection des ressources en eau dans les milieux karstiques (Geophysical methods applied to water exploration and protection in karst environment). PhD Thesis, Université Pierre et Marie Curie, Paris, France.

Chen S. (1988). The use of selected frequency of natural field in finding water. *Site Investigation Science and Technology*, **2**, 53–55.

Colley G. C. (1963). The detection of caves by gravity measurements. *Geophys. Prospect.*, **11**, 1–9.

Dahlin T. and Zhou B. (2004). A numerical comparison of 2-D resistivity imaging with 10 electrode arrays. *Geophysical Prospecting*, **52**, 379–398.

DeGroot-Hedlin C. and Constable S. C. (1990). Occam's inversion to generate smooth, two dimensional models from magnetotelluric data. *Geophysics*, **55**, 1613–1624.

Dewandel B., Lachassagne P., Wyns R., Maréchal J. C. and Krishnamurthy N. S. (2006). A generalized 3-D geological and hydrogeological conceptual model of granite aquifers controlled by a single or multiple weathering. *J. Hydrol.*, **330**, 260–284.

Dey A. and Morrison H. F. (1979a). Resistivity modelling for arbitrary shaped two dimensional structures. *Geophysical Prospecting*, **27**(1), 106–136.

Dey A. and Morrison H. F. (1979b). Resistivity modelling for arbitrarily shaped three dimensional shaped structures. *Geophysics*, **44**(4), 753–780.

Dillon P. (2005). Future management of aquifer recharge. *Hydrogeol. J.*, **13**, 313–316.

Dillon P., Pavelic P., Page D., Beringen H. and Ward J. (2009). Managed aquifer recharge: An Introduction. Waterlines Report Series No. 13, National Water Commission, Canberra ACT 2600, Australia.

Dutta S., Krishnamurthy N. S., Arora T., Rao V. A. and Ahmed S. (2006). Localization of water bearing fractured zones in a hard rock area using integrated geophysical techniques in Andhra Pradesh, India. *Hydrogeology Journal*, **14**(5), 760–766.

Ezersky M., Bruner I., Keydar S., Trachtman P. and Rybakov M. (2006). Integrated study of the sinkhole development site on the western shores of the dead sea using geophysical methods. *Near Surf. Geophys.*, **4**, 335–343.

Farrell D., Mayers B. L., Moseley L., Sandberg S., Mwansa J., Sutherland A. and Barnes H. (2003). Geophysical imaging of seawater intrusion into karst limestones to support groundwater modelling along the west coast of Barbados [On-line]. Available: http://scitec.uwichill.edu.bb/OAS/Reports/Reports.htm

Gale I. and Dillon P. (2005). Strategies for Managed Aquifer Recharge (MAR) in semi-arid areas. United Nations Environment Programme, Division of Technology, Industry and Economics, UNESCO IHP, Paris, France.

Gan F. P., Yu L. P. and Lu C. J. (2011). Geophysical prospecting and analyzing on different karst waterbearing structures. *Geology and Exploration*, **47**(435), 663–672.

Gan F., Chen Y., Zhao W., Chen Y. and Liu W. (2013). Integrated geophysical methods for groundwater exploration in a karst area with or without thin cover – a case study from Tai'an City, Shandong Province, China. 13th Multidisciplinary Conference on Sinkholes and the Engineering and Environmental Impacts of Karst, Carlsbad, New Mexico, United States.

Ghosh D. P. (1971). The application of linear filter theory to the direct interpretation of geoelectrical resistivity sounding measurements. *Geophysical Prospecting*, **19**(2), 192–217.

Griffiths D. H., Turnbull J. and Olayinka A. I. (1990). Two dimensional resistivity mapping with a computer-controlled array. *First Break*, **8**(4), 121–129.

Guérin R., Baltassat J. M., Boucher M., Chalikakis K., Galibert P. Y., Girard J. F., Plagnes V. and Valois R. (2009). Geophysical characterisation of karstic networks – application to the ouysse system (Poumeyessen, France). *Applied Geophysics*, **341**, 810–817.

Jacob T., Chéry J., Bayer R., Le Moigne N., Boy J. P., Vernant P. and Boudin F. (2009). Time-lapse surface to depth gravity measurements on a karst system reveal the dominant role of the epikarst as a water storage entity. *Geophys. J. Int.*, **177**, 347–360. doi:10.1111/j.1365-246X.2009.04118.x

Jardani A., Revil A., Akoa F., Schmutz M., Florsch N. and Dupont J. P. (2006). Least squares inversion of self-potential (SP) data and application to the shallow flow of ground water in sinkholes. *Geophysical Research Letters*, **33**(19), L19306. doi:10.1029/2006GL027458.

Kaspar M. and Pecen J. (1975). Finding the caves in a karst formation by means of electromagnetic waves. *Geophys. Prospect.*, **23**, 611–621.

Koefoed O. (1979). Geosounding Principles 1: Resistivity Sounding Measurements. Elsevier Science Publishing Company, Amsterdam.

LaBrecque D. J. and Casale D. (2002). Experience with anisotropic inversion for electrical resistivity tomography: Proceedings of the Symposium on the Application of Geophysics to Engineering and Environmental Problems (SAGEEP) '02, compact disk paper 11ELE7.

Leucci G and De Giorgi L. (2005). Integrated geophysical surveys to assess the structural conditions of a karstic cave of archaeological importance. *Natural Hazards and Earth System Sciences*, **5**, 17–22.

Li G. Z. and Wang X. (2009). Application effect of water prospecting in granite region by comprehensive geophysical methods. *Site Investigation Science and Technology*, **4**, 55–57.

Li Y. and Oldenburg D. W. (1992). Approximate inverse mappings in DC resistivity problems. *Geophysical Journal International*, **109**(2), 343–362.

Loke M. H. and Barker R. D. (1996a). Rapid least-squares inversion of apparent resistivity pseudosections using a quasi-Newton method. *Geophysical Prospecting*, **44**(1), 131–152.

Loke M. H. and Barker R. D. (1996b). Practical techniques for 3D resistivity surveys and data inversion. *Geophysical Prospecting*, **44** (3), 499–523.

Loke M. H., Chambers J. E., Rucker, D. F., Kuras O. and Wilkinson P. B. (2013). Recent developments in the direct-current geoelectrical imaging method. *Journal of Applied Geophysics*, **95**, 135–156.

Loke M. H., Dahlin T. and Rucker D. F. (2014). Smoothness-constrained time-lapse inversion of data from 3-D resistivity surveys. *Near Surface Geophysics*, **12**(1), 5–24.

Metwaly M., Elawadi E. and Moustafal S. R. (2012). Groundwater exploration using geoelectrical resistivity technique at Al-Quwy'yia area central Saudi Arabia. *International Journal of Physical Sciences*, **7**(2), 317–326.

Meyerhoff S. B., Karaoulis M., Fiebig F., Maxwell R. M., Revil A., Martin J. B. and Graham W. D. (2012). Visualization of conduit-matrix conductivity differences in a karst aquifer using time-lapse electrical resistivity, *Geophysical Research Letters*, **39**(24), L24401. doi: 10.1029/2012GL053933.

Moore D. L. and Stewart M. T. (1983). Geophysical signatures of fracture traces in a karst aquifer (Florida, U.S.A.). *J. Hydrol.*, **61**, 325–340.

Neumann R. (1965). La gravimètrie de haute precision: Application aux recherches de cavités (high precision gravimetry: Application to cavities research). *Geophys. Prospect.*, **15**, 116–134.

Saydam A. S. and Duckworth K. (1978). Comparison of some electrode arrays for their IP and apparent resistivity responses over a sheet like target. *Geoexploration*, **16**(4), 267–289.

Singha, K., Day-Lewis, F. D., Johnson, T., and Slater, L. D. (2014). Advances in interpretation of subsurface processes with time-lapse electrical imaging. Hydrological Processes, DOI: 10.1002/hyp.10280.

Szalai S. and Szarka L. (2008). On the classification of surface geoelectric arrays. *Geophysical Prospecting*, **56**(2), 159–175.

Turberg P. and Barker R. (1996). Joint application of radio-magnetotelluric and electrical imaging surveys in complex subsurface environments. *First Break*, **14**, 105–112.

Vargemezis G., Tsourlos P. and Stampolidis A. (2012). A focusing approach to ground water detection by means of electrical and EM methods: The case of Paliouri, Northern Greece. *Studia Geophysica et Geodaetica*, **56**(4), 1063–1078.

Vlahović T. and Munda B. (2011). Karst aquifers on small islands—the island of Olib, Croatia. *Environ Monit Assess.*, **184**, 6211–6228.

Zaidi F. K. and Kassem O. M. K. (2012). Use of electrical resistivity tomography in delineating zones of groundwater potential in arid regions: A case study from Diriyah region of Saudi Arabia. *Arabian Journal of Geosciences*, **5**(2), 327–333.

Zhou W., Beck B. F. and Adams A. L. (2002). Effective electrode array in mapping karst hazards in electrical resistivity tomography. *Environmental Geology*, **42**(8), 922–928.

Chapter 19

Numerical and analytical models for natural water treatment systems in the Indian context

Christoph Sprenger, Bertram Monninkhoff, Christian Tomsu and Wolfram Kloppmann

19.1 WHY MODELLING INDIAN NATURAL TREATMENT SYSTEMS?

Modelling either by numerical or analytical approaches is a widespread approach in hydrogeology, but requires careful considerations of the model type and level of effort to be spent which in turn depends on the purpose of the study. Before starting to model, Anderson and Woessner (1992) proposed to answer the following generic questions:

- Is the model to be constructed intended for prediction, system interpretation, or a generic modelling exercise?
- What do you want to learn from the model? What questions do you want the model to answer?
- Is a modelling exercise the best way to answer those question(s)?
- Can an analytical model provide the answer or must a numerical model be constructed?

In this Chapter we will outline the use of groundwater models focusing on the strategic phases of Natural Treatment Systems (NTS) (feasibility in a given watershed, NTS design and implementation, NTS operation) as well as on the model selection (types of models). Generic numerical models are then developed for two NTS types, riverbank filtration (RBF) and Management of Aquifer Recharge (MAR) via infiltration ponds. The purpose of this modelling exercise is to determine important NTS performance parameters such as recovery rates, infiltration rates or travel times of the infiltrate in the subsurface. Investigation of modelling scenarios for each of the NTS types is another aim. Furthermore results of numerical modelling and selected analytical solutions are compared.

Real-world applications of different types of models within the case studies of the Saph Pani project as outlined in previous Chapter complete and illustrate this more generic approach (Chapter 14).

The specificity of NTSs is that they rely on natural processes. Those are complex interactions of surface water, wastewater and groundwater and the contaminants they may contain with the aquifer matrix, with microorganisms and plants. In contrast to completely engineered and controlled systems, the functioning of NTS needs to be understood first to be able to predict their performance. Once a conceptual model of the relevant processes has been established on the basis of a variety of measurements, analytical or numerical models can be set up to simulate the behaviour of the NTS. As for any numerical flow and transport model this is done through an iterative process of model setup and model calibration.

Both a completely natural and a partly engineered NTS can be, at the very beginning of their implementation, considered as a black box, contrary to engineered treatment systems. Their potential impact or performance will largely depend on the local climatic, hydrological, geological, and biological (including macrophytes, microphytes, microbia) conditions, on land use and, last but not least, on the quality and quantity of water to be treated. Measurements and monitoring are indispensable to get an insight into this black box system and are a prerequisite to any establishment of any model, even a conceptual one. The data situation will largely determine the complexity and the reliability of NTS models. Setting up a highly sophisticated model on a weak data basis will only create artificially precise simulations and predictions. In sum, modelling is inseparably linked to monitoring. A series of key questions need to be addressed before, during and after implementing of an NTS facility:

- What will be the impact of a NTS at local scale in terms of water availability and water quality improvement (or deterioration)?
- What will be the radius of influence on water quality and quantity of an individual NTS?
- Will its performance be durable over time?
- How will it behave in cases of changing boundary conditions (e.g. climatic, hydrologic, land use) or in the case of extreme events (droughts, floods)
- How can it be improved through adapted configuration (e.g. position of wells with respect to a river, pumping rates), or by adding engineered components to the system such as in-situ or pre- or post- treatment measures for water quality improvement?
- What will be the impact of NTS at the basin scale when a large number of individual systems are implemented and if different systems are combined within a watershed?

These questions will be asked from the very beginning of the planning phase and over the entire lifetime of the NTS project. Coupled surface-groundwater models, potentially with contaminant transport will provide the unique possibility to preview the feasibility of an NTS in the regional context, to optimize the choice of the site and the configuration and to optimize operating conditions in a way so as to meet fixed quantity and quality targets. Those targets are most frequently quantified through criteria like water quality acceptable for given uses such as drinking water, irrigation or industry, groundwater level evolution and salinity.

Especially in India, groundwater is an important resource, accounting for approximately 60% of irrigation water and 85% of drinking water and it is estimated that 60% of groundwater sources will be in a critical state of quantitative degradation within the next twenty years (Worldbank, 2010 and referenced therein). MAR is identified as a strategy to cope with dwindling water resources in India and "The National Groundwater Recharge Master Plan" is developed to assess the nationwide feasibility of MAR (CGWB, 2005). Modelling can accompany the implementation of NTS over different generic phases, common to all NTS types and regional contexts (Table 19.1).

Table 19.1 Planning phases of Natural Treatment Systems (NTSs) and case studies in India (Saph Pani project), see Chapter 14.

Phase	Examples from Saph Pani
Phase 1: Initial feasibility study in the regional context and choice of the NTSs	• Choice of NTSs (MAR) for saline intrusion management in the coastal Arani and Koratalaiyar watershed, Chennai, Tamil Nadu
Phase 2: Estimation of the radius of influence and positive/negative impact of an individual NTS	• Simulation of the behaviour of individual MAR-SAT* percolation tanks, Maheshwaram, Telangana
Phase 3: Planning of NTS implementation at watershed scale	• Implementation of MAR check dams in the coastal Arani and Koratalaiyar watershed, Chennai, Tamil Nadu
Phase 4: Estimation of the impacts on water quality and quantity at aquifer and watershed scale	• Scenarios of wetland impacts on water balance in the Musi watershed, Hyderabad, Telangana Simulation of contaminant transport/attenuation in an alluvial aquifer: RBF at Yamuna River, New Delhi
Phase 5: Optimisation of individual and watershed scale solutions	• Optimisation of well technology and exploitation schemes assisted by flow modelling for RBF in Haridwar, Uttarakhand

*Managed aquifer recharge (MAR) combined with soil-aquifer treatment (SAT).

19.2 WHAT MODELS FOR INDIAN NATURAL TREATMENT SYSTEMS?

The variety and degree of sophistication of models is large and, as stated above, has to be adapted to the problem to be solved and to the available data situation. Geometry of groundwater models range from 1D to full 3D and the chosen spatial resolution will determine the calculation times. Processes used for natural water treatment mainly take place at the interfaces of different compartments of the local or regional water cycle (surface flow, unsaturated flow, groundwater flow, seawater intrusion) so that there is need for integration of different types of models (river models, unsaturated-saturated groundwater models, density driven flow models). These models have revealed a major challenge for the simulation of the behaviour e.g. constructed wetlands at basin scale (Musi river study site, Chapter 14).

A complete response to the questions listed above (Chapter 19.1.1), that also addresses contaminant transport and water quality in general, may require the use of reactive transport models or even of state of the art bio-geochemical reaction modelling. Even simple models (analytical models) can provide sufficient information at least for a preliminary design or evaluation of NTS's but, most frequently, numerical models will be used. Standard numerical models are nowadays able to simulate up to full 3D advective and dispersive flow and transport of water and solutes. Supplementary features may be needed in a given context. Such features could involve, in the order of increasing complexity, the following processes:

- Density driven flow (in the context coastal aquifers salinisation), e.g. the Chennai case study (Chapter 14).
- Sorption and (bio-) degradation of solutes (e.g. through sorption isotherms, degradation factors) e.g. the New-Delhi RBF case study investigating ammonium transport (Chapter 14).
- Variable saturation flow (in the case of a significant thickness of the unsaturated zone, in particular if the latter plays an important role for water quality improvements in SAT systems) e.g. the Maheshwaram case study, which looked into infiltration processes when using infiltration ponds/tanks for MAR-SAT (Managed aquifer recharge combined with soil-aquifer treatment) (Chapter 14).
- Geochemical reactions through the combined use of flow-transport models and thermodynamic equilibrium models or thermo-kinetic models taking into account the reaction kinetics e.g. the Maheshwaram case study dealing with Fluoride mobilisation upon MAR (Chapter 14).
- Biologically mediated geochemical reactions (specific models available).

In this Chapter we will outline, through simulation of generic benchmark tests, the use of different types of groundwater models available for NTSs, in particular MAR and RBF. These generic model runs will allow selecting modelling approaches adapted to the problem to be treated and to the available calculation capacity and modelling tools. Applications of different types of models within the case studies of Saph Pani will illustrate this synthesis in Chapter 14.

19.3 SOME ANALYTICAL SOLUTIONS FOR NT SYSTEMS

Analytical solutions are simplifications and generally assume hydraulic properties to be homogenous and isotropic. Boundary conditions are often simplified and assumed to be constant. Nevertheless, analytical solutions for NT systems often provide a straight-forward approximation of important performance parameters such as recovery rates, infiltration rates or travel times of the infiltrate in the subsurface.

19.3.1 Bank filtration

Bank filtration (BF) systems in India are often utilized as the sole purification treatment along with limited post-treatment such as chlorination (Sandhu *et al.*, 2010). The purification capacity of the BF systems depends to a large extend on hydraulic parameters such as mixing ratio between native groundwater with induced surface water (bank filtrate) and the travel time of the bank filtrate to the abstraction well.

Simple analytical solutions for BF systems were developed by Rhebergen and Dillon (1999) and Dillon *et al.* (2002) to approximate travel time of bank filtrate from the surface water to the abstraction well. The authors assume an initially horizontal water table and do not consider riverbed clogging. Both river and well are fully penetrating the aquifer. All the water which is pumped is assumed to come from the surface water body in the final steady state condition. The minimum travel time ($t_{\min}$) of bank filtrate to the abstraction well is calculated according to:

$$t_{\min} = \frac{2\pi D n_e a^2}{3Q} \quad \left(\text{Dillon } et\ al., 2002\right) \tag{19.1}$$

where:

$t_{\min}$ = minimum travel time [d]
D = average saturated thickness [m]
n_e = effective porosity of the aquifer [–]
a = distance of the well from the bank [m]
Q = abstraction rate [m^3/d]

The minimum travel time ($t_{\min}$) is calculated under steady-state conditions and underestimates travel time for intermittent pumping conditions (Dillon *et al.* 2002). The analytical approach overestimates the proportion of bank filtrate in the abstraction well because rivers in nature are usually only partially penetrating the aquifer and analytical solution which assume full penetration will overestimate the infiltration from the river (Chen, 2001).

The share of bank filtrate in the abstraction well (q/Q) of the above example changes in time and can be calculated according to a generalized solution developed by Glover and Balmer (1954) based on the equations developed by Theis (1941):

$$\frac{q}{Q} = erfc\left(\frac{a}{\sqrt{4\alpha t}}\right) \quad \text{(Glover and Balmer, 1954)} \tag{19.2}$$

where:

q = rate of induced infiltration from the river (bank filtrate) [m^3/d]
α = aquifer diffusivity = transmissivity/storage coefficient, for unconfined conditions it can be calculated with K = hydraulic conductivity [m/d] according to KD/n_e [m^2/d]
t = time of pumping [d]
$erfc$ = complementary error function
q/Q = the proportion of water derived from the stream for transient pumping conditions.

As t increases to values where steady-state conditions can be assumed (approx. one year in the test cases), the solution approaches one (equal to 100% bank filtrate). Therefore, eq. 19.2 overestimates the share of bank filtrate because the influence of the stream (fully penetrating) is overestimated. As discussed above the critical parameters travel times and share of bank filtrate were worst case estimations, travel time is underestimated and the share of bank filtrate is overestimated. This makes it a conservative approach since both parameters, essential for the NTS purification capacity, are assumed to be better in reality: Longer travel times will lead to longer contact time with the aquifer material and a lower share of bank filtrate in the recovery well(s) results in stronger dilution so that the overall system performance will be higher than estimated by the analytical model.

In addition to the above example, Hunt (1999) derived a solution which takes into account a situation in which a river only penetrates partially into the aquifer system, the river has a semi pervious sediment layer and the river is not necessary located at the boundary of the model. The system Hunt describes gives a non-stationary solution for a phreatic aquifer system in which a well extracts groundwater and this extraction causes inflow from the river into the groundwater (Figure 19.1). Hunt presents a solution for the drawdown of the groundwater, both in space and in time, as well as a solution for the discharges from the river into the groundwater. The equations he provides for these two solutions are given in eq. 19.3 (drawdown $\omega(x,y,t)$ [m]) and eq. 19.4 (ratio between the infiltration and the extraction rate $\Delta Q/Q_w$):

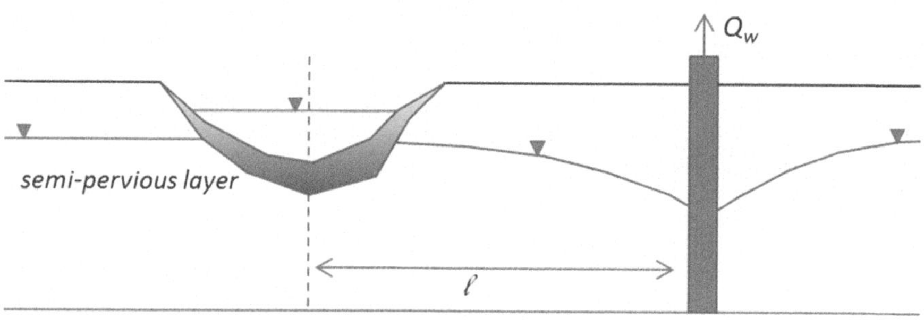

Figure 19.1 The problem considered by Hunt (1999).

$$\omega(x,y,t) = \frac{Q_w}{4\pi T}\left\{W\left[\frac{(\ell - x)^2 + y^2}{4Tt/S}\right] - \int_0^{\infty} e^{-\theta} W\left[\frac{(\ell + |x| + 2T\theta/\lambda)^2 + y^2}{4Tt/S}\right] d\theta\right\} \tag{19.3}$$

where:

λ = constant of proportionality between the seepage flow rate per unit distance (in the y direction) through the streambed and the difference between river and groundwater levels at $x = 0$ (location of the river) [m/d]
W = Theis well function (for example in Barry, 2000)
S = porosity [–]
T = transmissivity of the aquifer [m^2/d]

$$\frac{\Delta Q}{Q_w} = erfc\left(\sqrt{\frac{S\ell^2}{4Tt}}\right) - \exp\left(\frac{\lambda^2 t}{4ST} + \frac{\lambda\ell}{2T}\right) erfc\left(\sqrt{\frac{\lambda^2 t}{4ST}} + \sqrt{\frac{S\ell^2}{4Tt}}\right) \tag{19.4}$$

As a simplification, Hunt (1999) also assumes that the water level in the river does neither change with time as a result of runoff variations or of the infiltration into the groundwater.

19.3.2 Surface spreading methods

Surface spreading methods comprise NTSs such as infiltration ponds, soil-aquifer treatment or surface flooding. During surface spreading, the source water such as river water or surface runoff is collected and diverted to the area of recharge. Recharge takes place by percolation through the unsaturated zone to the groundwater table. In India, surface spreading is often operated without managed abstraction and the artificially recharged groundwater is consumed by the local community mostly for agricultural purposes (Gale *et al.* 2006). Important hydraulic parameters during surface spreading are the infiltration rate or the development of the groundwater mound beneath the recharge area, reducing the thickness of the unsaturated zone. Infiltration rates during surface spreading are subject to large temporal and spatial variations. This is caused by geological heterogeneities, by varying filling and extension of ponds but also by operational needs such as dry/wet cycles. The hydraulic performance of an infiltration system is therefore best expressed in long-term infiltration rates or hydraulic loading rates (Bouwer, 2002).

The Green-Ampt equation was developed to calculate the infiltration rate (V_i) from a ponded surface (e.g. infiltration basin) into a deep homogeneous porous media with uniform initial water content. The Green-Ampt model has been found to apply best to infiltration into initially dry, coarse textured media which exhibit a sharp wetting front.

The Green-Ampt solution was developed in 1911 and is based on Darcy's law:

$$V_i = K\left(\frac{H_W + L_f - h_{we}}{L_f}\right) \quad \text{(Green and Ampt, 1911)} \tag{19.5}$$

where:

V_i = infiltration rate or hydraulic loading rate [m/s],
K = hydraulic conductivity [m/s],
H_w = depth of water in the pond or infiltration facility [m]
L_f = depth of the wetting front below the bottom of the pond [m]
h_{we} = suction or negative pressure head at the wetting front [m]. Approximately equal to the air entry pressure or bubbling pressure

Unsaturated K values are lower than saturated K (K_{sat}) values, because of the entrapped air. Bouwer (1978) refers to K/K_{sat} ratios of 0.5 for sandy soils and 0.25 for clays. Values of h_{we} describe the suction at the wetting front (negative pressure head). Typical values of h_{we} along with other important hydraulic properties for various soils can be found in Table 19.2.

Table 19.2 Hydraulic properties for various soils (modified after Rawls *et al.*, 1983).

Texture	Effective Porosity n_e [–]	Suction Head h_{we} [cm]	Hydraulic Conductivity K [cm/hr]
Sand	0.417	–4.95	11.78
Loamy Sand	0.401	–6.13	2.99
Sandy Loam	0.412	–11.01	1.09
Loam	0.434	–8.89	0.34
Silt Loam	0.486	–16.68	0.65
Sandy Clay Loam	0.330	–21.85	0.15
Clay Loam	0.309	–20.88	0.10
Silty Clay Loam	0.432	–27.30	0.10
Sandy Clay	0.321	–23.90	0.06
Silty Clay	0.423	–29.22	0.05
Clay	0.385	–31.63	0.03

Operators of infiltration ponds may also be interested in the height of groundwater mound which is created by a MAR facility. A minimum thickness of the unsaturated zone may be important to ensure a sufficient degree of natural treatment with respect to required water quality standards (Figure 19.2).

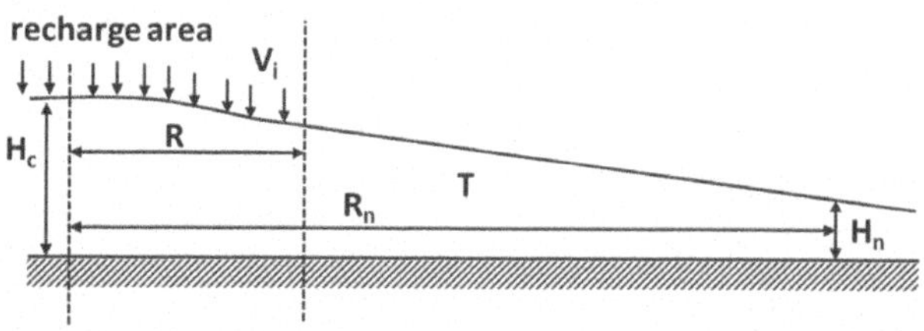

Figure 19.2 Cross sectional view illustrating geometry and parameters used for calculation of groundwater mound (Bouwer, 2002).

Bouwer *et al.* (1999) developed an analytical solution for round or square recharge ponds, where the groundwater flows radially away from the point of recharge. The ultimate or steady-state height of the groundwater mound right below of the centre of the recharge pond is calculated according to:

$$H_c - H_n = \frac{V_i R^2}{4T}\left(1 + 2\ln\frac{R_n}{R}\right) \quad \left(\text{Bouwer } et\ al.\text{, 1999}\right) \tag{19.6}$$

where:
R = radius or equivalent radius of the recharge area [m]
R_n = distance from the centre of the infiltration pond to the control area [m]
H_c = height of groundwater mound in the centre of recharge area [m]
H_n = height of water table in control area [m]
V_i = average infiltration rate (total recharge divided by total area) [m/s]
T = transmissivity of the aquifer [m²/s]

Control area is here defined as the area where the groundwater table is stable. The value of transmissivity in eq. 19.6 must reflect the average transmissivity of the aquifer at the steady-state stage of the mound.

In numerical groundwater models, infiltration rates or exchange fluxes (q) between the ground- and surface water are usually calculated by introducing a transfer or leakage coefficient ϕ_h [1/d]:

$$q = \phi_h (h_{\text{ref}} - h_{gw}) \tag{19.7}$$

in which:
q = Darcy flux [md^{-1}] of fluid (positive from river to groundwater) and
h_{ref}, h_{gw} = heads [m] in the river and groundwater respectively.

Assuming a simple surface spreading infiltration system represented by an initially fully rectangular canal with no other in- or outflow than the fluxes to or from the connected groundwater, the conservation of mass equation for such a unit can be written as follows:

$$\frac{\delta h_{\text{ref}}}{\delta t} = -\frac{Q_o}{A_r} \tag{19.8}$$

In which t [d] represents time and A_r represents the cross section area of the canal [m²].

It is assumed that the groundwater is initially way below the bottom of the canal and even after the canal has been drained completely, the groundwater still has no direct contact to the surface water. Substituting eq. 19.7 in eq. 19.8, taking into account that the width of the canal (B_r [m]) is water level independent, infiltration takes place along the complete wetted perimeter of the canal (bottom and lateral infiltration) and the lowest value of h_{gw} is limited to the bottom of the canal (constraining the infiltration), the time T_e [d] which is needed to empty the canal from a water depth wd_r1 to a depth wd_r2 can be calculated by:

$$T_e = -\left[\frac{B_r(\ln(wd_r) - \ln(B_r\phi_h + 2\phi_h wd_r))}{B_r\phi_h}\right]_{wd_{r1}}^{wd_{r2}} \tag{19.9}$$

For a canal with a triangular shape and a constant slope of the banks $1/\eta$ [–], the solution of T_e can be expressed in a slightly more convenient way:

$$T_e = -\left[\frac{\eta}{\phi_h\sqrt{(1+\eta^2)}}\ln(wd_r)\right]_{wd_{r1}}^{wd_{r2}} \tag{19.10}$$

As these equations also describe non stationary infiltration processes with varying water levels in the infiltration unit, they can be used to verify the results of numerical groundwater models simulating infiltration processes related to surface-spreading MAR structures.

19.4 USE OF NUMERICAL MODELS FOR NATURAL TREATMENT SYSTEMS

In contrast to analytical solutions, numerical models can be adapted to a wide range of site-specific conditions and problem statements. A large number of numerical models have been used to analyse various MAR systems ranging from basic hydraulic problems (Neumann *et al.* 2004) to complex temperature-dependant redox zonation and associated contaminant removal (Henzler *et al.* 2014; Greskowiak *et al.* 2006). Most numerical models, no matter if based on finite elements or finite differences, are generally capable to simulate three-dimensional advective, diffusive and dispersive flow and solute transport. In the following, we provide a brief description of commonly used codes:

MODFLOW (Harbaugh, 2005) is a modular, finite-difference flow model developed since the 1980's. The code is public domain free software, but there are several commercial and non-commercial graphical user interfaces available. MODFLOW can be combined with several packages to simulate fate and transport (MT3DMS, Zheng & Wang, 1999 and PHT3D, Prommer, 2006), density driven flow (SEAWAT, Langevin *et al.* 2007). It also provides modules for parameter estimation and uncertainty analysis (e.g. PEST, Welter *et al.* 2012). With this set of packages MODFLOW is a very powerful, robust and flexible modelling tool.

FEFLOW (Diersch, 2014) provides an advanced 3D graphically based modelling environment for performing complex groundwater flow, contaminant transport, and heat transport modelling. Regarding contaminant transport, multiple species as well as kinetic reactions between the species can be modelled. Both saturated and unsaturated flow regimes can be described. It uses a Galerkin-based finite element numerical analysis approach with a selection of different numerical solvers and tools for controlling and optimizing the solution process. FEFLOW is a completely integrated system from simulation engine to graphical user interface including a public programming interface for user code. By this interface, also integrated surface water–groundwater interactions can be modelled, for example using the plug-in IfmMIKE11. This module couples FEFLOW to the surface water modelling engine MIKE11 (DHI, 2014a). Its scope of application ranges from simple local-scale to complex large-scale simulations. Special features like biodegradation, density dependent flow and random walk analyses enable the use of FEFLOW also in very specific cases. With FEPEST FEFLOW offers a powerful tool for auto calibration and parameter uncertainty analyses.

MARTHE v7.0 is a complete numerical hydrosystem code designed for hydrodynamic and hydrodispersive modelling of groundwater flow and mass and energy transfer in porous media (Thiéry, 1990, 1993, 1995, 2010a). This code allows the three-dimensional simulation of flow and transport under saturated conditions and in the vadose zone (Herbst *et al.* 2005) using a finite volume method for hydraulic calculations (Thiéry, 2010b) and integrates a hydroclimatic balance (precipitation, evapotranspiration, runoff, infiltration, recharge) using the GARDENIA scheme (Thiéry, 2010c) as well as density driven flow. Full interaction between a surface water network and groundwater is implemented in MARTHE v7.0 and has been applied to river basins to predict the influence of climatological changes on river flows and to predict floods (Thiéry and Amraoui, 2001; Habets *et al.* 2010; Thiéry, 2010d). Reactive transport (flow and solute transport coupled to geochemistry) uses the REACT solver from Berkeley LBNL TOUGHREACT code (Xu *et al.* 2011). Coupling with the thermokinetic PHREEQC code for reactive transport modelling is currently being implemented. Furthermore, MARTHE was applied in the context of MAR systems (Gaus *et al.* 2007; Kloppmann *et al.* 2012). However in the release of MARTHE used for earlier studies, all surface water bodies are connected to the river network allowed to flow out of the system, but surface water cannot be stored in topographic depressions and re-infiltrate to the aquifer. For this reason, a specific module (LAC), allowing for a complete water balance (rainfall, evapotranspiration, infiltration, storage, water level changes and lateral extension) of MAR structures has been implemented and tested on the Saph Pani project study site of Maheshwaram watershed (Picot-Colbeaux *et al.* 2013).

MIKE SHE (DHI, 2014b) is a fully distributed, process-based hydrological model and includes process models for evapotranspiration, overland flow, unsaturated flow, groundwater flow, and channel flow and their interactions including solute transport. Each of this process is described by its governing equation or by a simpler conceptual representation and a user can tailor the model structure by choosing processes to be included and the solving methods. MIKE SHE is a comprehensive catchment modelling framework with applications ranging from aquifer management and remediation to wetland management, flooding and flood forecasting. MIKE SHE is dynamically coupled to MIKE11, which is a one-dimensional surface water model that simulates fully dynamic channel flows and is therefore able to represent river processes and river management. The process-based approach allows different model structures to be applied within the same modelling framework. In MIKE SHE, governing partial differential equations are solved by discrete numerical approximations in space and time using finite differences.

19.4.1 Calculating mixing proportions by water balance modelling

Water balance modelling or water budget calculation is the easiest procedure to analyse e.g. shares between the different recharge components. Typical problem statements in MAR systems comprise characterisation of sources of water which is abstracted in a well. By specifying sub-regions in the model domain (i.e. boundary conditions), the flow between each of the adjacent zones is calculated. This method may be applied in two- or three-dimensions using any groundwater flow model that includes water balance calculations (for example MODFLOW, MARTHE or FEFLOW). In case of more complex flow patterns, e.g. multiple wells and transient boundary conditions, it may be necessary to assess the water budget by more demanding particle tracking or solute transport approaches.

19.4.2 Calculating mixing proportions and travel times by particle tracking

Particle tracking is used to trace flow lines by simulating the movement of imaginary particles in a given flow field. When advective transport is the dominant process controlling solute mobility, particle tracking in groundwater flow models can be an alternative to more demanding solute transport models. Particle-tracking using model packages such as MODPATH (Pollock, 1994) or FEFLOW (DHI-WASY, 2013) provide a tool to calculate e.g. travel time of water between two points. In MAR and RBF systems, particle tracking was used e.g. by Abdel-Fattah *et al.* (2008) to investigate travel time of bank filtrate, riverbed infiltration zone length and well capture zone. During particle tracking, it is assumed that solute movement is controlled entirely by advection and that density-dependent flow, dispersion and diffusion are negligible. Random-Walk Particle Tracks, however, incorporate diffusion and dispersion, bringing field-line analysis a large step closer to the capabilities of a full advection-dispersion solution. As this option does not require the setup of a complete transport problem, pre- processing effort and computational cost remain comparably low. Random-Walk particle tracking is available in FEFLOW (DHI-WASY, 2013) and MARTHE (Thiéry, 2010a). Flow models can be transient or steady state and particle tracking can be calculated forward or backward in time.

Important performance parameters of MAR systems such as the share of bank filtrate in the abstraction well can be approximated by backward modelling of particles released around a well screen. The angle between the uppermost and lowermost streamline gives an approximation of the share of bank filtrate in the abstraction well (Chen, 2001). River water between these two lines flows to the abstraction well, while the water outside of these two lines does not flow to the well. The angle is then measured by visual inspection close to the well screen and compared to a complete full circle. This method will give only a rough approximation for simple models (e.g. single layer models). It has to be taken into account that in heterogeneous, multi-layer models, particles for each layer must be weighted according to the layer specific flow. This leads to laborious calculations and other methods may apply better. Travel times of bank filtrate to the abstraction well can be approximated by e.g. end point calculations in MODPATH. Travel time of particles from the abstraction well to their point of termination is calculated backward in time. Termination points are model boundaries e.g. the river.

19.4.3 Calculating mixing proportions and travel times by solute transport

In transient models, conservative solute transport can be used to approximate travel times and mixing proportions in MAR systems. Among other transport options, conservative transport can be simulated with MT3DMS (Zheng & Wang, 1999) and MARTHE (Thiéry, 1990, 1993, 1995, 2010a). In FEFLOW, a new feature called "Groundwater Age" offers a method to calculate groundwater ages, mean lifetime expectancies and mean exit probabilities also for non-conservative transient transport processes involving 1st order decay and linear retardation processes (DHI-WASY, 2013). By this, valuable information to estimate risk vulnerabilities or evaluate outlet capture zones and the origin of water, also under density dependent dominated conditions, can be provided. In MARTHE, transport in the aquifer considers advective, diffusive and dispersive components as well as exponential decrease for a given compound, chain degradation, a retardation factor or partition coefficient (Kd for adsorption-desorption), double porosity (equilibrium or with kinetics), Freundlich and Langmuir isotherms.

In MAR systems the point of recharge (e.g. the river or lake during bank filtration, infiltration pond or injection well) is assigned to species concentration C_0 of 1, while the rest of the model domain is assigned to species concentration of 0. The resulting breakthrough curve for continuous injection is shown schematically in Figure 19.3.

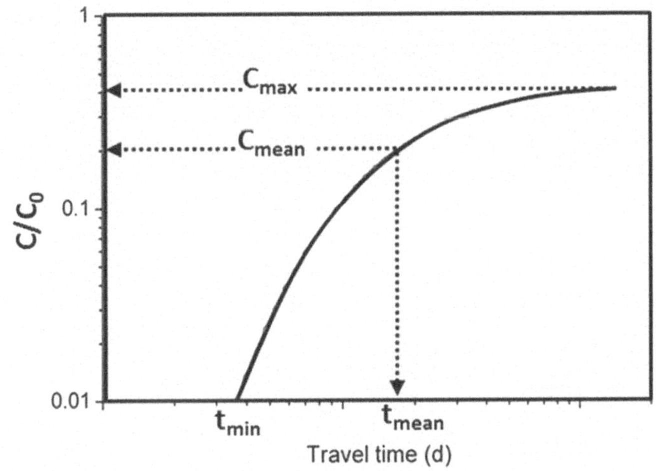

Figure 19.3 Log-log scale of an exemplary breakthrough curve of an ideal tracer and calculation of mean, minimum travel time and bank filtrate share in the abstraction well.

The proportion or share of bank filtrate in the abstraction well water is defined as the maximum concentration (C_{max}) during late, quasi steady-state conditions. The mean travel time (or dominant travel time) of e.g. bank filtrate to the abstraction well can be calculated by differentiating the cumulative breakthrough curve and retrieving the time at which its mean value is reached. Minimum travel time can, for example, be defined when one percent ($C/C_0 = 0.01$) of the maximum concentration is reached.

19.5 COMPARISON OF ANALYTICAL AND NUMERICAL SOLUTIONS

19.5.1 Bank filtration

Model descriptions

The BF model domain and associated boundary conditions are illustrated in Figure 19.4.

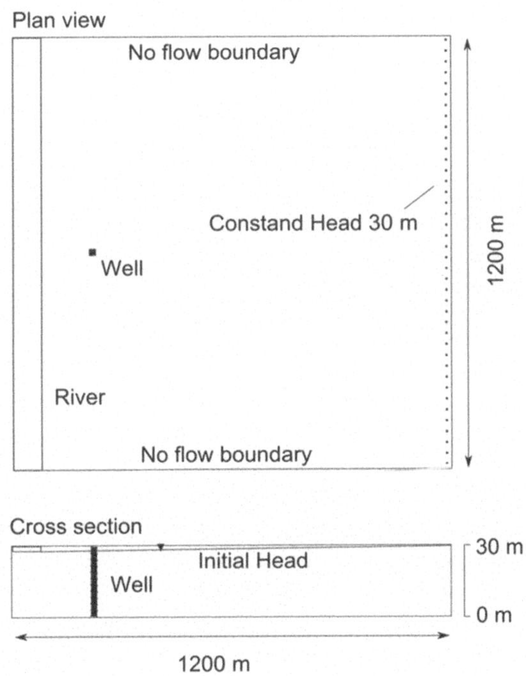

Figure 19.4 Model domain and boundary conditions for the bank filtration scenarios.

The cell size in the model domain is 20 × 20 m without any grid refinement. The abstraction well is fully penetrating. River stage is kept constant 1 m below the constant head boundary and vertical hydraulic conductivity is 1/10 of horizontal hydraulic conductivity. Two exemplary model scenarios have been used to compare analytical and numerical solutions for BF systems. Differences in the two BF scenarios are shown in Table 19.3. These scenarios represent two arbitrary, basic and simple capture zone characteristics, which can be approximated by using the presented analytical solutions. It was not intended to analyse the differences between the two scenarios in this study.

Table 19.3 Differences in model parameter between scenarios.

Parameter	Scenario 1	Scenario 2
Hydraulic conductivity [m/s]	1×10^{-4}	1×10^{-3}
Effective porosity [–]	0.15	0.25
Pumping rate [m³/d]	1000	2000
Riverbed conductance [m²/s]	0.04	0.4
Distance of pumping well from riverbank [m]	60, 80, 100	60, 80, 100

In FEFLOW, the riverbed conductance is applied as transfer rates T_{in} and T_{out}, which are calculated as follows:

$T_{in} = T_{out}$ = Riverbed conductance [m²/s]/Element area [m²] = 0.04/20 = 2×10^{-4} [1/s]

In each of the scenarios, the riverbed conductance was adjusted to the hydraulic conductivity of the aquifer in order to ignore any riverbed clogging effects. The river is represented in MODFLOW as well as in FEFLOW by a head-dependent flux boundary. This head is kept constant during all model scenarios. Water is flowing from the river to the groundwater when the head in the nearby cell is lower than the river stage. The flux between river and aquifer (q_{riv}) is calculated with riverbed conductance (C_{riv}) and the head difference between the river stage and the adjacent groundwater head (Δh):

$$q_{riv} = C_{riv} \times \Delta h \tag{19.11}$$

Clogging of the riverbed is expressed by riverbed conductance (C_{riv}) according to:

$$C_{riv} = \frac{K \times L \times W}{M} \tag{19.12}$$

where:
C_{riv} = riverbed conductance [m²/s]
K = hydraulic conductivity of riverbed [m/s]
L = river length [m] in cell
W = river width [m] in cell
M = thickness of clogging layer [m]

Both equations are solved individually for each model time step at each model grid cell, which is identified as a river cell. This approach enables consideration of the temporally and spatially variable extent of the interactions between the groundwater and the surface water.

The BF model scenarios were first run in steady-state mode to calculate the water budget and particle tracking. In a last step, the flow model was coupled to a MT3DMS solute transport model using a third-order total-variation-diminishing (TVD) scheme for solving the advection term in transient mode. The TVD scheme is mass conservative and does not produce excessive numerical dispersion or artificial oscillation (Zheng & Wang, 1999). In FEFLOW, the particle tracking analyses taking into account dispersion as well as diffusion was performed using classic mass transport simulations as well as the calculation of lifetime expectancies with the Groundwater Age feature.

In a second example, the shown benchmark of Hunt (1999) was simulated using a coupled FEFLOW and MIKE11 setup using the plug-in IfmMIKE11 (Monninkhoff *et al.*, 2009). The model setup is shown in Figure 19.5.

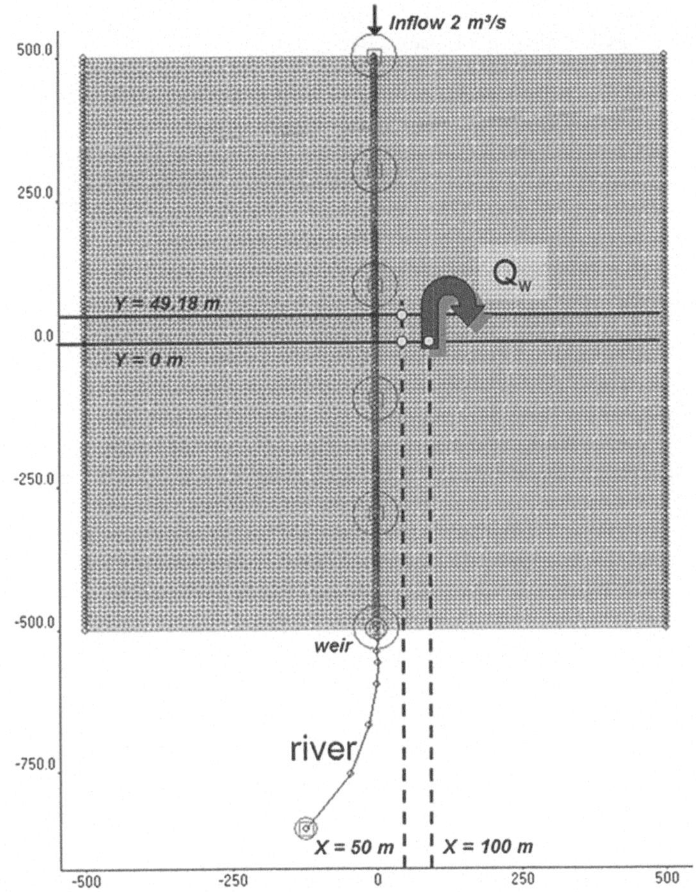

Figure 19.5 The problem considered by Hunt (1999) described by coupled groundwater and surface water system.

For this example, the MIKE11 model had to be built in a way that the river water level would not, or hardly, change as a result of the infiltration. For that purpose, a rather wide (200 m) and very smooth river bed (Manning–Strickler roughness coefficient Kst = 80 $m^{1/3}/s$) was defined. Furthermore, a constant upstream inflow rate of 2 m^3/s and a weir level of approximately 10 m at the downstream end of the coupled region ensured that the water level in the river along the FEFLOW model was infiltration rate-independent as well as constant along the river itself. MIKE SHE was verified using the Hunt benchmark as well (Illangasekare, 2001). For that verification λ in eq. 19.4 was set equal to $1*10^{-5}$ m/s. With a 200 m wide rectangular channel with a constant water depth of 2 m (the river bed was set at 8 m) an identical value for λ can be achieved using a global transfer coefficient equal ϕ_h of $42.4*10^{-4}$/d. By fixing a porosity of 0.2, a transmissivity of 0.001 m^2/s, a distance ℓ between river and well of 100 m and an extraction rate Q_w equal to 10'000 m^3/a, exactly the same conditions could be tested with IfmMIKE11.

Results

Differences of calculated minimum travel time by analytical and numerical approaches were found for the two BF scenarios (Table 19.4). These can be partly explained by the differences in the numerical approximation and discretization (MODFLOW and FEFLOW) and by the differences in conceptualization (in the numerical models the river was not fully penetrating the aquifer and a clogging effect has been taken into account by the FEFLOW model). Furthermore, it has to be noted that the calculation methods for deriving minimum travel times using particle tracking and solute transport are fundamentally different. In case of solute transport, a concentration threshold at the well determines the minimum travel time (which can be defined as $C/C_0 = 0.01$), taking into account dispersion, diffusion and mixing processes within the well capture zone. Using particle tracking, however, the minimum travel time from different starting points within the capture zone is derived. With the simple model setup at hand, the minimum travel time path always coincides with the shortest distance between the well and the river.

Table 19.4 Minimum travel time [d] calculated by different analytical and numerical approaches (longitudinal dispersivity = 5 m, transversal dispersivity = 0.5 m for solute transport).

	Well Distance from Riverbank [m]	Dillon *et al.* (2002)	Particle Tracking (Advection only)		Solute Transport (Advection + Dispersion)	
			MODFLOW	FEFLOW	MODFLOW (TDV)	FEFLOW[1]
Scenario 1	100	94	104	91	75	60
	80	60	67	57	39	43
	60	34	39	30	25	18
Scenario 2	100	79	132	109	101	91
	80	50	79	62	64	50
	60	28	43	31	29	24

[1]: Classic FEFLOW mass transport simulation; TDV = Total Variation Diminishing solution.

Despite these differences in model setup and calculation methods, the results clearly show that, compared to particle tracking results, on average the analytical solution from Dillon *et al.* (2002) produces lower travel times which confirms the conservative approach of the analytical solution, especially for scenario 2. If compared to the solute transport solutions, the analytical solution yields substantial higher travel times and underestimates dispersive effects during subsurface passage.

In Table 19.5, a comparison between FEFLOW simulations using particle tracking, the classic mass transport simulation and the Age Problem Class (Life Time Expectancy) is shown. The minimum travel time in the classic mass transport simulation is based on the explanation in Figure 19.3 and is therefore derived from the concentrations in the well. The evaluation of travel times using life time expectancies is based on mean travel times including dispersion and diffusion. From the resulting travel times, the minimum travel time between the river and the well is selected. Dispersion and diffusion in reality causes both longer and shorter travel times compared to mere advection based simulations. In this, the proportion of the longer travel times has the tendency to easily shift the mean towards older values, causing on average longer single travel times compared to pure advection based analyses (particle tracking). This also results in larger minimum travel times using the Age Problem Class. Like in the particle tracking, the minimum travel time is mostly located at the minimum distance between the riverbank and the well. The results show that according to the choice of definition of minimum travel time, significantly different results can be obtained. It is therefore important to determine which definition is most appropriate for the problem statement under consideration.

Table 19.5 Minimum travel time [d] calculated by different approaches in FEFLOW (longitudinal dispersivity = 5 m, transversal dispersivity = 0.5 m for solute transport simulations).

	Well Distance from Riverbank [m]	Minimum Travel Times [d]		
		Particle Tracking (Advection only)	Life Time Expectancy	Classic Mass Transport
Scenario 1	100	91	103	60
	80	57	68	43
	60	30	41	18
Scenario 2	100	109	120	91
	80	62	73	50
	60	31	41	24

In Figure 19.6 an exemplary result of Scenario 1 is shown with a well distance of 100 m from the riverbank, calculated with FEFLOW, taking into account advection and dispersion using the Age Problem Class (Life Time Expectancy).

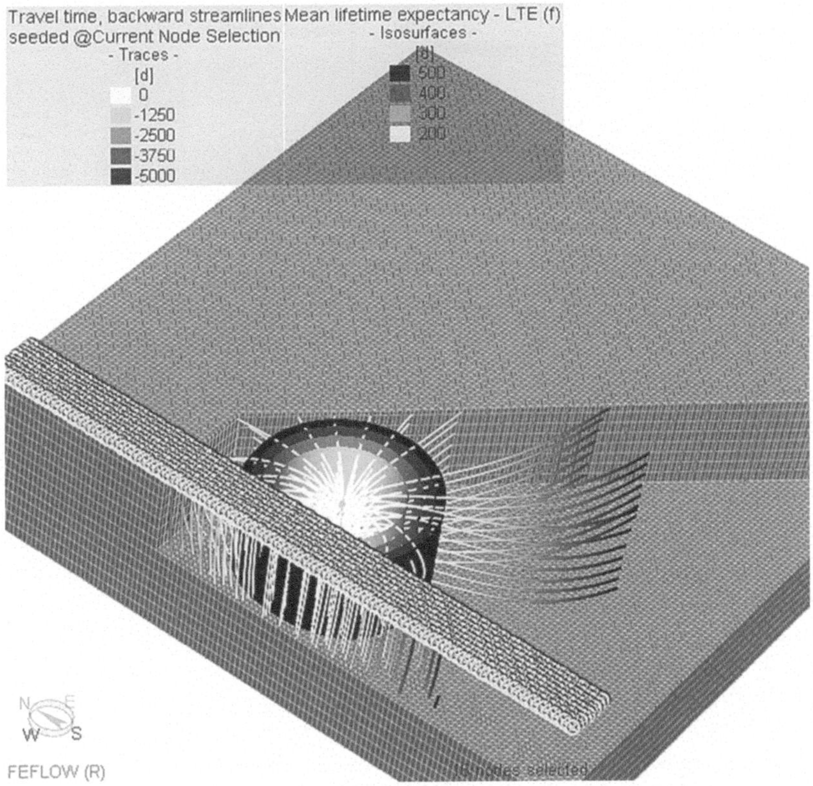

Figure 19.6 Result of a FEFLOW bank filtration simulation showing Lifetime Expectancies (advection and dispersion), Backward Streamlines (purely advection) and the river boundary nodes in slice 1. In FEFLOW, the Lifetime Expectancy (days) is defined as the time required for the water molecules to reach an outlet boundary of the aquifer.

Calculations of the share of bank filtrate in the abstraction well based on zone budget, particle tracking and solute transport come to almost identical results, whereas analytical solutions by Glover and Balmer (1954) largely overestimate the share of bank filtrate (%) compared to numerical solutions (Table 19.6).

Table 19.6 Percentage share of bank filtrate calculated by analytical and numerical (advection only) approaches.

	Well Distance from Riverbank [m]	Glover and Balmer (1954)*	Zone Budget (Water Balance)	Modflow	Feflow
Scenario 1	100	86	61	61	62
	80	86	63	63	64
	60	86	66	66	67
Scenario 2	100	94	17	17	17
	80	94	22	22	22
	60	94	27	27	28

*Calculated with t_{min} derived from eq. 19.1.

Minimum travel time and share of bank filtrate in the abstraction well can also be calculated based on breakthrough curves of conservative solute transport calculated for the abstraction well for different well distances (like shown in Figure 19.7 for the MODFLOW simulations).

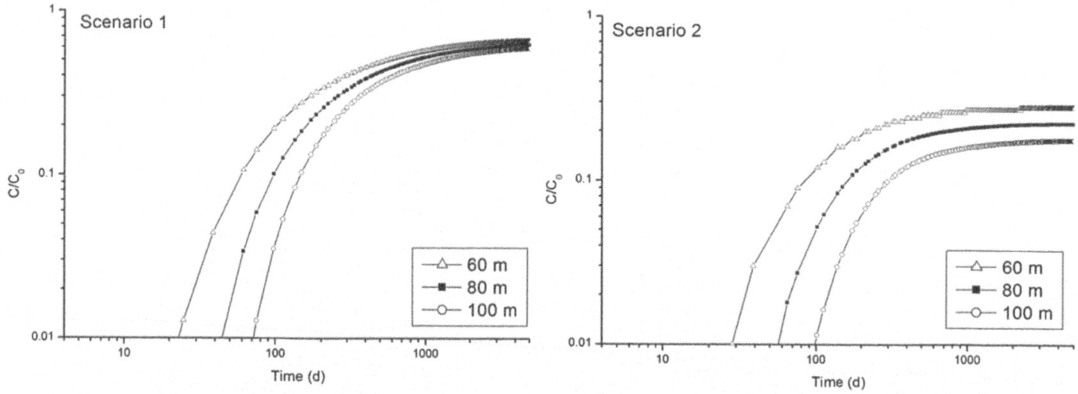

Figure 19.7 Log-log scale breakthrough curves of tracer indicating travel time (d) and share of bank filtrate (C/C_0) calculated by solute transport for 60 m, 80 m and 100 m distance of the abstraction well from the riverbank, MODFLOW simulations.

During these additional solute transport simulations, longitudinal dispersivity was kept constant with 10 m, which was approximated by one tenth of the maximum flow distance from the riverbank to the abstraction well according to (Adams & Gelhar, 1992) and transversal dispersivity was neglected.

In Figure 19.8 and Figure 19.9 the results for the benchmark of Hunt (1999) using IfmMIKE11 are shown. These figure show both the simulation results for the infiltration rate along the coupled river (Figure 19.8) as well as the drawdown along the line $y = 0$ and $y = 49.18$ m in Figure 19.5 at day 23 of the simulation (Figure 19.9). Within the figures, the analytical solutions presented by Hunt (1999) are also included. The analytical solutions and the IfmMIKE11 results are nearly identical. It is therefore concluded that these kind of MAR applications can be simulated using a coupled setup of MIKE11 and FEFLOW. The same setup has been successfully benchmarked by an OPENMI based coupling between FEFLOW and MIKE SHE (Yamagata *et al.*, 2012).

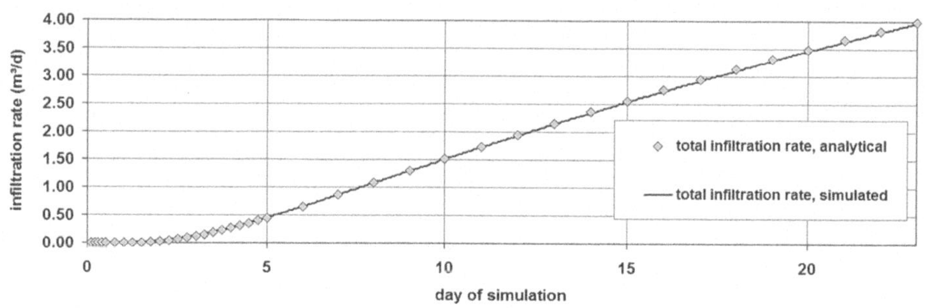

Figure 19.8 Comparison between the analytically solved and numerically simulated total infiltration rates over time along the coupled river branch using FEFLOW and MIKE11.

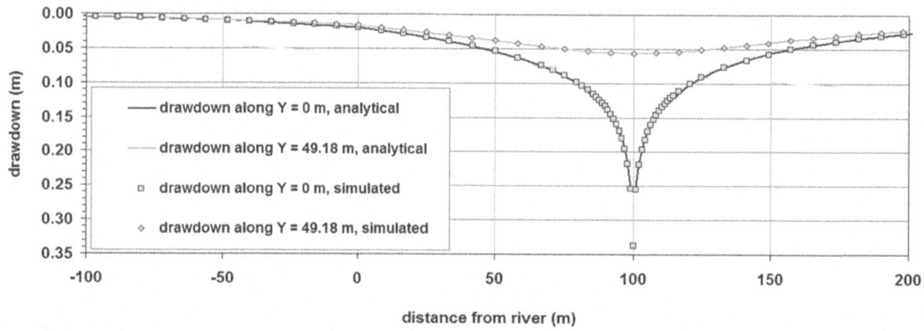

Figure 19.9 Comparison between the analytically solved and numerically simulated drawdown along $y = 0$ and $y = 49.18$ m at day 23 of the simulation using FEFLOW and MIKE11.

19.5.2 Infiltration pond

Model description

A simple model has been constructed to compare analytical and numerical solutions for infiltration ponds. The infiltration pond model domain and associated boundary conditions are illustrated in Figure 19.10.

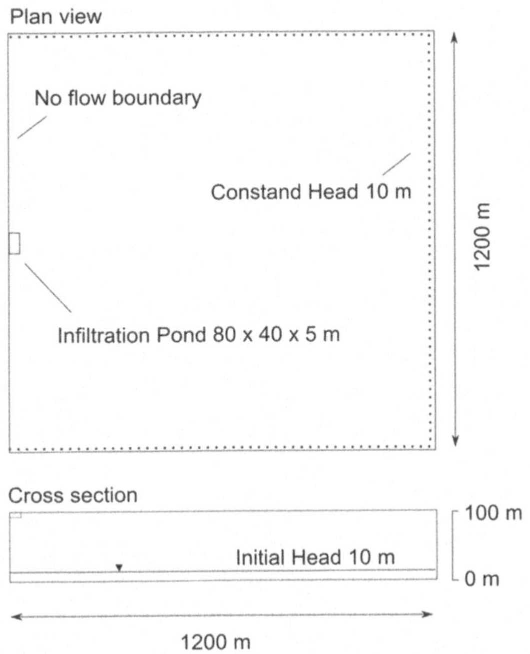

Figure 19.10 Model domain and boundary conditions for infiltration pond scenarios.

The infiltration pond is a square type with 80 × 80 m area, but, in order to save computational time, only half of the infiltration pond is represented by the model. The development of the groundwater mound beneath the infiltration pond was simulated using the unsaturated-zone flow package coupled to MODFLOW 2005 under steady-state conditions. Unsaturated flow is calculated based on a simplified Richard's equation (Niswonger *et al.*, 2006). In FEFLOW, the fully integrated 3D Richard's equation is used, applying a simplified Van Genuchten scheme. Model parameters used for the different pond scenarios are shown in Table 19.7.

Table 19.7 Differences in model parameter for infiltration pond scenarios.

Hydraulic Loading Rate [m/d]	Hydraulic Conductivity [m/s]	Effective Porosity [–]
1	1×10^{-4}	0.15
2	1×10^{-4}	0.15
3	1×10^{-4}	0.15
1	1×10^{-3}	0.25
2	1×10^{-3}	0.25
3	1×10^{-3}	0.25

Results

Analytical solutions from Bouwer *et al.* (1999) and numerical solutions for the development of a groundwater mound beneath a recharge pond for different hydraulic loading rates and hydraulic conductivities of the aquifer are shown in Figure 19.11.

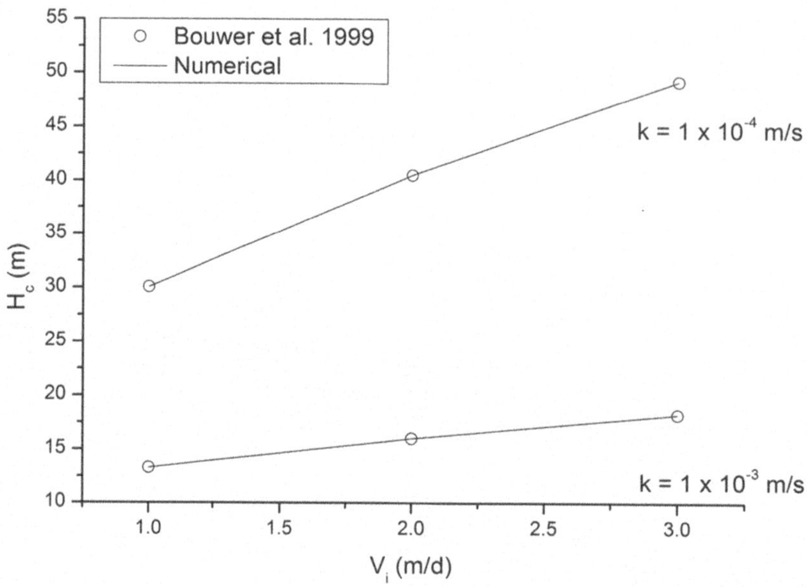

Figure 19.11 Comparison of numerical and analytical solutions of mounding height (H_c) below infiltration ponds using MODFLOW.

Calculations with MODFLOW and FEFLOW show that a sandy aquifer with $k = 1 \times 10^{-4}$ m/s and a 10 m groundwater thickness at the control area creates a mounding height of approx. 50 m ($V_i = 3$ m/d). Please note that these values are steady-state solutions and it may take years (up to 10 years) to develop equilibrium between percolation from the recharge structure and radial flow away from the recharge area. If the calculated groundwater mounding heights exceeds the thickness of the unsaturated zone (i.e. the zone below the recharge area is totally saturated), the groundwater mound must be controlled e.g. by pumping or by reducing the long-term infiltration rate. In Figure 19.12 an example of the FEFLOW setup and the simulated groundwater mounding beneath the infiltration pond is shown.

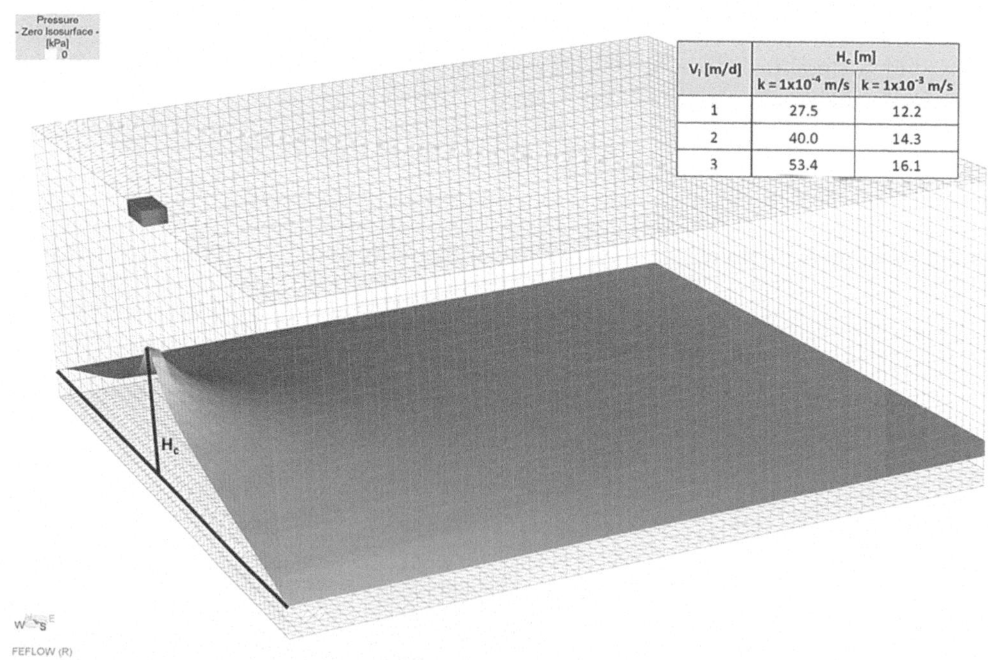

Figure 19.12 Results of FEFLOW simulations for different infiltration pond scenarios.

The value of transmissivity (T) in eq. 19.6 must reflect the average transmissivity of the aquifer during steady-state mound height. If T of the entire aquifer is used, eq. 19.6 underestimates the mounding height and if T only for the initially saturated thickness is used, eq. 19.6 overestimates the mound rise. Hence, the challenging part is to find a representative aquifer transmissivity. Bouwer *et al.* (1999) proposed to run pilot infiltration areas and to calculate T from that mound rise.

Eq. 19.9 and eq. 19.10, describing a transient infiltration process out of a rectangular and triangular shaped infiltration pond, have also been verified using a coupled numerical model under FEFLOW and MIKE11. The initial water depth of the river in both models is 10 m, the river width of the rectangular cross section is 25 m and the slope of the bank of the triangular cross section amounts 1:2. The transfer coefficient is equal to 0.1/d and the length of the river is 500 m. Using a maximum allowed time step of FEFLOW of 0.05 d, both the exchange discharges and the water depths of the latest version of IfmMIKE11 fit perfectly to the analytical solutions (Figure 19.13). As IfmMIKE11 is using an explicit coupling approach, also the influence of the maximum allowed time step of FEFLOW was tested. It was found that the lack of an iterative coupling causes discrepancies between the analytical and numerical solution if the FEFLOW time step exceeds a length of approximately 0.25 d.

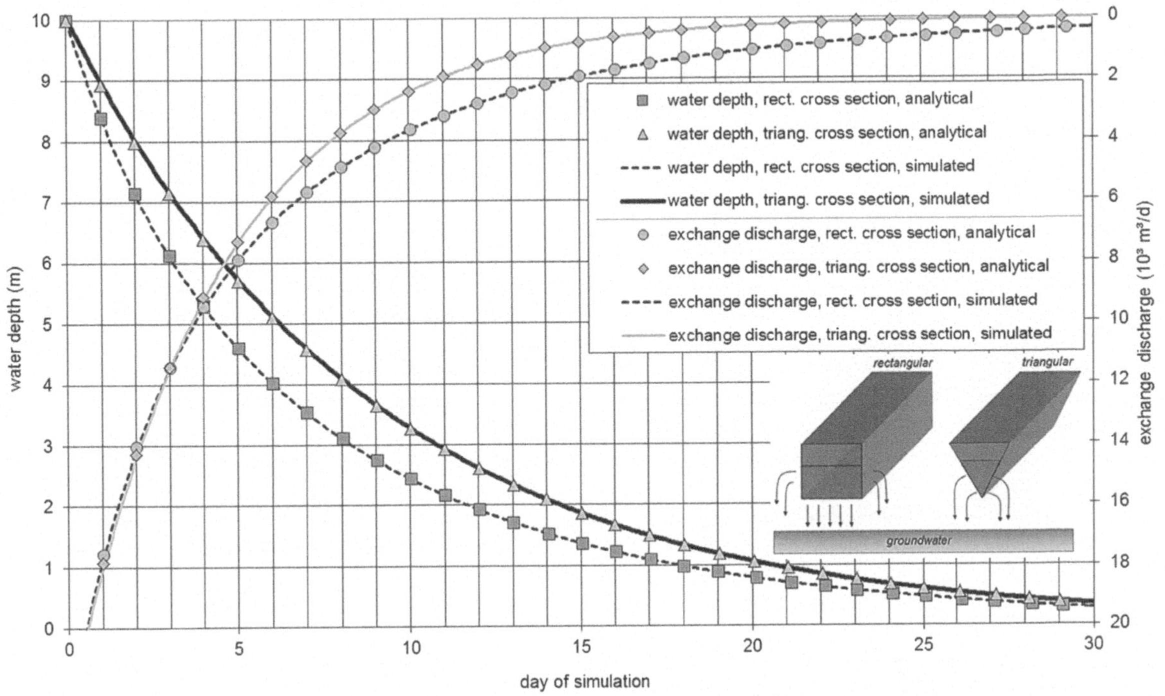

Figure 19.13 Comparison between the analytical solutions and simulated results for a rectangular as well as a triangular river cross section using a maximum FEFLOW time step of 0.05 d.

These simulations show that, besides bank filtration processes, also surface spreading MAR structures can be simulated using numerical coupled ground and surface water models. Furthermore, with the coupling of MIKE11 and FEFLOW also integrated multi-species and non-conservative mass transport processes can be described (Monninkhoff *et al.* 2011).

19.6 CONCLUSIONS

The examples presented in this Chapter show that by using numerical modelling a variety of different natural treatment techniques (MAR, RBF) can be accurately described. However, it has also been shown that particle-based, purely advective numerical calculations may yield substantial longer travel times of infiltrated source water to the abstraction well compared to solute transport. It is therefore recommended to use classical mass transport simulations to evaluate the minimum travel times, taking into account both dispersion- and diffusion-driven processes.

Besides the numerical tools presented in this Chapter, the included analytical solutions may provide a first approximation of important performance parameters of NTSs, but limitations and assumptions have to be taken into account. Integrated analyses of surface water and groundwater interactions, especially for geometrically complex MAR structures, in diverse geological environments or at a regional catchment levels, can only be performed using numerical modelling tools.

19.7 REFERENCES

Abdel-Fattah A., Langford R. and Schulze-Makuch D. (2008). Applications of particle-tracking techniques to bank infiltration: A case study from El Paso, Texas, USA. *Environmental Geology*, **55**, 505–515.

Adams E. E. and Gelhar L. W. (1992). Field study of dispersion in a heterogeneous aquifer 2. Spatial Moments Analysis. *Water Resour. Res*, **28**(12), 3293–3307. doi: 10.1029/92WR01757.

Anderson M. P. and Woessner W. W. (1992). Applied groundwater modeling. Simulation or Flow and Advective Transport. Elsevier, San Diego. ISBN – 10:0-12-059485-4.

Barry D. A., Parlange J.-Y. and Li L. (2000). Approximation for the exponential integral (Theis well function), *Journal of Hydrology* **227**, 287–291.

Bouwer H. (1978). Groundwater hydrology. McGraw-Hill, New York.

Bouwer H. (2002). Artificial recharge of groundwater: Hydrogeology and engineering. *Hydrogeology Journal*, **10**, 121–142.

Bouwer H., Back J. T. and Oliver J. M. (1999). Predicting infiltration and ground water mounding for artificial recharge. *J Hydrol Eng*, **4**, 350–357.

CGWB (2005). Central Ground Water Board, Master Plan for Artificial Recharge to Ground Water in India.

Chen X. (2001). Migration of induced-infiltrated stream water into nearby aquifers due to seasonal ground water withdrawal. *Ground Water*, **29**(5), 721–729.

DHI (2014a). MIKE 11, Reference and User Manual, MIKE by DHI, Copenhagen.

DHI (2014b). MIKE SHE user manual, volume 2: Reference guide, MIKE by DHI, Copenhagen.

DHI-WASY (2013). FEFLOW 6.2, Finite Element Subsurface Flow & Transport Simulation System, user manual, Berlin.

Diersch H.-J. G. (2014). FEFLOW – Finite element modeling of flow, mass and heat transport in porous and fractured media, Springer, Heidelberg/Berlin.

Dillon P. J., Miller M., Fallowfield H. and Hutson J. (2002). The potential of riverbank filtration for drinking water supplies in relation to microsystin removal in brackish aquifers. *Journal of Hydrology*, **266**, 209–221.

Gale I. N., Macdonald D. M. J., Calow R. C., Neumann I., Moench M., Kulkarni H., Mudrakartha S. and Palanisami K. (2006). Managed aquifer recharge: An assessment of its role and effectiveness in watershed management. British Geological Survey Commissioned Report, CR/06/107N, Nottingham, United Kingdom.

Gaus I., Cikurel H., Picot G., Aharoni A., Guttman Y., Azaroual M. and Kloppmann W. (2007). "Alternative SAT of secondary effluents using a combination of UF and short term SAT for increasing infiltration rates at the Shafdan site (Israel)" Proc., Int. Symp. on Managed Aquifer recharge ISMAR 6, Phoenix, USA, 544–557.

Glover R. E. and Balmer G. G. (1954). River depletion resulting from pumping a well near a river. *Transactions of the American Geophysical Union*, **35**, 468–470.

Green W. H. and Ampt G. (1911). Studies of soil physics, part I – the flow of air and water through soils. *J. Ag. Sci*, **4**, 1–24.

Greskowiak J., Prommer H., Massmann G. and Nuetzmann G. (2006). Modeling seasonal redox dynamics and the corresponding fate of the pharmaceutical residue phenazone during artificial recharge of groundwater. *Environmental Science & Technology*, **40**, 6615–6621.

Harbaugh A. W. (2005). MODFLOW-2005. The U.S. Geological Survey Modular Ground-Water Model—the Ground-Water Flow Process USGS.

Henzler A. F., Greskowiak J. and Massmann G. (2014). Modeling the fate of organic micropollutants during riverbank filtration (Berlin, Germany). *Journal of Contaminant Hydrology*, **156**, 78–92.

Habets F., Gascoin S., Korkmaz S., Thiéry D., Zribi M., Amraoui N., Carli M., Ducharne A., Leblois E., Ledoux E., Martin E., Noilhan J., Ottlé C. and Viennot P. (2010). "Multi-model comparison of a major flood in the groundwater-fed basin of the Somme River (France)", *Hydrol. Earth Syst. Sci.*, **14**, 99–117.

Herbst M., Fialkiewicz W., Chen T., Putz T., Thiery D., Mouvet C., Vachaud, G. and Vereecken H. (2005). Intercomparison of flow and transport models applied to vertical drainage in cropped lysimeters. *Vadose Zone Journal*, **4**(2): 240–254.

Hunt B. (1999). Unsteady stream depletion from ground water pumping, Ground Water 37, no. 1, 98–102.

Illangasekare T. H. (2001). MIKE SHE Code Verification and Validation for RFETS Site-wide Water Balance Model, Colorado School of Mines, Golden, Colorado. In collaboration with Robert Prucha, RMC Consultants and DHI-Water and Environment for Kaiser-Hill Company, L.L.C. http://www.integratedhydro.com/reviews.html (accessed 9 September 2015).

Kloppmann W., Aharoni A., Chikurel H., Dillon P., Gaus I., Guttman Y., Kraitzer T., Kremer S., Masciopinto C., Pavelic P., Picot-Colbeaux G., Pettenati M. and Miotlinski K. (2012). "Use of groundwater models for prediction and optimisation of the behaviour of MAR sites." in "Water Reclamation Technologies for Safe Managed Aquifer Recharge – Chapter 18", C. Kazner, T. Wintgens and P. Dillon (eds.), Chapter book: IWA Publishing. ISBN 9781843393443.

Langevin C. D., Thorne D. T. J., Dausman A. M., Sukop M. C. and Guo W. (2007). SEAWAT Version 4: A Computer Program for Simulation of Multi-Species Solute and Heat Transport. U.S. Geological Survey Techniques and Methods Book 6, United States.

Monninkhoff B. L. and Hartnack J. N. (2009). Improvements in the coupling interface between FEFLOW and MIKE11, Conference proceedings of FEFLOW 2009, Berlin, Germany.

Monninkhoff B. L., Kaden S. and Mayer J. (2011). Integrated analyses of MAR techniques in Shandong, China, International Conference on Integrated Water Resources Management (IWRM 2011). Management of Water in a Changing World: Lessons Learnt and Innovative Perspectives, 12–13 October, Dresden.

Neumann I., Barker J., MacDonald D. and Gale I. (2004). Numerical approaches for approximating technical effectiveness of artificial recharge structures. British Geological Survey Commissioned Report, CR/04/265N, Nottingham, United Kingdom.

Niswonger R. G., Prudic D. E. and Regan R. S. (2006). Documentation of the Unsaturated-Zone Flow (UZF1) Package for Modeling Unsaturated Flow Between the Land Surface and the Water Table with MODFLOW-2005. U.S. Geological Survey Techniques and Methods 6, United States.

Picot-Colbeaux G., Thiéry D., Sarah S., Boisson A., Pettenati M., Perrin J., Dewandel B., Maréchal JC., Ahmed S. and Kloppmann W. (2013) Modelling artificial recharge capacity of fractured hard rock under semi-arid conditions in Southern India: Implementing storage basin dynamics into MARTHE code. *ISMAR8, Beijing is the 8th International Symposium on Managed Aquifer Recharge*, Beijing, China, October 15–19, 2013.

Pollock D. W. (1994). User's guide for modpath/modpath-plot, version 3: A particle tracking post-processing package for MODFLOW. The U.S. geological survey finite- difference ground-water flow model, U.S. Geologic Survey, Washington DC, United States.

Prommer H. (2006). A Reactive Multicomponent Transport Model for Saturated Porous Media, User's Manual Version 1.46.

Rawls W. J., Brakensiek D. L. and Soni B. (1983) Agricultural management effects of soil water processes: Part I – Soil water retention and Green ampt infiltration parameters. *Trans., Amer. Soc. of Agric. Engin.*, **26**(6), 1752–1753.

Rhebergen W. and Dillon P. (1999). Riverbank filtration models for assessing viability of water quality improvement. Centre for Groundwater Studies Report No. 90, Glen Osmond, South Australia.

Sandhu C., Grischek T., Kumar P. and Ray C. (2010). Potential for riverbank filtration in India. *Clean Technologies and Environmental Policy*, **13**, 295–316.

Theis C. V. (1941). The effect of a well on the flow of a nearby stream. *Transactions of the American Geophysical Union*, **22**, 734–738.

Thiéry D. (2010a). "Groundwater Flow Modelling in Porous Media Using MARTHE", in "Modelling Software Volume 5, Chapter 4, Environmental Hydraulics Series". Tanguy J.M. (ed.), Wiley/ISTE London, pp. 45–60. ISBN: 978-1-84821-157-5.

Thiéry D. (2010b). "Presentation of the Finite Volume Method", in "Numerical Methods Volume 3, Chapter 8.6, Environmental Hydraulics Series". Tanguy J.M. (ed.), Wiley/ISTE London, pp. 195–211. ISBN: 978-1-84821-155-1.

Thiéry D. (1995). "Modélisation 3D du transport de masse avec le logiciel MARTHE version 5.4". Report BRGM R 38149 DR/HYT 95, pp. 171.

Thiéry D. (1993). "Modélisation des aquifères complexes – Prise en compte de la zone non saturée et de la salinité. Calcul des intervalles de confiance". *Revue Hydrogéologie* **4**, 325–336.

Thiéry D. (1990). "Software MARTHE. Modelling of Aquifers with a Rectangular Grid in Transient state for Hydrodynamic calculations of heads and flows. Release 4.3". Report BRGM 4S/EAU n° R32548.

Thiéry D. (2010c). "Reservoir Models in Hydrogeology", in "Mathematical Models Volume 2, Chapter 13, Environmental Hydraulics Series". Tanguy J.M. (ed.), Wiley/ISTE London, pp. 409–418. ISBN: 978-1-84821-154-4.

Thiéry D. (2010d). "Interaction between Surface and Subsurface Flows: Somme Basin", in "Practical Applications in Engineering: Volume 4, Chapter 13, Environmental Hydraulics Series". Tanguy J.M. (ed.), Wiley/ISTE London, pp. 143–156. ISBN: 978-1-84821-156-8.

Thiéry D. and Amraoui N. (2001). "Hydrological modelling of the Saone basin sensitivity to the soil model". *Phys. Chem. Earth, Pt B*, **26**(5–6), 467–472.

Welter D. E., Doherty J. E., Hunt R. J., Muffels C. T., Tonkin M. J. and Schreüder W. A. (2012). Approaches in highly parameterized inversion: PEST++, a Parameter ESTimation code optimized for large environmental models. U.S. Geological Survey Techniques and Methods 7, United States.

Worldbank (2010). Deep wells and prudence: Towards Pragmatic Action for Addressing Groundwater Overexploitation in India. The International Bank for Reconstruction and Development/The World Bank, Washington, D.C.

Yamagata K., Butts B., Grooss J., Clausen T., Clausnitzer V., Gründler R. and Monninkhoff B. L. (2012). Investigating an OpenMI coupling of FEFLOW and MIKE SHE, Conference proceedings of FEFLOW 2012, Berlin, Germany, September.

Xu T. F., Spycher N., Sonnenthal E., Zhang G. X., Zheng L. E. and Pruess K. (2011). TOUGHREACT Version 2.0: A simulator for subsurface reactive transport under non-isothermal multiphase flow conditions. *Computers & Geosciences*, **37**(6): 763–774.

Zheng C. and Wang P. P. (1999). A Modular Three-Dimensional Multispecies Transport Model for Simulation of Advection. Dispersion and Chemical Reactions of Contaminants in Groundwater Systems. Documentation and User's Guide, Vicksburg, United States.

IWA Publishing's authorised EU representative for General Product Safety Regulations is Diane D'Arras, 15 rue Duret, 75116 Paris, France, e-mail: safety@iwap.co.uk.

Printed and bound by CPI Group (UK) Ltd, Croydon, CR0 4YY
16/06/2026
02140377-0001